The Science of Health Disparities Research

The Science of Health Disparities Research

Edited by

Irene Dankwa-Mullan
IBM Watson Health
IBM Corporation
Bethesda, MD, USA

Eliseo J. Pérez-Stable
National Institute on Minority Health and Health Disparities
National Institutes of Health
Bethesda, MD, USA

Kevin L. Gardner
Department of Pathology and Cell Biology
Columbia University Medical Center
New York, NY, USA.

Xinzhi Zhang
Division of Clinical Innovation, National Center for Advancing Translational Science
National Institutes of Health
Bethesda, MD, USA

Adelaida M. Rosario
Commissioned Corps Headquarters
Office of the Surgeon General
Rockville, MD, USA

WILEY Blackwell

This edition first published 2021

Registered Office
John Wiley & Sons, Inc., 111 River Street, Hoboken, NJ 07030, USA

Editorial Office
9600 Garsington Road, Oxford, OX4 2DQ, UK

For details of our global editorial offices, customer services, and more information about Wiley products visit us at www.wiley.com.

Wiley also publishes its books in a variety of electronic formats and by print-on-demand. Some content that appears in standard print versions of this book may not be available in other formats.

Library of Congress Cataloging-in-Publication Data

Names: Dankwa-Mullan, Irene, editor. | Pérez-Stable, Eliseo J., editor. | Gardner, Kevin L., M.D., editor. | Zhang, Xinzhi, 1971– editor. | Rosario, Adelaida M., 1977– editor.
Title: The science of health disparities research / edited by Irene Dankwa-Mullan, Eliseo J. Pérez-Stable, Kevin L. Gardner, Xinzhi Zhang, Adelaida M. Rosario.
Description: Hoboken, NJ : Wiley-Blackwell, 2021. | Includes bibliographical references and index.
Identifiers: LCCN 2020024358 (print) | LCCN 2020024359 (ebook) | ISBN 9781119374817 (cloth) | ISBN 9781119374831 (adobe pdf) | ISBN 9781119374848 (epub)
Subjects: MESH: Healthcare Disparities | Translational Medical Research | Socioeconomic Factors | Social Determinants of Health | Minority Health
Classification: LCC RA563.M56 (print) | LCC RA563.M56 (ebook) | NLM W 76.1 | DDC 362.1089–dc23
LC record available at https://lccn.loc.gov/2020024358
LC ebook record available at https://lccn.loc.gov/2020024359

Cover Design: Wiley
Cover image: Barbershop Photo Courtesy of Stephen B. Thomas, Hula Dance Photo Courtesy of Joseph Kaholokula, Happy Baby Girl Playing With Toys In Playroom © Monkey Business Images/Shutterstock

Set in 9.5/12.5pt STIXTwoText by SPi Global, Pondicherry, India

SKY10026176_040821

Contents

List of Contributors

Adebowale Adeyemo
Center for Research on Genomics and Global Health, National Human Genome Research Institute,
National Institutes of Health
Bethesda, MD
USA

Margarita Alegria
Department of Medicine, Disparities of Research Unit
Massachusetts General Hospital
Boston, MA
USA;
Departments of Medicine and Psychiatry
Harvard Medical School
Boston, MA
USA

Naomi Ali
Department of Medicine, Disparities of Research Unit
Massachusetts General Hospital
Boston, MA
USA

Jeroan Allison
University of Massachusetts
Medical School
Worcester, MA
USA

Jennifer Alvidrez
National Institute on Minority Health and Health Disparities
National Institutes of Health
Bethesda, MD
USA

Hortensia Amaro
Herbert Wertheim, College of Medicine
Florida International University
Miami, FL
USA;
Robert Stempel, College of Public Health and Social Work
Florida International University
Miami, FL
USA

Noa Appleton
Department of Population Health
New York University School of Medicine
New York, NY
USA

Inna Arnaudova
Department of Psychiatry
University of California
Los Angeles, CA
USA

Ligia Artiles
National Institute on Minority Health and Health Disparities
National Institutes of Health
Bethesda, MD
USA

Leslie J. Baier
Phoenix Epidemiology and Clinical Research Branch, National Institute of Diabetes and Digestive and Kidney Diseases
National Institutes of Health
Phoenix, AZ
USA

Tracy Bastain
Department of Preventive Medicine, Keck School of Medicine
University of Southern California
Los Angeles, CA
USA

Sara E. Baumann
Department of Behavioral and Community Health Sciences
University of Pittsburgh Graduate School of Public Health
Pittsburgh, PA
USA

Amy R. Bentley
Center for Research on Genomics and Global Health, National Human Genome Research Institute
National Institutes of Health
Bethesda, MD
USA

Rick A. Berzon
National Institute on Minority Health and Health Disparities
National Institutes of Health
Bethesda, MD
USA

Bernadette Boden-Albala
Department of Population Health and Disease Prevention
University of California
Irvine, CA
USA;
Department of Epidemiology
University of California
Irvine, CA
USA

Nancy Breen
National Institute on Minority Health and Health Disparities
National Institutes of Health
Bethesda, MD
USA

Carrie Breton
Department of Preventive Medicine, Keck School of Medicine
University of Southern California
Los Angeles, CA
USA

Esteban Burchard
Department of Bioengineering and Therapeutic Sciences
University of California
San Francisco, CA
USA

Jung S. Byun
National Institute on Minority Health and Health Disparities
National Institutes of Health
Bethesda, MD
USA

Tamela Cannady
Choctaw Nation Health Service
Talihina, OK
USA

Olveen Carrasquillo
Division of General Internal Medicine, Department of Medicine, Miller School of Medicine
University of Miami
Miami, FL
USA

Patricia Chalela
Institute for Health Promotion Research
University of Texas Health Science Center
San Antonio, TX
USA

Guanjie Chen
Center for Research on Genomics and Global Health, National Human Genome Research Institute
National Institutes of Health
Bethesda, MD
USA

Marshall Chin
Department of Medicine
The University of Chicago
Chicago, IL
USA

Kelvin Choi
National Institute on Minority Health and Health Disparities
National Institutes of Health
Bethesda, MD
USA

Benjamin Cook
Harvard Medical School
Boston, MA
USA;
Cambridge Health Alliance
Cambridge, MA
USA

Lisa A. Cooper
Johns Hopkins Center for Health Equity
Johns Hopkins University
Baltimore, MD
USA

Nancy J. Cox
Vanderbilt Genetics Institute, Department of Medicine, Division of Genetic Medicine
Vanderbilt University Medical Center
Nashville, TN
USA

Timothy Creedon
Cambridge Health Alliance
Cambridge, MA
USA

Rada K. Dagher
National Institute on Minority Health and Health Disparities
National Institutes of Health
Bethesda, MD
USA

Irene Dankwa-Mullan
IBM Watson Health
International Business Machines (IBM) Corporation
Bethesda, MD
USA

Stuti Dang
Division of Geriatrics, Department of Medicine, Miller School of Medicine
University of Miami
Miami, FL
USA

Rina Das
National Institute on Minority Health and Health Disparities
National Institutes of Health
Bethesda, MD
USA

Alison F. Davis
Scientific Workforce Diversity, Office of the Director
National Institutes of Health
Bethesda, MD
USA

Karissa DiMarzio
Department of Medicine, Disparities of Research Unit
Massachusetts General Hospital
Boston, MA
USA

Ayo P. Doumatey
Center for Research on Genomics and Global Health, National Human Genome Research Institute
National Institutes of Health
Bethesda, MD
USA

Sherine El-Toukhy
National Institute on Minority Health and Health Disparities
National Institutes of Health
Bethesda, MD
USA

Michael Engelgau
National Heart, Lung, and Blood Institute
National Institutes of Health
Bethesda, MD
USA

Tilda Farhat
National Institute on Minority Health and Health Disparities
National Institutes of Health
Bethesda, MD
USA

Shohreh Farzan
Department of Preventive Medicine, Keck School of Medicine
University of Southern California
Los Angeles, CA
USA

Morgan Foreman
Center for Computational Health & Integrated Care Research, IBM Research
International Business Machines (IBM) Corporation
Cambridge, MA
USA

Sarah Forrester
University of Massachusetts
Medical School
Worcester, MA
USA

Federico Fuentes
Kidney Diseases Branch, National Institute of Diabetes and Digestive and Kidney Diseases,
National Institutes of Health
Bethesda, MD
USA

Samantha Garcia
Program in Public Health
University of California
Irvine, CA
USA

Kevin L. Gardner
Department of Pathology and Cell Biology
Columbia University Medical Center
New York, NY
USA

Tiffany L. Gary-Webb
Department of Behavioral and Community Health Sciences
University of Pittsburgh Graduate School of Public Health
Pittsburgh, PA
USA;
Department of Epidemiology
University of Pittsburgh Graduate School of Public Health
Pittsburgh, PA
USA

Page Goddard
Department of Bioengineering and Therapeutic Sciences
University of California
San Francisco, CA
USA

Rima Habre
Department of Preventive Medicine, Keck School of Medicine
University of Southern California
Los Angeles, CA
USA

Benyam Hailu
National Institute on Minority Health and Health Disparities
National Institutes of Health
Bethesda, MD
USA

Ross A. Hammond
Center on Social Dynamics and Policy
Brookings Institution
Washington, DC
USA;
Brown School
Washington University in St. Louis
St. Louis, MO
USA

Nora Haney
The James Buchanan Brady Urological Institute
Johns Hopkins Hospital
Baltimore, MD
USA

Mark L. Hatzenbuehler
Department of Psychology
Harvard University
Cambridge, MA
USA

Bertha Hidalgo
Ryals School of Public Health
The University of Alabama at Birmingham
Birmingham, AL
USA

Carl V. Hill
Office of Special Populations, National Institute of Aging
National Institutes of Health
Bethesda, MD
USA

Lydia A. Isaac
Department of Health Policy and Management, Milken Institute School of Public Health
The George Washington University
Washington, DC
USA

Chandra L. Jackson
Epidemiology Branch, Social and Environmental Determinants of Health Equity, National Institute of Environmental Health Sciences
National Institutes of Health
Cary, NC
USA

Tvli Jacob
Center for Health Sciences
Oklahoma State University
Tulsa, OK
USA

Anil Jain
IBM Watson Health
International Business Machines (IBM) Corporation
Cleveland, OH
USA

Regina James
2M Research
Arlington, VA
USA

Valarie Blue Bird Jernigan
Center for Health Sciences
Oklahoma State University
Tulsa, OK
USA

Jill Johnston
Department of Preventive Medicine, Keck School of Medicine
University of Southern California
Los Angeles, CA
USA

Monica P. Jolles
Suzanne Dworak-Peck, School of Social Work
University of Southern California
Los Angeles, CA
USA;
Cecil G. Sheps Center for Health Services Research
University of North Carolina
Chapel Hill, NC
USA

Matt Kasman
Center on Social Dynamics and Policy
Brookings Institution
Washington, DC
USA

Sonjia Kenya
Division of General Internal Medicine, Department of Medicine, Miller School of Medicine
University of Miami
Miami, FL
USA

Jeffrey B. Kopp
Kidney Diseases Branch, National Institute of Diabetes and Digestive and Kidney Diseases,
National Institutes of Health
Bethesda, MD
USA

Heather Kuczynski
School of Global Public Health
New York University
New York, NY
USA

Thomas A. LaVeist
Tulane University School of Public Health and Tropical Medicine
New Orleans, LA
USA

Phuong-Tu Le
National Institute on Minority Health and Health Disparities
National Institutes of Health
Bethesda, MD
USA

Cynthia Lebron
School of Nursing and Health Studies
University of Miami
Coral Gables, FL
USA

Maria F. Lima
CUNY School of Medicine
The City College of New York
New York, NY
USA

Benmei Liu
National Cancer Institute
National Institutes of Health
Bethesda, MD
USA

Jacqueline J. Lloyd
National Institute on Drug Abuse
National Institutes of Health
Rockville, MD
USA

Jill A. Marsteller
Johns Hopkins Center for Health Equity
Johns Hopkins University
Baltimore, MD
USA

Linda Sprague Martinez
School of Social Work
Boston University
Boston, MA
USA

David O. Meltzer
Department of Medicine
The University of Chicago
Chicago, IL
USA

Priscah Mujuru
National Institute on Minority Health and Health Disparities
National Institutes of Health
Bethesda, MD
USA

Anup K. Nair
Phoenix Epidemiology and Clinical Research Branch, National Institute of Diabetes and Digestive and Kidney Diseases
National Institutes of Health
Phoenix, AZ
USA

Emily Nangle
Department of Epidemiology, School of Global Public Health
New York University
New York, NY
USA

Anna M. Nápoles
National Institute on Minority Health and Health Disparities
National Institutes of Health
Bethesda, MD
USA

Uchenna J. Ndulue
Boston Public Health Commission, Child Adolescent and Family Health Bureau
Boston, MA
USA

Tung T. Nguyen
School of Medicine
University of California
San Francisco, CA
USA

Abigail Ortiz
Southern Jamaica Plain Health Center
Jamaica Plain, MA
USA

John E. Pachankis
School of Public Health
Yale University
New Haven, CT
USA

William V. Padula
Department of Pharmaceutical and Health Economics, School of Pharmacy
University of Southern California
Los Angeles, CA
USA;
Leonard D. Schaeffer Center for Health Policy and Economics
Los Angeles, CA
USA

Jane Pang
Pacific Islander Health Partnership
Huntington, Beach, CA
USA

Nina S. Parikh
Department of Epidemiology, School of Global Public Health
New York University
New York, NY
USA

Eliseo J. Pérez-Stable
National Institute on Minority Health and Health Disparities
National Institutes of Health
Bethesda, MD
USA

Robert L. Phillips
American Board of Family Medicine
Washington, DC
USA

Latrice C. Pichon
Social and Behavioral Sciences
University of Memphis
Memphis, TN
USA

Tanjala S. Purnell
Johns Hopkins Center for Health Equity
Johns Hopkins University
Baltimore, MD
USA

Nishadi Rajapakse
National Institute on Minority Health and Health Disparities
National Institutes of Health
Bethesda, MD
USA

Amelie G. Ramirez
Institute for Health Promotion Research
University of Texas Health Science Center
San Antonio, TX
USA

Neda Ratanawongsa
Division of General Internal Medicine, Department of Medicine at Zuckerberg, San Francisco General Hospital
University of California-San Francisco
San Francisco, CA
USA;
Center for Vulnerable Populations
University of California-San Francisco
San Francisco, CA
USA;
San Francisco Department of Public Health
San Francisco, CA
USA;
Richard Fine People's Clinic
San Francisco, CA
USA

William T. Riley
Office of Behavioral and Social Sciences Research
National Institutes of Health
Bethesda, MD
USA

Erik J. Rodriquez
National Heart, Lung, and Blood Institute
National Institutes of Health
Bethesda, MD
USA

Adelaida Rosario
Commissioned Corps Headquarters
Office of the Surgeon General
Rockville, MD
USA

Charles N. Rotimi
Center for Research on Genomics and Global Health, National Human Genome Research Institute
National Institutes of Health
Bethesda, MD
USA

Melanie D. Sabado-Liwag
National Institute on Minority Health and Health Disparities
National Institutes of Health
Bethesda, MD
USA

Mariana Sanchez
Robert Stempel College of Public Health and Social Work
Florida International University
Miami, FL
USA

Milda R. Saunders
Department of Medicine
The University of Chicago
Chicago, IL
USA

Dean Schillinger
Division of General Internal Medicine, Department of Medicine at Zuckerberg, San Francisco General Hospital
University of California-San Francisco
San Francisco, CA
USA;
Center for Vulnerable Populations
University of California-San Francisco
San Francisco, CA
USA;
Richard Fine People's Clinic
San Francisco, CA
USA

Daniel Shriner
Center for Research on Genomics and Global Health, National Human Genome Research Institute
National Institutes of Health
Bethesda, MD
USA

Sarah C. Stallings
Vanderbilt University Medical Center
Nashville, TN
USA;
Meharry-Vanderbilt University
Nashville, TN
USA

Anita L. Stewart
Institute for Health and Aging, Center for Aging in Diverse Communities
University of California
San Francisco, CA
USA

Andrea Williams Stubbs
St. Jude Children's Research Hospital
Memphis, TN
USA

Meryl Sufian
National Institute on Minority Health and Health Disparities
National Institutes of Health
Bethesda, MD
USA

Derrick C. Tabor
National Institute on Minority Health and Health Disparities
National Institutes of Health
Bethesda, MD
USA

Sora Park Tanjasiri
University of California
Irvine, CA
USA

Claudia Toledo-Corral
Department of Preventive Medicine, Keck School of Medicine
University of Southern California
Los Angeles, CA
USA

Hannah A. Valantine
Scientific Workforce Diversity, Office of the Director
National Institutes of Health
Bethesda, MD
USA

Elizabeth Vang
Department of Preventive Medicine, Keck School of Medicine
University of Southern California
Los Angeles, CA
USA

Salina P. Waddy
Department of Neurology, Atlanta Veterans Administration
Decatur, GA
USA

Nina Wallerstein
College of Population Health
University of New Mexico
Albuquerque, NM
USA

Joan Wasserman
Uniformed Services University of the Health Sciences
Bethesda, MD
USA

Kristina Weeks
Johns Hopkins Center for Health Equity
Johns Hopkins University
Baltimore, MD
USA

Roy E. Weiss
Miller School of Medicine
University of Miami
Miami, FL
USA

Marquitta White
School of Medicine
University of California
San Francisco, CA
USA

Consuelo H. Wilkins
Vanderbilt University Medical Center
Nashville, TN
USA;
Meharry-Vanderbilt University
Nashville, TN
USA

Kesi Williams
National Institute on Minority Health and Health Disparities
National Institutes of Health
Bethesda, MD
USA;
Center for Research to Advance Community Health (REACH)
UT Health San Antonio
San Antonio, TX
USA

Jessamine Winer-Jones
IBM Watson Health
International Business Machines (IBM) Corporation
Herndon, VA
USA

Cheryl A. Winkler
Frederick National Laboratory for Cancer Research
National Cancer Institute
National Institutes of Health
Frederick, MD
USA

Kent Woo
NICOS Chinese Health Coalition
San Francisco, CA
USA

Shumei Yun
Resolve to Save Lives and Vital Strategies
New York, NY
USA

Xingyou Zhang
US Bureau of Labor Statistics
Washington, DC
USA

Xinzhi Zhang
Division of Clinical Innovation, National Center for Advancing Translational Sciences
National Institutes of Health
Bethesda, MD
USA

Foreword

January 10, 1992, the National Institutes of Health (NIH) published within its *Guide for Grants and Contracts*, special instructions to applicants regarding implementation of new policies concerning the inclusion of women and minorities in clinical research study populations. Many of the senior authors of chapters in this seminal volume and I were among those who advocated strongly for this, then, new policy. It spoke to the under-representation of racial and ethnic minorities in the science of the times, and strongly argued for closer attention to their inclusion in NIH-sponsored research as a means of improving said science. The introductory chapter to this volume chronicles the circumstances that led to this policy change, as well as the ensuing response on the part of the NIH and Department of Health and Human Services to its implementation.

Unfortunately, despite such a remarkable advance, the vigilance necessary to realize the promise embodied in this change waned. Indeed, as a long-standing member of several established NIH study sections charged with determining the scientific merit of thousands of grant applications submitted for prospective funding, I witnessed firsthand the gradual attenuation of this policy of inclusion. Study sections struggled with its operationalization; we lacked carefully crafted examples of how adherence to this policy could be accomplished in a pragmatic fashion and thus improve our science. As a consequence, reviewers gradually drifted from a rigorous application of the policy to a *pro forma* determination of the *adequacy* of including under-represented racial ethnic minorities, roughly equivalent to whether or not sampling plans assured their representation equal to that characteristic of the settings in which this work was to be conducted. One seldom read arguments, for example, for oversampling special populations in a manner that would truly test the hypotheses underpinning the proposed research, and thereby address race, ethnicity, or disadvantage as possibly important sources of variation in the phenomena under study.

Again, numerous contributors to this volume who also served on NIH study sections shared this concern and lamented the absence of careful attention to the implications for the science. Understanding the original arguments for why, *scientifically*, such individuals should be included in the work sponsored by NIH gradually faded; inattention diluted the promise of this remarkable policy. Arguments for the inclusion of under-represented racial and ethnic minorities and the disadvantaged in NIH-sponsored research eventually shifted from the benefit to science to the underlying importance of health equity and social justice. While the latter are important motives for undertaking such work, they are, in my opinion, necessary, but not sufficient conditions for promoting knowledge acquisition and ensuring the scientific merit of such efforts.

One and one-half decades later, the now National Academy of Medicine reviewed the NIH's strategic research plan to reduce and ultimately eliminate health disparities. This assessment essentially focused on the impact of the National Center on Minority Health and Health Disparities as of 2006, just prior to elevation to its current Institute status. In the Academy's report, the committee, of which several authors in this volume and I were members, concluded that the unfinished business of the Center and NIH was to revitalize a focus on the relevant science and to offer a coherent thematic framework for its pursuit. An emphasis on the social determinants of health emerged and gained increasing currency in anticipating and focusing this work. As the present volume amply demonstrates, this particular framework has played a critical role in organizing attention to the key domains and related constructs that bear importantly on the science underpinning minority health and health disparities research today.

This volume, and its companion piece in the special issue of the *American Journal of Public Health*, provide a roadmap to guide the science in this area. The chapters contained herein illustrate the importance and *feasibility* of systematic, rigorous inquiry for understanding the specifics of minority health and health disparities. They also convey the importance of the lessons learned for science in general: for discovery, for generalizability, for advancing theory, for enhancing measurement, for improving investigative methods, for promoting attention to neglected areas of research, and for diversifying the scientific work workforce. *The Science of Health Disparities Research* returns us to the spirit of 1992, and conviction, albeit now *empirically* demonstrable, that work of this nature can be exacting, meritorious, innovative, and broadly relevant. I applaud my colleagues' efforts in this regard and am confident those who follow can more effectively integrate health equity, social justice, and good science in service of improving the health of racial and ethnic minorities, as well as the disadvantaged, and humankind in general.

Spero M. Manson, PhD
Distinguished Professor of Public Health and Psychiatry
Colorado Trust Chair in American Indian Health
Colorado School of Public Health
University of Colorado Anschutz Medical Campus

Acknowledgements

We would like to thank all of the scientists, clinicians, and community-based researchers who have contributed to the scientific literature on minority health and health disparities research. Their vast contributions to the field serve as the foundation for this text. We also would like to thank chapter authors for their commitment to this project and their submissions. Further, we are eternally grateful to the staff of the National Institute on Minority Health and Health Disparities for their support, suggestions, and advice that helped us refine our ideas. Finally, we extend special thanks to Ligia Artiles and Richard C. Palmer for the immeasurable support they provided in successful completion of this project.

1

Definitions, Principles, and Concepts for Minority Health and Health Disparities Research

Eliseo J. Pérez-Stable[1], Jennifer Alvidrez[1], and Carl V. Hill[2]

[1] *National Institute on Minority Health and Health Disparities, National Institutes of Health, Bethesda, MD, USA*
[2] *Office of Special Populations, National Institute on Aging, National Institutes of Health, Bethesda, MD, USA*

1.1 Introduction

In 1985, Department of Health and Human Services (DHHS) Secretary Margaret Heckler commissioned a report on minority health at the urging of African American health leaders. The Heckler Report on Black and minority health examined the health status of Americans by race/ethnicity and identified the gaps in disease rates, mortality, and other outcomes among Blacks compared to Whites [1]. The report provided a foundation for the scientific field of minority health research and legitimized a perspective that had been developing for several decades. At the time, the public health paradigm was to evaluate health differences in populations from a socioeconomic perspective and access to care on the assumption that these were the main drivers of health outcome differences. The Heckler Report introduced the notion at a national level that race and ethnicity may be an independent contributor to health outcomes, which merited scientific study and targeted intervention programs. In 1987, the DHHS Office of Minority Health was founded, led by Herb Nickens, MD.

In 1990, the Office of Minority Programs was founded at the National Institutes of Health (NIH) under DHHS Secretary Louis Sullivan, MD. In 1993 the name was changed to the Office of Minority Health Research. Through congressional legislation, this office was transformed in 2000 into the Center on Minority Health and Health Disparities, and in 2010 to the National Institute on Minority Health and Health Disparities (NIMHD). John Ruffin, PhD was the director from 1990 until his retirement in 2014.[1]

In 1999, DHHS Deputy Secretary David Satcher, MD cited the unacceptability of demonstrated healthcare disparities by race in commenting on a study published in the *New England Journal of Medicine,* showing that Blacks were less likely to be referred for cardiac

1 https://www.nimhd.nih.gov/about/overview/history/

The Science of Health Disparities Research, First Edition. Edited by Irene Dankwa-Mullan, Eliseo J. Pérez-Stable, Kevin L. Gardner, Xinzhi Zhang, and Adelaida M. Rosario.

evaluation when presenting with classic chest pain symptoms compared to Whites [2]. Two years later, the Institute of Medicine (IOM) published the Unequal Care report summarizing a legacy of unequal healthcare and more adverse results for most leading causes of death and disability in the United States among African Americans compared to Whites [3]. Remarkably, little data were contained in this Report about the status of other race/ethnic groups in the United States. The IOM report broke down silos and provided the field with unifying principles about healthcare disparities. These events brought together scientific disciplines from population health, social science, and clinical care to focus on minority health and health disparities research.

In the twenty-first century, data collection and availability have dramatically improved. Scientific advances in understanding basic biological mechanisms have transformed our understanding of etiological pathways and potential interventions to improve minority health and reduce health disparities. The creation of a critical mass of interdisciplinary investigators has made possible further development of the science of minority health and health disparities. Collaboration among all health-related disciplines will make it possible for the next generation of minority health and health disparities researchers to advance the science. In that spirit, NIMHD is producing this book as it celebrates its tenth anniversary as an NIH Institute to further advance the science and lay the foundation for future research.

1.2 NIMHD Mission

NIMHD is charged with coordinating and leading the NIH's vision and programs on minority health and health disparities research by funding research to improve minority health and reduce health disparities. The topics are broad and include the epidemiology, etiology, prevention, and treatment for all diseases across the life course for all health disparity populations. Research that advances understanding and improvement of health and disease in minority racial/ethnic groups in the United States is a primary area of interest, requiring a basic understanding of race and ethnicity in the context of science. Research to understand the causes of health and healthcare disparities, leading to interventions to reduce these disparities, is NIMHD's mandate. The training and development of a diverse scientific workforce is also part of a broad NIH mandate embraced by NIMHD. NIMHD envisions an America in which all populations will have an equal opportunity to live long, healthy, and productive lives.

1.3 Definitions and Concepts of Minority Health and Health Disparities

1.3.1 Racial/Ethnic Minority Populations

In 1997, the Office of Management and Budget (OMB) issued the Revisions to the Standards for the Classification of Federal Data on Race and Ethnicity.[2] These standards are used for federal data collection purposes in the decennial US census, national household surveys,

2 https://www.whitehouse.gov/wp-content/uploads/2017/11/Revisions-to-the-Standards-for-the-Classification-of-Federal-Data-on-Race-and-Ethnicity-October30-1997.pdf

many administrative forms, and in medical and clinical research. Race and ethnicity categories have been modified over the decades to reflect the evolving demography of the United States, although full understanding of the meaning of the constructs captured by these categories is lacking and the categories need ongoing refinement and study. Currently, the five categories for race are defined as: American Indian or Alaska Native, Asian, Black or African American, Native Hawaiian and Other Pacific Islander, and White. In addition, a multirace category was included for the first time in the 2000 census. There are two categories for ethnicity: "Hispanic or Latino" and "Not Hispanic or Latino" and the question of ethnicity is posed before the race categories. The 2020 census plans to ask about family background for all respondents and continue to collect information on nation or US territory of origin for Hispanics/Latinos, Other Pacific Islanders and Asians, as well as tribal affiliation for American Indians/Alaska Natives. We consider that standardization of race and ethnic categories is an essential component of minority health and health disparities research and recommend using the census nomenclature and definitions as the starting point for specificity and standardization in the field.

1.3.2 Minority Health and Minority Health Research

NIMHD defines minority health research as the study of all aspects of health and disease in one or more OMB-defined racial/ethnic minority populations. Minority health research can include comparative research to examine and understand better or worse health outcomes in a racial/ethnic minority group relative to other groups. For example, greater prostate cancer mortality in African American men compared to White men, longer life expectancy in Asian and Hispanic/Latino populations compared to the general population, and higher rates of diabetes in all minority groups. Minority health research also encompasses within-group variation in health, such as asthma prevalence and morbidity in Puerto Ricans compared to Mexican Americans, and variation in cancer rates among American Indians from different regions. Minority health research can also address health conditions or risk and resilience factors specific to or disproportionately found in specific racial/ethnic minority groups, such as pain management in African American sickle cell patients and the use of native or traditional medicines or health practices by American Indians/Alaska Natives and other population groups.

An overarching common theme for all racial/ethnic minorities in the United States is to share a common experience of having been subject to some level of discrimination or social exclusion, which vary across groups and by socioeconomic status (SES) and need to be placed in historical and current contexts. The historical trauma experienced by American Indians as they were displaced from their lands and restricted to reservations, and the legacy of slavery for Black Americans carry a special burden. Moreover, societal discriminatory practices, like redlining to support residential segregation, have affected racial/ethnic minorities and are a fundamental cause that must be understood in order to reduce health disparities.

1.3.3 Health Disparities and Health Disparities Research

NIMHD defines a health disparity as a health difference, based on one or more health outcomes, that adversely affects defined disadvantaged populations. According to the legislation that created NIMHD, a health disparity population is characterized by a pattern of poorer health outcomes, indicated by the overall rate of disease incidence, prevalence, morbidity, mortality, or survival in the population as compared to the health status of the general population. NIH-designated health disparity populations were defined in this legislation and the authority to modify these designations was given to the director of NIMHD in consultation with the director of the Agency for Healthcare Research and Quality (AHRQ). Currently designated health disparity populations include the racial/ethnic minority groups mentioned above, populations of less privileged SES or poor persons from any race/ethnic group, underserved rural populations, and sexual and gender minorities. Health disparities research is devoted to (i) understanding determinants that cause, sustain, or mitigate health disparities; and (ii) how this knowledge is translated into interventions to reduce disparities.

NIMHD's definition of health disparities emphasizes an adverse difference in populations with a social disadvantage. From the NIMHD perspective, a defined difference in health outcomes between populations is not necessarily a disparity. We define disparity as a difference between a disadvantaged population and a more advantaged referent population or the general population.

By comparison, in *Healthy People 2020*, the contributions of social determinants to an individual's ability to achieve good health also include sex, age, and disability.[3] *Healthy People 2020* defines a health disparity as "a particular type of health difference that is closely linked with social, economic, and/or environmental disadvantage." The NIMHD definition emphasizes the disadvantage of population groups, and thus all adverse health outcomes within the designated population groups are considered health disparities. The NIMHD definitions also emphasize the importance of the historical or current link to discrimination or exclusion in determining health and healthcare disparities. The contributions of biological mechanisms, healthcare access and quality, and interventions to ameliorate disparities also need to be considered.

1.3.4 Is It Minority Health or Health Disparities?

There is substantial overlap in minority health research and health disparities research, in particular, research that focuses on worse health outcomes among particular racial/ethnic minority groups compared to Whites or other populations (Figure 1.1). For example, the

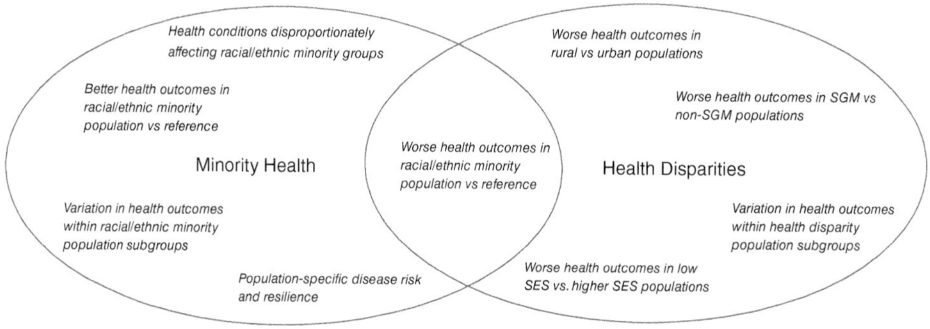

Figure 1.1 Overlapping but distinct constructs of Minority Health and Health Disparities Research.

3 https://www.healthypeople.gov/2020/about/foundation-health-measures/Disparities

fact that African American men have a higher prevalence and mortality for lung cancer is both a minority health and health disparities issue [4]. Blacks are a disparity population, smoke at a similar or lower rate than other racial/ethnic groups, and yet experience up to 50% more lung cancer for the same cigarette smoking intensity [4].

Another example is type 2 diabetes, which is more common and has more severe manifestations in all racial/ethnic minority groups studied in the United States compared to Whites [5]. However, within a staff model healthcare system, the rates of myocardial infarction or heart attacks in patients with diabetes were lower for all minority race/ethnic groups compared to Whites, while the rates of end-stage renal disease were higher. Understanding the factors that lead to these substantial differences in outcomes by race/ethnicity in a well-characterized disease such as diabetes are likely to advance knowledge about mechanisms of how the condition progresses [6].

There are conditions where some racial/ethnic minority groups may have better health outcomes than the reference population, placing the study of these conditions within the domain of minority health research. Latinos or Hispanics have the longest life expectancy by gender of any other demographic group in the United States. This longer life expectancy is a consequence of lower overall rates of cardiovascular disease, cancer, and cerebrovascular disease [7]. Suicide and opioid overdoses are other examples of conditions with lower rates for African Americans, Latinos, and Asians, but higher rates among American Indians/Alaska Natives and lower SES and rural Whites [8]. Research related to the last two populations would fall squarely into the category of health disparities research.

1.3.5 Standardized Measures of Minority Health- and Health Disparities-Related Constructs

Evaluation of health disparities by SES requires a consistent and reliable approach. Household income is a powerful predictor of overall mortality in the United States, with those at the poverty level (defined as ≤ $25 000 household income per year) being three times more likely to die from any cause compared to the mortality rate among persons in a household with an income ≥ $115 000 per year (Figure 1.2) [9]. Although household income is a useful measure of SES, formal years of education may be simpler to obtain,

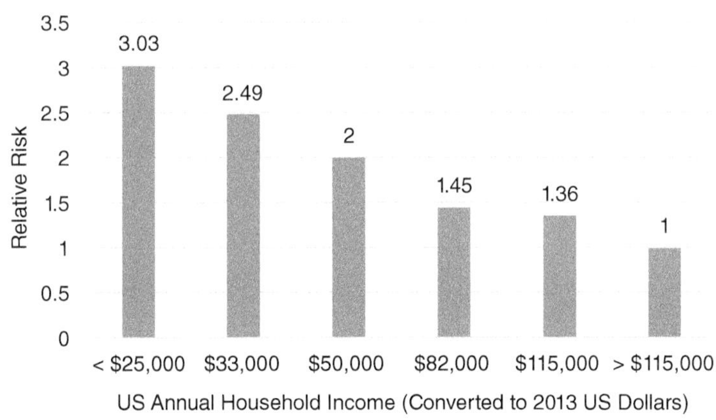

Figure 1.2 Relative risk of all-cause mortality by US annual household income level.

more reliable and stable over the life course, and overall more efficient. Education may also be incorrectly reported, and years of schooling do not equate to quality of education. Thus, depending on the research question at hand, multiple indicators of SES may be the best strategy [10]. This relationship of SES measures and health outcomes is most robust for Whites and Blacks, but there is evident interaction between SES and race/ethnicity that needs to be inherent in most studies focused on minority health and health disparities.

Although health disparities are usually related to SES and traditional minority race/ethnic groups, people living in rural areas are also a minority in their own way. There is increasing evidence of disparities from the leading causes of death among persons living in the most rural areas compared to those living in cities and these disparities merit increased research attention [11]. The operational definition of rurality used in data reported from the Centers for Disease Control and Prevention categorized about 18% of the US population as residing in rural counties. NIMHD endorses this definition and encourages researchers to examine the intersectionality of rural residence, less privileged SES, and race/ethnic minorities.

The addition of sexual and gender minorities (SGM) as a health disparity population is expected to lead to more innovative research to examine health determinants that contribute to disparities. SGM populations share the experience of discrimination with other disparity populations. This implies that there may be shared mechanisms of health determinants to specific conditions that can be examined. Although sexual orientation and gender identity questions are more recent additions to national surveys, it is clear that how these questions are asked significantly affects how respondents identify themselves. More research is needed to understand the best ways to assess sexual orientation and gender identity.

In comparing outcomes across populations, it is important to use standard terminology. Disease or condition rates, important aspects of morbidity, are typically described in terms of incidence and/or prevalence and are important components of population morbidity. Population health is often measured by mortality, so researchers will frequently evaluate not only rates, but also whether a population has premature and/or excessive mortality for specific conditions. Using a general index that reflects population health—such as a global burden of disease measured by disability-adjusted life years (DALYS) or premature years of life lost—can provide insights into understanding mortality patterns in areas where populations vary [12].

For clinicians, rates of risk factors such as level of blood pressure, health-related behaviors such as cigarette smoking, and biomarkers linked to disease outcomes such as glycosylated hemoglobin for diabetes are important components of population morbidity directly in the causal pathway of disease incidence. For clinicians and social scientists, the assessment of how patients or people feel and function using standardized measures provides important data and outcomes to consider. Such outcomes could include psychometrically tested symptom scores for specific conditions, quality of life measures, and activities of daily living. The interdisciplinary nature of minority health and health disparities science will benefit from concurrence on standardized terminology and measures.

1.4 The NIMHD Research Framework: Health Determinants in Action

There is no single identifiable mechanism or etiologic pathway for observed health or healthcare disparities. The robust association of SES with overall mortality underlies all other factors studied. If SES could be made equal across society, would health disparities be eliminated? If access to care were similar for all groups in terms of accessibility, cost and quality, would healthcare outcomes differ by SES or race/ethnicity? Opportunity in education and employment and access to quality healthcare will go a long way to addressing health and healthcare disparities, but other factors will continue to affect outcomes. Understanding complex interactions that may lead to health disparities, such as biological processes and genetics, cultural influences on individual behaviors and lifestyle, effects of the physical and social environment, and access to and interaction with the healthcare system all may play significant roles depending on the populations and conditions. To depict the wide array of determinants that promote or worsen minority health and/or cause, sustain, or reduce health and healthcare disparities, NIMHD scientific staff developed the NIMHD Research Framework (also referred to here as the Framework).[4]

The Framework reflects a hybrid of two models: the socioecological model [13] and the National Institute on Aging (NIA) Health Disparities Research Framework [14]. The socioecological model posits that health and human development are influenced by factors at multiple levels, from the individual to the macro or societal level. The NIA Framework organizes many of these factors into levels of analysis of health disparities relevant to aging research into several domains, including the biological, behavioral, sociocultural, and environmental.

The NIMHD Research Framework (Figure 1.3) has two axes, with the *Y* axis depicting domains of influence on health (biological, behavioral, physical/built environment, sociocultural environment, healthcare system) and the *X* axis depicting levels of influence on health (individual, interpersonal, community, societal). These 20 cells portray a set of determinants, any one of which may be relevant for a particular minority health outcome or health disparity. Examples of factors within each cell are provided in Figure 1.3; this list is not intended to be comprehensive.

In addition to characterizing determinants of health disparities, the Framework describes how health outcomes span multiple levels (individual, family/organizational, community, and population). Health outcomes beyond the individual level reflect collective or aggregate outcomes. Also included in the Framework are designated health disparity populations, as well as demographic factors relevant to understanding the impact of the intersectionality of population group membership—such as sex/gender, disability status, and geographic region—on minority health and health disparities. Finally, the Framework identifies the importance of a life-course perspective when examining determinants across domains of influence, including consideration of early adverse events, chronic and

4 https://www.nimhd.nih.gov/about/overview/research-framework.html

National Institute on Minority Health and Health Disparities Research Framework

Domains of Influence (Over the Life course)	Levels of Influence*			
	Individual	**Interpersonal**	**Community**	**Societal**
Biological	Biological Vulnerability and Mechanisms	Caregiver–Child Interaction Family Microbiome	Community Illness Exposure Herd Immunity	Sanitation Immunization Pathogen Exposure
Behavioral	Health Behaviors Coping Strategies	Family Functioning School/Work Functioning	Community Functioning	Policies and Laws
Physical/Built Environment	Personal Environment	Household Environment School/Work Environment	Community Environment Community Resources	Societal Structure
Sociocultural Environment	Sociodemographics Limited English Cultural Identity Response to Discrimination	Social Networks Family/Peer Norms Interpersonal Discrimination	Community Norms Local Structural Discrimination	Social Norms Societal Structural Discrimination
Healthcare System	Insurance Coverage Health Literacy Treatment Preferences	Patient–Clinician Relationship Medical Decision Making	Availability of Services Safety Net Services	Quality of Care Healthcare Policies
Health Outcomes	**Individual Health**	**Family/ Organizational Health**	**Community Health**	**Population Health**

National Institute on Minority Health and Health Disparities, 2018
*Health Disparity Populations: Race/Ethnicity, Low SES, Rural, Sexual/Gender Minority
Other Fundamental Characteristics: Sex/Gender, Disability, Geographic Region

Figure 1.3 The NIMHD Research Framework. *Source:* National Institute on Minority Health and Health Disparities. NIMHD Research Framework. 2017. Public Domain Available at: https://www.nimhd.nih.gov/about/overview/research-framework.html. Accessed March 2, 2018. (See color representation of this figure in color inserts section.)

cumulative social and environmental exposures, transgenerational transmission of risk and resilience, and critical periods for developing risk or resilience.

The NIMHD Research Framework is intended to convey the complexity of minority health and health disparities and the reality that focusing research exclusively on one cell of the Framework may produce incomplete knowledge. Much of the early research on minority health and health disparities used a unidimensional approach, focusing, for example, on individual behaviors and lifestyle comparisons in an effort to understand observed epidemiological differences and develop tailored interventions; or to take another example, examining the roles of language fluency and culture to understand how immigrants interacted with the healthcare system.

The role of racism and discrimination has for decades been postulated as a major contributor to adverse health outcomes, and the volume of research on this topic has steadily increased [15]. However, racism and discrimination have not been routinely included in health studies that primarily emphasize other types of determinants or etiological factors. For any particular minority health or health disparity issue, multiple factors may be relevant, from lifestyle choices, cultural beliefs, and racism and discrimination to place-based factors such as community violence exposure, availability of green space, public transportation, and access to healthy foods. Social factors, such as social support and size and cohesiveness of social networks and macro-level factors, such as health and other policies, are all relevant as well. The Research Framework is intended to stimulate consideration and research about potential factors and determinants of health outcomes that transcend traditional disciplinary silos or service sector boundaries.

Another principle conveyed by the NIMHD Research Framework is the importance of the interaction between biological, sociocultural, and environmental processes. Early work in minority health and health disparities often focused primarily on biological or social-behavioral factors. The explosion of information describing and understanding biological processes and genetics has provided several plausible models for observed differences in specific conditions. The genetic contribution to the higher prevalence, earlier onset, and higher morbidity of prostate cancer among African Americans is not fully understood and warrants exploration of the social and environmental factors that may play a role in these outcomes [16].

The fact that all racial/ethnic minority groups studied in the United States have a higher prevalence of diabetes would suggest a common pathway that triggers metabolic abnormalities based on chronic stress, nutritional habits, and/or environmental interactions with genetic susceptibility. Major differences in the numbers of suicides by race/ethnicity may be explained by the interactions of underlying neurological and psychological factors with cultural and social factors that have not been measured. Minority health and health disparities researchers, who typically are aware of the social determinants of health, need also to understand and follow developments in the research on biological mechanisms, such as genetic stability, cellular function, and physiological indicators. Such integrative research requires interdisciplinary collaborations and team science approaches.

Additionally, the Framework highlights the importance of healthcare as a significant determinant of minority health and health disparity outcomes. Access to healthcare, usually defined as having health insurance coverage, an identifiable place to obtain care, and a clinician who can manage and coordinate healthcare [17] is an essential component of improving health for health disparity populations. As an increasing proportion of the population develops chronic diseases, the role that the healthcare system and clinicians have in incorporating environmental and sociocultural factors to influence health outcomes intensifies. Differences in intensity of blood pressure control, ordering and following-up cancer screening tests, recommending and administering vaccines, and coordinating with consultants all affect health disparities. Communication and shared medical decision making between patients and clinicians are also factors relevant to health disparities. More research is needed on how race/ethnicity, SES, sexual orientation and gender identity, health literacy, limited English proficiency, and other social determinants influence patient–clinician communication and how this process may affect minority health and health disparities. At the macro-level, local, state, and federal policies regarding availability, eligibility, and cost of healthcare can also result in disparities in access to, utilization of, and quality of care.

1.5 Inclusion of Diverse Participants in Clinical Research

The inclusion of diverse participants in clinical studies is related to minority health research but is a distinct topic. The inclusion of women, children, and minorities in clinical research was mandated by Congress in 1993.[5] Although significant progress has been made

5 https://grants.nih.gov/grants/funding/women_min/guidelines.htm

in engaging women as participants in clinical research, participation from racial/ethnic minority groups remains low. Every clinical study cannot be powered to stratify outcomes by race/ethnicity, but when the data indicate a significant disparity in rates or outcomes, there is a scientific responsibility to address diversity.

One example where this was prospectively addressed is the Diabetes Prevention Trial that evaluated results by race/ethnicity [18]. The lack of diverse genotype data on minority populations in the United States has been highlighted [19], and the randomized trials evaluating efficacy of prostate cancer screening, for example, have had limited minority participation despite the fact that prostate cancer is two to three times more common among African Americans [20–22]. Future clinical research needs to ascertain greater granularity of social and demographic determinants in addition to age, gender, and race/ethnicity, such as SES, birthplace, language proficiency, sexual orientation/gender identity, and geographic residence.

Many investigators find it challenging to recruit minorities to participate in clinical studies. Successful recruitment may require different strategies, more resources, and, frequently, unfamiliar skills. Research indicates that recruitment of minorities may be more successful when active strategies (e.g., more in-person contact, personalized messages, and active outreach) are used rather than passive strategies (e.g., mass mailings, advertisements), which are the norm for many studies [23]. Further research is needed to understand the best mechanisms to recruit and retain minority and other health disparity populations in clinical research, especially clinical trials to identify the most appropriate prevention, detection, and treatment strategies for everyone. Given that minorities constitute close to 40% of the US population, diversity of participants in clinical studies is not merely a legal mandate but critical to the conduct of ethical, equitable, and rigorous science.

1.6 Conclusions

Although significant challenges remain, increasing public interest and scientific advancements point to opportunities to better understand the causes of, and potentially eliminate, certain disparities. It is important for minority health and health disparities researchers to construct consistent/standardized terminology and extend cross-collaborations and networks with diverse research communities using interdisciplinary approaches. The science of minority health and health disparities is a platform not only to reduce inequities but to advance knowledge in health and healthcare for all populations.

The scientists who have addressed minority health and health disparities come from a variety of disciplines. Historically, scientists conducting research on minority health and health disparities have come from epidemiology, sociology, psychology, economics, anthropology, demography, and other social sciences. Increasingly, clinician investigators, health services researchers, and translational scientists have become engaged in these topics, leveraging the power of computational techniques and advances in data collection methods. More recently, scientific advances have led to increased collaboration between social and clinical scientists and basic biomedical and data scientists. The importance of engaging communities, service providers, and policymakers is also increasingly recognized. Clearly, the broad spectrum of the field makes the task of scientific orientation daunting but necessary. The objective of this book is to create a starting point and reference for emerging and early career investigators, including graduate students, postdoctoral scholars, and junior faculty. This book is intended to lay the foundation and be a guide for scientific inquiry in minority health and health disparities research.

1.7 Key Points

- Constructing consistent/standardized terminology on minority health and health disparities research.
- Encouraging collaborations and networks among diverse research communities.
- Promoting an interdisciplinary approach with three major focus areas: clinical and health services research, integrative biological and behavioral sciences, and community health and population health sciences.

Disclaimer

The views and opinions expressed in this chapter are those of the authors only and do not necessarily represent the views, official policy, or position of the U.S. Department of Health and Human Services or any of its affiliated institutions or agencies.

References

1 Heckler, M.M. (1985). *Report of the Secretary's Task Force on Black and Minority Health.* Washington, DC.: US Department of Health and Human Services.

2 Schulman, K.A., Berlin, J.A., Harless, W. et al. (1999). The effect of race and sex on physicians' recommendations for cardiac catheterization. *New England Journal of Medicine* 30: 618–626.

3 National Academy of Medicine (2003). *Unequal Treatment: Confronting Racial and Ethnic Disparities in Health Care* (eds. B.D. Smedley, A.Y. Stith and A.R. Nelson). Washington, DC: National Academies Press.

4 Haiman, C.A., Stram, D.O., Wilkens, L.R. et al. (2006). Ethnic and racial differences in the smoking-related risk of lung cancer. *New England Journal of Medicine* 354: 333–342.

5 Kanaya, A.M., Adler, A., Moffet, H.H. et al. (2011). Heterogeneity of diabetes outcomes among Asians and Pacific Islanders in the US. The Diabetes Study of Northern California (DISTANCE). *Diabetes Care* 34: 930–937.

6 Karter, A.J., Ferrara, A., Liu, J.Y. et al. (2002). Ethnic disparities in diabetic complications in an insured population. *Journal of the American Medical Association* 287: 2519–2527.

7 Centers for Disease Control (2015). Vital signs: leading causes of death, prevalence of diseases and risk factors, and use of health services among Hispanics in the United States — 2009–2013. *Morbidity and Mortality Weekly Report* 64 (17): 469–478.

8 Rudd, R.A., Seth, P., David, F., and Scholl, L. (2016). Increases in drug and opioid-involved overdose deaths––United States, 2010–2015. *Morbidity and Mortality Weekly Report* 65: 1–8.

9 Chetty, R., Stepner, M., Abraham, S. et al. (2016). The association between income and life expectancy in the United States, 2001–2014. *Journal of the American Medical Association* 315 (16): 1750–1766.

10 Cubbin, C., Pollack, C., Flaherty, B. et al. (2011). Assessing alternative measures of wealth in health research. *American Journal of Public Health* 101: 939–947.

11 Garcia, M.C., Faul, M., Massetti, G. et al. (2017). Reducing potentially excess deaths from the five leading causes of death in the rural United States. *MMWR Surveillance Summaries* 66 (2): 1–7.

12 Global Burden of Cardiovascular Diseases Collaboration (2018). The burden of cardiovascular disease among US States, 1990-2016. *JAMA Cardiology* 3 (5): 375–389.

13 Bronfenbrenner, U. (1977). Toward an experimental ecology of human development. *American Psychologist* 32 (7): 513–531.

14 Hill, C.V., Perez-Stable, E.J., Anderson, N.A., and Bernard, M.A. (2015). The National Institute on Aging Health Disparities Research Framework. *Ethnicity & Disease* 25 (3): 245–254.

15 Paradies, Y., Ben, J., Denson, N. et al. (2015). Racism as a determinant of health: a systematic review and meta-analysis. *PLoS One* 10 (9): e0138511. https://doi.org/10.1371/journal.pone.0138511.

16 Bhardwaj, A., Sanjeev, K., Alsam Khan, M. et al. (2017). Racial disparities in prostate cancer: a molecular perspective. *Frontiers in Bioscience, Landmark* 22: 772–782.

17 Starfield, B. (1998). *Primary Care: Balancing Health Needs, Services and Technology*. New York: Oxford University Press.

18 Diabetes Prevention Program Research Group (2002). Reduction in the incidence of type 2 diabetes with lifestyle intervention or metformin. *New England Journal of Medicine* 346 (6): 393–403.

19 Popejoy, A.B. and Fullerton, S.M. (2016). Genomics is failing on diversity. *Nature* 538 (7624): 161–164.

20 Martin, R.M., Donovan, J.L., Turner, E.L. et al. (2018). Effect of a low-intensity PSA-based screening intervention on prostate cancer mortality. The CAP randomized clinical trial. *Journal of the American Medical Association* 319 (9): 883–895.

21 Schröder, F.H., Hugosson, J., Roobol, M.J. et al. (2014). Screening and prostate cancer mortality: results of the European Randomised Study of Screening for Prostate Cancer (ERSPC) at 13 years of follow-up. *Lancet* 384 (9959): 2027–2035.

22 Andriole, G.L., Crawford, E.D., Grubb, R.L. et al. (2009). Mortality results from a randomized prostate-cancer screening trial. *New England Journal of Medicine* 360 (13): 1310–1319.

23 Sturgeon, K.M., Hackley, R., Fornash, A. et al. (2018). Strategic recruitment of an ethnically diverse cohort of overweight survivors of breast cancer with lymphedema. *Cancer* 124 (1): 95–104.

2

Getting Under the Skin

Pathways and Processes that Link Social and Biological Determinants of Disease

Chandra L. Jackson[1], *Rada K. Dagher*[2], *Jung S. Byun*[3], *Tilda Farhat*[4], *and Kevin L. Gardner*[5]

[1] *Epidemiology Branch, Social and Environmental Determinants of Health Equity, National Institute of Environmental Health Sciences, National Institutes of Health, Cary, NC, USA*
[2] *National Institutes of Health, National Institute on Minority Health and Health Disparities, Division of Scientific Programs, Bethesda, MD, USA*
[3] *National Institutes of Health, National Institute on Minority Health and Health Disparities, Division of Intramural Research, Bethesda, MD, USA*
[4] *National Institutes of Health, National Institute on Minority Health and Health Disparities, Office of Science Policy, Planning, Evaluation, and Reporting, Bethesda, MD, USA*
[5] *Department of Pathology and Cell Biology, Columbia University Medical Center, New York, NY, USA*

2.1 Introduction

A fundamental aspect of health disparities research is to understand the differential influences through which "where" and "how" we live affects our daily lives, and to provide everyone with an opportunity to reach and maintain their greatest health potential. "Where and how" we live encompasses, for instance, how we feed, how we sleep, and how we feel. It is also characterized by the differing ways in which we each respond to the vicissitudes of life. These influences on health are complex, multifactorial, interact with each other, accrue throughout the life course, and differentially affect various populations.

The domains that have the greatest influence on health and health disparities include the healthcare system, the sociocultural environment, the physical environment, and behavioral and biological determinants. Of note, modifiable environmental exposures and social factors, combined with a certain level of susceptibility may trigger epigenetic changes, thus producing biological processes contributing to health disparities. These domains are inextricably linked and may combine to produce synergistic effects on health, but their impacts are not distributed uniformly throughout the population. Several diseases and conditions can occur with increased frequency and severity (producing worse outcomes) in certain population groups. Understanding how these domains interact and combine to determine health outcomes (along the dimensions of race, ethnicity, socioeconomic status, and lifestyle) requires a transdisciplinary approach [1]. A transdisciplinary perspective cuts across different areas of expertise while filling in gaps of knowledge and understanding. Through cooperative and integrated efforts, subject matter experts must work together across

The Science of Health Disparities Research, First Edition. Edited by Irene Dankwa-Mullan, Eliseo J. Pérez-Stable, Kevin L. Gardner, Xinzhi Zhang, and Adelaida M. Rosario.

disciplines and with communities to understand the causes/determinants and environmental or social pathways, as well as biological mechanisms of health disparities in order to implement interventions to eliminate or reduce disparities. A precision perspective in this effort requires the recognition of heterogeneity in the environmental, social, and behavioral influences. It also requires an understanding of how population-specific differences in the magnitude of environmental, social, and behavioral influences may be fundamental causes of disparities in disease incidence, progression, and outcomes. Effective interventions—whether social, behavioral, or therapeutic—to eliminate or mitigate health disparities may depend on how thoroughly we understand the biological underpinnings of the contributions of the environment and society (including their influences on behavior) to disease. Developing such interventions will require the combined efforts of interdisciplinary teams, including, for instance, social, behavioral, environmental, and translational scientists. Moreover, understanding the biology of disease and how it may be exacerbated by environmental or behavioral influences will also provide new insights that will assist in identifying population-specific differences in disease susceptibility across the life course, as well as effective interventions at the biological, behavioral, societal, and environmental levels.

In this chapter, we discuss the pathways that influence how environmental, social, behavioral, and biological determinants independently and synergistically influence health disparities. We will begin by discussing the concept of allostatic load and how it provides a framework and perspective through which we can understand how environmental, social, behavioral, and biological determinants combine to influence health outcomes and potentially contribute to health disparities. By focusing on studies of social, racial, and ethnic differences in sleep, nutrition, and depression, we will provide examples from the literature that illustrate how such mechanisms converge to influence disparities in health outcomes.

2.2 Allostasis and Allostatic Load

One of the most significant universal biological imperatives is the maintenance of homeostasis. From the Greek *homeo*, meaning "similar" and *stasis*, meaning "stand" or "state," homeostasis refers to the body's ability to maintain internal stability. Humans and mammals have evolved numerous physiological systems to maintain homeostasis (e.g., body temperature, serum electrolytes, blood pressure, and pH). Similarly, from the Greek *allo,* meaning "other" or "variable," allostasis refers to change or adaptation to maintain stability and preserve homeostasis, particularly in response to environmental challenges (Figure 2.1a). Allostasis, as initially coined by Sterling and Eyer [2] and popularized by McEwen [3], defines the central role of the brain in stress responses through the neuroendocrine and autonomic nervous system. This system is referred to as the hypothalamic-pituitary-adrenal (HPA) axis and, along with elements of the brain governing behavioral and physiological changes in the cardiovascular, metabolic, and autonomic nervous system, is a central component of the global stress response system [4]. Traditionally referred to as the "fight or flight" response, many components of the HPA have adapted to environmental threats or change to promote survival (Figure 2.1b) [5].

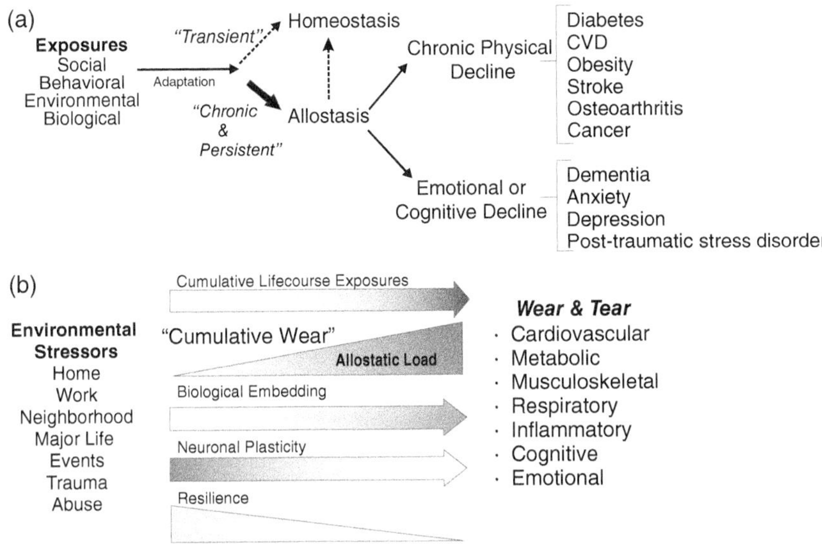

Figure 2.1 Schematic representation of the concept of Allostatic Load: (a) Graphic illustration of conceptual linkages between allostasis and environmental stress and chronic disease; (b) Graphic representation of the role and components of allostatic load over the life course illustrating influences of biological embedding, neuronal plasticity, and cumulative "wear and tear" in response to environmental stressors. (See color representation of this figure in color inserts section.)

However, persistent or chronic overuse of the stress response systems leads to cumulative "wear and tear," or cellular, physiological, cognitive, and emotional dysfunction, that eventually becomes maladaptive. Over time, this "weathering," the disproportionate deterioration as a result of cumulative wear and tear that begins at an earlier age and is patterned by race, can result in disease [6]. The chronic conditions or diseases that result from persistent allostasis are referred to as allostatic load (Figure 2.1b) [3].

Allostatic load provides a conceptual bridge to understanding how the basic cellular and molecular biology underlying human physiology interacts with behavior and environmental exposures to affect health. Furthermore, it provides a framework to help us understand how certain aspects of human lived experience (e.g., social isolation and racism) and environmental exposure can become "embedded" or "baked in" to influence behavioral patterns and biological events across the life course [3]. The central thesis of allostatic load is that cumulative chronic stress may "get under the skin," so that past events, occurring as distantly as early childhood or even prenatally, can have persistent effects far into adulthood [7]. These concepts form the foundation that supports how differences in societal experience can be the root cause of disparities in health outcome [8]. The concept of allostatic load enables an exploration of gene-environment and epigenetic interactions that will ultimately provide insights into intervention.

In summary, allostasis and the influence of allostatic load occur and accumulate throughout life with consequences that ultimately result in chronic physical, emotional, and cognitive decline [3]. Understanding the forces through which societal, behavioral, and environmental determinants combine with biological susceptibility will be the subject of this chapter.

2.3 The HPA Axis

The HPA axis wields a pervasive influence on health behaviors, including diet, physical activity, and sleep [4, 5]. The paraventricular nucleus (PVN) is a core component of the HPA axis. In response to stress, its neurons produce corticotropin releasing factor (CRF) to stimulate release of pituitary adrenal corticotropic hormone (ACTH), which stimulates production of glucocorticoids (predominantly cortisol) from the adrenal cortex (Figure 2.2) [4, 5]. The net result of the release of circulating cortisol is to mobilize energy through gluconeogenesis, cause peripheral inhibition of glucose and amino acid uptake, and impair or blunt the immune inflammatory response [5]. The result is a rapid increase in blood glucose that will persist under conditions of chronic stress. Cortisol also has significant influence on the central nervous system through feedback inhibition of pituitary ACTH secretion and production of CRF from the PVN, in addition to significant influence on neurogenesis in the amygdala, hippocampus, and mesocorticolimbic regions (Figure 2.2).

CRF also acts at the level of the locus coeruleus (LC) to potentiate the secretion of norepinephrine (NE), which works in combination with cortisol to enhance consolidation

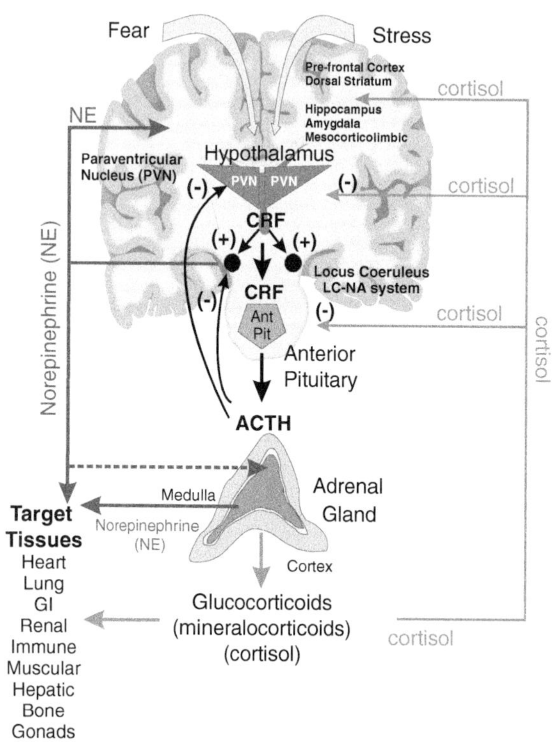

Figure 2.2 Schematic presentation of stress pathway outputs and inputs to the hypothalamic-pituitary-adrenal (HPA) axis in response to stress. NE, norepinephrine; CRF, corticotropin releasing factor; PVN, paraventricular nucleus; ACTH, adreno-corticotropin. *Source:* Derived in part from Spencer and Deak [9]. (See color representation of this figure in color inserts section.)

and retrieval of highly emotional events through their combined influence on neurogenesis in the hippocampus, amygdala, and the mesocorticolimbic system (Figure 2.2) [3]. LC-derived NE also enhances peripheral levels of NE through stimulation of the adrenal medulla, resulting in significant changes in the cardiovascular, pulmonary, renal, gastrointestinal, immune, and hepatic tissues [4, 5].

2.3.1 How We Feed: The Role of the Hypothalamus in Pathways Controlling Feeding and Nutrition

Multiple nuclei within the hypothalamus are involved in complex gating and signaling that result from an intricate mixture of feedback and feed forward involving signals that originate in the complex neurotransmitter and neuroendocrine systems. Often, these interactions involve feedback communication amongst nuclei and signaling molecules from the periphery that indicate nutrient availability (Figure 2.3) [4, 5]. Following a meal, the two major harbingers of nutrient availability, leptin and insulin, play a major role in driving the signaling cascade in the hypothalamus (Figure 2.3) [10]. Elevation of blood glucose levels following a meal results in the activation and secretion of insulin from the beta-islet cells of the pancreas. This occurs in concert with the leptin secretion from adipose tissue, which conveys the signal that energy is available for storage. The arcuate nucleus of the

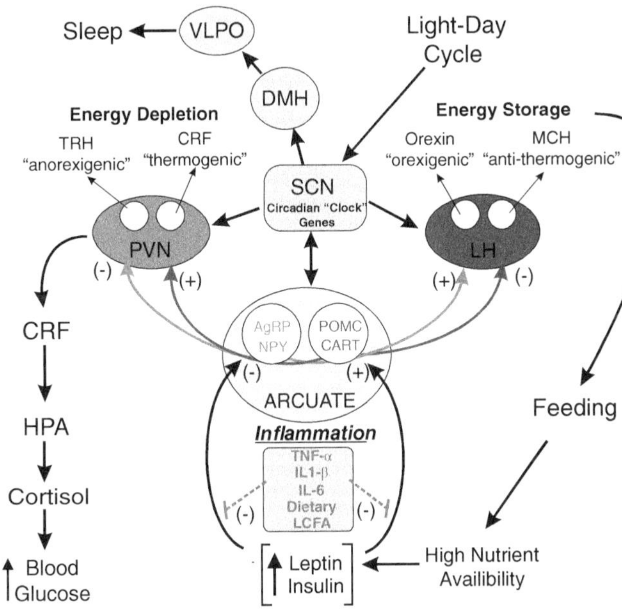

Figure 2.3 Schematic representation of pathways important in hypothalamic control of energy balance and sleep. VLPO, ventrolaterol preoptic nucleus; DMH, dorsomedial hypothalamus; TRH, thyrotropin releasing hormone; CRF, corticotrophin releasing factor; PVN, paraventricular nucleus; SCN, suprachiasmic nucleus; LH, lateral hypothalamus; HPA, hypothalamus-pituitary-adrenal axis; AgRP, agouti-related peptide; NPY, neuropeptide Y; POMC, pro-opiomelanocortin; CART, cocaine and amphetamine related transcript; MCH, melanin-concentrating hormone; TNF-α, tumor necrosis factor alpha; IL1-β, interleukin 1 beta; IL-6, interleukin-6; LCFA, long chain fatty acids. (See color representation of this figure in color inserts section.)

hypothalamus has receptors for both insulin and leptin and responds differently depending on which region of the arcuate receives input (Figure 2.3). Leptin and insulin repress signaling from neurons of the arcuate that produce the orexigenic agouti-related peptide (AgRP) and neuropeptide Y (NPY), each of which functions to repress neurons in the PVN responsible for producing anorexigenic thyrotropin releasing hormone (TRH), and CRF. AgRP and NPY also stimulate neurons in the lateral hypothalamus (LH) that produce the orexigenic orexin and the orexigenic and anti-thermogenic melanin concentrating hormone (MCH) [5, 11]. The net effect of leptin and insulin on AgRP- and NPY-producing neurons is the repression of hunger. In contrast, the function of leptin and insulin on arcuate neurons is to stimulate the expression of anorexigenic pro-opiomelanocortin (POMC) and cocaine and amphetamine regulated transcript (CART). POMC- and CART-active neurons in the PVN express anorexigenic and thermogenic TRH and CRH, while repression neurons in the LH express orexigenic orexin and MCH. Therefore, in contrast to AgRP and NPY, POMC and CART repress hunger. Thus AgRP/NPY and POMC/CART have significant influence on modulating feeding, nutrient sensing, and the flux of signaling through the HPA axis [11, 12]. Finally, the hypothalamus is subject to modulation by local inflammatory events. Studies have shown that normal physiological control of feeding by the hypothalamus can be disrupted by local inflammation, some of which is influenced by high levels of dietary long chain fatty acids (LCFA) [4, 11]. This local inflammation blunts the response of the hypothalamus to leptin and insulin, thus implicating a role for inflammation in the evolution of obesity (Figure 2.3) [11].

2.3.2 How We Sleep: Light–DayCycle, Circadian Clock, and Hypothalamic Linkages to Metabolic Control and Sleep

The 24-hour rotation of the Earth creates a light-dark cycle that has conditioned the biology and temporal behavior of humans and other mammals to synchronize activities with the 24-hour clock. This control is exerted in the hypothalamus at the level of the suprachiasmatic nucleus (SCN) (Figure 2.3). In response to temporal cues (or *zeitgeber* for "time giver"), the SCN executes control of central clock gene function through both endocrine and autonomic outputs. The SCN inputs to the arcuate nucleus, PVN, and LH to cyclically regulate feeding, energy storage, and body temperature [13, 14]. Signals from the SCN are also relayed to the dorsomedial hypothalamus (DMH), which sends significant output to the ventrolateral preoptic nucleus (VLPO), which directly promotes sleep. Its activity is highest during sleep and inhibited by NE and serotonin arising from the arousal network. Thus, timing of sleep, feeding, nutrient intake, times of arousal, and energy expenditure are regulated cyclically (e.g., the early rise of cortisol at the beginning of the day just before activity) [14]. In addition to the SCN, most peripheral cells and tissues express autonomous clocks that function in synchrony with the SCN through the action of a regulatory cassette of clock genes. This regulatory cassette consists of a class of transcriptional activators (Clock/BMAL1) that, in addition to controlling the expression of numerous genes involved in carbohydrate and lipid metabolism, induce the expression of their own repressors (Per1-3/Cry1-2) [13].

Besides receiving signals from the SCN, these peripheral clocks are influenced by surrogate zeitgebers other than light, including food availability, glucocorticoids, and temperature. Because of these linkages, it is easy to understand how disruption of this network by environmental changes, such as shift work, jet lag, light pollution, changes in eating patterns, and sleep deprivation, may lead to metabolic imbalance and weight gain. These changes all have the common effect of altering hypothalamic control of processes involved in energy storage and expenditure. Indeed, studies of shift work provide important insights into the linkages between circadian rhythm and metabolism. Studies of individuals experiencing circadian misalignment reveal significantly higher rates of hypoleptinemia, insulin resistance, inverted cortisol peaks, and increased blood pressures. All of these measures are frequently used biomarkers of allostatic load and consistent with the association between poor sleep health and metabolic syndrome, diabetes, and obesity (Figure 2.3) [13, 14].

Notably, these influences appear bidirectional, as patients with diabetes show dampened circadian oscillation of both glucose and insulin. Thus, synchronization of activity, feeding behavior, and sleep with light–dark cycles has significant influence on metabolic imbalance and obesity. As mentioned above, many aspects of these regulatory circuits are likely to have bidirectional effects, since glucocorticoids have a strong influence on the expression of multiple clock genes in the SCN and peripheral tissues [13].

2.3.3 How We Feel: Stress and the Role of HPA Axis in Memory and Mood

A critically significant adaptive function for promoting and maintaining survival is the memory system. The active process of recollection and embedding of past adverse experiences provides a means of assuring that similar events can be appropriately avoided or circumnavigated. The release of glucocorticoids through the HPA axis in response to stress also influences pathways and mechanisms that govern multiple aspects of memory [15]. Cortisol has access to many areas of the brain that express both glucocorticoid receptor (GR) and mineralocorticoid receptor (MR) [9]. While GRs are ubiquitous throughout the brain, MRs are mostly concentrated in the limbic regions, including the hippocampus and the amygdala. GR activation is associated with consolidation, the transition of short-term to long-term memory. MR activation is associated with evaluation and response to stressful experiences.

Research indicates that glucocorticoids have complicated influences on memory consolidation and retrieval. Moreover, chronic elevation of glucocorticoid in combination with NE has a strong influence on neurogenesis through their influence on neuronal plasticity by promoting dendritic remodeling at the cellular level [5]. Such influences are complex and sometimes paradoxical. While glucocorticoids play a role in the consolidation of memory of emotionally arousing experiences, they can also impair retrieval of primary learned or memorized events. Learning and memory processes take place in complex networks of regions in the brain that may act in parallel or competition. Studies suggest that emotional stress can promote rerouting of systems of memory and learning.

For example, new findings have demonstrated that stress can create a shift in the hippocampal cognitive learning to "reflexive" stimulus-response or habitual learning governed through the dorsal striatum (DR) (Figure 2.4). A role for the amygdala in mediating this switch has been suggested [15]. The activation of stress pathways in the regulation of emotional memory is highly adaptive and important for survival. However, its chronic

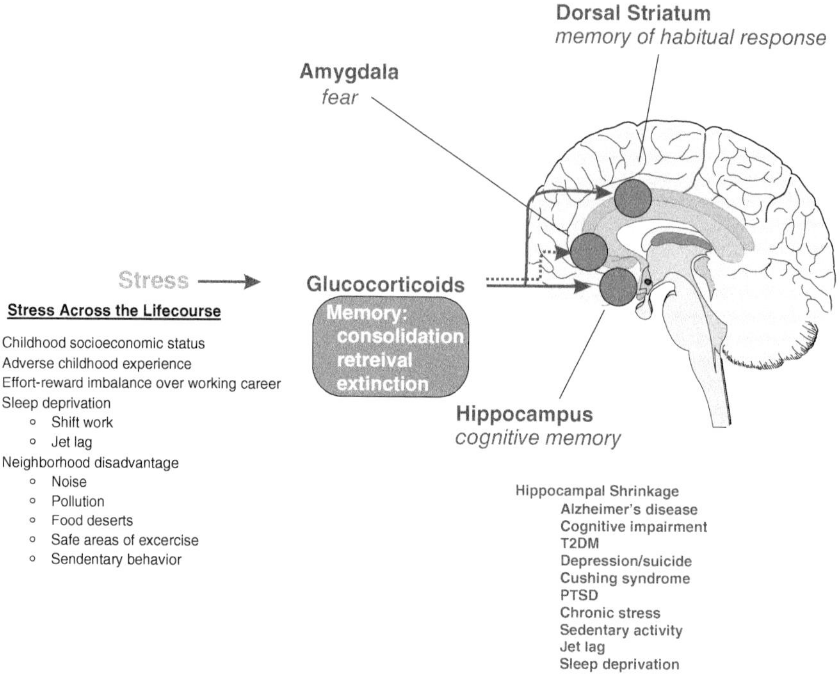

Figure 2.4 Schematic diagram of the influence of glucocorticoids and stress on memory, learning, and mood. T2DM, Type 2 diabetes mellitus; PTSD, post-traumatic stress syndrome. (See color representation of this figure in color inserts section.)

activation occurs at the cost of an "allostatic load," which becomes the underlying cause of emotional conditions or disease characterized by fear, anxiety, depression, and post-traumatic stress disorders. The hippocampus is a central region controlling disorders of mood, including anxiety and depression [16]. Patients with major depression develop hippocampal atrophy [17]. The neuronal population within the hippocampus is maintained by a population of neural progenitors that are exquisitely sensitive to glucocorticoids [18, 19]. In experimental systems, excess glucocorticoid exposure has been associated with decreased neuronal growth and differentiation and increased cell death in the hippocampus [16, 18, 20, 21]. An example of this "cost" of chronically elevated cortisol level is well documented by the mood disorders associated with patients suffering from Cushing syndrome, a condition characterized by high levels of cortisol due to chronic activation of the HPA by a pituitary tumor [20].

2.4 Anticipatory Biology and Behavior: The Embedding of Exposures Across the Life Course

The life course perspective provides a useful framework to understand the dimensions of human behavior; that is, how chronological age, life transitions, and social and environmental contexts and changes shape human lives, behaviors, and health, from birth through the lifespan. The framework builds on evidence that lived experiences set people on trajectories that influence health and well-being over the life course. Each of these experiences in time contribute to the health status so that repeated periods of stress, nutritional deficiency, and/or social deprivation beginning in utero and continuing through childhood and adult life set up a sequence of poor development at each stage. The cumulative stress or the accumulation of these deprivations over the lifespan can eventually manifest in poor health and premature death (Figure 2.1b). This is a biological imperative that persists throughout the life course and involves mechanisms in which the changes and sequelae of adaptation become "baked-in" or "get under the skin" and accrue across the life course—and, in some cases, are transmitted trans-generationally [3, 22].

This imperative and its impact on allostatic load are supported through many concepts that examine processes and mechanisms of adaptation. One concept originally proposed by Neel [23] and referred to as the "thrifty gene" concept purports that certain genes linked to metabolisms could have evolved in response to specific environmental conditions characterized by fluctuations in nutrient availability. Thus, genes and gene systems that are energy conserving would have selective advantage. A simple example provided by Power and Schulkin [24] would be the adaptation of the genetic pathways linked to metabolic energy conservation, storage, and adiposity that could have evolved in response to environments that were episodically exposed to periods of famine. Chronic use of these adaptive pathways in environmental conditions that are nutrient rich would thus predispose to an allostatic load, which would be reflected in metabolic imbalance, obesity, and diabetes [24].

The mechanisms underlying this type of adaptation or selection need not be restricted to Darwinian time scales, but could involve mechanisms that evolve in a developmental context through epigenetic regulation. A classic example is the cohort of individuals that were exposed to prenatal famine in the Dutch Hunger Winter of 1944–1945, and who later, in adulthood, experienced higher incidence of depression, schizophrenia, hyper-responsiveness to stress, metabolic imbalance, and diabetes. This cohort was later found to show epigenetic alterations at genes important in controlling metabolism, including loss of imprinted deoxyribonucleic acid (DNA) methylation at the IGF2 gene [25]. Thus, the earliest of exposures became embedded across the life course at loci that were likely to influence disease risk decades later.

Mechanisms of epigenetic regulation are likely to have a pervasive influence on how traits and susceptibility become embedded and accrue across the life course [26]. Epigenetic changes and regulatory pathways regulating the hypothalamic expression of the fat mass and obesity-associated gene (FTO), a regulator of food intake and energy balance, has been widely studied [27]. Exploration of epigenetic control of the genes governing pathways that regulate feeding, sleep, energy balance, and metabolism has only begun [22]. In this regard, GR and MR are nuclear receptors for glucocorticoids that influence a broad array of epigenetic regulatory events linked to changes in chromatin modification and DNA methylation. Thus, the stress response system is a prime target for the process in which its overuse can become stably embedded in the genome through alterations in epigenetic control.

Epigenetic biomarkers may reflect causative mechanisms or provide quantitative measures of cumulative biological influences of allostatic load. For example, studies are beginning to show that life's stressors accelerate aging, which can be reflected in DNA methylation patterns, some of which drive or reflect the aging phenotype [22]. In other instances, these stressors shorten telomeres found in peripheral blood lymphocytes, which have been significantly associated with allostatic load and impoverished neighborhood resources [28]. Changes that occur to chromatin with aging are a new, fast-growing area of study [22].

2.4.1 Studies of Stress and Allostatic Load Across the Life Course

The concept of allostatic load and the biomarkers used to study it have been widely accepted in studies of how behavioral, social, and biological exposures combine to influence the wear and tear on the human body and predispose it to disease. Numerous biomarkers are now in frequent use in many studies and are gradually providing a clearer picture of the role of stress and allostatic load in health disparities [3, 29]. For example, a review by Rodriquez et al. describes markers of allostatic load (e.g., systolic and diastolic blood pressure; hemoglobin A_{1c}; C-reactive protein; dehydroepiandrosterone; cortisol; telomeres) representing multiple biological systems (e.g., cardiovascular, metabolic, inflammatory, neuroendocrine) that are derived from research among minority and disparity populations (see Table 2.1) [30]. Research is also showing that measures of allostasis differ by racial/ethnic group and, therefore, need to have a range. For example, cardiometabolic risk factors for Whites with a body mass index (BMI) of 25 appeared at a much lower BMI than in patients from other groups: 22.9 for African Americans, 21.5 for Hispanics, 20.9 for Chinese, and 19.6 for South Asians. After controlling for lifestyle factors such as smoking, alcohol intake, exercise, and diet, only 21% of Whites had cardiometabolic risk factors but normal BMIs, whereas 39% of Hispanic Americans and 44% of South Asians had a normal weight but a higher risk for chronic diseases that are usually associated with being overweight [31].

Several cohort studies have documented clear associations between adverse childhood experiences and allostatic load for both men and women, with major influences on health behaviors, BMI, and socioeconomic factors in adult life [29]. Studies measuring allostatic load across ethnic and racial populations demonstrate how a multitude of social factors, including discrimination, can contribute to overall allostatic load, significantly alter health aging, and act as an early indicator of subsequent disease burden [32].

In the following sections, we provide examples of how such studies are increasing our understanding of causal mechanisms and how they impact health disparities and provide new and important insights for future interventions.

2.5 Sleep

Sleep is an essential human need for maintaining biological homeostasis. It plays an important role as both a mediator and downstream measure of allostatic load. Defined as "a recurring, reversible neuro-behavioral state of relative perceptual disengagement from and unresponsiveness to the environment" [33] with complex behavioral and physiological processes, it normally begins in a light stage of Non-Rapid Eye Movement (NREM) sleep and progresses into deeper periods of NREM over approximately 90-minute cycles. NREM sleep typically comprises 75% of the sleeping period [34]. The remaining period consists of Rapid Eye Movement (REM) sleep, which is characterized by increased brain activity causing the eyes to dart back and forth [35]. Based on polysomnographic measures of the

Table 2.1 Representative list of biomarkers for allostatic load [3, 22].

Biomarkers of allostatic load		
Biomarker	**Function**	**Typical measures**
Cardiovascular		
Systolic blood pressure	Cardiovascular activity	Mean of 3, resting, (mm Hg)
Diastolic blood pressure	Cardiovascular activity	Mean of 3, resting, (mm Hg)
Heart rate	Cardiovascular activity	Mean of 3, resting, (bpm)
Peak expiratory flow	Lung function	Highest of 3, Peak flow, ($l\,min^{-1}$)
Metabolic		
HDL, $mmol\,l^{-1}$	Low levels, unfavorable	Serum, nonfasting, ($mmol\,l^{-1}$)
LDL	Low levels, favorable	Serum, nonfasting, ($mmol\,l^{-1}$)
Triglyceride	High levels, unfavorable	Serum, nonfasting, ($mmol\,l^{-1}$)
Total cholesterol/HDL	Low levels, favorable	Ratio <5.9
BMI	Measure of adiposity	BMI >30
WHR (waist: hip)	Abdominal adiposity	WHR >9.94
HbA_{1c}	Measure of insulin resistance	$mmol\,mol^{-1}$
Inflammation		
IGF-1	Inflammation	Serum nonfasting, ($nmol\,l^{-1}$)
Fibrinogen	Blood clotting, thrombosis risk	Serum, nonfasting, ($g\,l^{-1}$)
IgE	Type I hypersensitivity	$KU\,l^{-1}$
CRP	Acute inflammation	Citrated plasma, ($mg\,l^{-1}$)
Neuroendocrine		
DHEA-S	HPA antagonist	Urinary, ($ng\,ml^{-1}$)
Cortisol t1	HPA function	Salivary, 45 after awakening, ($nmol\,l^{-1}$)
Cortisol t2	HPA function	Salivary, 3 after awakening, ($nmol\,l^{-1}$)
Epigenetic		
DNA methylation	Embedded cellular aging	Circulating white cells
Telomere shortening	Cellular aging	Circulating leukocytes

HDL, high density lipoprotein; LDL, low density lipoprotein; WHR, waist to hip ratio; CRP, C-reactive protein; DHEA-S, dihydroepiandosterone sulfate; HPA, hypothalamus pituitary adrenal axis.

neurophysiological signal's frequency and amplitude, the various sleep stages are associated with particular health benefits. For instance, slow wave (also known as deep sleep or "N3") is considered the most physiologically restorative stage of sleep [35]. In addition to other physiological processes, cellular and tissue growth as well as repair occurs in this stage. The metabolically active hormones referred to previously are also released. During this stage, blood pressure also customarily drops. Breathing tends to slow. Regarding REM sleep, this stage has been associated with restoration of cognition, which is important for learning, consolidating memories, and regulating emotions [36].

While sleep needs vary by age group, the National Sleep Foundation recommends that adults need between seven and nine hours of quality, uninterrupted sleep with minimal arousals for sleep to be fully restorative, and for maintaining and promoting optimal health [37]. Some data suggest that the mean sleep duration of persons in the United States has decreased over time. Estimates indicate that American adults are getting an average of 6.1 hours of sleep per day independent of quality [38] and sleep deprivation annually costs the US economy approximately $411 billion [39]. An estimated 60 million prescriptions for sleeping pills are written per year and more than 300,000 automobile accidents occur, with 6400 of them being fatal, that are largely attributed to the under-recognized phenomenon of drowsy driving, which results from sleep deprivation [40]. Unfortunately, approximately 50 to 70 million US adults have sleep or wakefulness disorders, such as obstructive sleep apnea (OSA), insomnia, narcolepsy, and restless leg syndrome. In 2006, sleep was considered an unmet public health need by the Institute of Medicine [41]. The Department of Health and Human Services also established *Healthy People 2020* goals to increase the proportion of Americans getting sleep in recommended amounts.

Poor sleep (or insufficient sleep quantity and/or inadequate quality relative to what is needed for optimal performance as well as physical and mental/emotional health) is a gateway to numerous conditions linked to rising allostatic load and increased risk of suboptimal chronic health conditions, such as obesity, hypertension, type 2 diabetes, cardiovascular disease, and premature mortality [42]. In terms of potential biological mechanisms, partial sleep deprivation is believed to influence hormonal and growth factor signaling to the hypothalamus. For instance, sleep deprivation has been shown to downregulate leptin (a hormone associated with satiety) and upregulate ghrelin, a hormone associated with increased hunger [42]. The combined action of these hormones has considerable influence on the development of a host of chronic health conditions (e.g., obesity, type 2 diabetes). Additionally, hyperarousal of certain brain regions related to sleep deprivation is associated with hedonistic behaviors like sweet/salty food cravings. Furthermore, sleep disturbances often precede the onset of mood disorders, like anxiety and depression, which, for instance, exacerbate risk factors for suicide and depression relapse [43].

2.5.1 Sleep Health Disparities and Allostatic Load

Large racial/ethnic disparities in sleep and sleep disorders exist. Blacks (especially men) are nearly twice as likely as Whites to report short sleep duration [44]. Blacks are also over 60% more likely to report being long sleepers [45]. Evidence also suggests that Latinos compared to non-Hispanic Whites may also be at higher risk for both a short and long duration of sleep [46].

Furthermore, sleep architecture appears to differ by race. For instance, Blacks consistently have been shown, based on objective sleep measures, to experience less deep (or N3) sleep and more light sleep (N1 and N2), take more time to fall asleep, and spend less time in bed actually sleeping. Although there are believed to be many undiagnosed cases of sleep apnea (especially among minorities), the prevalence is highest among Blacks, and Black children are alarmingly, four to six times more likely than White children to have OSA [47].

Moreover, Blacks also experience more severe health consequences as a result of significantly higher rates of obesity and hypertension at the time of sleep apnea diagnosis [48]. Compared to Whites, Blacks, and Hispanics, Chinese individuals appear to have a stronger association of increasing BMI and waist circumference with sleep-disordered breathing [49].

Regarding insomnia, the results are mixed, but Whites seem to have the highest prevalence. These estimates may, however, be based on methodological issues related to varying insomnia definitions, the use of self-reported data, and the need for a formal diagnosis that is highly influenced by healthcare access and utilization, which often varies by race and ethnicity. Sleep irregularities may also contribute to disproportionately high rates of chronic conditions, such as hypertension among Blacks. This may be due, at least in part, to a higher likelihood of nondipping blood pressure during deep or N3 sleep. Disproportionately poorer sleep may also contribute to disparities in metabolic health conditions, such as type 2 diabetes, by contributing to greater levels of insulin resistance and cardiovascular disease through pathways linked to higher levels of the circulating inflammatory cytokines described above. Therefore, preventing or minimizing the impact of modifiable environmental and social disturbances on sleep duration, quality, and timing could help mitigate racial/ethnic and socioeconomic disparities in common chronic health conditions associated with rising levels of allostatic load [12].

Understanding the fundamental cause(s) of racial/ethnic differences in sleep, its contribution to allostatic load, and the combined influence on chronic health conditions across the lifespan is important. The ultimate goal of generating new knowledge in these areas is to provide strategies for intervention and disease risk reduction that will most efficiently reduce widespread health-associated inequities. The upstream factors that could contribute significantly to racial disparities include physical and social environments and/or epigenetic influences.

Racial groups have historically lived and continue to live in different environments—both physically and socially—and these differences in exposures to adverse conditions lead to differential levels of physiological dysregulation at multiple points across the life course. These differences likely contribute to an increased risk of acute and chronic health conditions. Experimental and observational studies demonstrate that the properties contributing to sleep are both intrinsic and extrinsic, and involve a broad complement of environmental as well as social factors, such as light, temperature, noise, air quality, diet/nutrition, medication use, physical activity, safety, and social cohesion. Adverse exposure to any of these and other factors can lead to insufficient or disturbed sleep [50]. Compared to Whites, Blacks and many other racial/ethnic groups have greater exposure to potentially health-damaging environments and less access to health-promoting goods and services that may influence their risk of both poor sleep and health. This could be due to living in less healthy urban neighborhoods or resource-limited rural regions with higher frequencies of food deserts or swamps, and where there is less access to safe areas for recreational activity.

2.5.2 Sleep Health Disparities and Genetics

Genetics is clearly another important factor to consider. Genes account for 31–55% of a person's sleep duration, with behavior and environment accounting for the remainder [51].

Although genetics is a major contributor to sleep in all individuals, modifiable environmental factors are likely to play a larger role than genetics in poor sleep and poorer health between races. For instance, racial residential segregation resulting in differential exposure to artificial light at night, noise, and safety concerns due to poverty, for example, may have a more profound impact on sleep than genetic differences. Moreover, single nucleotide polymorphisms with minor allele frequencies (which generally need to be functional and may still confer small risks) are likely to play a much smaller role than the environment in shaping racial variation in health and disease. It is now widely accepted that race, as distinct from ancestry, is a social construct whose distinguishing phenotypes (e.g., skin and eye color) are not strongly linked to lifestyle-related health conditions that cause the majority of premature deaths in the United States [52]. In fact, the human genome project found that all humans are 99.9% the same (meaning that genetic variants overlap). There is more genetic diversity within (especially among people of African ancestry) than between racial categories—which are, again, socially constructed and have changed over time.

Despite the large overlap of genetic variants across racial groups, studies should investigate how much (or how little) ancestral differences in genetic architecture (e.g., haplotypes, structural variation) contribute to observed differences in various sleep characteristics that affect health outcomes. These studies should also acknowledge that behaviors such as sleep are complicated and strongly influenced by nongenetic environmental factors. In fact, race could serve as a proxy/marker for relative advantage or disadvantage by identifying groups with more or less social and economic obstacles to health.

Finally, gene expression can be modified without altering an individual's actual genetic code (known as epigenetics, or gene-by-environment interactions). Environmental factors such as diet/nutrition, as well as toxicants (e.g., phthalates, flame retardants, surfactants, bisphenol-A, and fungicides like vinclozolin), can promote the epigenetic transgenerational inheritance of disease and phenotypic variation [53]. Epigenetics (especially in early life) may play an important role in racial differences in the relationship between sleep and many health outcomes influenced by sleep, since Blacks and Whites have historically been and currently are exposed to different physical and social environments across the life course.

Behaviors and biological mechanisms along the pathway between sleep and suboptimal health outcomes are likely affected by the aforementioned upstream environmental and social factors, contributing to the often-observed racial differences in risk, prevalence, and severity of both poor sleep characteristics and their downstream health sequelae.

2.5.3 Methodologies in Sleep Research

Research focused on sleep and sleep disparities can advance the field by addressing important methodological challenges. For instance, studies should include more prospective designs to establish temporal ordering of sleep and health outcomes. Objective sleep measures, as obtained by wrist actigraphy and polysomnography (the gold standard), are needed to minimize hard-to-report sleep-related behavior misclassification (e.g., sleep latency, snoring among individuals without a bed partner). Researchers should keep in mind that addressing sleep disparities at a population level to prevent subsequent poor health outcomes will require more racial/ethnic minority inclusion in studies, as well as scientific evidence (likely from multilevel research including individual, interpersonal/household, and neighborhood/community factors), that helps coordinate efforts that connect policies, systems, and environmental interventions.

2.6 How We Feed: Nutrition and Nutrition-related Health Disparities

Nutrition-related disparities are well documented, with low nutritional value foods and low-quality diets more prevalent among disadvantaged than nondisadvantaged populations [54]. This could be related to food deserts and food insecurity, which are more likely to be concentrated in neighborhoods with a high proportion of minority and low-income populations. According to the United States Department of Agriculture (USDA), food deserts include census tracts that are low-income (i.e., where a portion of residents make below a threshold income) and low access (i.e., where a significant portion of the residents is far from a supermarket). The USDA defines food insecurity as "the limited or uncertain availability of nutritionally adequate and safe foods or limited or uncertain ability to acquire acceptable foods in socially acceptable ways" [55].

Residents of minority and low-income neighborhoods have limited availability of affordable and healthy foods, and must rely on the nonhealthy/non-nutritious foods available in their proximity or incur significant tangible (e.g., monetary) or intangible (e.g., long distance travel) costs to buy more expensive healthy options [56]. Low-income neighborhoods have lower availability of fruits and vegetables and nutrient-dense foods [56]. Such neighborhoods also include more small stores than large supermarkets. This is important because prices of staple food items such as milk, eggs, fruits, and vegetables are higher in small stores compared to large ones, with one study finding the difference ranging from 10% to 50% higher [56]. Smaller stores also have fewer fresh items and sell smaller units of packaged items than supermarkets, resulting in higher prices [56, 57]. One study estimated that a resident of a poor Black neighborhood lived 1.1 miles farther from the closest supermarket than a resident of a poor White neighborhood [57]. Further, low-income neighborhoods have a high concentration of unhealthy food restaurants (e.g., fast food), thus adding to the limited availability of healthy food options [57]. As a result, and in contrast to the diet of higher-income populations, the diet of low-income populations generally costs less than a healthier, nutrient-dense diet because it is of low nutritional value and low quality [54].

Disparities in intake of nutritious food are associated with disparities in health and well-being. Populations whose diet is mostly of low nutritional quality experience poor dietary patterns, including low fruit and vegetable intake. They also experience lower levels of serum carotenoids and higher rates of obesity and high blood pressure, which are associated with diabetes, cardiovascular disease, and other chronic diseases [54, 56, 57]. Even after controlling for home food-environment factors, an association persisted between neighborhood food-environment factors such as food deserts and obesity [58].

Healthy, nutrient-dense food is often neither available nor affordable in communities where health disparity populations live. Consequently, those populations experience a high prevalence of nutrition-related diseases and conditions. Possible biological pathways linking nutritional stressors to health disparities include allostatic load and changes to the microbiome. Both pathways are responses to local inflammations caused by nutritional stressors, including inadequate intake of key nutrients or a high intake of calorie-dense foods, such as frequent fast food meals. The hypothalamus, which plays a central role in regulating feeding and nutrition, is subject to modulation by local inflammatory events. Studies have shown that normal physiological control of feeding by the hypothalamus can be disrupted by local inflammation, some of which is influenced by high levels of dietary LCFA [4, 11]. This local inflammation blunts the response of the hypothalamus to leptin and insulin, thus implicating a role for inflammation in the evolution of obesity.

Alteration of the microbiome can also result from nutritional stressors. The microbiome, mostly located in the gut, is made of microorganisms that are largely beneficial but can sometimes be harmful. Both types exist in harmony in healthy bodies, but the balance can be disrupted by factors such as infectious disease (and frequent use of antibiotics), chronic stress, or malnutrition. Health disparity populations often experience one or more of these factors, thus resulting in frequent detrimental alterations to their microbiome, thus increasing their susceptibility to disease.

Nutrition-related disparities contribute to disparities in the prevalence, incidence, and burden of disease through their impact on allostatic load and the human microbiome. Efforts to decrease those disparities and their effect entail modifying the environments that promote inadequate nutrition and improving understanding of the mechanisms through which inadequate nutrition contributes to health disparities. In terms of environmental changes, moving populations to better environments, although possible (e.g., Moving to Opportunity program [59]), is not scalable or sustainable. Interventions have been tested to improve the physical and social environments to support healthier food choices. This evidence needs to be effectively disseminated and scaled up. Future research needs to examine the feasibility, scalability, and sustainability of interventions to improve the physical and social environments that promote and sustain healthy nutrition among health disparity populations. With regard to mechanisms, more research is needed to understand how nutritional stressors contribute to allostatic load and affect the microbiome, and the extent to which changes in the microbiome can contribute to health disparities. In particular, linking environmental, social, and behavioral nutrition-related data to data on the microbiome would enable examining how these contextual factors affect the microbiome in various racial/ethnic groups [60].

2.7 How We Feel: Mood and Depression

Depression is ranked by the World Health Organization (WHO) as the leading cause of disability around the world [61]. It is also the major contributor to deaths by suicide, approximately 800,000 per year [61]. Worldwide, depression prevalence was 4.4% in 2015 (about 322 million people) with higher prevalence among females (5.1%) compared to males (3.6%) [61]. In 2015 in the United States, 6.7% of adults had experienced at least one major depressive episode (MDE) in the previous year and 4.3% had experienced an MDE with severe impairment. This national survey defined MDE using the standard criteria from the *Diagnostic and Statistical Manual of Mental Disorders–4th edition (DSM-IV)* as "adults who had a period of 2 weeks or longer in the past 12 months when they experienced a depressed mood or loss of interest or pleasure in daily activities, and they had at least some additional symptoms, such as problems with sleep, eating, energy, concentration, and self-worth." Such adults were considered to have experienced an MDE, while those whose depression caused "severe problems with their ability to manage at home, manage well at work, have relationships with others, or have a social life" were considered to have an MDE with severe impairment [62].

Major depressive disorder[1] (MDD) inflicts serious economic and social costs on society [63]. In addition, it has been associated with low educational attainment, marital disruption, unstable employment, and increased risk of teenage pregnancy among those who experience early onset [64]. Moreover, it has been associated with significant decreases in role functioning, including low marital quality, low job performance, and low income. MDD has also been implicated in increased risk of onset, persistence, and severity of many chronic physical disorders, as well as increased early mortality due to suicide and a variety of physical disorders [65].

Emotional stress, whether acute or chronic, is an established causal factor in the onset and progression of depression [66]. Stress arises when individuals experience demands (physical or psychological) or threats for which they do not have sufficient social and material resources to respond to and thereby ameliorate the resulting health effects [67]. In contrast, when sufficient resources are available to individuals, these demands or threats are perceived as challenges that promote growth and resilience [68]. Stress hormones, such as cortisol and adrenaline, released in response to stressors are usually adaptive and beneficial in the process of allostasis, or maintaining stability during times of change [69]. However, when these stress hormones are not turned off when not needed, are overused, or are not turned on when needed, allostatic load ensues, a wear and tear on the body and the brain that is evident in major depressive illness [70].

Individuals of low socioeconomic status (SES) and racial/ethnic minorities (especially Blacks, Latinos, and American Indians/Alaska Natives) usually live and work in environments with limited material and social resources and increased probability of exposure to threats, and thus report higher levels of stress than high-SES and White individuals [68]. Moreover, racial and ethnic minorities have a greater likelihood of being exposed to discrimination and discrimination-related stressors, threats to safety and financial security, violence, and barriers to occupational advancement across their life course than Whites [71]. In fact, WHO has deemed chronic stress as one of the main social determinants of health inequities [68]. The consistent adverse exposures that racial and ethnic minorities experience starting as early as the prenatal period and throughout their adult lives result in the accumulation of chronic stress; this repeated adaptation to stressors and overuse of the stress response systems results in cumulative "wear and tear" or physiological deterioration that contributes to allostatic load, the chronic conditions or diseases caused by persistent allostasis [6]. Depression is one of the chronic diseases whose development has been shown to be associated with disturbance of allostasis and inflammatory activation of the immune system [72]. For example, the HPA axis plays a key role in mediating the

1 The main difference between MDD and MDE is that the latter includes episodes of depression that are part of a bipolar disorder, while MDD does not include bipolar depression. However, because most of lifetime MDE is MDD and different epidemiological surveys may focus on either one, we use whichever term is used in the study cited.

relationship between stress and depression through the excessive production of glucocorticoids [73]. Over time, prolonged exposure to adversity and excessive levels of glucocorticoids may have negative effects on areas of the developing brain that pertain to depression [74]. A number of studies have shown an association between elevated allostatic load and depression [75–77]. Moreover, a case control study showed that the effects of childhood physical abuse on adult depression were mediated by allostatic load [78]. Given that racial/ethnic minorities may experience more adversity and stressful circumstances than Whites throughout their life course and thus accumulate higher allostatic load, one would expect that they experience higher prevalence of chronic conditions, including MDDs. However, while racial and ethnic disparities in many physical illnesses have been clearly established, the disparities in depression prevalence present a more complex picture. For example, data (2013–2016) from the National Health and Nutrition Examination Survey (NHANES) show no statistically significant differences in depression prevalence for Hispanic, non-Hispanic White, and non-Hispanic Black adults, yet lower depression prevalence for Asian adults [79]. Earlier national studies showed lower rates of lifetime prevalence of depression and lower lifetime risk for depression among Latinos, Blacks, Asian, and Pacific Islanders compared to Whites, and higher rates of lifetime prevalence and risk for American Indians/Alaska Natives [70]. However, compared to Whites, Blacks and Latinos experience higher rates of dysthymia, a persistent depressive disorder characterized by chronic negative mood, and are more likely to characterize their own depressive symptoms as severe [80, 81]. Moreover, once present, depression has been shown to follow a chronic and persistent pattern among Blacks [82]. In addition, the consequences of depression have been shown to be worse and more disabling in Blacks compared to Whites, including lower quality of life, higher severity of symptoms when depression is present, and lower rates of treatment [83]. The underlying assumptions about the rates of depression in racial/ethnic minorities, as established in earlier national studies, coupled with physician–patient communication problems, may lead to discrimination in the detection and diagnosis of depression among minorities. For example, in two mental healthcare studies, clinicians responded with less alacrity to variation in depression severity in minority patients compared to White patients [84]. Moreover, two other studies found that Hispanics and Blacks were less frequently given depression diagnoses than Whites [85].

Given that culture can influence the experience of depression and how patients communicate their symptoms, diagnosis recommendations for depression specific to Latinos state that a somatic presentation pattern of depression is more common in this minority group [83]. Similarly, diagnosis recommendations for Blacks state that somatic and neurovegetative symptoms are more common in this group than mood or cognitive symptoms of depression [86]. These cultural differences may be playing a role in underdiagnosing MDD among racial/ethnic minorities. Interestingly, the literature consistently finds lower lifetime risk for depression among immigrants than the US-born population, pointing to a potential role of environmental factors and stressors specific to the US context and culture [87]. For example, using data from the Collaborative Psychiatric Epidemiology Surveys, one study examined American-born and foreign-born Blacks, Hispanics, Asians, and Afro-Caribbeans, as well as American-born Whites and found that immigrants had 37% lower odds of lifetime MDD than those who were American-born and that foreign-born Afro-Carribeans and foreign-born Asians had the lowest rates of lifetime depression [88]. In addition, some studies point to lower allostatic load among immigrants compared to

US-born racial/ethnic minorities. For instance, one study that used the 2001–2010 NHANES data found a health advantage for foreign-born Blacks in terms of allostatic load as compared to US-born Blacks [89]. Two other studies found lower allostatic load among foreign-born Mexican immigrants compared to US-born Mexicans [90, 91].

Studies that examine allostatic load as a mediator of racial/ethnic disparities in depression prevalence have shown mixed findings. For example, a study by Rodriquez et al. that used 2005–2012 NHANES data on individuals aged 40–79 years showed that while African Americans and Latinos had higher levels of allostatic load and higher rates of depressive disorder than Whites, there was no association between allostatic load and depressive disorder in either racial/ethnic group [92]. Another study that used 2005–2010 NHANES data on Black and White adults aged 18–64 found that the relationship between allostatic biomarkers and depression varied with gendered race, such that different biomarkers were associated with depression among Black men, Black women, White men, and White women [93]. An Australian study that examined the relationship of two indices of sustained stress, cortisol, and allostatic load with depressive symptoms in Australian Aboriginal and Torres Strait Islander populations found that neither of these two hypothesized mediators were related to depressive symptoms [94]. A review article on 58 studies of allostatic load, health, and health disparities highlighted the heterogeneity in the operationalization of allostatic load and the measurement of its biomarkers, which makes it difficult to compare different studies in this area of research [95]. It also highlighted the importance of moving beyond the simple count-based index to more complex scoring methods for creating allostatic indices, such as recursive partitioning, canonical correlation analyses, and Grade of Membership (GOM) multivariate methods.

The field of racial/ethnic disparities in depression prevalence still has many unanswered questions. Underdiagnosing depression in racial/ethnic minorities by the healthcare system and cultural differences in experiences of depression may provide some explanations as to why current depression prevalence rates in racial/ethnic minorities are not in line with assumptions of the "weathering hypothesis." However, studying mediators of the association between stress and depression using the concept of allostatic load shows promise in clarifying the mechanisms through which early life adversity and chronic exposure to stress may affect depression outcomes in different racial/ethnic groups. In addition, future research should examine mechanisms of resilience in immigrant versus US-born populations given the consistent finding of lower depression prevalence among immigrants. Moreover, research on allostatic load and depression in racial/ethnic minorities may gain from a more unified approach to operationalizing allostatic load and measurement of its biomarkers to enhance comparability of findings. Finally, a focus on identifying key developmental periods in the life course to target the antecedents of allostatic load, in addition to delineating key areas of vulnerability and resilience, may yield effective interventions to decrease depression rates among vulnerable populations.

2.8 Summary

In this chapter, we outline major biological pathways through which environmental, behavioral, and social exposures influence health. We summarize the mechanisms through which disparate exposures to societal inequities and inequalities are manifested in the form of health disparities. The central theme in this discussion has been the concept of allostatic load and how alterations in the most basic aspects of the human experience (where and how we feel, sleep, and feed) provide the conduit through which unequal exposure to behavioral social stressors (e.g., poverty, social isolation, discrimination, and racism) culminates in disparate disease frequency and outcome. Though its linkage to specific diseases is beyond the scope of this chapter, the influence of allostatic load is one of the major gateways to chronic disease and disease conditions (cardiovascular, metabolic, and cancer). Understanding these biological physiological pathways identifies important areas for future investigation that will augment strategies for intervention, reveal mechanisms of differential disease susceptibility, and elucidate how these factors interact with behavioral, social, and environmental determinants to influence disease incidence and outcome.

2.9 Key Points

- The domains that have the greatest impact on health disparities include environmental, social, behavioral, and biological determinants. These domains are inextricably linked and their combinations may have synergistic impacts on health disparities.
- The concept of allostatic load provides a conceptual framework from which to understand the root or fundamental causes of health disparities in sleep, diet/nutrition, and depression.
- The concept of allostatic load provides the mechanistic underpinnings revealing the biological factors that link the influence of societal conditions (e.g., socioeconomic status, social isolation, racism) to disparities in health and disease outcomes.
- Understanding the biological pathways through which environmental and behavioral determinants affect health disparities will provide scientific insights that could inform interventions to decrease health disparities.
- Sleep is an essential human need for maintaining biological homeostasis.
- Poor sleep is a gateway to higher allostatic load and numerous mental and physical health conditions such as anxiety and depression, obesity, hypertension, type 2 diabetes, cardiovascular disease, and premature mortality.
- Large racial/ethnic disparities in sleep and sleep disorders exist. Therefore, preventing or minimizing the impact of modifiable environmental and social disturbances on sleep duration, quality, and timing could help mitigate racial/ethnic and socioeconomic disparities in common chronic health conditions associated with rising levels of allostatic load.
- Addressing sleep disparities to prevent subsequent poor health outcomes will require more racial/ethnic minority inclusion in scientific studies.
- Disparities in the consumption of nutritious food are associated with disparities in health and well-being.
- Food deserts and food insecurity are more likely to be concentrated in neighborhoods with a high proportion of minority and low-income populations, which likely contributes to health disparities.
- While racial and ethnic disparities in many physical illnesses have been clearly established, the depression disparities have not been clearly delineated.

- Individuals of low SES and racial/ethnic minorities (especially Blacks, Hispanics/Latinos, and Native Americans) usually live and work in environments with limited material and social resources and increased probability of exposure to threats—and, thus, report higher levels of stressors and stress than high-SES and White individuals. Nonetheless, Blacks and Hispanics/Latinos have been less frequently given depression diagnoses than Whites.
- Research should focus on identifying key developmental periods in the life course to target the antecedents of allostatic load which, in addition to delineating key areas of vulnerability and resilience, may yield effective interventions to decrease depression rates among vulnerable populations.

Disclaimer

The views and opinions expressed in this chapter are those of the authors only and do not necessarily represent the views, official policy, or position of the U.S. Department of Health and Human Services or any of its affiliated institutions or agencies.

References

1 Gardner, K. (2018). The science of cancer health disparities: a young discipline with an old heritage. *The American Journal of Pathology* 188 (2): 268–270.

2 Sterling, P. and Eyer, J. (1981). Biological basis of stress-related mortality. *Social Science & Medicine Part E* 15 (1): 3–42.

3 Seeman, T.E., McEwen, B.S., Rowe, J.W. et al. (2001). Allostatic load as a marker of cumulative biological risk: MacArthur studies of successful aging. *Proceedings of the National Academy of Sciences of the United States of America* 98 (8): 4770–4775.

4 Chrousos, G.P. and Gold, P.W. (1992). The concepts of stress and stress system disorders. Overview of physical and behavioral homeostasis. *Journal of the American Medical Association* 267 (9): 1244–1252.

5 Nicolaides, N.C., Kyratzi, E., Lamprokostopoulou, A. et al. (2015). Stress, the stress system and the role of glucocorticoids. *Neuroimmunomodulation* 22 (1-2): 6–19.

6 Geronimus, A.T., Hicken, M., Keene, D. et al. (2006). "Weathering" and age patterns of allostatic load scores among blacks and whites in the United States. *American Journal of Public Health* 96 (5): 826–833.

7 McEwen, B. and Lasley, E.N. (2003). Allostatic load: when protection gives way to damage. *Advances in Mind-Body Medicine* 19 (1): 28–33.

8 Williams, D.R. and Sternthal, M. (2010). Understanding racial-ethnic disparities in health: sociological contributions. *Journal of Health and Social Behavior* 51 (Suppl): S15–S27.

9 Spencer, R.L. and Deak, T. (2017). A users guide to HPA axis research. *Physiology & Behavior* 178: 43–65.

10 Nicolaides, N.C., Charmandari, E., Kino, T. et al. (2017). Stress-related and circadian secretion and target tissue actions of glucocorticoids: impact on health. *Frontiers in Endocrinology (Lausanne)* 8: 70.

11 Le Thuc, O., Stobbe, K., and Cansell, C. (2017). Hypothalamic inflammation and energy balance disruptions: spotlight on chemokines. *Frontiers in Endocrinology (Lausanne)* 8: 197.

12 Saper, C.B., Scammell, T.E., and Lu, J. (2005). Hypothalamic regulation of sleep and circadian rhythms. *Nature* 437 (7063): 1257–1263.

13 Huang, W., Ramsey, K.M., Marcheva, B. et al. (2011). Circadian rhythms, sleep, and metabolism. *The Journal of Clinical Investigation* 121 (6): 2133–2141.

14 Plano, S.A., Casiraghi, L.P., Moro, P.G. et al. (2017). Circadian and metabolic effects of light: implications in weight, homeostasis and health. *Frontiers in Neurology* 8: 558.

15 de Quervain, D., Schwabe, L., and Roozendaal, B. (2017). Stress, glucocorticoids and memory: implications for treating fear-related disorders. *Nature Reviews Neuroscience* 18 (1): 7–19.

16 Kino, T. (2015). Stress, glucocorticoid hormones, and hippocampal neural progenitor cells: implications to mood disorders. *Frontiers in Physiology* 6: 230.

17 Sheline, Y.I., Wang, P.W., Gado, M.H. et al. (1996). Hippocampal atrophy in recurrent major depression. *Proceedings of the National Academy of Sciences of the United States of America* 93 (9): 3908–3913.

18 Spijker, A.T. and van Rossum, E.F.C. (2012). Glucocorticoid sensitivity in mood disorders. *Neuroendocrinology* 95 (3): 179–186.

19 McEwen, B.S. (2005). Glucocorticoids, depression, and mood disorders: structural remodeling in the brain. *Metabolism: Clinical and Experimental* 54 (5): 20–23.

20 Starkman, M.N., Gebarski, S.S., Berent, S. et al. (1992). Hippocampal-formation volume, memory dysfunction, and cortisol-levels in patients with Cushings-syndrome. *Biological Psychiatry* 32 (9): 756–765.

21 Mann, J.J. and Currier, D. (2007). A review of prospective studies of biologic predictors of suicidal behavior in mood disorders. *Archives of Suicide Research* 11 (1): 3–16.

22 Gassen, N.C., Chrousos, G.P., Binder, E.B. et al. (2017). Life stress, glucocorticoid signaling, and the aging epigenome: Implications for aging-related diseases. *Neuroscience and Biobehavioral Reviews* 74 (Pt B): 356–365.

23 Neel, J.V. (1962). Diabetes mellitus: a "thrifty" genotype rendered detrimental by "progress"? *American Journal of Human Genetics* 14: 353–362.

24 Power, M.L. and Schulkin, J. (2012). Maternal obesity, metabolic disease, and allostatic load. *Physiology & Behavior* 106 (1): 22–28.

25 Heijmans, B.T., Tobi, E.W., Stein, A.D. et al. (2008). Persistent epigenetic differences associated with prenatal exposure to famine in humans. *Proceedings of the National Academy of Sciences of the United States of America* 105 (44): 17046–17049.

26 Tobi, E.W., Goeman, J.J., Monajemi, R. et al. (2014). DNA methylation signatures link prenatal famine exposure to growth and metabolism. *Nature Communications* 5: 5592.

27 Frayling, T.M., Timpson, N.J., Weedon, M.N. et al. (2007). A common variant in the FTO gene is associated with body mass index and predisposes to childhood and adult obesity. *Science* 316 (5826): 889–894.

28 Zalli, A., Carvalho, L.A., Lin, J. et al. (2014). Shorter telomeres with high telomerase activity are associated with raised allostatic load and impoverished psychosocial resources. *Proceedings of the National Academy of Sciences of the United States of America* 111 (12): 4519–4524.

29 Barboza Solis, C., Kelly-Irving, M., Fantin, R. et al. (2015). Adverse childhood experiences and physiological wear-and-tear in midlife: findings from the 1958 British birth cohort. *Proceedings of the National Academy of Sciences of the United States of America* 112 (7): E738–E746.

30 Rodriquez, E.J., Kim, E.N., Sumner, A.E. et al. (2019). Allostatic load: importance, markers, and score determination in minority and disparity populations. *Journal of Urban Health* 96 (Suppl 1): 3–11.

31 Tomiyama, A.J., Hunger, J.M., Nguyen-Cuu, J. et al. (2016). Misclassification of cardiometabolic health when using body mass index categories in NHANES 2005–2012. *International Journal of Obesity* 40 (5): 883–886.

32 Chyu, L. and Upchurch, D.M. (2011). Racial and ethnic patterns of allostatic load among adult women in the United States: findings from the National Health and Nutrition Examination Survey 1999–2004. *Journal of Women's Health (2002)* 20 (4): 575–583.

33 Keenan, S. and Hirshkowitz, M. (2011). *Chapter 141 - Monitoring and staging human sleep.* In: *Principles and Practice of Sleep Medicine*, 5e (eds. M.H. Kryger, T. Roth and W.C. Dement), 1602–1609. Philadelphia, PA: W.B. Saunders.

34 Van Cauter, E., Holmback, U., Knutson, K. et al. (2007). Impact of sleep and sleep loss on neuroendocrine and metabolic function. *Hormone Research* 67 (Suppl 1): 2–9.

35 Andrade, L., Caraveo-Anduaga, J.J., Berglund, P. et al. (2003). The epidemiology of major depressive episodes: results from the International Consortium of Psychiatric Epidemiology (ICPE) Surveys. *International Journal of Methods in Psychiatric Research* 12 (1): 3–21.

36 Walker, M.P. (2009). The role of sleep in cognition and emotion. *Annals of the New York Academy of Sciences* 1156: 168–197.

37 Hirshkowitz, M., Whiton, K., Albert, S.M. et al. (2015). National sleep foundation's updated sleep duration recommendations: final report. *Sleep Health* 1 (4): 233–243.

38 Ford, E.S., Cunningham, T.J., and Croft, J.B. (2015). Trends in self-reported sleep duration among US adults from 1985 to 2012. *Sleep* 38 (5): 829–832.

39 Hafner, M., Stepanek, M., Taylor, J. et al. (2017). Why sleep matters-the economic costs of insufficient sleep: a cross-country comparative analysis. *Rand Health Quarterly* 6 (4): 11.

40 Tefft, B.C. (2018). Acute sleep deprivation and culpable motor vehicle crash involvement. *Sleep* 41 (10): 144.

41 Institute of Medicine Committee on Sleep Medicine Research (2006). *Sleep Disorders and Sleep Deprivation: An Unmet Public Health Problem* (eds. H.R. Colten and B.M. Altevogt), 424. Washington, DC: The National Academies Press.

42 Gallicchio, L. and Kalesan, B. (2009). Sleep duration and mortality: a systematic review and meta-analysis. *Journal of Sleep Research* 18 (2): 148–158.

43 Bersani, F.S., Iannitelli, A., Pacitti, F. et al. (2012). Sleep and biorhythm disturbances in schizophrenia, mood and anxiety disorders: a review. *Rivista di Psichiatria* 47 (5): 365–375.

44 Krueger, P.M. and Friedman, E.M. (2009). Sleep duration in the United States: a cross-sectional population-based study. *American Journal of Epidemiology* 169 (9): 1052–1063.

45 Hale, L. and Do, D.P. (2007). Racial differences in self-reports of sleep duration in a population-based study. *Sleep* 30 (9): 1096–1103.

46 Hale, L. and Do, D. (2007). Racial differences in self-reports of sleep duration in a population-based study. *Sleep* 30: A123–A124.

47 Boss, E.F., Smith, D.F., and Ishman, S.L. (2011). Racial/ethnic and socioeconomic disparities in the diagnosis and treatment of sleep-disordered breathing in children. *International Journal of Pediatric Otorhinolaryngology* 75 (3): 299–307.

48 Meetze, K., Gillespie, M.B., and Lee, F.S. (2002). Obstructive sleep apnea: a comparison of black and white subjects. *Laryngoscope* 112 (7 Pt 1): 1271–1274.

49 Chen, X., Wang, R., Lutsey, P.L. et al. (2016). Racial/ethnic differences in the associations between obesity measures and severity of sleep-disordered breathing: the Multi-Ethnic Study of Atherosclerosis. *Sleep Medicine* 26: 46–53.

50 Knutson, K.L. and Van Cauter, E. (2008). Associations between sleep loss and increased risk of obesity and diabetes. *Annals of the New York Academy of Sciences* 1129: 287–304.

51 Watson, N.F., Buchwald, D., Vitiello, M.V. et al. (2010). A twin study of sleep duration and body mass index. *Journal of Clinical Sleep Medicine* 6 (1): 11–17.

52 Martin, A.R., Gignoux, C.R., Walters, R.K. et al. (2017). Human demographic history impacts genetic risk prediction across diverse populations. *American Journal of Human Genetics* 100 (4): 635–649.

53 Waterland, R.A., Travisano, M., Tahiliani, K.G. et al. (2008). Methyl donor supplementation prevents transgenerational amplification of obesity. *International Journal of Obesity* 32 (1379): 1373–1379.

54 Darmon, N. and Drewnowski, A. (2015). Contribution of food prices and diet cost to socioeconomic disparities in diet quality and health: a systematic review and analysis. *Nutrition Reviews* 73 (10): 643–660.

55 United States Department of Agriculture and Economic Research Service (2018). *Food Security in the US: Overview* September 05, 2018 [cited 2019 May 14, 2019]; https://www.ers.usda.gov/topics/food-nutrition-assistance/food-security-in-the-us. (accessed June 16, 2020).

56 Caspi, C.E., Pelletier, J.E., Harnack, L.J. et al. (2017). Pricing of staple foods at supermarkets versus small food stores. *International Journal of Environmental Research and Public Health* 14 (8): 915.

57 Tach, L. and Amorim, M. (2015). Constrained, convenient, and symbolic consumption: neighborhood food environments and economic coping strategies among the urban poor. *Journal of Urban Health* 92 (5): 815–834.

58 Chen, D., Jaenicke, E.C., and Volpe, R.J. (2016). Food environments and obesity: household diet expenditure versus food deserts. *American Journal of Public Health* 106 (5): 881–888.

59 US Department of Housing and Urban Development (2019). *Moving to opportunity for fair housing*. https://www.hud.gov/programdescription/mto (accessed 16 June, 2020).

60 Findley, K., Williams, D.R., Grice, E.A. et al. (2016). Health disparities and the microbiome. *Trends in Microbiology* 24 (11): 847–850.

61 World Health Organization (2017). *Depression and Other Common Mental Disorders: Global Health Estimates*, 1–24. Geneva: World Health Organization.

62 Center for Behavioral Health Statistics Quality. (2016). *Key substance use and mental health indicators in the United States: results from the 2015 National Survey on drug use and health*. SAMHSA: Rockville, MD.

63 Greenberg, P.E., Fournier, A.-A., Sisitsky, T. et al. (2015). The economic burden of adults with major depressive disorder in the United States (2005 and 2010). *The Journal of Clinical Psychiatry* 76 (2): 155–162.

64 Kessler, R.C. (2012). The costs of depression. *Psychiatric Clinics of North America* 35 (1): 1–14.

65 Kessler, R.C. and Bromet, E.J. (2013). The epidemiology of depression across cultures. *Annual Review of Public Health* 34 (1): 119–138.

66 Hammen, C. (2005). Stress and depression. *Annual Review of Clinical Psychology* 1 (1): 293–319.

67 American Psychological Association and APA Working Group on Stress and Health Disparities (2017). *Stress and Health Disparities: Contexts, Mechanisms, and Interventions Among Racial/Ethnic Minority and Low-Socioeconomic Status Populations*, 1–76. Washington: American Psychological Association.

68 Bartley, M., Wilkinson, R., and Marmot, M. (eds.) (2003). *Social Determinants of Health: The Solid Facts*, 2e. Denmark: World Health Organization.

69 McEwen, B.S. (2003). Mood disorders and allostatic load. *Biological Psychiatry* 54 (3): 200–207.

70 Breslau, J., Aguilar-Gaxiola, S., Borges, G. et al. (2007). Risk for psychiatric disorder among immigrants and their US-born descendants: evidence from the national comorbidity survey replication. *The Journal of Nervous and Mental Disease* 195 (3): 189–195.

71 Sternthal, M.J., Slopen, N., and Williams, D.R. (2011). Racial disparities In health: how much does stress really matter? *Du Bois Review: Social Science Research on Race* 8 (1) 95–113.

72 Oglodek, E., Szota, A., Just, M. et al. (2014). The role of the neuroendocrine and immune systems in the pathogenesis of depression. *Pharmacological Reports* 66 (5): 776–781.

73 Lupien, S.J., McEwen, B.S., Gunnar, M.R. et al. (2009). Effects of stress throughout the lifespan on the brain, behaviour and cognition. *Nature Reviews Neuroscience* 10 (6): 434–445.

74 Anacker, C., O'Donnell, K.J., and Meaney, M.J. (2014). Early life adversity and the epigenetic programming of hypothalamic-pituitary-adrenal function. *Dialogues in Clinical Neuroscience* 16 (3): 321–333.

75 Juster, R.P., Marin, M.-F., Sindi, S. et al. (2011). Allostatic load associations to acute, 3-year and 6-year prospective depressive symptoms in healthy older adults. *Physiology & Behavior* 104 (2): 360–364.

76 Kobrosly, R.W., Seplaki, C.L., Cory-Slechta, D.A. et al. (2013). Multisystem physiological dysfunction is associated with depressive symptoms in a population-based sample of older adults. *International Journal of Geriatric Psychiatry* 28 (7): 718–727.

77 Kobrosly, R.W., van Wijngaarden, E., Seplaki, C.L. et al. (2014). Depressive symptoms are associated with allostatic load among community-dwelling older adults. *Physiology & Behavior* 123: 223–230.

78 Scheuer, S., Wiggert, N., Brückl, T.M. et al. (2018). Childhood abuse and depression in adulthood: the mediating role of allostatic load. *Psychoneuroendocrinology* 94: 134–142.

79 Brody, D.J., L.A. Pratt, and J.P. Hughes, *Prevalence of Depression Among Adults Aged 20 and Over: United States, 2013–2016*. NCHS Data Brief, 2018(303): p. 1–8.

80 Kessler, R.C., Birnbaum, H., Demler, O. et al. (2005). The prevalence and correlates of nonaffective psychosis in the national comorbidity survey replication (NCS-R). *Biological Psychiatry* 58 (8): 668–676.

81 Riolo, S.A., Nguyen, T.A., Greden, J.F. et al. (2005). Prevalence of depression by race/ ethnicity: findings from the national health and nutrition examination survey III. *American Journal of Public Health* 95 (6): 998–1000.

82 Walsemann, K.M., Gee, G.C., and Geronimus, A.T. (2009). Ethnic differences in trajectories of depressive symptoms: disadvantage in family background, high school experiences, and adult characteristics. *Journal of Health and Social Behavior* 50 (1): 82–98.

83 Lewis-Fernandez, R., Das, A.K., Alfonso, C. et al. (2005). Depression in US Hispanics: diagnostic and management considerations in family practice. *The Journal of the American Board of Family Practice* 18 (4): 282–296.

84 Balsa, A.I., McGuire, T.G., and Meredith, L.S. (2005). Testing for statistical discrimination in health care. *Health Services Research* 40 (1): 227–252.

85 Borowsky, S.J., Rubenstein, L.V., Meredith, L.S. et al. (2000). Who is at risk of nondetection of mental health problems in primary care? *Journal of General Internal Medicine* 15 (6): 381–388.

86 Das, A.K., Olfson, M., McCurtis, H.L. et al. (2006). Depression in African Americans: breaking barriers to detection and treatment. *The Journal of Family Practice* 55 (1): 30–39.

87 Breslau, J., Aguilar-Gaxiola, S., Kendler, K.S. et al. (2006). Specifying race-ethnic differences in risk for psychiatric disorder in a USA national sample. *Psychological Medicine* 36 (1): 57–68.

88 Budhwani, H., Hearld, K.R., and Chavez-Yenter, D. (2015). Depression in racial and ethnic minorities: the impact of nativity and discrimination. *Journal of Racial and Ethnic Health Disparities* 2 (1): 34–42.

89 Doamekpor, L.A. and Dinwiddie, G.Y. (2015). Allostatic load in foreign-born and US-born blacks: evidence from the 2001–2010 national health and nutrition examination survey. *American Journal of Public Health* 105 (3): 591–597.

90 Peek, M.K., Cutchin, M.P., Salinas, J.J. et al. (2010). Allostatic load among non-Hispanic whites, non-Hispanic blacks, and people of Mexican origin: effects of ethnicity, nativity, and acculturation. *American Journal of Public Health* 100 (5): 940–946.

91 Kaestner, R., Pearson, J.A., Keene, D. et al. (2009). Stress, allostatic load and health of Mexican immigrants. *Social Science Quarterly* 90 (5): 1089–1111.

92 Rodriquez, E.J., Livaudais-Toman, J., Gregorich, S.E. et al. (2018). Relationships between allostatic load, unhealthy behaviors, and depressive disorder in US adults, 2005–2012 NHANES. *Preventive Medicine* 110: 9–15.

93 Bey, G.S., Jesdale, B.M., Ulbricht, C.M. et al. (2018). Allostatic load biomarker associations with depressive symptoms vary among US black and white women and men. *Healthcare (Basel)* 6 (3): 105.

94 Berger, M., Taylor, S., Harriss, L. et al. (2019). Hair cortisol, allostatic load, and depressive symptoms in Australian Aboriginal and Torres Strait Islander people. *Stress*: 1–9.

95 Beckie, T.M. (2012). A systematic review of allostatic load, health, and health disparities. *Biological Research for Nursing* 14 (4): 311–346.

3

Racial/Ethnic, Socioeconomic, and Other Social Determinants

Tiffany L. Gary-Webb[1, 2], Sara E. Baumann[1], Erik J. Rodriquez[3], Lydia A. Isaac[4], and Thomas A. LaVeist[5]

[1] Department of Behavioral and Community Health Sciences, University of Pittsburgh Graduate School of Public Health, Pittsburgh, PA, USA
[2] Department of Epidemiology, University of Pittsburgh Graduate School of Public Health, Pittsburgh, PA, USA
[3] Division of Intramural Research; National Heart, Lung and Blood Institute, National Institutes of Health, Bethesda, MD, USA
[4] The George Washington University, Milken Institute School of Public Health, Department of Health Policy and Management, Washington, DC, USA
[5] Tulane University School of Public Health and Tropical Medicine, New Orleans, LA, USA

3.1 Introduction

Race and ethnicity are among the most frequently used concepts in research conducted by public health, nursing, and medical scientists. It is well known that most racial/ethnic minority groups experience higher rates of morbidity and mortality compared to Whites. Likewise, the association between lower socioeconomic status (SES) and lower educational levels with poor health outcomes is one of the most established relationships in the literature. This chapter will explore the concepts and measurement of race and ethnicity and aspects of SES, provide some epidemiological data for racial/ethnic groups in the United States, as well as discuss studies that have used innovative methods to explain the intersection of race and ethnicity, SES, and social context.

3.2 Introduction to the Topic, Including Key Definitions

There is increasing recognition that social determinants are major contributors to health above and beyond access to and use of healthcare services [1]. Social conditions are so closely linked to health that health disparities are an indicator of social and economic inequalities [2]. Since a discussion of the full spectrum of social determinants of health, while important, is beyond the scope of this chapter, we focus on the concepts and measurement of race and ethnicity and aspects of SES, provide epidemiological data for racial/ethnic groups in the United States, and explore studies that have used innovative methods to study the

The Science of Health Disparities Research, First Edition. Edited by Irene Dankwa-Mullan, Eliseo J. Pérez-Stable, Kevin L. Gardner, Xinzhi Zhang, and Adelaida M. Rosario.

intersection of race and ethnicity, SES, and social context. This chapter will set the foundation for subsequent chapters in this text to investigate the complexity of racial/ethnic and socioeconomic health disparities.

3.3 Used and Recommended Measures and Research Methods

3.3.1 Conceptual Model

In this section, we discuss a conceptual model developed by LaVeist [3], modified to include ethnicity and ethnic differences. It aims to explain the influence of social and behavioral factors on racial/ethnic disparities in health (see Figure 3.1). Race and ethnicity are considered latent factors in this model, with skin color being the most commonly used indicator. Along the left side of the model is the social pathway, which illuminates how an individual engages the social world. Through this pathway, he or she is assigned to a racial/ethnic group through a process called physiognomy, literally defined as the "art of judging human character from facial features" [4]. After an individual has been labeled as a member of a particular racial/ethnic group, he or she is exposed to the social health risks associated with that group. Social health risks are variables that influence health outcomes but are commonly outside the direct control of an individual. For example, occupational health hazards, poor quality housing, exposure to discrimination or racism, and poorer quality medical care are all factors that influence the health of an individual but are rarely manageable by an individual. These social health risks have been found to differ by race and ethnicity.

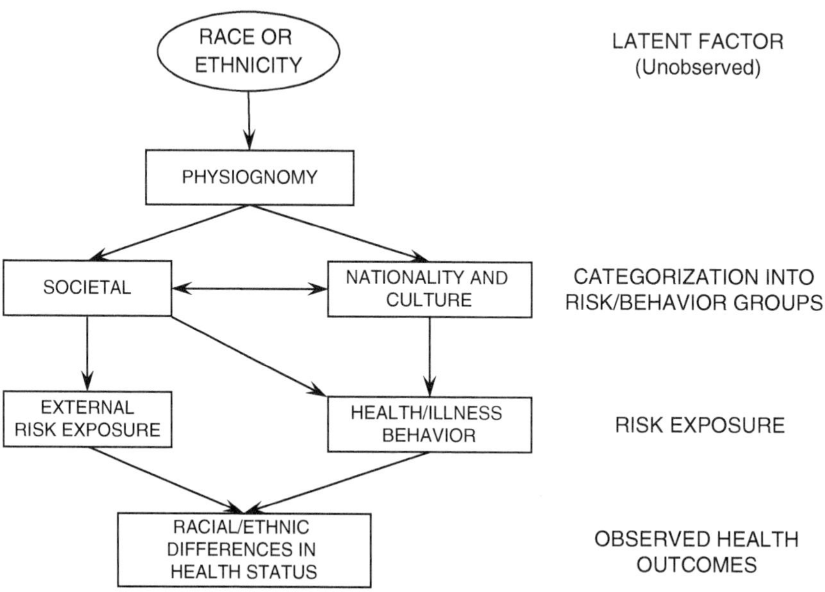

Figure 3.1 Conceptual model of race or ethnicity. *Source:* LaVeist et al. [3]. © 2005, John Wiley & Sons.

The right side of the model outlines the behavioral pathway, showing that there may be characteristics of the nationality or culture of an ethnic group that influence health or illness behaviors and, thus, health status downstream. These factors may account for some degree of health disparities.

The third pathway through which racial/ethnic differences in health status are produced is demonstrated by the arrow linking societal factors to health/illness behavior. Specifically, societal factors limit the ability of an individual to address health issues or adopt behaviors that protect one's health. For example, a person's race may lead to lower SES, which may lead to the under-utilization of health services. This model explains that illness behavior is not directly associated with race or ethnicity; rather, it is associated with social class, which is an important distinction. However, considering the way that race and ethnicity are commonly explored, it is possible to inaccurately attribute illness behaviors to one's race and ethnicity rather than understanding the effects of social class on behavior. These errors lead to the assumption that there is something about a person's skin color, rather than their societal context, that makes them engage in risky behavior [3].

3.3.2 US Census Definitions

Race, ethnicity, and nationality are often used interchangeably. Ethnicity refers to cultural commonality, whereas race refers to one's physical characteristics, predominantly skin color. Nationality refers to one's country of origin. It is important to note that there has been a lack of clarity when it comes to defining race and ethnicity. In examining representative medical and allied health dictionaries for definitions of race, one finds significant variability. This lack of clarity has had important implications for the collection of data by race and ethnicity. It was not until 1977, when the US Office of Management and Budget (OMB) issued OMB Directive No. 15, that standards were established for the collection of data on race and ethnicity, which allowed for consistency and comparable data for a variety of government programs. Since then, federal data have been routinely reported for Latinos and Asian Americans. Detailed census definitions are provided in chapter 1 [3].

3.3.3 Race Versus Ethnicity

Compared to race, which often refers to one's physical characteristics, ethnicity refers to an individual's belonging to a social group that shares a common national or cultural background. The largest minority group in the United States, Latinos, is an ethnic group that combines persons from multiple countries and ethnic identities. Latinos sometimes share a similar history, culture, and religion, but not always. They mainly speak Spanish, but often different dialects. It is also important to note that throughout much of Latin America there are indigenous groups that do not speak Spanish as their primary language.

Health risks can arise due to ethnicity-specific characteristics regardless of race as well as due to race regardless of ethnicity. For example, consider a Cuban of African descent and an African American who share physical characteristics. This may result in both persons being at increased health risks due, for example, to living in poor-quality housing. However, they are from different ethnic groups, which may result in different health behaviors and health risk profiles [5]. Thus, while both may be at increased risk of exposure to housing discrimination (due to their race – as measured by skin color), the individual identifying as Cuban may be at an increased risk of encountering barriers to accessing quality healthcare due to his or her status as an immigrant, Spanish being their first language, and/or holding ethnicity-specific health beliefs.

Consider the experiences of US immigrants from East Asia, Southeast Asia, and the Pacific Islands, who originate from dozens of ethnic groups, some with mixed indigenous or European descent. Upon arrival in the United States, they are viewed as the single racial group of Asian or Pacific Islander. A similar example can be made of immigrants during the nineteenth and twentieth centuries from the European nations of Ireland, Italy, and the former Soviet Union and Russia. Despite coming from distinct cultural backgrounds, they were categorized as White in the United States. Rarely is this source of variation taken into account in health research. The same can be said for indigenous populations in South America and Africa. The Yoruba of Brazil and Nigeria share a common cultural background, yet they differ in nationality. Within the American cultural context, they both would be viewed as the single racial group of Black or African American, thus adopting the health risk profile associated with that group.

We often try to capture the concepts of ethnicity and race in one summary term, though doing so can result in some degree of measurement error. More importantly, the lack of conceptual clarity leaves a great deal of room for erroneous interpretations of research findings and, consequently, for ineffective public policy, public health programming, and even medical practice. OMB Directive No. 15 instructs that, where possible, race and ethnicity be captured by separate questions: the first for the individual's race (American Indian or Alaskan Native, Asian or Pacific Islander, Black or African American, or White), the second for ethnicity (Hispanic or Latino origin or not of Hispanic or Latino origin) [3]. These questions provide the opportunity to identify an individual's self-reported race and ethnicity.

Notably, some mixed-race groups, such as Latinos, often don't self-report race outside their ethnicity. A Cuban of African descent may only view themselves as Hispanic or Latino, and not as an individual of Hispanic or Latino ethnicity and Black or African American race based on their skin color. Due to this issue, "other" is a rapidly growing response to the census race question. The US Census Bureau is now considering changing to a single question to address this issue. However, it is worth pointing out that if this change happens, it will be essentially impossible to conduct intersectional research on groups like Black Latinos [3].

3.4 How and Why this Topic is Important to Minority Health and Health Disparities Research

3.4.1 The Epidemiological Profile of Racial/Ethnic Minorities

3.4.1.1 Mortality/Morbidity

In the United States, as in other industrialized nations, chronic conditions such as heart disease, cancer, and stroke are the leading causes of death. However, there are differences in the leading causes of death for different racial/ethnic groups.

Table 3.1 Rankings of the ten leading causes of death for each racial/ethnic group, United States, 2016.

Cause of death	Total	White	Black or African American	Hispanic or Latino	Asian or Pacific Islander	American Indian or Alaskan Native
Heart disease	1	1	1	2	2	1
Cancer	2	2	2	1	1	2
Chronic lower respiratory disease	3	3	6	8	8	6
Unintended injuries	4	4	4	3	4	3
Stroke	5	5	3	4	3	7
Alzheimer's disease	6	6	9	6	6	—
Diabetes	7	7	5	5	5	4
Influenza and pneumonia	8	8	—	10	7	10
Nephritis, nephritic syndrome, and nephrosis	9	10	8	9	9	9
Suicide	10	9	—	—	10	8
Homicide	—	—	7	—	—	—
Septicemia	—	—	10	—	—	—
Chronic liver disease and cirrhosis	—	—	—	7	—	5

Source: National Center for Health Statistics [6].

As reported by the National Center for Health Statistics for 2015 (see Table 3.1), the five leading causes of death for Whites, in descending order, are heart disease, cancer, chronic lower respiratory disease, unintended injuries, and stroke. The leading causes for Blacks are the same but in a slightly different order and instead of chronic lower respiratory disease, diabetes is among the top five. For Hispanics or Latinos, the five leading causes of death in descending order include cancer, heart disease, unintentional injuries, stroke, and diabetes. Asians or Pacific Islanders have the same top five leading causes of death as Latinos, although the order is slightly different. American Indians or Alaska Natives have similar trends to Blacks, except instead of stroke, chronic liver disease and cirrhosis are among the top five leading causes of death.

Notably, suicide is within the top 10 leading causes of death for Whites, Asians or Pacific Islanders, and American Indians or Alaska Natives, but not for Blacks and Latinos. It is also important to note that Blacks are the only group that have homicide among the 10 leading causes of death (seventh) [3].

3.4.1.2 Life Expectancy

Disparities can be measured in the United States in years of potential life lost before age 75, defined as an estimation of the average time an individual would have lived if he or she had

Table 3.2 Years of potential life lost before age 75 (per 100 000 population under age 75) for all causes of death by race and ethnicity, United States, selected years 1980–2015.

	Crude	Age adjusted					
	2015	**1980**	**1990**	**2000**	**2010**	**2014**	**2015**
White	7090.6	9554.1	8159.5	6949.5	6342.8	6390.1	6514.8
Black or African American	9764.6	17873.4	16593.0	12897.1	9832.5	9490.6	9702.3
Hispanic or Latino	4452.8	—	7963.3	6037.6	4795.1	4676.8	4750.4
Asian or Pacific Islander	3073.6	5378.4	4705.2	3811.1	3061.2	2954.4	3049.7
American Indian or Alaska Native	6895.3	13390.9	9506.2	7758.2	6771.3	6954.0	7176.2

Source: National Center for Health Statistics [6].

not died early due to race and ethnicity [7]. Years of potential life lost for Blacks is higher than for all other groups (see Table 3.2). In 2015, the crude estimate for Blacks is 9764.6 years of potential life lost per 100 000 population under age 75. The lowest figure is observed among Asians or Pacific Islanders: 3073.6 years of potential life lost. Though the highest figures of years of potential life lost are among Blacks, there has been a decreasing trend, from 17 873.4 years in 1980 to 9702.3 years in 2015.

For Latinos and Whites, the years of potential life lost also show a decreasing trend, with the exception of a slight increase between 2013 and 2014 (4668.1 years in 2013 to 4676.8 years in 2014 for Latinos and 6338.2 years in 2013 to 6390.1 years in 2014 for Whites). There were large improvements for American Indians or Alaska Natives between 1980 (13 390.9 years) and 1990 (9506.2 years), and small improvements between 1990 and 2015 (9506.2 years and 7176.2 years, respectively). Despite these improvements, a study in 2015 of mortality found that overall life expectancy in the United States has decreased slightly from 78.9 years in 2014 to 78.8 years in 2015 [8]. This is noteworthy, as it is the first time the United States has seen a drop in life expectancy in decades.

3.4.1.3 Socioeconomic Status

Lower SES is consistently associated with poorer health outcomes. Adapted from Brown and colleagues [9], a conceptual framework of the distal mediators and moderators of the relationship between SES and health outcomes and behaviors is located in Figure 3.2. This model shows how socioeconomic factors influence multiple levels of distal and proximal mediators and moderators that impact health behaviors and outcomes. SES can be measured at the individual level, such as by annual household income, the highest level of educational attainment, and wealth (e.g., earnings, home ownership, and the net value of assets) by questionnaire. A full list of measures and references is located elsewhere [10].

Lower SES is known to affect health through psychosocial factors such as stress and depression, and health behaviors such as cigarette smoking and physical inactivity [11]. In recent literature, there has been a rise in the use of the term Socioeconomic Position (SEP), which is used instead of SES as both a "resource-based and prestige-based measure, as linked to both childhood and adult social class position" [12]. Resource measures are linked

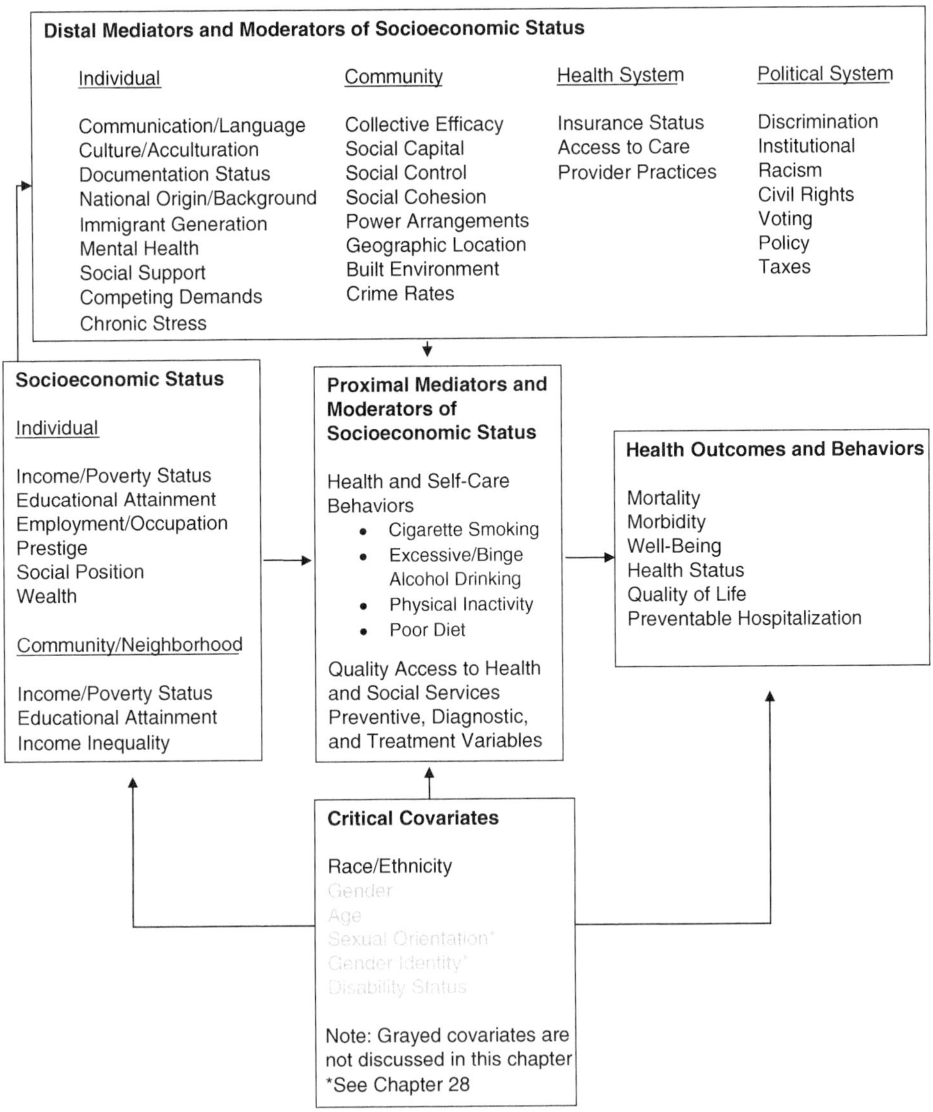

Figure 3.2 Conceptual framework of the distal mediators and moderators of the relationship between socioeconomic status and health outcomes and behaviors. *Source:* Based on Brown et al. [9].

to income, wealth, and educational attainment, whereas prestige measures are linked to social hierarchy [12].

SES can also be measured at the community and neighborhood levels. Recent studies on SES include investigations of income inequality across cities or countries, as well as neighborhood poverty and access to resources in neighborhoods (see Other Social Determinants in Section 3.4.2 and examples of neighborhood environment in Section 3.5.3). Among developed nations, the United States has the highest income inequality, with the top 0.1%

of the population owning nearly the same share of wealth as the bottom 90% [13]. The United States also has poorer health outcomes compared to other industrialized nations as measured by life expectancy, mental illness, obesity, infant mortality, teenage births, homicides, imprisonment, educational attainment, distrust, and social mobility.

The most commonly used measure of income inequality is the Gini coefficient, "formally defined as half of the arithmetic average of the absolute differences between all pairs of incomes within the sample, with the total then being normalized on mean income. If incomes are distributed completely equally, the value of the Gini will be zero. If one person has all the income, representing complete inequality, the Gini will assume a value of 1" [14].

Research on income inequality and health, including the role of violence, strongly supports a causal link [15]. Furthermore, a growing body of literature linking neighborhoods and health has shown that residing in a disadvantaged neighborhood is associated with increased rates of chronic disease [16]. Data suggest that these neighborhoods have fewer resources available to promote healthy behaviors and facilitate good health (further discussed in Section 3.4.2).

While this SES–health relationship has been robust across many health outcomes and at multiple levels, more work is needed to understand the complexity of the components of SES and how they affect health disparities [11]. These components do not always act independently but, instead, are affected by one another. Determining the biological mechanisms linking these social factors to disease outcomes has been a major priority for the field of social epidemiology.

Racial and ethnic minority populations, who have disproportionately lower SES, are especially affected by stressors related to their minority status (i.e., discrimination and racism) that exacerbate health disparities. Disentangling the effects of race and ethnicity from those of SES is extremely difficult due to the complex historical inequities and ongoing social issues. Rigorously designed studies to address this question are needed.

3.4.2 Other Social Determinants

3.4.2.1 Acculturation

In addition to the health disparities that racial and ethnic minority groups experience, the health of immigrant populations is influenced by their level of acculturation. The term acculturation refers to changes in the values, attitudes, and behaviors of an ethnic minority individual due to their continuous interaction with individuals from a different ethnic group. This is one example of an individual mediator and moderator of SES (Figure 3.2). Lower mortality rates for lower acculturated Latinos, despite their relatively low SES—also known as the Latino Paradox—is an example of the relationship between acculturation and health. However, an immigrant's risk of chronic disease and disability has been shown to increase with a greater level of acculturation. Although the reasons for such relationships are complex, there are a variety of measures to assess acculturation level, ranging from a short multiple-item measure to multidimensional assessments (see Table 3.3).

The biggest advantages of a short multiple-item measure are its brevity and general ability to explain a majority of the variance when modeling health outcomes or behaviors. Although the psychometric properties of multidimensional assessments have been identified for some immigrant populations, the short multiple-item measure is widely used across immigrant subgroups. National surveys such as the National Health Interview Survey, the American Community Survey, the National Health and Nutrition Examination Survey, the Current Population Survey, and the National Survey of Children's Health ask at least one or more of the questions contained in the short multiple-item measure mentioned below. When there is concern that a survey instrument may take too long for a participant to complete in a predetermined timeframe, the short multiple-item measure can be used to quickly assess an individual's acculturation level.

In order to assess acculturation level using a brief measure, researchers should consider collecting data on nativity and the number of years an individual has lived in the United States. Nativity is either asked as "In what country or territory were you born?" or "Were you born in the United States?" Although this question may be problematic in some instances because of fears of deportation, most respondents understand that the question does not ask about documentation status. Years lived in the United States is often phrased as "From the time that you first moved to the United States to today, about how many years have you lived in the mainland United States (50 states + DC)?" and is skipped for US-born individuals. Responses can either be collected in an open-ended fashion or as a predetermined count, ranging from zero to the current age of the individual. For some immigrants, their time in the United States may be segmented because they returned to their home country or moved to another country for a period of time. When this is likely, supplemental questions can be used to obtain a clearer measure of the exact number of years an individual has spent living in the United States. Such questions may include "Except for short vacations, do you return to your native country for part of each year?" Individuals who respond in the affirmative are then asked "For how many months per year?"

One of the most commonly used and easily assessable measures of acculturation level is language preference. This item can be asked in combination with nativity, years lived in the United States, or both to create a short multiple-item measure. It should be noted that the preferred method would be to ask about language preference and at least one of the other two questions. Typically, language is assessed by asking, "What language(s) do you usually speak at home?" Similar to years in the United States, possible responses can be open-ended or predetermined and listed for selection by the individual. When language preference cannot be collected by self-report, language in which the survey or interview was conducted is often used to determine language preference. Sometimes, additional questions are asked about language preferences in other settings, such as "In general, what language(s) do you read and speak?" "What was the language(s) you used as a child?" "In which language(s) do you usually think?" and "What language(s) do you usually speak with your friends?" Potential response options for these questions can be: Only non-English language (e.g., Spanish); non-English language better than English; Both equally; English better than non-English language; and Only English, utilizing Likert-type scale values of 1 to 5, respectively. When combined with an item such as nativity, these questions can be used to identify if an individual is more acculturated (e.g., US-born and prefers English), less acculturated (e.g., foreign-born and prefers a language other than English), or bicultural (e.g., foreign-born and prefers English or US-born and prefers a language other than English). An example of how acculturation level, accessed using a short multiple-item measure, is associated with risk factors for cardiovascular disease in the Hispanic Community Health Study/Study of Latinos (HCHS/SOL) can be found in Section 3.5.2.

Table 3.3 Selected list of the most frequently cited acculturation scales.

Title	Authors	Year	Population(s)	Description
Acculturation rating scale for Mexican Americans-II [17]	Israel Cuellar, Bill Arnold, Roberto Maldonado	1995	Mexican Americans, also used among other Latinos	Multidimensional measure based on language use and preference, ethnic identity and classification, cultural heritage and ethnic behaviors, and ethnic interaction
Short acculturation scale for Hispanics [18]	Gerardo Marín, Fabio Sabogal, Barbara Vanoos Marín, Regina Otero-Sabogal, Eliseo J. Pérez-Stable	1987	Latinos	Bidimensional measure based on language use, media, and ethnic social relations
Suinn-Lew Asian self-identity acculturation scale [19]	Richard M. Suinn, Kathryn Rickard-Figueroa, Sandra Lew, Patricia Vigil	1987	Asian Americans	Bidimensional measure based on language, identity, friendship choice, behaviors, generation/geographic history, and attitudes
Vancouver index of acculturation [20]	Andrew G. Ryder, Lynn E. Alden, Delroy L. Paulhus	2000	Multicultural	Bidimensional measure based on heritage and mainstream cultural identification

These supplemental language preference questions sometimes contribute to multidimensional assessments of acculturation level. A general list of acculturation scales will not provide a sense of which are the most frequently cited by researchers. A list of scales to assess acculturation level among different immigrant populations is located in Table 3.3. Due to space constraints, descriptions of these scales have been reserved for the table.

3.4.2.2 Social and Physical Environments

It is important to examine the context through which the social determinants interact to influence health and the mechanisms by which these factors lead to poor health outcomes. As mentioned in Section 3.4.1.3, Figure 3.2 is a conceptual model that illustrates distal and proximal mediators and moderators of SES at multiple levels that influence health behaviors and outcomes. In addition to some of the determinants discussed in this chapter, SES is affected by social context. For example, geographic location (e.g., urban, suburban, and rural) and how structures within these locations are built to impede or enhance health is determined by access to resources as well as the political, community, and economic will for the built environment to support health. The health and political systems are also depicted in Figure 3.2 to highlight their important influence on health.

Social context can be divided into two environments: the social environment and the physical environment. Each can separately or interactively exert pressure on the ability of individuals and communities to access necessary resources to support a healthy lifestyle. Some of the components of these environments are related to SES and can be measured at the individual and community/neighborhood levels (e.g., income and educational attainment). The term social environment refers to socioeconomic factors (e.g., employment and educational attainment), social relations (e.g., within a community or workplace), and power arrangements (e.g., political empowerment, individual and community control and influence) [21]. At the community level, other aspects of the social environment include social capital (i.e., a measure of the quality of relationships among community residents), social cohesion (i.e., conditions of mutual trust and solidarity among neighbors), collective efficacy (i.e., a measure of the willingness to help out for the common good), and social control (i.e., the capacity of a group to regulate its members according to desired principles). SES is the component of the social environment that has been studied the most [22].

Social capital has been used as a measure of the social environment in order to elucidate aspects of that environment other than SES that can influence health outcomes and health behaviors. In 2014, the Office of National Statistics in the United Kingdom produced a report that aimed to unify the different measures of social capital [23]. The authors organized social capital into four major areas: (i) Personal Relationships, (ii) Social Network Support, (iii) Civic Engagement, and (iv) Trust and Cooperative Norms. Measures of personal relationships include "Have at least one close friend" and "Meeting with friends or family members at least once a week." For social networks, measures include, "Have someone to rely on if they have a serious problem," and "Borrow things and exchange favors with neighbors."

The next two areas go beyond personal relationships and explore community-level factors of social capital, including the measure of "Voted in last national election" as a proxy for civic engagement. For trust and cooperative norms, the more salient measures focus on agreeing with the following statements: "Most people can be trusted," "People in my neighborhood can be trusted," and "People in my neighborhood are willing to help each other." Using these measures to understand the type of personal and community relationships from which people can draw support allows for the creation of programs and policies that can leverage these resources for improving health.

Another social construct, collective efficacy [24], measures the ability of communities to harness social environmental factors. This construct uses the concepts of informal social control and social cohesion and trust. Informal social control can be measured by asking respondents about the likelihood that their neighbors could be counted on to intervene in various ways if (i) children were skipping school and hanging out on a street corner, (ii) children were spray-painting graffiti on a local building, (iii) children were showing disrespect to an adult, (iv) a fight broke out in front of their house, and (v) the fire station closest to their home was threatened with budget cuts, where respondents were given response options from very likely to very unlikely on a five-point scale. Also on a five-point scale, the social cohesion and trust construct asks respondents how strongly they agree that "People around here are willing to help their neighbors," "This is a close-knit neighborhood," "People in this neighborhood can be trusted," "People in this neighborhood generally don't get along with each other," and "People in this neighborhood do not share the same values." Collective efficacy and social cohesion have been found to be associated with a variety of social factors and health behaviors, including rates of violence, smoking, and obesity.

Factors in the physical environment that are important to health include harmful substances (e.g., air pollution or proximity to toxic sites), physical surroundings (e.g., neighborhood and work conditions), access to various health-related resources (e.g., healthy or unhealthy foods, recreational resources, and medical care), and community design and the built environment (e.g., land use mix, street connectivity, and transportation systems) [25]. The built environment includes all of the physical parts of where people live and work (e.g., homes, buildings, streets, open spaces, and infrastructure) [16]. The built environment can be measured subjectively or objectively. Objective measurements include directly observing the presence of litter, the number of abandoned buildings and cars, walkability, presence of green space, and the number or duration of broken windows. Walkability is the measure that is most often used in examining the impact of the built environment on health behaviors. Measurements of walkability include proximity and accessibility to key destinations, availability of sidewalks, availability of trails and parks, safety, and number and location of complex intersections (see example of walkability data in Section 3.5.3).

Subjective measurements include self-reported perceptions of the accessibility, usability, and condition of buildings and structures in the environment in which one lives. Whether measured separately or together, both subjective and objective measurements of the built environment have been found to affect various health behaviors, including walking, physical activity, and access to healthy foods and health services.

3.5 Selected Examples of the State of the Science in the Field to Illustrate Best Practices

3.5.1 Exploring Health Disparities in Integrated Communities Study

The Exploring Health Disparities in Integrated Communities (EHDIC) study examines health disparities within racially integrated communities among urban Blacks and Whites with similar SES in order to address the issue of confounding between race, SES, and segregation. Confounding by race and SES can be problematic, as health status varies by both factors [26]. The overlap between these factors complicates efforts to understand if it is "race and class" that produces health disparities or "race or class."

This cohort study collected data through a structured questionnaire and blood pressure measurements from adults in two urban, low-income, racially integrated contiguous census tracts in southwest Baltimore, Maryland. In order to identify racially comparable communities for the study, the team used the following set of criteria: (i) at least 35% each of Black and White residents, (ii) a ratio of Black to White median income between 0.85 and 1.15, and (iii) a ratio of the percentage of Black to White high school graduates age 25 and above between 0.85 and 1.15. Using these criteria allowed for the examination of health disparities with minimal confounding between race and SES measures, such as income and level of educational attainment [27].

After two contiguous census tracts were selected, every occupied dwelling was identified in the study area, and a letter was mailed to all occupied dwellings to notify residents about the study. Various recruitment methods were used, including door-to-door visitation, walk-ins at the study's administrative office, and community-based health fairs [27].

The results indicated that Blacks had lower rates of both smoking and fair or poor self-rated health as compared to Whites. However, no race differences were observed in obesity, drinking, or physical inactivity (see Table 3.4). The results suggest that considering race differences in exposure to social conditions reduces or eliminates some health-related disparities. Further analyses from the EHDIC study sought to determine if racial health disparities, which are typically reported in national-level studies, persist when Blacks and Whites live in integrated settings [26]. The study results indicated that nationally reported disparities, including hypertension, diabetes, obesity among women, and use of health services, lessened or disappeared altogether in racially integrated communities. The one exception was smoking, which was higher among White residents. This study demonstrates that racial differences in social environments explain a significant portion of disparities reported in national data, and that when social factors are leveled, racial disparities are reduced.

Table 3.4 Summary of previously published findings from the Exploring Health Disparities in Integrated Communities – Southwest Baltimore Study (EHDIC-SWB) compared to findings from national surveys.

	EHDIC-SWB		National survey		
Health-related outcome	**Odds ratio[a]**	**95% CI**	**Odds ratio[a]**	**95% CI**	**Major finding**
Hypertension	1.42	1.09, 1.86	2.01	1.63, 2.48	Racial disparity related to hypertension prevalence smaller in EHDIC-SWB than in NHANES, but statistically significant
Diabetes	1.07	0.71, 1.58	1.61	1.26, 2.04	Racial disparity related to diabetes found in NHIS, but not in EHDIC-SWB
Obesity	1.25	0.90, 1.75	1.99	1.71, 2.32	Racial disparity related to obesity found in NHIS, but not in EHDIC-SWB
Cigarette smoking					Blacks have lower odds of being current smokers and smoke fewer cigarettes per day compared to Whites in EHDIC-SWB, but no racial disparity in smoking in NHIS
Lifetime smoker	0.55	0.41, 0.72	0.62	0.49, 0.79	
Current smoker	0.71	0.56, 0.90	0.93	0.72, 1.21	
Cigarettes smoked per day	0.68	0.61, 0.75	0.86	0.74, 1.01	
Use of health services	1.44	1.00, 1.87	0.74	0.51, 1.07	No disparity in having a medical care visit in MEPS, but more likely to have a visit in EHDIC-SWB

CI = confidence interval, NHANES = National Health and Nutrition Examination Survey, NHIS = National Health Interview Survey, MEPS = Medical Expenditure Panel Survey.

[a] All odds ratios are for the odds of the health-related outcome among Blacks compared to the odds of the health-related outcome among Whites [26].

Source: LaVeist et al. [26]. © 2011, The People-to-People Health Foundation, Inc.

Though the EHDIC study advances our understanding of health disparities, the study was conducted in only two census tracts in an urban population. As a result, the generalizability of the study findings is limited. Conducting this study in nonurban environments or higher SES communities may yield different results. Future studies plan to examine health disparities in different environments.

3.5.2 Hispanic Community Health Study/Study of Latinos

The Hispanic Community Health Study/Study of Latinos (HCHS/SOL), colloquially referred to as SOL, is a multisite, community-based, longitudinal study of the health of Latino populations in the United States. Participants were selected to be representative of the target population in four communities: the Bronx, NY; Chicago, IL; Miami, FL; and San Diego, CA. The study enrolled participants from diverse national backgrounds, including Cuban, Dominican, Puerto Rican, Mexican, Central American, and South American.

Using data from HCHS/SOL, Kershaw and colleagues used two easily assessable measures of acculturation to investigate the relationship between acculturation level and favorable levels of cardiovascular risk factors [28]. They used nativity and years lived in the United States to derive a three-level acculturation variable: (i) low acculturation: foreign-born and lived in the United States less than 10 years, (ii) moderate acculturation: foreign-born and lived in the United States equal to or more than 10 years, and (iii) high acculturation: US-born. Among cardiovascular risk factors, the study assessed cholesterol levels, systolic and diastolic blood pressure, fasting plasma glucose, current smoking, electrocardiograph abnormalities, and stress due to ethnic discrimination.

Results from multivariable-adjusted models showed that low-acculturated women had significantly higher odds of favorable levels of cardiovascular risk factors compared to moderate- or high-acculturated women. Further, favorable levels of cardiovascular risk factors differed significantly among men by their national background. For example, moderate-acculturated Dominican men had significantly higher odds of favorable levels of cardiovascular risk factors compared to their low-acculturated counterparts.

One of the limitations of this study is its cross-sectional design and, therefore, inability to examine changes in health behaviors over time and determine temporality. The HCHS/SOL study has longitudinal follow-up and will eventually be able to assess these factors over time. Another limitation is that the study may have overadjusted for covariates that may be on the causal pathway between acculturation level and favorable levels of cardiovascular risk factors.

Despite these limitations, the present study exhibits how acculturation level can influence established risk factors for chronic health conditions and how that relationship can differ among subpopulations of immigrants.

3.5.3 Pittsburgh Hill/Homewood Research on Neighborhoods and Health Study

The Pittsburgh Hill/Homewood Research on Neighborhoods and Health (PHRESH) study evaluates a natural experiment of neighborhood revitalization on health in Black communities. To reduce inequalities in food access, policy solutions have concentrated on addressing the issue of "food deserts," or neighborhoods with inadequate access to healthy, affordable foods. One proposal has been to place full-service supermarkets in neighborhoods with limited access to healthy food. To better understand how this solution may lead to changes in diet, Dubowitz and colleagues conducted a rigorous study in Pittsburgh, PA that surveyed two predominately Black sociodemographically similar neighborhoods. The PHRESH study uses a quasi-experimental longitudinal design. One of the study neighborhoods received a new supermarket, and the other acted as a control in order to understand if the introduction of the supermarket led to improved dietary outcomes. Data were collected using surveys with randomly selected residents, which included two 24-hour dietary recalls administered between seven to fourteen days apart [29].

The study results showed net positive changes in overall dietary quality; average daily intakes of kilocalories and added sugars; and percentage of kilocalories from solid fats, added sugars, and alcohol in the neighborhood that received a full-service supermarket [29]. However, the only positive outcome in the recipient neighborhood specifically associated with regular use of the new supermarket was improved perceived access to healthy food. The researchers did not observe a significant increase between the neighborhoods in fruit and vegetable intake, whole grain consumption, or body mass index. Though the introduction of supermarkets in food deserts is appropriate, the mechanisms for understanding how the introduction of such supermarkets may influence dietary behaviors and weight status requires further study. Policymakers considering the approach of introducing full-service supermarkets in food deserts should include assessments of other community-level mechanisms that may affect both store users and nonusers. Finally, future assessments should consider longer-term follow-ups, which may allow for changes in weight status to be detected [29].

Another PHRESH study examined neighborhood walkability. Previous studies have found that Black adults are not likely to get the recommended amount of daily moderate to vigorous physical activity (MVPA) [30], which is known to be an important factor for lowering diastolic blood pressure [31] and for reducing the risk of cardiovascular disease and risk factors [32]. In a study conducted by Richardson and colleagues using data from the PHRESH study, it was hypothesized that low levels of MVPA (where accelerometry was used to measure physical activity, which is superior to self-reporting) among Black adults may be associated with crime rates, the amount of green space, and walkability in one's neighborhood [30]. In addition to data from PHRESH, the study examined crime data available from the City of Pittsburgh and used ArcGIS to calculate street network distances from households to crime locations. The team hypothesized that crime occurring in the previous year would influence perceptions of safety. The team studied the relationship between these three variables (i.e., neighborhood green space, walkability, and crime) and their influence on MVPA, as well as a potential interaction effect.

The results suggested that, although all residents engaged in little MVPA, women under the age of 65 living in more walkable neighborhoods engaged in significantly more MVPA. The variables green space and crime were not associated with MVPA in this study. However, there may have been limited power to detect associations between neighborhood characteristics and MVPA. Future studies are required to increase knowledge regarding the impact of lifestyle and neighborhood factors (e.g., walkability, green space, and crime) and their interactions on MVPA to inform public health policy [30].

3.6 Challenges and Future Opportunities

3.6.1 Biological Links to Social Determinants

One of the major challenges that researchers face is understanding how the social determinants of health get "under the skin" to impact health. This requires studies to link specific social determinants of health to biological mechanisms. Several developing research areas aim to investigate such links as epigenetic markers, and how heritable and potentially modifiable changes in gene expression can be affected by nonbiological and environmental exposures [33]. Other examples include biological markers of stress, such as cortisol or telomere length, as a marker for cellular aging. These issues will add to the ongoing complexity of health disparities research. Nonetheless, continuing to develop new methods to elucidate mechanisms will be crucial in the quest to eliminate health disparities.

3.7 Summary

This chapter focused on summarizing the state of the literature, measurement issues, and research opportunities related to race, SES, and other social determinants. The chapter references two conceptual frameworks in order to provide an understanding of the factors involved in the relationships between race/ethnicity, SES behavioral mediators/moderators, and health outcomes. The use of widely accepted measures, such as those used by the US Census Bureau, is critical to minority health and health disparities research because they make it possible to compare findings across studies and over time.

The differences between race and ethnicity are distinct, although they can be difficult to distinguish if one is not intimately familiar with them. As the United States becomes more diverse, future health research will need to disaggregate the larger racial/ethnic categories to understand the causes of known differences in health status, behaviors, and outcomes between subgroups. Among the key measures of health are mortality, morbidity, and life expectancy, which can be used to gauge progress in improving minority health and in reducing health disparities.

One of the most studied social determinants, SES, is a strong predictor of health and can be measured on an individual or community/neighborhood level. Other social determinants, including acculturation, social environment, and physical environment, can also influence health independent of SES. Assessing such factors, in addition to SES, can elucidate relationships that have previously gone unobserved. The examples above represent some of the leading research related to race, ethnicity, and social determinants of health.

In addition to disaggregating racial/ethnic groups, the opportunity to link social determinants of health to biological mechanisms will advance research addressing minority health and health disparities. Future research should also consider integrating race/ethnicity, SES, and other social determinants into the healthcare setting, as well as into the emerging areas of precision medicine, clinical trials, sexual and gender minority research, and workforce diversity.

3.8 Key Points

- Race and ethnicity are complex constructs and incorporate aspects of skin color, societal influences, nationality, and culture, which all impact racial/ethnic differences in health status.
- Racial/ethnic minority populations in the United States experience significant morbidity, mortality, and decreased life expectancy compared to the White population.
- Lower SES, income inequality, and other factors such as poorer physical and social environments are social determinants that impact health, particularly for racial/ethnic minority groups.
- Several innovative studies have been designed and conducted, especially to understand complex issues in health disparities research and intricate aspects of the culture and life of racial/ethnic populations.
- More research is needed to link the social determinants of health to biological mechanisms and to understand the nuances of social determinants in understudied and underserved populations.

Disclaimer

The views and opinions expressed in this chapter are those of the authors only and do not necessarily represent the views, official policy, or position of the U.S. Department of Health and Human Services or any of its affiliated institutions or agencies.

References

1 Marmot, M., Friel, S., Bell, R. et al. (2008). Closing the gap in a generation: health equity through action on the social determinants of health. *The Lancet* 372 (9650): 1661–1669.

2 Marmot, M. and Allen, J.J. (2014). Social Determinants of Health Equity. *American Journal of Public Health* 104 (Suppl4): S517–S519.

3 LaVeist, T.A. (2005). Conceptual issues in race/ethnicity and health. In: *Minority Populations and Health*, 15–29. San Francisco, CA: Jossey-Bass.

4 Houghton Mifflin Harcourt Publishing Company (2017). *American Heritage Dictionary of the English Language*, 5e. Boston, MA: Houghton Mifflin Harcourt.

5 LaVeist-Ramos, T.A., Galarraga, J., Thorpe, R.J. et al. (2012). Are black Hispanics black or Hispanic? Exploring disparities at the intersection of race and ethnicity. *Journal of Epidemiology and Community Health* 66 (7): e21.

6 National Center for Health Statistics (2016). *Health, United States, 2015: With Special Feature on Racial and Ethnic Health Disparities*. Hyattsville, MD: National Center for Health Statistics.

7 Gardner, J.W. and Sanborn, J.S. (1990). Years of potential life lost (YPLL) – what does it measure? *Epidemiology* 1 (4): 322–329.

8 Xu, J., Murphy, S.L., Kochanek, K.D., and Arias, E. (2016). *Mortality in the United States, 2015*. Hyattsville, MD: National Center for Health Statistics.

9 Brown, A.F., Ettner, S.L., Piette, J. et al. (2004). Socioeconomic position and health among persons with diabetes mellitus: a conceptual framework and review of the literature. *Epidemiologic Reviews* 26: 63–77.

10 Berkman, L.F., Kawachi, I., and Glymour, M.M. (2014). *Social Epidemiology*. Oxford: Oxford University Press.

11 Chen, E. and Miller, G.E. (2013). Socioeconomic status and health: mediating and moderating factors. *Annual Review of Clinical Psychology* 9: 723–749.

12 Krieger, N., Williams, D.R., and Moss, N.E. (1997). Measuring social class in US public health research. *Annual Review of Public Health* 18: 341–378.

13 Saez, E. and Zucman, G. (2014). *Wealth Inequity in the United States since 1913: Evidence from Capitalized Income Tax Data*. Cambridge, MA: National Bureau of Economic Research.

14 Kondo, N., Sembajwe, G., Kawachi, I. et al. (2009). Income inequality, mortality, and self rated health: meta-analysis of multilevel studies. *The BMJ* 339: b4471.

15 Pickett, K.E. and Wilkinson, R.G. (2015). Income inequality and health: a causal review. *Social Science & Medicine* 128: 316–326.

16 Diez Roux, A.V. and Mair, C. (2010). Neighborhoods and health. *Annals of the New York Academy of Sciences* 1186: 125–145.

17 Cuéllar, I., Arnold, B., and Maldonado, R. (1995). Acculturation rating-scale for Mexican Americans-II: a revision of the original ARSMA scale. *Hispanic Journal of Behavioral Sciences* 17 (3): 275–304.

18 Marín, G., Sabogal, F., Marín, B.V. et al. (1987). Development of a short acculturation scale for Hispanics. *Hispanic Journal of Behavioral Sciences* 9 (2): 183–205.

19 Suinn, R.M., Rickard-Figueroa, K., Lew, S., and Vigil, P. (1987). The Suinn-Lew Asian self-identity acculturation scale: an initial report. *Educational and Psychological Measurement* 47 (2): 401–407.

20 Ryder, A.G., Alden, L.E., and Paulhus, D.L. (2000). Is acculturation unidimensional or bidimensional? A head-to-head comparison in the prediction of personality, self-identity, and adjustment. *Journal of Personality and Social Psychology* 79 (1): 49–65.

21 Lillie-Blanton, M. and Laveist, T. (1996). Race/ethnicity, the social environment, and health. *Social Science & Medicine* 43 (1): 83–91.

22 Williams, D.R. (1999). Race, socioeconomic status, and health. The added effects of racism and discrimination. *Annals of the New York Academy of Sciences* 896: 173–188.

23 Siegler, V. (2014). *Measuring National Well-being: Measuring Social Capital Consultation Response*. London: Office for National Statistics.

24 Sampson, R.J., Raudenbush, S.W., and Earls, F. (1997). Neighborhoods and violent crime: a multilevel study of collective efficacy. *Science* 277 (5328): 918–924.

25 National Research Council (2013). *US Health in International Perspective: Shorter Lives, Poorer Health*. Washington, DC: National Academies Press.

26 LaVeist, T., Pollack, K., Thorpe, R. et al. (2011). Place, not race: disparities dissipate in southwest Baltimore when blacks and whites live under similar conditions. *Health Affairs* 30 (10): 1880–1887.

27 LaVeist, T., Thorpe, R., Bowen-Reid, T. et al. (2008). Exploring health disparities in integrated communities: overview of the EHDIC study. *Journal of Urban Health* 85 (1): 11–21.

28 Kershaw, K.N., Giacinto, R.E., Gonzalez, F. et al. (2016). Relationships of nativity and length of residence in the U.S. with favorable cardiovascular health among Hispanics/Latinos: The Hispanic Community Health Study/Study of Latinos (HCHS/SOL). *Preventive Medicine* 89: 84–89.

29 Dubowitz, T., Ghosh-Dastidar, M., Cohen, D.A. et al. (2015). Diet and perceptions change with supermarket introduction in a food desert, but not because of supermarket use. *Health Affairs* 34 (11): 1858–1868.

30 Richardson, A.S., Troxel, W.M., Ghosh-Dastidar, M.B. et al. (2017). One size doesn't fit all: cross-sectional associations between neighborhood walkability, crime and physical activity depends on age and sex of residents. *BMC Public Health* 17 (1): 97.

31 Carson, V., Ridgers, N.D., Joward, B.J. et al. (2013). Light-intensity physical activity and cardiometabolic biomarkers in US adolescents. *PLoS One* 8 (8): e71417.

32 Andersen, L.B., Harro, M., Sardinha, L.B. et al. (2006). Physical activity and clustered cardiovascular risk in children: a cross-sectional study (The European Youth Heart Study). *Lancet* 368 (9532): 299–304.

33 Gary-Webb, T.L., Suglia, S.F., and Tehranifar, P. (2013). Social epidemiology of diabetes and associated conditions. *Current Diabetes Reports* 13 (6): 850–859.

4

Behavioral Determinants in Population Health and Health Disparities Research

Amelie G. Ramirez[1], *Patricia Chalela*[1], *Melanie D. Sabado-Liwag*[2], *and Kelvin Choi*[2]

[1] *The University of Texas Health Science Center, Institute for Health Promotion Research, San Antonio, TX, USA*
[2] *National Institute on Minority Health and Health Disparities, National Institutes of Health, Bethesda, MD, USA*

4.1 Introduction

Americans spent more than $3.2 trillion on healthcare in 2015 [1], yet that expenditure accounts for as little as 10% of the variability in life expectancy in the overall population [2]. As much as 30% of US healthcare spending does nothing to improve health [3]. Behaviors, in contrast, may explain as much as 50% overall, and more than 75% for certain diseases, of the variability in life expectancy. Both health-promoting and health-related risk behaviors are shaped by biological and genetic factors, social interactions and cultural norms, psychosocial determinants, physical environment (e.g., urban vs. rural, barrio vs. enclave), and healthcare.

Behavior acts as a credit or debit on the balance of health assets; that means following evidence-based recommendations can pay dividends. The American Institute for Cancer Research and the World Cancer Research Fund, for example, estimate that engaging in health-promoting behaviors such as eating a nutritious diet, limiting alcohol intake, keeping the body at a healthy weight, and incorporating physical activity daily could prevent approximately 375 000 cases of the most common cancers in the United States annually [4]. Conversely, risky behaviors, such as the intake of high-calorie foods and a lack of exercise, can contribute to people becoming overweight or obese and causing hypertension, coronary heart disease, diabetes mellitus, and certain types of cancer [5]. Smoking, alcohol abuse, poor nutrition, and lack of exercise are known causes of these and other chronic diseases [6]. We can add cardiomyopathy, neuropsychiatric disorders, and increased risk of injury or accidents to the risks of chronic disease imposed by long-term drinking [7].

In trying to understand behavioral health determinants, we strive to address these behaviors through interventions and policy changes to create better health outcomes. The structural forces, societal issues and socioeconomic factors, environment, and culture that influence behavior must also be part of the equation. Without a broad perspective of what governs behavior, efforts to improve health and reduce long-standing health disparities

The Science of Health Disparities Research, First Edition. Edited by Irene Dankwa-Mullan, Eliseo J. Pérez-Stable, Kevin L. Gardner, Xinzhi Zhang, and Adelaida M. Rosario.

may fail [8]. The presence of readily available tobacco, alcohol, and drugs and a social climate that endorses their use affects their adoption, use, and misuse. Cultural factors can affect adoption or rejection of safe sex practices or birth control. Social influences may encourage or discourage use of tanning salons, safety equipment such as seatbelts, and adoption of practices that detect cancer before it is a threat. Lack of social support, negative health-seeking behaviors influenced by family and friends, and negative previous patient–provider experiences may significantly affect the utilization of healthcare and mental health services for chronic illness and/or diagnosis.

In health disparities research, US investigators study these factors within groups that have systematically experienced greater barriers to health because of social or economic disadvantage and characteristics long connected to trust, discrimination, and/or exclusion. These unfair differences are based on: racial or ethnic group; ethnic identity and acculturation; socioeconomic position; sex; religion; age; sexual orientation or gender identity; immigration/generation status; geographic location; mental health; health literacy; cultural understanding; use of cultural-traditional health services; and cognitive, sensory, or physical disability [9]. These populations shoulder illness and poorer health outcomes disproportionately [9, 10].

Let's begin with the basic measure of mortality, whose causes, including heart disease and diabetes, as well as statistics on other diseases, demonstrate the differences. All-cause death rates in 2014 were highest for African Americans (males, 1034.0/100 000; females, 713.3/100 000) than for any other racial or ethnic group, and all-cause death rates in infants (younger than a year old) were more than twice as high for African Americans than for Whites in both boys (1125.4 vs. 551.3) and girls (956.3 vs. 457.6; all rates per 100 000) [11]. African Americans outpaced their counterparts in rates of heart disease and hypertension [11]. Though Latinos are less likely than Whites to die from most of the top 10 causes of death of Whites, the death rate associated with diabetes is about 50% higher in Latinos than in Whites [12]. Unlike other major racial and ethnic groups, Latinos are more likely to die of cancer than heart disease [13]. Childhood obesity, which is found disproportionately in communities with high poverty rates and in communities of color (especially among Latino children), flourishes where there are few safe places to be physically active and access to healthy foods and beverages is limited [14]. Many Latino families suffer a lack of access to and knowledge about proper nutrition and active spaces for physical activity. They also lack economic support, educational opportunities, and access to healthcare and health insurance. One-third of US Latino families lives in poverty, while nearly 27% report not having access to a regular healthcare provider. Lack of access to early childhood education has led to gaps in cognitive development in Latino children [15]. All of these circumstances impact Latinos later in life.

Changes in the types and dimensions of recognized disparities, the increasing proportions and diversity of minority populations, and the compounding influence of chronic disease compel researchers to uncover the multiple factors responsible for poorer health outcomes. Often, people think of health disparities as only differences between racial and ethnic groups, so it is important to make clear the variety of types of recognized health disparities. Through research that pays attention to disparities, investigators have shown why it is important to measure their multilevel dimensions. Have disparities in health widened or narrowed over time? If so, what do these changes tell us about intervening to improve health? What new groups are experiencing disparities? The answers will tell us if we are making progress in erasing health disparities. They have also encouraged investigators to move out of the corridors of academia and into the streets, where, using community-based participatory research (CBPR), investigators and community members work together in collaborative, nonhierarchical partnerships to develop more effective, culturally tailored, and theory-based health initiatives.

Efforts to reduce health disparities have expanded, most active within the scope of specific diseases or the domain of health services research. Fueling the expansion has been the recognition of the interrelationships between health and biology, genetics, and behavior, as well as the influences of socioeconomic position, literacy, the physical environment, mental health, health services, and racism and discrimination. These factors affect the health not only of individuals but also of populations. Over a lifespan, behavioral determinants can affect outcomes, so the earlier a disparity occurs, the greater the opportunity to compound its negative effects. Conversely, consider the child who escapes lung cancer by not adopting her parents' smoking habit, the adolescent who avoids being overweight and the risks of diabetes by substituting physical activity for screen time, and the adult or elder who sits less and walks more to avoid chronic disease. Physical activity in adults can decrease the risk of disease and early death, reduce symptoms of psychological distress (e.g., depression, stress), improve control of body weight, help control blood pressure and blood glucose, enhance one's quality of sleep, and promote independent living [16].

4.2 Importance of Behavioral Determinants to Minority Health and Health Disparities Research

Behavioral determinants of health are important because they change health status and outcomes of individuals and populations. In an epidemiological shift, the major causes of death worldwide and in the United States have become noncommunicable diseases. Most prominent are cardiovascular disease, most cancers, chronic respiratory disease, and type 2 diabetes (Figure 4.1) [17]. Eighty-five percent of the noncommunicable disease mortality rate in some parts of the world, represented by these diseases, can be attributed to four behavioral determinants: tobacco use, physical inactivity, an unhealthy diet, and harmful and excessive use of alcohol. Smoking accounts for most cases of lung cancer, a disease responsible overall for one-quarter of all US cancer deaths annually (and 40% of cancer deaths in the South) [18], and for chronic respiratory disease, itself linked to lung cancer and heart disease. Diets low in fruits and vegetables and high in trans-fatty acids and the three *s*'s—sugar, salt, and saturated fats—exacerbated by a sedentary lifestyle and resulting in obesity are associated with such metabolic and physiologic changes as high blood pressure, overweight and obesity, hyperglycemia, and hyperlipidemia, and with decreasing productivity along with shrinking opportunity, including ineligibility for US military service [18].

Health disparities can grow in minority, rural, and other communities where education, supportive institutions, employment, health engagement, and care access and utilization are in short supply. Organizations, communities, states, or nations—any individuals who band together—in contrast, may have options for structural change through evidence-based advocacy and legislation.

One area of growing disparities relates to the rapidly evolving opioid epidemic in the United States. An estimated 30 000 people die each year from opioid overdoses [19].

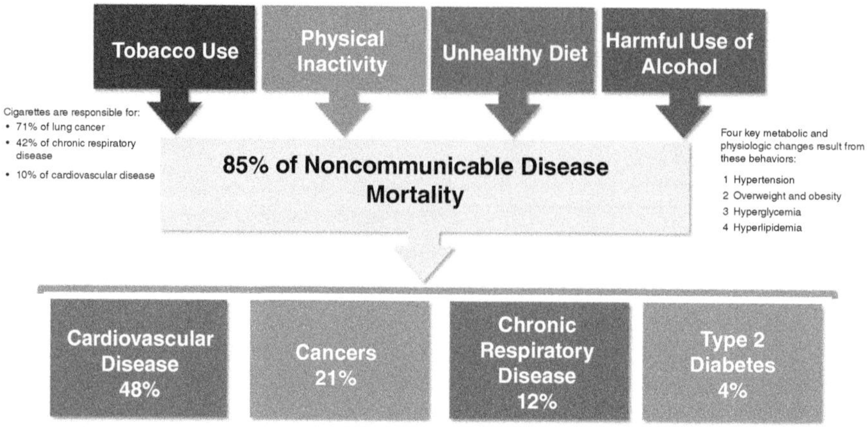

Figure 4.1 In this illustration of behavioral determinants for noncommunicable disease mortality in low- and middle-income countries, identified by the World Health Organization, four risk factors are linked to 85% of the noncommunicable disease mortality, which is caused by cardiovascular disease, cancer, chronic respiratory disease, and type 2 diabetes. *Source:* World Health Organization [17]. (See color representation of this figure in color inserts section.)

According to the Centers for Disease Control and Prevention, the rate of heroin usage among young white adults increased by 114% from the period 2002–2004 to 2011–2013, and more than 600 000 people died from drug overdoses in the years 2000 to 2016 [20]. Unlike drug epidemics of the past, minority populations have seen less of a rise in drug addiction and death during the opioid epidemic than have young White adults, both male and female [19]. The rise in opioid addiction can be traced to the increased use of prescription pain relievers. In 2012, the number of prescriptions for opioid drugs reached an all-time high of 259 million [21]. Minorities, especially African Americans and Latinos, are historically underprescribed pain medication [19]. Research has shown that a patient's race or ethnicity is one factor that affects physicians' decisions to prescribe opioids. In a study of emergency room visits, non-Latino Blacks had significantly lower odds of receiving opioids during visits for back pain and abdominal pain than did non-Latino Whites [21]. This is backed up by a 2008 study that found that, overall, members of minority groups were less likely than Whites to receive opioids for pain in an emergency department [22].

Failing to address the rapidly increasing prevalence of behavior-related conditions has financial and health consequences; the challenge of childhood overweight and obesity is a good example. In the past 40 years, the rate of obesity has quadrupled among US children aged 6 to 11 and tripled among adolescents aged 12 to 19 [14]. Most of the burden falls on Latino and African American children, who overall (ages 2 to 19) and in every age segment measured from 2011 to 2014 (ages 2 to 5, ages 6 to 11, and ages 12 to 19) had the highest rates of obesity (15.6–25%), while Asians had the lowest (5.2–9.8%) (Figure 4.2) [11]. Children who are overweight or obese for an extended period are more likely to become overweight or obese adults, who are then more susceptible to heart disease, high blood pressure, and type 2 diabetes. Researchers have found associations between pre-adulthood obesity and higher risk in adulthood of certain types of cancer. Such findings have prompted calls for more research, including analyses that are more robust across the lifespan [23].

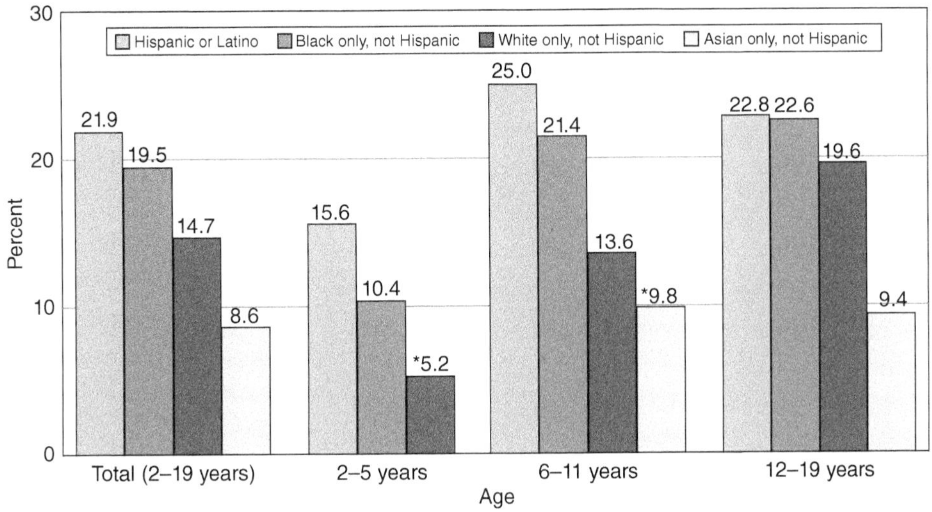

Figure 4.2 Children and adolescents with obesity, overall (2–19 years of age; top, left), and in (following, left to right), generally, preschool, elementary, and secondary school age groups. *Source:* US Centers for Disease Control and Prevention [11]. (See color representation of this figure in color inserts section.)

Though annual costs related to childhood obesity are estimated to be $14 billion, annual costs related to adult obesity rise to 10 times that or more ($147–$210 billion). Stigmatizing people with obesity, though a prejudice, is not generally recognized as one and remains largely acceptable in the young and old. Despite its adverse effects on its targets and the generation of health disparities, such prejudice further contributes to chronic burden and medical attention deterrence [24]. Though rapid increases have ended and rates have leveled off, indicators suggest in certain parts of the country disparities are widening because most reductions in obesity rates have occurred in students who are White and from families of higher than average socioeconomic position [14].

We address behavior and other health determinants, including socioeconomics, cultural environment, physical environment, biological and genetic influences, and healthcare, because we value good health for all. By recognizing disparities, we seek to create effective interventions that take into account these influences and to identify and enact policies that support the correction of disparities and the wise use of resources. That is not to suggest that these influences are static. They are dynamic, changing across the lifespan and across generations, and their influence waxes and wanes according to such factors as education, vulnerabilities, environmental stability, and structural drivers (power, income, and other resources) at individual, local, national, and global levels.

4.3 Relevant Metrics and Research Methods

4.3.1 Relevant Metrics

Table 4.1 shows the most common metrics relevant to health disparities in behavioral health determinants and specific examples [9, 25]. Investigations will continue to be undertaken in a wide array of designs and methods, but interdisciplinary teams and approaches,

Table 4.1 Types of disparity measurements and examples from behavioral determinants of health.

Indicator	Grouping		Method of comparison	
Measurement of health disparities				
• Health outcomes (e.g., life expectancy, mortality, chronic disease rates) • Healthcare access/quality of care • Social/environmental/behavioral risk factors (e.g., neighborhood characteristics, discrimination)	Social grouping by level of disadvantage or advantage (racial/ethnic groups, income groups by poverty level, level of educational attainment)		Comparison of rates on single indicator for two groups (usually top and bottom), absolute difference between rates for two groups, other methods (e.g., slope and relative index of inequality)	
Example 1: Disparities in health outcomes				
Life expectancy at birth by US state, both sexes, 2014	Estimated lowest and highest life expectancy by state, 2014 (years)		Absolute difference	Ratio (Mississippi/Hawaii)
	Mississippi	Hawaii	(years)	(%)
	74.91	81.15	−6.24	92.3
Example 2: Disparities in healthcare				
Change in percentage making visit to clinic including tobacco screening, 2009–2011	Percentage of men and women, 18 years and older, 2009–2011		Absolute difference	Ratio
	NH Black	NH White	(Percentage points)	(NH Black/White)
	−14.8	−0.1	−14.7	148
Example 3: Disparities in social, environmental, and behavioral risk factors				
Children, 3–11 years of age, exposed to secondhand smoke	Percentage exposed by family poverty status		Absolute difference	Ratio
	<100% poverty threshold	≥500% poverty threshold	(Percentage points)	(<100%/≥500%)
	62.1	14.1	48.0	4.4

Abbreviation: NH, non-Hispanic.

Sources: Types of disparity measurements adapted from Appendix 11 of the US Department of Health and Human Services [9]. Data from Example 1 are from the *Global Health Data Exchange: US Health Map*, and data for Examples 2 and 3 are from *Data 2020*, a resource linked to *Healthy People 2020*.

new methodologies, new technology, and multilevel applications of methodologies will take these in new directions.

Certain characteristics have long had a place in quantitative research on behavioral determinants of health and health disparities, including, for example, race or ethnicity, gender, education, and socioeconomic position. Place of residence can be an indicator of socioeconomic position, or it may reveal disparities resulting from a remote, rural location, such as identification as a medically underserved area or one without public transportation, broadband service, or other amenities common in cities. All these characteristics contribute to limited access to and utilization of healthcare or health information.

Place-based approaches to improving population health outcomes have a lengthy history and are gaining momentum [10]. Among them are multilevel strategies that may involve working with nonprofit organizations or local, state, or national governments, performing epidemiological surveys, evaluating physical environments and nutrition status, and even instituting sustainable brick and mortar clinical care [10].

Inequalities of health, as we too well know, lie not only between countries but within them. One of the most pervasive of these is the common finding that the lower the socioeconomic position of a group, the poorer their health. One of the opportunities that lie before us is the chance to explicate the intricate biological and nonbiological mechanisms driving behavior and thereby understand them more comprehensively within social, cultural, and environmental contexts. Some think of this approach as a scientific one and an opportunity to reimagine old paradigms and models and seek how they may align or not with new theories, models, and measures [26].

4.3.2 Research Methods

Researchers conducting studies that seek to understand or change health behavior today think globally and act locally. Recognized more frequently and more broadly today as central to achieving public health goals, behavioral determinants of health hold high places in the hierarchy of health variables capable of changing a population's health trajectory. Researchers' concerns, including obesity and smoking, are often shared worldwide, and today's mobile and electronic health (mhealth/ehealth) technologies not only make it possible for researchers to share findings but also to incorporate novel communication strategies to deliver in-time messages about behavior to study participants. Smartphones and web applications belonging to former drug users in recovery programs may alert their owners to risks, warning them to avoid neighborhoods where they previously bought drugs. Similarly, weight-conscious individuals may actively use health applications to help monitor daily activity, food and calorie intake, and heart rate.

The methods researchers employ to promote healthy behaviors, or to engage in prevention to reduce disease, include:

- surveillance, epidemiologic study, or other traditional public health tools;
- randomized controlled trials, ecological study, or another tested design;
- re-analysis of existing data; and
- meta-analyses or systematic reviews of completed investigations.

What the best studies have in common:

- attention directed at a prioritized problem (e.g., adolescent smoking);
- a commitment to the most rigorous design applicable (e.g., the randomized controlled trial);
- thoughtful statistical approach (e.g., particularly determining the size of the study group so that findings have statistical meaning);
- a team of dedicated and well-trained investigators who may represent many disciplines (e.g., epidemiology, statistics, public health, medicine, nursing, and social work);
- recruitment that creates a representative subject population able to produce generalizable results; and
- a plan for prompt dissemination of findings and/or replication of successful interventions.

Some samples include:

- a randomized trial to test a physical activity intervention for Latinas;
- a statistical study to identify and characterize a minority population with high rates of colorectal cancer;
- a re-analysis of national survey data to identify differences in behavioral factors and breast cancer status according to race and ethnicity; and
- an experiment with computer simulation to create a reliable model of behavior change.

A foundational premise must be a commitment to appreciating and respecting all aspects of the beliefs, values, and traditions of the communities of interest, and being responsible in obtaining the cultural competence to work with them. Any researcher needs to be concerned about the effect of participation on the community and whether necessary and sufficient support is available to communities to sustain any changed behavior introduced by an intervention.

This is especially true in terms of precision medicine and precision prevention. Making precision medicine and prevention available to large populations requires the participation and inclusion of a diverse patient population, and not just those who can afford it and have ready access to care. Most research data are extracted from uniform study populations composed largely of White men, though the minority population of the United States is approaching 40%. Members of these groups cannot fully benefit from advances in cancer treatment, especially in this era of precision medicine. Minority representation in clinical trials is less than 5%, and according to a review of clinical trials from 2000 to 2010, only 2.2% of participants were Latino [27]. This calls for a large area of improvement to empower minority communities toward positive attitudes and beliefs in research and clinical practices through increased promotion and knowledge of clinical trials, active participation in research protocols, and involvement in transdisciplinary and translational partnerships for better minority health outcomes.

4.4 State of the Science: Promising Practices

Promising practices in behavioral determinants research, described below, rely on both new and tried-and-true methods. Rather than toiling within the confines of traditional academic practice with a single team, some researchers are creating consortia and combining multiple teams to address significant health behaviors. An evolving approach is incorporating technology into interventions. Health technology is transforming everyday tools for research, or for directly promoting health-enhancing behavior.

These include more than just uses for the electronic health record (EHR) and smartphones. Promising practices also include respectful responses to underserved constituencies that build trust and the creative concept of public health advocates influencing legislation by bringing together a broad coalition of partners across the state to create a sustainable investment in community-level population health strategies to advance the cycle of better care, better health, and lower costs [28, 29].

CBPR also should be included on the list of promising practices. In addressing the disproportionate and growing burden of health disparities among minority and lower socioeconomic populations, CBPR has shown great promise. Researchers, practitioners, and community partners who have participated in CBPR projects have been able to point to the success that results when everyone is able to fully integrate their talents, resources, and expertise toward a common goal [30].

4.4.1 *Salud America!*

In response to the high prevalence of obesity in Latino children and adolescents, researchers at the Institute for Health Promotion Research at the University of Texas Health Science Center at San Antonio, buoyed by support from the Robert Wood Johnson Foundation, led the creation of *Salud America!* The National Research Network to Prevent Obesity among Latino Children. In its first few years, *Salud America!* formed a national advisory committee, selected 20 pilot study investigators, and presented a Latino childhood obesity research priority agenda [31]. Investigators received an average of $75 000 to develop and implement promising policy-relevant strategies to prevent and reduce obesity among Latino children. Awards were made across the country. Investigators believed the need for research and intervention was urgent because: childhood and adolescent overweight and obesity portend a propensity to adult overweight and obesity; healthcare costs rise dramatically for obese adults who were not at a healthy weight in childhood; and Latino children are vulnerable to marketers hawking unhealthy food to youth through television, social media, and other outlets.

One pilot investigator's approach included Photovoice, which encouraged parents and children to use photography to identify patterns or themes within their communities that were barriers to physical activity and root causes of inactivity. Among barriers participants noted were lack of access to activity programs outside school hours, lack of transportation, and communication problems. This involvement prompted them to make proposals for solutions within the community.

Other initiatives brought together mothers and daughters for physical activity; taught families about nutrition and how to evaluate grocery purchases; profiled understudied migrant farmworkers' children, almost half of whom were at unhealthy weights; and linked greater physical activity with higher performance in math. Others assessed the built environment, and still others addressed crime as an inhibitor of physical activity [10].

A second phase shifted *Salud America!* efforts into advocacy and policy development based on the evidence [31]. The program developed a Policy Contribution Spectrum to identify and describe the stages of policy development. Program leaders then developed an online multipronged strategy—research reviews, culturally relevant digital content curation, stories, videos, and action items in key aspects of Latino child and family health [10]—to inspire people to start and support healthy changes to policies, systems, and environments where Latino children and families can equitably live, learn, work, and play. The *Salud America!* program uses this content to fuel its social and online network of more than 125 000 moms and dads, providers, researchers, and community and school leaders to push for healthy changes in schools and communities for Latinos and all kids.

Overall, the Institute for Health Promotion Research team has created more than 100 national, statewide, local, and private research programs. The work not only produced results but also has generated new investigators, new funds, wider interest, and awards. The Institute has pursued expansion of Latino health initiatives, including research in childhood obesity [31], work in tobacco cessation among young adults [32], action on hepatocellular cancer [33], and attention to cancer health disparities in a recently launched national initiative [27].

4.4.2 Health Technology—New Vital Signs

EHRs are proving to be significant partners in research. The Committee on the Recommended Social and Behavioral Domains and Measures for Electronic Health Records with the Board on Population Health and Public Health Practice published *Capturing Social and Behavioral Domains and Measures in Electronic Health Records: Phase 2* [34]. The effort conceptualizes the EHR as a kind of nervous system with the potential to unify policy and business frameworks and to produce coordinated action on improving healthcare in light of social and behavioral research. Guided by the National Academies of Science, the committee recommended expanding the four domains commonly collected (tobacco use, alcohol use, race/ethnicity, residential address) with eight more (education level, financial resource strain, stress, depression, physical activity, social isolation, intimate partner violence, and neighborhood median household income [deduced from address]).

These variables are known as "psychosocial vital signs" and are considered means to ensure that all clinical, public health, and community resources needed were made available to patients [35]. Behavioral determinants of health—physical activity joins tobacco use and alcohol use—figure prominently. This information can help link patients who need support to improve physical activity, alter alcohol habits, change diet, or access other resources in the community known to healthcare personnel.

Many of the domains are directly aligned with those in the World Health Organization PROGRESS mnemonic (*p*lace of residence [rural or urban], *r*ace/ethnicity, *o*ccupation, *g*ender, *r*eligion, *e*ducation level, *s*ocioeconomic position, and *s*ocial capital or resources) [25]. Standardizing this information may facilitate better care, improve population health, and promote more informative research, but also help patients overcome barriers to prevention and wellness efforts by identifying modifiable determinants of health when support can be established.

4.4.3 Health Technology—Smartphone Use

The frequent use and broad reach of messaging on mobile phones have made them attractive tools in studies of behavior and health. Researchers appreciate their ubiquity, efficiency, cost-effectiveness, potential for message tailoring, just-in-time communication, and likelihood of intake and responsiveness: within three minutes of receipt, 90% of text messages are read and 99% are opened. Receivers are rewarded with a dopamine release associated with text message receipt. A meta-analysis and systematic review of text messaging interventions directed at behavior change and health improvement amassed from 15 carefully selected studies found that messaging interventions were effective in confronting behavioral determinants of health, such as medication adherence for antiretroviral therapy, diabetes management, smoking cessation, physical activity, stress and coping, and weight loss.

Of particular interest to health disparities researchers is that Latinos are significantly more likely than Whites to send or receive text messages (83% vs. 70%) and that both Latinos (83%) and African Americans (76%) use messaging more than Whites [36]. Likewise, Latinos (73%) and African Americans (67%) have been shown to be more likely than Whites (58%) to use their smartphones for health information [35].

Taking advantage of this use is *Quitxt*, a Latino-attuned subscription service that promotes tobacco cessation by providing young adults real-time support, interactive messages, and motivating videos and music. These create a culturally relevant path toward quitting and tools for handling related stress. *Quitxt* enrolled 798 participants in its initial launch; at the 7-month follow-up, investigators found that 21% (171) of enrollees reported abstinence. This finding is consistent with high rates of success found in studies of telephone counseling for young adults, and confirms that text and mobile media services specifically designed for young adults provide a feasible and potentially cost-effective approach to promoting cessation [32].

Simply having access to mobile broadband can be an issue in remote areas [37], which means the lack of Internet access must be added to the limitations in that health setting. The cost of smartphone data and phone plans, as well as technology literacy and comfort among different age groups with technology, are also barriers that may further widen the digital divide and affect minority health behavior interventions and outcomes.

4.4.4 Underresearched Constituencies

Questions on sexual and gender minorities (SGMs) were included on 10 recurring federal surveys, according to an Institute of Medicine report in 2011 [38] (an expansion and follow-up after a 1999 volume on lesbian health) [39], but none measured gender identity. The committee producing the report met in open session three times to hear from invited speakers and members of the public, ask questions, and provide a forum to share knowledge. The committee recommended that researchers continue to collect data on sexual orientation and gender identity on federal surveys, and that the National Institutes of Health (NIH) foster more research in the SGM community.

The lack of data on the SGM community, like the sparse data on American Indian populations and other underresearched populations, cripples efforts to craft basic and effective interventions to enhance health and prevent health-adverse consequences, including behavior-related lung disease and energy imbalance when calorie consumption and expenditure are not equal. The lack of data means public health officials cannot construct an evidence-based hierarchy of risks for the SGM community and prioritize those risks. The NIH has created the Sexual and Gender Minority Research Coordinating Committee, whose activities are led in part by the National Institute on Minority Health and Health Disparities. Among its goals are holding listening sessions with stakeholders to identify SGM health priorities and developing an NIH SGM strategic plan for health research [39].

4.4.5 Advancing the Cycle of Better Care, Better Health, Lower Costs

Acknowledging that the continual increase in the cost of healthcare is unsustainable—the share of the US economy devoted to healthcare in 2015 was 17.8%, an increase of 5.8% over 2014—and that prevention is more cost-effective than treatment, the Massachusetts Public Health Association helped create a \$60 million trust fund for community-level health strategies from those who would benefit from the savings—health plans and large healthcare systems [28]. Its aims included increasing healthy behaviors, reducing health disparities, reducing rates of the state's most costly preventable health conditions, developing a more robust evidence base of effective prevention programs, and promoting implementation of workplace programs that promote wellness. Public health advocates made public health part of the debate of a healthcare cost-containment legislative initiative and built a robust coalition of businesses, civic leaders, and local government officials.

This multilevel approach has served more than 372 000 Massachusetts citizens, and evaluators estimate cost savings in targeting diet and nutrition, leisure-time exercise, and stress reduction to be between \$760 000 and \$4.1 million [29]. Over a lifetime, hypertension programs are expected to avert healthcare costs \$9 million to \$16 million, while tobacco use programs could save from \$622 118 to \$5.6 million. Other advances are expected in building capacity, improving affordable housing environments, and engaging hard-to-reach populations.

Another promising initiative, especially in reaching the Latino community, has been the use of *promotores de salud* (community health workers/promoters). These bilingual, bicultural health advocates work directly in communities, often being directly employed from the neighborhoods that they serve, and link communities to healthcare providers, healthcare services, education, and resources [15]. In Latino communities, *promotores* have helped to reduce stigmas associated with certain medical conditions and have increased use of mental health services through community advocacy and healthcare navigation [15].

4.5 Challenges and Future Opportunities

In the twentieth century, Americans witnessed an increase in life expectancy of 30 years, 80% of which can be attributed to public health advances [40]. These included reducing the rate of adult tobacco use by half, improving food safety and nutrition, broadening access to clean water and improving sanitation, eradicating smallpox and creating and using vaccines against other infectious diseases, and making family planning methods accessible and effective [41]. What we learn from this, and even more from the statistics indicating how far from good health our behaviors take us as well as the cost of treatment, is that "the potential of behavioral and social interventions for increasing life expectancy may be orders of magnitude greater than traditional medical interventions" [2].

In the twenty-first century, chronic disease has been identified as the predominant emerging challenge to global health [42]. It is a growing challenge in the United States and in subpopulations characterized by health disparities. Chronic conditions, which are responsible for most ill health, disability, and death as well as most of the cost of healthcare in the United States [43], represent 7 of the top 10 causes of death—6 of which have related behavioral determinants—and are responsible for about 75% of all deaths [11]. Heart disease is first on the list of chronic conditions. Rates for hypertension are highest in non-Latino Blacks (43.3%) and lowest in non-Latino Asians (26.5%). Years lost under age 75 per 100 000 population because of diabetes (seventh on the list) is 321 per 100 000 non-Latino Blacks. The rate for non-Latino Asians is 4.3 times lower (74 years/100 000). Though healthcare spending since 1960 has outpaced the growth of the gross domestic product five-fold [44], the United States has little to show for it. While ranking as the biggest global healthcare spender, it posts a life expectancy that puts it behind 25 other countries [43].

Not unlike understanding biological pathways in disease, understanding how an individual's interaction with his/her environment can result in chronic disease can help researchers identify and prevent social and behavioral determinants of chronic disease. Multidisciplinary investigators at Michigan State University and Washington State University investigated ways to jumpstart physical activity and more healthful eating in Flint, Michigan, and began a search for successful hypertension interventions that could work with American Indians, Pacific Islanders, and Alaska Natives in the Pacific Northwest [45].

In Michigan, Vicki Johnson-Lawrence, PhD, and other investigators working in part on evidence that Blacks, Latinos, and women of lower socioeconomic position are at greater risk of obesity [46], conceived a multilevel—individual, community, and structural—approach to reduce obesity and cardiovascular disease. Investigators believe economically disadvantaged communities need interventions that work on multiple levels to produce complementary, sustainable, longitudinal results. In Flint, a community already dealing with a lead-contaminated water supply [47], the county health department partnered with local churches to devise an intervention incorporating classes in cooking and African American dietary history, peer support, and church-based food pantry access. The pursuit of environmental policy goals, including health-positive land use (e.g., making way for grocery stores with fresh food and creating bike lanes and sidewalks), is part of an effort that will involve educational institutions, government, community organizations, and individuals.

In Washington, Dedra Buchwald, MD, and colleagues are working with American Indians, who are prone to strokes at younger ages and are three times more likely to die of a stroke if it occurs before they turn 65. These investigators are addressing blood pressure control to reduce cardiovascular disease and stroke. Health team members who work with these populations rely on technology at hand—grocery shopping applications, text messaging, and patient-friendly blood pressure and physical activity monitors—to increase activity, improve diet, and monitor and prevent cardiovascular events.

Working under the umbrella of the Transdisciplinary Collaborative Centers for Health Disparities Research on Chronic Disease Prevention program, funded by National Institute on Minority Health and Health Disparities (NIMHD), these and other researchers are undertaking multilevel investigations that will identify what works in these populations characterized by health disparities. They and others are seeking ways to control energy balance, reduce hypertension and cardiovascular disease, and solve other health problems affecting populations across the United States.

4.6 Summary

The very high cost Americans pay for healthcare, almost $10 000 per person annually, does not buy longevity [1]. Investigators attribute 50% of health overall and more than 75% in certain diseases to behavior. Though health-adverse behaviors may contribute to a majority of the causes of mortality, we have no comprehensive explanation of why and how the behaviors begin or are sustained. Structural drivers, including social context, environment, money, and power, are not neutral forces. Populations at risk include racial and ethnic minority groups; those of low socioeconomic position; those who live in rural, underserved areas; and members of the SGM communities. To seek elimination of health disparities is to seek health equity. Promising practices include creating consortia to make advances against a specific priority and create a new cadre of investigators; adapting and adopting readily available technology to health-enhancing purposes; listening to underresearched communities and acting on their needs; and promoting positive health behaviors through multilevel approaches that advance the cycle of better care, better health, and lower costs. Understanding mechanisms and pathways responsible within the social and environmental context for behavior may afford new approaches to health disparities. Inequalities in health are a measure of any country, and failing to eliminate those disparities risks allowing them to widen.

The potential fallout from failing to change the status quo is not inconsequential. President and CEO of the Joint Center for Political and Economic Studies Ralph B. Everett said in the organization's report on managing chronic disease by mobile phone, "What is clear . . . is that . . . innovations are vitally important and needed as part of a comprehensive strategy to reduce and eventually eliminate health inequities" [37].

We cannot fail to accept the challenge posed by the significant contribution of behavioral determinants to health. We must act on what the research tells us about how to recognize, attack, and eliminate health disparities. Just as the public health advances of the twentieth century advanced longevity, public health initiatives of the twenty-first century must work to ensure that those gains are not eroded but extended, and that the health quality improvement achieved is not only sustained but enhanced.

4.7 Key Points

- Behavioral determinants of health influence health status and outcomes of individuals and populations. In an epidemiological shift, for example, the major causes of death worldwide are no longer communicable diseases but noncommunicable diseases largely determined by such behaviors as tobacco use, inactivity, unhealthy diet, and alcohol overuse.

- Health disparities can grow in minority, rural, and other communities where education, supportive institutions, employment, health engagement, and care access are in short supply.
- In addressing behavior and other health determinants, including socioeconomics, cultural environment, physical environment, biological and genetic influences, psychosocial factors, and healthcare, we must act on the belief that—recognizing disparities—we value good health for all.
- It is imperative that we act on research results to implement evidence-based interventions and that we continue to work on recognizing, addressing, and eliminating health disparities.

Disclaimer

The views and opinions expressed in this chapter are those of the authors only and do not necessarily represent the views, official policy, or position of the U.S. Department of Health and Human Services or any of its affiliated institutions or agencies.

References

1 Centers for Medicare and Medicaid Services (2015). *National Health Expenditures 2015 Highlights*. Baltimore, MD: Centers for Medicare and Medicaid Services. https://www.healthaffairs.org/doi/full/10.1377/hlthaff.2016.1330 (accessed July 6, 2017).

2 Kaplan, R.M. (2014). Behavior change and reducing health disparities. *Preventive Medicine* 68: 5–10.

3 Committee on Core Metrics for Better Health at Lower Cost (2015). *Vital Signs: Core Metrics for Health and Health Care Progress*. Washington, DC: Institute of Medicine of the National Academies. https://www.nap.edu/catalog/19402/vital-signs-core-metrics-for-health-and-health-care-progress (accessed August 15, 2017).

4 American Institute for Cancer Research (2014). *Cancer Research Update*. Arlington, VA: American Institute for Cancer Research.

5 Branca, F., Nikogosian, H., and Lobstein, T. (eds.) (2007). *The Challenge of Obesity in the WHO European Region and the Strategies for Response: Summary*. Copenhagen, Denmark: World Health Organization. http://www.euro.who.int/__data/assets/pdf_file/0008/98243/E89858.pdf (accessed August 17, 2017).

6 Cawley, J. and Ruhm, C. (2011). *The Economics of Risky Health Behaviors*. Cambridge, MA: National Bureau of Economic Research.

7 Piano, M.R. and Phillips, S.A. (2014). Alcoholic cardiomyopathy: pathophysiologic insights. *Cardiovascular Toxicology* 14 (4): 291–308.

8 McGovern, L., Miller, G., and Hughes-Cromwick, P. (2014). *Health Policy Brief: The Relative Contribution of Multiple Determinants to Health Outcomes*. Bethesda, MD: Project HOPE. http://healthaffairs.org/healthpolicybriefs/brief_pdfs/healthpolicybrief_123.pdf (accessed 21 August 2017).

9 US Department of Health and Human Services (2008). *Phase I Report: Recommendations for the Framework and Format of Healthy People 2020*. Washington, DC: The Secretary's Advisory Committee on National Health Promotion and Disease Prevention Objectives for 2020.

10 Dankwa-Mullan, I. and Perez-Stable, E.J. (2016). Addressing health disparities is a place-based issue. *American Journal of Public Health* 106 (4): 637–639.

11 US Department of Health and Human Services (2017). *Health, United States, 2015: With Special Feature on Racial and Ethnic Health Disparities*. Hyattsville, MD: National Center for Health Statistics. https://www.cdc.gov/nchs/data/hus/hus15.pdf#019 (accessed July 9, 2017).

12 US Centers for Disease Control and Prevention (2015). *Hispanic Health: CDC Vital Signs*. Atlanta, GA: US Centers for Disease Control and Prevention. https://www.cdc.gov/vitalsigns/hispanic-health/index.html (accessed August 14, 2017).

13 Heron, M. (2017). Deaths: leading causes for 2015. *National Vital Statistics Reports* 66 (5): 1–76.

14 Robert Wood Johnson Foundation (2016). *Declining Childhood Obesity Rates: Where Are We Seeing Signs of Progress?* Princeton, NJ: Robert Wood Johnson Foundation. http://www.rwjf.org/en/library/research/2016/06/declining-childhood-obesity-rates.html (accessed August 13, 2017).

15 Ramirez, A.G., Gallion, K.J., Aguilar, R., and Gamse, C. (2017). *Building Support for Latino Families*. Princeton, NJ: Robert Wood Johnson Foundation. https://salud-america.org/building-support-for-latino-families-research (accessed August 13, 2017).

16 US Department of Health and Human Services (2014). *Healthy Places, Healthy People: A Progress Review on Nutrition and Weight Status and Physical Activity*. Washington, DC: US Department of Health and Human Services. https://www.cdc.gov/nchs/ppt/hp2020/hp2020_nws-and_hp_progress_review_presentation.pdf (accessed August 15, 2017).

17 Johnson, S.R. (2016). The racial divide in the opioid epidemic. *Modern Healthcare* 46 (9): 12, 14–15.

18 World Health Organization (2011). *Global Status Report on Noncommunicable Diseases 2010*. Geneva: World Health Organization. http://www.who.int/nmh/publications/ncd_report_full_en.pdf (accessed August 30, 2017).

19 Siegel, R.L., Miller, K.D., and Jemal, A. (2017). Cancer statistics, 2017. *CA: a Cancer Journal for Clinicians* 67 (1): 7–30.

20 US Centers for Disease Control and Prevention (2017). *Opioid Overdose*. Atlanta, GA: US Centers for Disease Control and Prevention. https://www.cdc.gov/drugoverdose/epidemic/index.html (accessed August 20, 2017).

21 Singhal, A., Tien, Y.Y., and Hsia, R.Y. (2016). Racial-ethnic disparities in opioid prescriptions at emergency department visits for conditions commonly associated with prescription drug abuse. *PLoS One* 11 (8): e0159224.

22 Pletcher, M.J., Kertesz, S.G., Kohn, M.A., and Gonzales, R. (2008). Trends in opioid prescribing by race/ethnicity for patients seeking care in US emergency departments. *Journal of the American Medical Association* 299 (1): 70–78.

23 Park, M.H., Falconer, C., Viner, R.M., and Kinra, S. (2012). The impact of childhood obesity on morbidity and mortality in adulthood: a systematic review. *Obesity Reviews* 13 (11): 985–1000.

24 Puhl, R.M. and Heuer, C.A. (2010). Obesity stigma: important considerations for public health. *American Journal of Public Health* 100 (6): 1019–1028.

25 World Health Organization (2013). *Health Inequality Monitoring: With a Special Focus on Low- and Middle-Income Countries*. Geneva: World Health Organization. http://apps.who.int/iris/bitstream/10665/85345/1/9789241548632_eng.pdf (accessed July 8, 2017).

26 Dankwa-Mullan, I. and Maddox, Y.T. (2015). Embarking on a science vision for health disparities research. *American Journal of Public Health* 105 (Suppl 3): S369–S371.

27 Ramirez, A.G. and Thompson, I.M. (2017). How will the 'cancer moonshot' impact health disparities? *Cancer Causes & Control* 28 (9): 907–912.

28 Batdorf-Barnes, A. (2013). Massachusetts prevention and wellness trust fund: cutting health care costs through prevention. In: *Improving Population Health: Policy, Practice, Research* (ed. D.A. Kindig). Madison, WI: University of Wisconsin Population Health Sciences.

29 Prevention and Wellness Trust Fund (2017). *Joining Forces: Adding Public Health Value to Healthcare Reform*. Boston, MA: Prevention Wellness Trust Fund.

30 Becker, A.B., Israel, B.A., Gustat, J. et al. (2012). Strategies and techniques for effective group process in community-based participatory research partnerships. In: *Methods in Community-Based Participatory Research for Health* (eds. B. Israel, E. Eng, A. Schulz and E. Parker), 52–72. San Francisco, CA: Jossey-Bass.

31 Ramirez, A.G., Gallion, K.J., Despres, C.E., and Adeigbe, R.T. (2013). SaludAmerica!: a national research network to build the field and evidence to prevent Latino childhood obesity. *American Journal of Preventive Medicine* 44 (3 Suppl 3): S178–S185.

32 Ramirez, A.G., Chalela, P., Akopian, D. et al. (2017). Text and mobile media smoking cessation service for young adults in South Texas: operation and cost-effectiveness estimation. *Health Promotion Practice* 18 (4): 581–585.

33 Ramirez, A.G., Munoz, E., Parma, D.L. et al. (2017). Lifestyle and clinical correlates of hepatocellular carcinoma in South Texas: a matched case-control study. *Clinical Gastroenterology and Hepatology* 15 (8): 1311–1312.

34 Institute of Medicine (2014). *Capturing Social and Behavioral Domains and Measures in Electronic Health Records: Phase 2*. Washington, DC: National Academies Press.

35 Anderson, M. (2015). *Racial and Ethnic Differences in How People Use Mobile Technology*. Washington, DC: Pew Research Center.

36 Zickuhr, K., Smith, A. (2012). Digital differences. https://www.pewresearch.org/internet/2012/04/13/digital-differences/ (accessed June 16, 2020).

37 Turner-Lee, N., Smedley, B., and Miller, J. (2012). *Minorities, Mobile Broadband and the Management of Chronic Diseases*. Washington, DC: Joint Center for Political and Economic Studies.

38 Institute of Medicine (2011). *The Health of Lesbian, Gay, Bisexual, and Transgender People: Building a Foundation for Better Understanding*. Washington, DC: National Academies Press.

39 National Institutes of Health (2017). *The Sexual and Gender Minority (SGM) Research Coordinating Committee (RCC)*. Bethesda, MD: National Institutes of Health.

40 Frieden, T.R. (2015). Shattuck Lecture: the future of public health. *The New England Journal of Medicine* 373 (18): 1748–1754.

41 Forsberg, V. and Fichtenberg, C. (2012). *The Prevention and Public Health Fund: A Critical Investment in our Nation's Physical and Fiscal Health*. Washington, DC: American Public Health Association.

42 Bauer, U.E., Briss, P.A., Goodman, R.A., and Bowman, B.A. (2014). Prevention of chronic disease in the 21st century: elimination of the leading preventable causes of premature death and disability in the USA. *Lancet* 384 (9937): 45–52.

43 United Health Foundation (2016). *America's Health Rankings: A Call to Action for Individuals and their Communities*. Minnetonka, MA: United Health Foundation.

44 Kaufman, K. (2012). *Health Affairs Blog—Bending the Health Care Cost Curve: More than Meets the Eye?* Project HOPE: Bethesda, MD.

45 Abbasi, J. (2016). Social determinants of chronic disease get NIH attention. *Journal of the American Medical Association* 316 (16): 1636.

46 Zenk, S.N., Mentz, G., Schulz, A.J. et al. (2017). Longitudinal associations between observed and perceived neighborhood food availability and body mass index in a multiethnic urban sample. *Health Education & Behavior* 44 (1): 41–51.

47 Bellinger, D.C. (2016). Lead contamination in Flint–an abject failure to protect public health. *The New England Journal of Medicine* 374 (12): 1101–1103.

5

Sociocultural Environments and Health Disparities Research

Frameworks, Methods, and Promising Directions

Hortensia Amaro[1,2], *Samantha Garcia*[3], *Inna Arnaudova*[4], *and Monica P. Jolles*[5, 6]

[1] *Herbert Wertheim, College of Medicine, Florida International University, Miami, FL, USA*
[2] *Robert Stempel, College of Public Health and Social Work, Florida International University, Miami, FL, USA*
[3] *Program in Public Health, University of California, Irvine, CA, USA*
[4] *Department of Psychiatry, University of California, Los Angeles, CA, USA*
[5] *Suzanne Dworak-Peck, School of Social Work, University of Southern California, Los Angeles, CA, USA*
[6] *Cecil G. Sheps Center for Health Services Research, University of North Carolina at Chapel Hill, Chapel Hill, NC, USA*

5.1 Introduction

Our evolving understanding of health disparities [1] reflects concepts and factors relevant to the reduction of those disparities. One such factor, sociocultural environments, manifests across levels of influence (i.e., individual, interpersonal, community, and society), and the human lifespan (see Chapter 1). Other domains of influence (i.e., biological, behavioral, physical/built environment, and healthcare systems) are thought to interrelate and have complex and iterative influences on health disparities (i.e., individual, family, community, and population; see Chapter 1).

In this chapter, we selectively discuss sociocultural environment elements that are key to the understanding and reduction of health disparities, as well as being prime areas for future research. We end with a broad synthesis of take-home points emphasizing challenges and future opportunities.

5.2 Selected Overarching Sociocultural Environment Factors

5.2.1 Culture

While not directly affecting health, cultural influences do affect the perceptions, beliefs, desires, norms, and values (e.g., at the individual, family, community, and population levels) that influence how we approach health and healthcare. For example, qualitative analysis of interviews with human immunodeficiency virus (HIV)-positive Latinos on

The Science of Health Disparities Research, First Edition. Edited by Irene Dankwa-Mullan, Eliseo J. Pérez-Stable, Kevin L. Gardner, Xinzhi Zhang, and Adelaida M. Rosario.

adherence to their prescribed treatment visit regimen, especially among more recent arrivals and those born outside the United States, showed that cultural factors (e.g., traditional male gender roles and value of personal privacy) and related stigma (e.g., social taboos against homosexuality) were barriers to disclosure of HIV status and HIV treatment adherence (e.g., regular clinic attendance) for fear of potential family conflict and rejection from community members [2]. A longitudinal study of Latino youth found that Latino cultural orientation in the 10th grade was associated with *reduced* intimate partner physical violence perpetration later on during early adulthood [3]. In contrast, greater US-oriented acculturation among 10th-grade boys was associated with greater likelihood of perpetration of physical intimate partner violence in adulthood. These findings show that cultural influences that affect perceptions, beliefs, attitudes, and norms can have negative and/or positive effects on health behaviors.

Overall, research on how culture influences health has received little empirical attention and suffers from poor conceptualization; simplistic and superficial measures; lack of historical, socioeconomic, and contextual grounding; and a static view of culture [4].

The following definition of culture differentiates what culture is from what it does: "Culture is an internalized and shared schema or framework that is used by group (or subgroup) members as a refracted lens to 'see' reality, and in which both the individual and the collective experience the world" [4]. The authors emphasize the dynamic and adaptive nature of culture as a system that responds to changing economic, social, and political conditions within and outside a group's societal ecology. Thus, culture is not immutable, but rather, a constantly evolving, multidimensional, multilevel process. What culture does is provide its members with tools (e.g., beliefs, attitudes, practices, and spiritual and emotional explanations used to create norms of behavior and codify roles) "for making sense of their world and to find meaning in and for life by providing a sense of safety and well-being, integrity, and belonging as a contributing member of one's social network" [4]. The research worthiness of aspects of culture vary by the degree to which they shape behavior [4].

Studies on health disparities need to further specify how culture affects health. Here are three suggestions for exploring that relationship: (i) Future research needs to examine the interactions of culture with other levels of influence (e.g., how changes in individual cultural orientation among immigrants positively or negatively affect health-enhancing interpersonal relationship features such as social support and networks) and other domains of influence (e.g., how immigration policies and related cultural values of the host country members impact stress and internalization of negative stereotypes among minorities and immigrant groups, and healthcare utilization) across population groups and the lifespan. (ii) Intervention research using an informed cultural lens may produce multiple health benefits [4]. (iii) The interdisciplinary approach inherent in cultural neuroscience provides conceptual and methodological tools to explore links between biology and sociocultural factors—an important area of health disparities research [5].

5.2.2 Acculturation

A common approach to assessing culture's effects on health has been to employ acculturation measures, which refers to the adoption of the attitudes, values, customs, beliefs, and behaviors of the dominant culture by the minority individual in multicultural societies, or by migrants following arrival [6]. While many studies have reported negative associations between acculturation and health, measures of acculturation used in many health studies have relied on inexact proxy variables (e.g., primary language spoken at home, place of birth, years in the host country), which have contributed to mixed findings [7]. Schwartz et al. [7] argue that understanding the relationship between acculturation and health requires using a multidimensional model comprised of direct measures of cultural practices, values, and ethnic identity. Using this model, Ro and Bostean [8] studied the relationship between duration of stay in the US and body mass index (BMI) among Latino and Asian immigrants and six hypothesized aspects of integration (i.e., household income, English proficiency, ethnic identity, family cohesion, acculturative stressors, and discrimination) and did not find evidence for a unidirectional relationship. Their findings suggest that the relationship between immigrant integration and health indices (in this case, BMI) is complex, involving pathways (e.g., mediators) that vary across groups and gender.

While the process of (im)migration and adaptation to a new culture and context may bring both challenges and opportunities that affect health, research is needed to clarify the conditions under which acculturation affects health outcomes and the pathways through which this occurs. The salutary and detrimental effects of acculturation likely differ across (im)migrant groups and the pathways may also vary among (im)migrant groups, genders, socioeconomic status (SES), and lifespan stages, as well as according to the receiving culture's orientation and resources it provides to immigrant groups. Future studies of these effects will provide a more accurate understanding of when and how health disparities develop and inform future intervention strategies.

5.2.3 Racism

Racism has been defined as beliefs, attitudes, institutional arrangements, and acts based on rankings of social group by phenotype or ethnic characteristics [9]. Williams and Mohammed's [9] framework details the negative consequences of racism (cultural transmission, stress, SES opportunities, societal resources) and proposes pathways (institutional racism, cultural racism, and experience of racism) through which racism may affect health. Previous work has documented how institutional racism (e.g., segregation in housing) and the experience of racism are associated with poorer health behaviors and outcomes [9]. Their framework highlighted how cultural racism (i.e., structurally unequal practices that rely on cultural differences rather than phenotype markers that assume White racial superiority) shapes the conscious and unconscious biases observed in many social transactions, including medical treatment [10]. Institutional and cultural racism promote internalized racism (harvesting negative beliefs about one's self or group), which has been shown to be associated with negative health outcomes and resulting health disparities [9]. While research might be able to measure experienced racism [11], racist beliefs might not be readily reported (see later section on implicit bias for potential solutions). Understanding how different types of racism place groups of individuals at risk for poorer health outcomes and how racism is related to other sociocultural factors is important for achieving health equity.

5.2.4 Interconnection of Race/Ethnicity and Socioeconomic Status

Although SES contributes to health inequities across racial and ethnic groups, disparities in health behavior, health status, and healthcare often remain even after controlling for SES in statistical analyses [1]. But controlling for SES does not capture the complex ways in which SES affects health across race and ethnicity. Such social constructs reflect multiple dimensions of social inequality, including the patterning of SES, segregation, and exposure to a host of social environment and physical life conditions. Promising approaches include examining the interaction of interconnected SES, race, and other sociocultural factors within the multilevel systems in which they occur [1].

Next, we discuss three sociocultural factors (social capital, healthcare systems, and implicit bias) that might help in understanding health disparities and have promising areas of research in the pursuit of health equity.

5.3 Social Capital and Health Disparities

Considered one of the social determinants of health and health disparities, social capital refers to resources—and, some argue, liabilities—embedded in and accrued via social groups and networks. Social capital is thought to have consequential impacts on life outcomes, including health.

In the late 1990s, research on social capital and its relation to health began to grow exponentially [12], resulting in a significant, promising, and complex body of work. Social capital or related elements began to be included more often in health disparities frameworks. For example, the National Institute on Minority Health and Health Disparities Research Framework (see Chapter 1) includes characteristics known to affect social capital (networks, norms, and discrimination), while other frameworks have directly identified social capital and related measures (social connection/social cohesion) as contributors to health inequities [1].

Below, we summarize research on (i) definitions and operationalization of social capital, (ii) its relationship to health disparities, (iii) related interventions, and (iv) challenges and opportunities for future research.

5.3.1 Definitions and Operationalization

Kawachi et al. [12] characterize research on social capital as belonging to two broad orientations: (i) the social cohesion school of social capital, which understands social capital as the collective or aggregated resources (e.g., trust, norms, sanctions, reciprocity) available to members of social groups (e.g., organization, neighborhoods, workplace); and (ii) the social network theory of social capital, which defines social capital as the resources (e.g., social support, information channels, social credentials) and attributes of both individuals (i.e., resources an individual can access through their social networks) and groups (i.e., features of connections between members of entire networks).

The social cohesion school understands social capital as an attribute of the group rather than of individual members, with a focus on identifying the influences of the group resources on the individual. In the social network theory, specific forms of social ties, such as bonding (i.e., strong ties within a network that strengthen common identities and function as a source of help and support among members), bridging (i.e., weaker ties that link people from different networks together and are important sources of information and resources), and linking network ties (i.e., vertical ties between people in different formal or institutionalized power hierarchies), provide distinct social capital resources (e.g., norms related to reciprocity and trust) at both the individual and collective levels.

The operationalization of various forms of social capital have produced measures ranging from individual-level perceptions of trust and reciprocity in relation to the socia grouping; sources; and types of social support from social ties, to community/group-level measures, such as the number of community-based organizations per capita; participation in neighborhood/civic activities; membership in religious or other social or civic organizations; voter turnout; and the extent, nature, and intensity of links in networks within a defined group.

Aggregated individual survey data are often used to characterize social capital at the collective or group level [12]. Social capital is considered multidimensional and both cognitive (what people feel or perceive) and structural (i.e., behavioral component; what people do with the resources they have access to).

5.3.2 Relationship to Health Disparities

A review of international evidence concluded that there is strong support for a negative correlation between various components of social capital and socioeconomic inequalities in health [13], moderated by SES. Such correlations are most often observed in studies on bonding and bridging social capital, and not found in one study that relied solely on structural measures of social capital. One study analyzed 413 low-income neighborhoods in 20 US cities and found community participation, home ownership, and neighborhood stability to be positively associated with bonding social capital [14]. While bonding social capital (measured through trust or strong social associations) may be positively associated to health, some research suggests social bonding among disadvantaged members (e.g., low-SES, high-crime communities) may be linked to negative health outcomes [12]. Future research should focus on understanding under what conditions various forms of social capital are protective as opposed to harmful to health in low-SES neighborhoods.

For disadvantaged populations, social capital may be an important health-promoting community factor. Research using self-reported health measures (i.e., general or overall current health status ratings) most frequently assessed at the individual level found evidence for a buffering effect of bonding social capital (e.g., community trust and involvement) on health that is strongest among lower-SES individuals [13].

Research also suggests people with lower SES may have unequal access to various forms of social capital, excluding them from potential benefits. For example, bonding and bridging social capital are less prevalent in economically disadvantaged communities. Still, ethnic enclaves may provide additional support and opportunities that promote social capital. A large analysis of elderly Mexican Americans found protective effects of living in high-density Mexican American neighborhoods on the prevalence of six medical conditions and all-cause survival over 7 years of follow-up, after adjusting for covariates [15]. This and other studies have found that for Mexican Americans, living in an ethnic enclave seems to buffer the negative effects of neighborhood poverty, possibly through strong social bonds fostered by culture. However, further research is needed to better understand how specific neighborhood features, such as ethnic enclaves, function to foster health through cultural values, social relations, or other mechanisms.

Social capital's effects on health have been overwhelmingly assumed to be positive. Despite warnings of social capital's potentially negative impact [12], the potential costs and downside of social capital in relation to health disparities has been underinvestigated. One review studied how bridging and linking ties for contextual social capital may negatively influence health for marginalized individuals, concluding that groups without access to bridging and linking ties may experience better health outcomes in communities where they are not socially excluded, even if social ties are weaker [13]. What remains unclear is the extent to which bonding ties can outweigh the lack of bridging and linking ties.

5.3.3 Social Capital Interventions

The knowledge base on the role of social capital and its influence on health disparities requires further development. A better understanding of how economically disadvantaged communities build and sustain social capital and how such community action affects health disparities is a key area for knowledge development. These lines of scientific inquiry are essential for informing interventions to address adverse health effects as well as promoting well-being [12]. Bonding social capital may buffer some of the negative effects of low SES on health [13].

Findings related to social capital's impact on health within an intervention setting have been mixed [12]. Previously studied social capital interventions varied in duration (from one and a half months to a year) and aimed at increasing social capital through promoting support between members of a community and encouraging social engagement. However, the buffer effect of social capital alone is unlikely to contribute to health equity for certain populations, such as individuals of lower SES [13] and women [16]. The potentially beneficial effects of social capital cannot outweigh the impact of structural drivers of health inequality [12] and therefore cannot be solely relied on to achieve health equity.

5.3.4 Future Directions for Social Capital Research

As with other socioeconomic factors discussed in this chapter, the effect of social capital on health outcomes might vary between individuals, genders, or ethnic groups. A better understanding of individual or group differences in accumulating or utilizing various forms of social capital is needed in order to develop more effective interventions. Social capital also might change over time with age or as changes occur in the environment. For example, it is unclear how gentrification of low-income neighborhoods affects the level of social capital and associated health disparities among low-income residents.

The distinction between the effects of cognitive and structural forms of social capital and the different types of social ties has been underinvestigated. This line of research has focused primarily on self-reported measures of health and mortality rates. Other objective health biomarkers might offer us insight into whether and how social capital has direct and/or indirect effects on health disparities. Finally, the role of social capital in a neighborhood's capacity to build and engage in community action to improve health conditions, resources, and outcomes—a strategy considered important in promoting health equity [17]—is poorly understood.

5.4 Implicit Bias of Healthcare Providers

According to the *National Institute on Minority Health and Health Disparities Research Framework*, discrimination is a sociocultural environment factor contributing to health disparities (see Chapter 1). Discrimination is thought to have detrimental health effects via various levels of influence, including: individual (e.g., internalization, stress response, resilience, resistance); interpersonal (e.g., discriminatory attitudes or behaviors from individual interactions, including with healthcare providers [HCPs]); community (e.g., socially patterned exposure to risks, inequitable living and working conditions, and environmental hazards); and societal (e.g., state-sanctioned discrimination and de facto discrimination via policies that structure opportunities and liabilities) [11]. These forms of discrimination have cumulative, negative health effects throughout the lifespan [11].

Discrimination by healthcare providers (HCPs) and healthcare systems constitute important contributors to health disparities. As gatekeepers of clinical services, HCPs play a critical role in determining patients' type and quality of services. Prejudicial beliefs of HCPs about patient groups (e.g., racial minorities) have been the focus of significant research in recent years, because HCPs might be unaware of their own biases and how they influence the services they provide.

Implicit attitudes or unconsciously held beliefs are commonly measured with the Implicit Association Task (IAT) [18]. The strength of the association between IAT scores and self-reported attitudes varies [19]; thus, it is important to discuss the contribution of both explicit and implicit bias to health disparities. In this section, we discuss the examination of HCPs' implicit bias, its relationship to overt discriminatory or biased action, and promising interventions for reducing implicit bias and its effects.

HCPs hold a variety of implicit biases, reflecting those found in the general public. Studies with the IAT have documented negative bias among HCPs toward racial or ethnic minorities [20] and people of lower social class [21]. The strength of the implicit bias varies, but implicit bias has been found regardless of the HCP's level of education, gender, and race [10].

5.4.1 Relationship Between Implicit Bias and Behavior

Implicit and explicit attitudes can both shape overt behavior [22], but researchers disagree about the relative strengths of their contributions. A review of the published literature found consistent evidence for the link between HCPs' implicit bias and the quality of medical communication, but mixed evidence regarding the link with patient health outcomes and clinical decision making in hypothetical patient vignettes [20]. The research suffers methodological limitations, such as using mostly convenience samples and cross-sectional designs. Improved methodology should be used in the future to uncover the pathways through which implicit bias might perpetuate health disparities [20].

Several personal (e.g., self-control), situational (e.g., time pressure), and behavioral (e.g., controllability of the action) factors have been identified as influencing overt action through implicit bias [23]. The roles of these factors in clinical decision making and health communication need further examination [10]. An unpublished meta-analysis [24] of laboratory studies showed that various experimental procedures (e.g., induction of affective states) might affect implicit and explicit bias more readily than overt actions. Further, when observed, changes in action or explicit attitudes were not related to implicit bias change [24]. These studies used largely unrepresentative and convenience samples (e.g., university students) and tested, for the most part, the effects within a single session, which does not answer the questions of the effectiveness of these procedures in HCP samples or the duration of the effects. Other research suggests that changes in implicit bias are not long lasting [25]. Thus, it may be more productive for interventions aiming to reduce health disparities related to implicit bias to focus on overt behaviors or factors [26], which might be more likely to cause implicit biases and lead to biased action.

Next, we briefly discuss interventions that might be effective in reducing bias and its effect on health disparities. Even though these interventions might not have a primary focus of reducing implicit bias, they might produce such a change as well.

5.4.2 Promising Interventions that Reduce Biased Behavior

Since the publication of the contact hypothesis [27], which proposed that under certain conditions, positive contact with members of stigmatized groups reduces prejudice, substantial evidence from correlational and laboratory studies has been accumulated in its support [28]. Increasing positive and meaningful contact between individuals from different groups can be a useful strategy for reducing discrimination [28], including unequal treatment of patients. While reviews [29] also suggested that contact interventions (e.g., face-to-face or imagined contact between individuals) might be helpful, it should be noted that most of the evidence comes from laboratory studies rather than field studies. We are not aware of any randomized controlled trials examining contact interventions with HCPs. Future research should evaluate their promise and feasibility [26].

A second line of research has conceptualized biases as "mental habits" [26] or difficult-to-change automatic cognitive responses to specific groups of people. According to this theory, habit-breaking interventions might be particularly effective in reducing the influence of bias on health disparities associated with differential HCP-patient communication and clinical care. Devine et al. developed a multicomponent intervention that includes training in changing discriminatory behavior, exposing individuals to well-regarded members from the discriminated group in imagination exercises, focusing on an individual's specific characteristics, looking at issues from the point of view of the representative from the discriminated group, and increasing contact [30]. They tested the intervention in randomized controlled trials and compared it to controls not receiving an intervention. This intervention increased awareness and concern about racial bias in students [30] and gender bias in academic faculty members [31], which the authors argue might be necessary for the long-term maintenance of bias reductions. Other outcomes of the intervention (e.g., implicit bias reduction) were inconsistent across studies. This habit-breaking intervention is delivered as a two-hour workshop and might be particularly suited for HCPs, who often have busy schedules.

A third, more recent, line of research has proposed that mindfulness-based interventions (MBI) might be helpful for overcoming healthcare disparities through reducing the effects of implicit bias on HCPs' behavior [32]. MBIs generally include meditation, yoga, or other mindfulness exercises, and help individuals learn to observe thoughts and sensations in a nonjudgmental manner, thus making individuals less reactive to stimuli such as racial/ethnic characteristics. This intervention might reduce the effects of some factors (e.g., stress), which increase the likelihood of implicit bias guiding action. Some studies have experimentally shown that brief meditation might also reduce implicit bias [32]. The effectiveness of MBIs in bias reduction with HCPs has not been tested. Studies of mindfulness-based approaches with HCPs are needed to assess their efficacy in reducing bias and its impact on HCP communication with patients, clinical decisions, and health outcomes. MBIs may be especially acceptable to HCPs because they do not need to be specifically focused on decreasing bias and might provide them with additional benefits, such as improving stress-coping skills as well as a host of health and mental health indicators [32].

5.5 Sociocultural Factors that Influence the Quality of Healthcare Provider-Patient Interactions and Communication

Provider-patient factors also contribute to health disparities. The *National Institute on Minority Health and Health Disparities (NIMHD) Research Framework* (see Chapter 1) considers the healthcare system as a separate domain of influence to the sociocultural domain and it is covered in its own chapter. In this section, we discuss health disparities stemming from sociocultural factors that affect the quality of provider-patient interaction.

Patients from racial and ethnic minority groups are less likely to receive high-quality care compared to Whites [33]. These disparities are influenced by the characteristics of individual interactions, including patient-provider interactions and communication, even after controlling for patient socioeconomic characteristics [34]. Higher quality social interactions during clinical encounters have been associated with higher patient satisfaction and health outcomes [35].

This section describes three interconnected sociocultural factors known to influence cross-cultural patient-provider interactions: (i) communication characteristics, (ii) cultural characteristics, and (iii) individual values and beliefs, and provides a review of interventions that have targeted patients, clinicians, context, or the three combined to improve the quality of these interactions. The selection of these factors was guided by the NIMHD Research Framework discussed in Chapter 1.

5.5.1 Communication Characteristics

Two recent literature reviews have shown that minority patients experience poor communication with their HCPs. Minority patients received less health information, were less engaged during the clinical encounter, and reported lower participation in shared decision making compared to White patients [33, 34]. A third review found that nonverbal communication measured as provider warmth, positive feelings, and facial expression were associated with patient satisfaction, information exchange, visit length, and expression decoding during cross-cultural clinical encounters [36]. Most of the studies in these literature reviews employed cross-sectional and qualitative research designs, which limit the understanding of causal factors and how these characteristics may change over time. Additional studies with stronger designs, such as mixed-method studies, longitudinal studies, and randomized controlled trials, are needed to arrive at more convincing conclusions.

A review of intervention studies between 1979 and 2011 showed that the two most commonly used interventions to address health disparities, patient disease education and self-management, did not specifically address patient-provider communication [37]. Yet, provider-based interventions aimed at reducing disparities in cancer screening rates among underserved minority patient populations, for example, have been successful in targeting communication issues. Three review studies have shown that interventions can improve adherence to cancer screenings among minorities. Specifically, they find evidence for interventions tailored to increase patient health education combined with patient navigation services and the use of lay health advisors, community-based partnerships, and physician training in communicating with minority patients possessing low health literacy [38]. Future research should assess a wider range of interventions in order to find the most effective approach to reducing disparities stemming from patient-provider communication.

5.5.2 Cultural Factors

The concept of cultural competency in health policy and health research arenas is receiving wide attention. This concept is defined as a set of skills, behaviors, and attitudes among professionals that facilitates their work with racially and culturally diverse patient populations [39]. As patient populations have grown increasingly diverse, this area of research has gained prominence in the health disparities literature.

Many interventions designed to improve minority patients' healthcare experiences and outcomes seek to increase providers' cultural competency. In a literature review on cultural competence interventions, 20 of the 25 included studies were randomized controlled trials, and most of the interventions reviewed related to trainings and curricula designed to change professionals' attitudes towards individuals with disabilities [40]. A second review assessing the effectiveness of interventions found mixed results. Most studies reviewed found that interventions designed to improve cultural competence of providers had a moderate improvement in healthcare access and utilization outcomes but weaker evidence for improvements in patient/client outcomes [41]. The lack of conclusive evidence is in part due to study differences in conceptual frameworks used to guide the research, the variables measured, and a lack of standardized measures [41].

5.5.3 Factors Related to Individual Values and Beliefs

Scholars have argued that service providers value certain patient knowledge, behaviors, or characteristics that may afford the patient certain advantages during the clinical encounter [42]. Culturally based values and beliefs often make patients appear passive or lacking interest during their interactions with service providers [43]. Culturally based beliefs can also influence patients' level of comfort with voicing concerns and opinions and assuming an active role as partners in clinical decision making. Less engaged patients may relay less information to professionals and, as a result of being less engaged, ask fewer questions. The result is a lower level of comprehension of their medical needs and treatment options.

Another type of intervention supported by policy makers seeks to redesign the healthcare system to deliver care that is patient centered and coordinated to address chronic health disparities in the United States. Patient-centered Care (PCC) is described as HCP delivery of care based on the ability to "identify, respect, and care about patient differences, values, preferences, and expressed needs; relieve pain and suffering; coordinate continuous care; listen to and continuously advocate disease prevention, wellness, and promotion of healthy lifestyles, including a focus on population health" [44].

A recent literature review of PCC across clinical settings (i.e., outpatient clinics, hospitals, and medical teaching settings) found that PCC is associated with higher provider understanding of multiculturalism as well as higher cultural knowledge and sensitivity toward patient values and beliefs [45]. Yet, this review found mixed evidence on the role of PCC in improving patient outcomes, such as patient satisfaction with services. The authors of this review conclude that high levels of attrition and weak research designs probably contributed to reported mixed results. Additional research with stronger research designs, such as randomized trials, stratified analyses across racial and ethnic minority groups, and the use of mixed methodology that examines, the role of PCC on improving minority patients' health outcomes over time, is warranted.

One PCC intervention that has focused on increasing patients' ability to engage in their healthcare is "patient activation." This term refers to patients' sense of self-efficacy in their self-management of chronic disease, and their ability to know when, how, and where to go for help to meet their healthcare needs [46]. A randomized controlled study examined the effectiveness of a psychoeducational intervention called "parent activation" tailored to Latina mothers advocating for mental health services for their children. The PCC intervention comprised four weekly group meetings covering various health modules and parent role-playing to apply learned skills [47]. This study found that, compared to a support group-only condition, Latinas receiving the PCC intervention showed significant increases in self-reported measures of health-related activation, school-based activation, and quality of school interactions.

Research on the efficacy of PCC in reducing health inequities is an emerging field. Guidance on how to implement PCC, as well as empirical studies to test its impact on increasing the quality of minority patient-provider interactions, is under way.

5.5.4 Conclusion

Research on how sociocultural factors affect healthcare disparities has increased in the past two decades. New research avenues incorporate interdisciplinary approaches at the patient, clinician, and healthcare context levels. Research needs include studies with rigorous designs and methods (including mixed methods) that test the efficacy of (i) patient-centered and -coordinated care models and (ii) culturally tailored interventions to improve the quality of HCP-patient interactions and communication and to reduce health inequities.

5.6 Synthesis

5.6.1 Culture

Culture is the collective and universal traits by which we understand and make sense of the human condition. These systems of understanding human experience are not static but, rather, are evolving along with a changing environment. Culture affects health outcomes indirectly by shaping views of health, related behaviors, systems, and social arrangements. Yet, the pathways through which culture affects health inequities and what those effects are have not been systematically investigated.

While discussions of culture are often focused on the "other," most commonly "disenfranchised" groups, culture also refers to beliefs, norms, values, and behaviors manifest in the larger public and society. Studies of culture are relevant in understanding how society structures health-related social arrangements based on individual and group characteristics such as race, ethnicity, financial resources, age, religion, and sexual orientation. A significant body of knowledge exists on how larger societal cultural factors, such as structural racism, cultural racism, income inequality, implicit bias, and access to certain social networks, yield health inequities. Advances have also been made in conceptualizing how these larger societal cultural factors manifest across health domains at the individual, intrapersonal, community, and societal levels and affect health and well-being. Key research questions pertain to how and when health disparities develop and how the various health domains and levels of influence interact to produce them.

5.6.2 Adaptations to New Cultural Contexts

Adaptations to new cultural contexts by newcomers and marginalized groups (e.g., acculturation) seem to lead to or be associated with worse health behaviors and health outcomes for marginalized groups. Understanding how, why, and when such cultural adaptations result in deleterious versus salubrious health effects is important for addressing health inequities. We noted the need for more cohesive theories, rigorous methods, and systematic measurement of the processes of cultural changes and adaptations in relation to health outcomes.

5.6.3 Social Capital

While theories and empirical studies have not always agreed on the influence of various types of social capital on health, significant evidence supports the buffering effect of some types of social capital on the deleterious effects of individual and neighborhood low SES and income inequality on health inequities. However, research on social capital's relationship to health has suffered from varying conceptualizations, inexact and unsystematic operationalization and measurement of types of social capital, and study designs that result in contradictory findings. The potential related costs or negative effects of social capital on well-being have been largely ignored. Social relationships and networks can bring with them health-enhancing benefits but also deleterious effects. Differential access to and embeddedness in such networks has obvious health relevance, the understanding of which requires rigorous approaches to conceptualization, study design, and measurement. Research that advances such knowledge development could inform interventions on social networks. Also important is the investigation of the effects of social network interventions and their limits in alleviating health disparities in the face of continued structural factors that shape economic disadvantage.

5.6.4 Implicit Bias

A significant body of research has documented the implicit bias widespread in healthcare. Its effects on behavior seem to be facilitated by factors such as cognitive load, stress, and lack of familiarity with the group(s) subjected to implicit bias. Research on approaches to diminish implicit bias and/or its effects on behaviors (e.g., HCP decision making) have yet to identify effective strategies for longer-term behavior change. Most promising to date are interventions (e.g., mindfulness-based stress reduction) that seek to reduce behaviors resulting from implicit bias by building meta-cognitive skills that disrupt habitual cognitive processes and behaviors involved in implicit bias. Such interventions have shown efficacy in enhancing emotion regulation, reactive responses, and perceived stress.

5.6.5 Patient-Provider Communication

Patient-provider communication contributes to healthcare disparities. Such communication is influenced by the way information is delivered, cultural competency of the provider, and beliefs and values of the patient and provider. Interventions aimed at training providers about effective information delivery and cultural competency as well as at involving patients in their own care are promising approaches for improving the quality of patient–provider interactions for racial and ethnic minorities. Research on the efficacy of PCC and its efficacy in reducing health inequities is an emerging field. Guidance on how to implement PCC, as well as empirical studies to test its impact on increasing the quality of minority patient-provider interactions, is under way.

5.7 Key Points

- Cultural influences affect how individuals, families, communities, and societies approach health and healthcare via perceptions, beliefs, desires, norms, and values. The pathways through which culture and cultural changes affect health disparities and what those effects are warrant empirical attention.
- SES, race, and ethnicity are closely intertwined. Research is needed on their interactions and interconnectedness with other sociocultural factors within the multilevel systems in which they occur.
- Racism places individuals and communities at risk for poorer health. Furthering our understanding of how racism is related to and impacts other sociocultural factors is important for informing interventions to achieve health equity.
- Various components of low social capital are associated with health disparities. How economically disadvantaged communities build and sustain impactful social capital and

how such community actions affect health disparities are a key area for knowledge development.

- Implicit bias affects overt action via several personal, situational, and behavioral factors. The roles of these factors in clinical decision making and health communication need further examination, and assessment of interventions that target these mechanisms of action are needed.
- Patient-provider communication contributes to healthcare disparities via the way information is delivered, provider cultural competency, and patient and provider beliefs and values. Provider training about effective information delivery and cultural competency, as well as training of patients to enhance involvement in their own care, are promising approaches for improving the quality of patient-provider interactions for racial and ethnic minorities.

Disclaimer

The views and opinions expressed in this chapter are those of the authors only and do not necessarily represent the views, official policy, or position of the U.S. Department of Health and Human Services or any of its affiliated institutions or agencies.

References

1 Adler, N.E. and Stewart, J. (2010). Health disparities across the lifespan: meaning, methods, and mechanisms. *Annals of the New York Academy of Sciences* 1186 (1): 5–23.
2 Levison, J.H., Bogart, L.M., Khan, I.F. et al. (2017). Where it falls apart: barriers to retention in HIV care in Latino immigrants and migrants. *AIDS Patient Care and STDs* 31 (9): 394–405.
3 Grest, C.V., Amaro, H., and Unger, J. (2017). Longitudinal predictors of intimate partner violence perpetration and victimization in Latino emerging adults. *Journal of Youth and Adolescence* 47 (3): 560–574.
4 Kagawa-Singer, M., Dressler, W., George, S. et al. (2016). Culture: the missing link in health research. *Social Science & Medicine* 170: 237–246.
5 Chiao, J.Y. and Ambady, N. (2007). Cultural neuroscience: parsing universality and diversity across levels of analysis. In: *Handbook of Cultural Psychology* (eds. S. Kitayama and D. Cohen), 237–254. New York: Guilford Press.
6 Abraído-Lanza, A.F., Chao, M.T., and Florez, K.R. (2005). Do healthy behaviors decline with greater acculturation? Implications for the Latino mortality paradox. *Social Science & Medicine* 61 (6): 1243–1255.
7 Schwartz, S.J., Unger, J.B., Zamboanga, B.L., and Szapocznik, J. (2010). Rethinking the concept of acculturation: implications for theory and research. *American Psychologist* 65 (4): 237–251.
8 Ro, A. and Bostean, G. (2015). Duration of US stay and body mass index among Latino and Asian immigrants: a test of theoretical pathways. *Social Science & Medicine* 144: 39.
9 Williams, D.R. and Mohammed, S.A. (2013). Racism and health I: pathways and scientific evidence. *American Behavioral Scientist* 57 (8): 1152–1173.

10 Sabin, J.A., Nosek, B.A., Greenwald, A., and Rivara, F.P. (2009). Physicians' implicit and explicit attitudes about race by MD race, ethnicity, and gender. *Journal of Health Care for the Poor and Underserved* 20 (3): 896–913.

11 Krieger, N. (2012). Methods for the scientific study of discrimination and health: an ecosocial approach. *American Journal of Public Health* 102: 936–945.

12 Kawachi, I., Subramanian, S.V., and Kim, D. (2008). Social capital and health: a decade of progress and beyond. In: *Social Capital and Health* (eds. I. Kawachi, S.V. Subramanian and D. Kim), 1–26. New York: Springer.

13 Uphoff, E.P., Pickett, K.E., Cabieses, B. et al. (2013). A systematic review of the relationships between social capital and socioeconomic inequalities in health: a contribution to understanding the psychosocial pathway of health inequalities. *International Journal for Equity in Health* 12 (1): 54.

14 Brisson, D.S. and Usher, C.L. (2005). Bonding social capital in low-income neighborhoods. *Family Relations* 54 (5): 644–653.

15 Eschbach, K., Ostir, G.V., Patel, K.V. et al. (2004). Neighborhood context and mortality among older Mexican Americans: is there a barrio advantage? *American Journal of Public Health* 94: 1807–1812.

16 Elgar, F.J., Davis, C.G., Wohl, M.J. et al. (2011). Social capital, health and life satisfaction in 50 countries. *Health & Place* 17 (5): 1044–1053.

17 National Academy of Sciences, Engineering, and Medicine (2017). *Communities in Action: Pathways to Health Equity*. Washington, DC: The National Academies Press.

18 Greenwald, A.G., McGhee, D.E., and Schwartz, J.L. (1998). Measuring individual differences in implicit cognition: the implicit association test. *Journal of Personality and Social Psychology* 74 (6): 1464–1480.

19 Greenwald, A.G., Poehlman, T.A., Uhlmann, E.L., and Banaji, M.R. (2009). Understanding and using the implicit association test: III. Meta-analysis of predictive validity. *Journal of Personality and Social Psychology* 97 (1): 17–41.

20 Hall, W.J., Chapman, M.V., Lee, K.M. et al. (2015). Implicit racial/ethnic bias among health care professionals and its influence on health care outcomes: a systematic review. *American Journal of Public Health* 105 (12): e60–e76.

21 Haider, A.H., Schneider, E.B., Sriram, N. et al. (2015). Unconscious race and social class bias among acute care surgical clinicians and clinical treatment decisions. *JAMA Surgery* 150 (5): 457.

22 Strack, F. and Deutsch, R. (2004). Reflective and impulsive determinants of social behavior. *Personality and Social Psychology Review* 8 (3): 220–247.

23 Friese, M., Hofmann, W., and Schmitt, M. (2009). When and why do implicit measures predict behavior? Empirical evidence for the moderating role of opportunity, motivation, and process reliance. *European Review of Social Psychology* 19 (1): 285–338.

24 Forscher, P. S., Lai, C. K., Axt, J. R., Ebersole, C. R., Herman, M., Devine, P. G., & Nosek, B. A. (2019). A meta-analysis of procedures to change implicit measures. *Journal of Personality and Social Psychology*, 117(3), 522–559.

25 Lai, C.K., Skinner, A.L., Cooley, E. et al. (2016). Reducing implicit racial preferences: II. Intervention effectiveness across time. *Journal of Experimental Psychology: General* 145 (8): 1001–1016.

26 Burgess, D.J., van Ryn, M., Dovidio, J., and Saha, S. (2007). Reducing racial bias among health care providers: lessons from social-cognitive psychology. *Journal of General Internal Medicine* 22 (6): 882–887.

27 Allport, G.W. (1954). *The Nature of Prejudice*. Cambridge, MA: Perseus Books.

28 Pettigrew, T.F. and Tropp, L.R. (2006). A meta-analytic test of intergroup contact theory. *Journal of Personality and Social Psychology* 90 (5): 751–783.

29 Dovidio, J.F., Love, A., Schellhaas, F.M.H., and Hewstone, M. (2017). Reducing intergroup bias through intergroup contact: twenty years of progress and future directions. *Group Processes & Intergroup Relations* 20 (5): 606–620.

30 Devine, P.G., Forscher, P.S., Austin, A.J., and Cox, W.T.L. (2012). Long-term reduction in implicit race bias: a prejudice habit-breaking intervention. *Journal of Experimental Social Psychology* 48 (6): 1267–1278.

31 Carnes, M., Devine, P.G., Baier Manwell, L. et al. (2015). The effect of an intervention to break the gender bias habit for faculty at one institution. *Academic Medicine* 90 (2): 221–230.

32 Burgess, D.J., Beach, M.C., and Saha, S. (2017). Mindfulness practice: a promising approach to reducing the effects of clinician implicit bias on patients. *Patient Education and Counseling* 100 (2): 372–376.

33 Saha, S., Arbelaez, J.J., and Cooper, L.A. (2003). Patient–physician relationships and racial disparities in the quality of health care. *American Journal of Public Health* 93 (10): 1713–1719.

34 Shen, M.J., Peterson, E.B., Costas-Muñiz, R. et al. (2017). The effects of race and racial concordance on patient-physician communication: a systematic review of the literature. *Journal of Racial and Ethnic Health Disparities* 5 (1): 117–140.

35 Horst, F., Van den Borne, B., Ryckman, R., and Crebolder, H. (2003). Provider–patient interaction in diabetes care: effects on patient self-care and outcomes: a systematic review. *Patient Education and Counseling* 51 (1): 17–28.

36 Lorié, Á., Reinero, D.A., Phillips, M. et al. (2017). Culture and nonverbal expressions of empathy in clinical settings: a systematic review. *Patient Education and Counseling* 100 (3): 411–424.

37 Clarke, A.R., Goddu, A.P., Nocon, R.S. et al. (2013). Thirty years of disparities intervention research: what are we doing to close racial and ethnic gaps in health care? *Medical Care* 51 (11): 1020–1026.

38 Naylor, K., Ward, J., and Polite, B.N. (2012). Interventions to improve care related to colorectal cancer among racial and ethnic minorities: a systematic review. *Journal of General Internal Medicine* 27 (8): 1033–1046.

39 Cross, T.L. (1989). *Towards a Culturally Competent System of Care: A Monograph on Effective Services for Minority Children Who are Severely Emotionally Disturbed*. Washington, DC: Georgetown University Child Development Center files.eric.ed.gov/fulltext/ED330171.pdf.

40 Butler, M., McCreedy, E., Schwer, N. et al. (2016). Improving cultural competence to reduce health disparities. *Journal of Nursing Scholarship* 35 (4): 325–331.

41 Truong, M., Paradies, Y., and Priest, N. (2014). Interventions to improve cultural competency in healthcare: a systematic review of reviews. *BMC Health Services Research* 14 (1): 99.

42 Shim, J.K. (2010). Cultural health capital: a theoretical approach to understanding health care interactions and the dynamics of unequal treatment. *Journal of Health and Social Behavior* 51 (1): 1–15.

43 Schouten, B.C., Meeuwesen, L., Tromp, F., and Harmsen, H.A. (2007). Cultural diversity in patient participation: the influence of patients' characteristics and doctors' communicative behavior. *Patient Education and Counseling* 67 (1): 214–223.

44 Greiner, A.C. and Knebel, E. (eds.) (2003). *Health Professions Education: A Bridge to Quality*. Washington, DC: Institute of Medicine, The National Academies Press.

45 Renzaho, A.M.N., Romios, P., Crock, C., and Sønderlund, A.L. (2013). The effectiveness of cultural competence programs in ethnic minority patient-centered health care—a systematic review of the literature. *International Journal for Quality in Health Care* 25 (3): 261–269.

46 Hibbard, J.H., Stockard, J., Mahoney, E.R., Tusler, M. (2004) Development of the Patient Activation Measure (PAM): Conceptualizing and measuring activation in patients and consumers. *Health Services Research*, 39 (4,p1): 1005–1026.

47 Thomas, K.C., Stein, G.L., Williams, C.S. et al. (2017). Fostering activation among Latino parents of children with mental health needs: an RCT. *Psychiatric Services* 68 (10): 1068–1075.

6

Physical Environment, and Minority Health and Health Disparities Research

Tracy Bastain[1], Carrie Breton[1], Shohreh Farzan[1], Rima Habre[1], Jill Johnston[1], Derrick C. Tabor[2], Claudia Toledo-Corral[1,3], and Elizabeth Vang[1]

[1] *Department of Preventive Medicine, Keck School of Medicine, University of Southern California, Los Angeles, CA, USA*
[2] *National Institute on Minority Health and Health Disparities, National Institutes of Health, Bethesda, MD, USA*
[3] *Department of Health Sciences, California State University Northridge, Northridge, CA, USA*

6.1 Introduction

The physical environment encompasses places, infrastructure, neighborhoods, and communities and affects a wide range of health conditions and quality of life outcomes that are distributed across populations. Positive attributes of the physical environment, such as clean water and air, healthy living spaces, safe schools, healthy workplaces, and safe and healthy communities, contribute generally to good population health. However, the physical environment may also contribute to disparities in health risks and may lead to health disparities, particularly when access or exposure to the positive and negative attributes of the physical environment varies across populations and places.

Health disparate populations, including African Americans, Latinos, American Indians/Alaska Natives, Asians, Native Hawaiians and Other Pacific Islanders, socioeconomically disadvantaged populations, underserved rural populations, and sexual and gender minorities, are disproportionately affected by higher rates of adverse health outcomes due to disparities in environmental exposures and illness. Attention to health disparities is increasing in environmental health sciences and epidemiological research due to a growing body of evidence indicating that the health of these populations is also disproportionately burdened by higher exposures to environmental chemicals, poorly designed and under-resourced communities, and other attributes of the physical environment that exacerbate adverse health outcomes.

In this chapter we aim to outline common methods used in environmental health research to measure the physical environment and to assess disparities in health risks, discuss characteristics of the built environment that are key to understanding how the physical environment produces or maintains health disparities across the course of individual lives and generations, and conclude with challenges and opportunities in environmental health and epidemiological research.

The Science of Health Disparities Research, First Edition. Edited by Irene Dankwa-Mullan, Eliseo J. Pérez-Stable, Kevin L. Gardner, Xinzhi Zhang, and Adelaida M. Rosario.

6.2 Methodologies and Measures

6.2.1 Exposure Assessment

Research methods for measuring and modeling exposure to adverse conditions in the physical environment and the health disparities that can result include approaches at the personal, micro-environmental, or area-wide levels. Methods and frameworks for conducting research at the intersection of environmental health and health disparities continue to evolve. A central concern is how disparities in environmental exposures contribute to observed health disparities. Many factors determine the optimal approach needed to answer research questions posed by environmental health and heath disparities researchers. These include the desired degree of spatial and temporal resolution in describing or estimating the exposure of interest, the exposure source, the relationship between the exposure and population health, the degree of measurement error or uncertainty that might be tolerated, and practical feasibility considerations, including participant burden, time, cost, and degree of and types of expertise involved.

The sections below describe the most common approaches to assess exposures in the physical environment that have been used to document the existence of exposure disparities: Geographic Information Systems (GIS), community- or area-wide monitoring, and personal monitoring. These methods apply to pollutants in air, water, soil, and other media.

6.2.1.1 Geographic Information Systems

Geographic Information Systems, also referred to as GIS, are software and hardware tools designed to facilitate the collection, management, analysis, and display or mapping of spatially referenced data. These include any information with location attributes, such as geographic coordinates (latitude, longitude, elevation), addresses, zip codes, census tracts, cities, states, or countries. GIS allows researchers to link information in space and time from several sources, analyze spatial relationships at various scales, and compute or model exposure metrics based on these relationships. Data can be represented as points (e.g., residences), lines (e.g., roads), polygons (e.g., census tract data), or rasters (e.g., images obtained through remote sensing) in GIS (Figure 6.1).

Other examples of commonly used spatial calculations include distance along the shortest route from a residence to a point of interest (for example, grocery stores), or to a road, or the sum of road lengths, or the number of healthy dining establishments within a certain radius around the residence. Examples of polygon data include census data or data from the CalEnviroScreen 3.0 environmental health screening tool [1].

Other types of spatial analyses enabled by GIS include investigating whether there are clusters or hot spots of a disease in space, or whether there is high (or low) spatial correlation between measurements taken at various points in space and time, as well as more sophisticated spatial and temporal modeling.

Many local, state, and national agencies host GIS data clearinghouses where they curate and publicly release data of interest to researchers. In addition, there are several free online GIS software packages as well as educational resources allowing the more widespread use of GIS for environmental health disparities research.

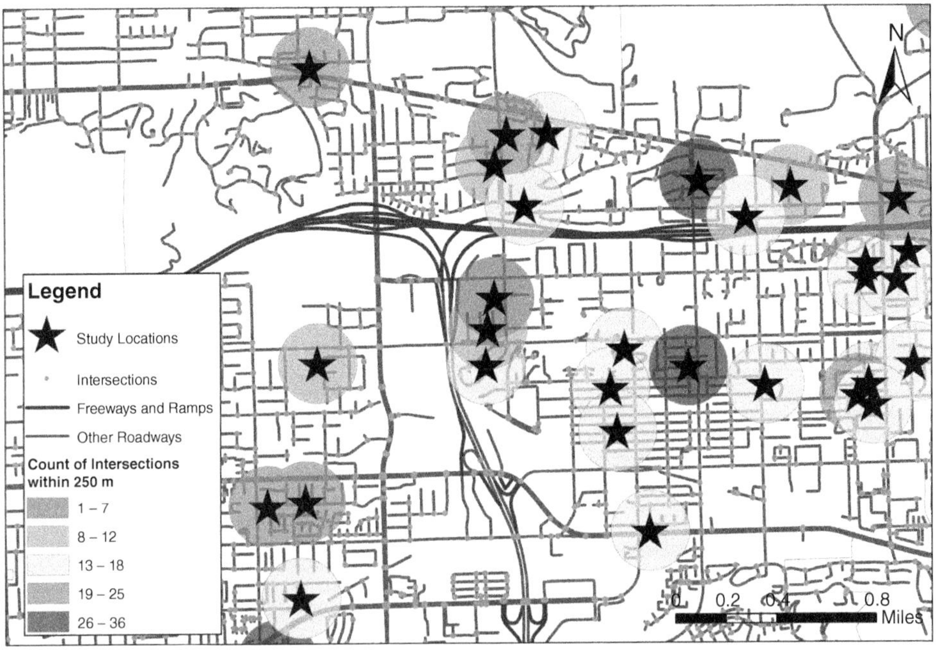

Figure 6.1 An illustration of the use of GIS to represent locations (points) in a study in relation to the road network (freeways, ramps, and other roadways as lines). The points of roadway intersections can be calculated and summarized within a 250 m buffer around a location. Color scales (green to red) can be used to represent the count of intersections on a continuous scale. *Source:* Figure created by Dr. Rima Habre, ScD, 2018, University of Southern California © OpenStreetMap (and) contributors, CC-BY_SA. (See color representation of this figure in color inserts section.)

The use of GIS is integral to recently developed and evolving research frameworks. One study utilized GIS in applying an exposome framework to understand the contribution of the environment to low birth weight and infant mortality disparities [2]. Whereas, another study used GIS methods to link Environmental Protection Agency (EPA) data on air toxins with census data on sociodemographics to understand disparities in cancer risk [3].

6.2.1.2 Community-level Sampling

The level or spread of environmental contamination can be sampled or measured at the community, neighborhood, or residence level. Depending on the expected variability of the exposure of interest in space and time, sampling or monitoring efforts should be designed to capture scales of interest. For example, a study investigated soil lead and children's blood lead data in pre and post Hurricane Katrina and matched these data by census tract communities. Their research teams started collecting samples of soil across New Orleans in 1998, and by 2000 they were able to develop a soil-lead map for the area. Ten years after Hurricane Katrina, the same research team remapped New Orleans and was able to compare time period differences and determine differences in children's blood lead levels [4].

Wind speed and direction also influence the spread and concentration of lead deposits from a source or facility. Any sampling scheme should take space and time into consideration, not only for emissions, but also in transport and dispersion patterns, by collecting at evenly spaced grid points in all directions around the source. Air and soil monitoring can be conducted side by side to capture airborne concentrations (inhalation as the exposure route of concern) as well as airborne contaminants deposited onto the soil, where the exposure routes of concern might include infant hand to mouth behavior, soil brought indoors on shoes or clothing, and food.

Sometimes, environmental health researchers are interested in characterizing an entire community's exposure to area-wide sources of pollution rather than point sources, such as general or background air pollution levels. The EPA mandates ambient air pollution monitoring for regulatory purposes and makes the data available to the public through the Air Quality System (https://www.epa.gov/aqs). These monitors are sited according to specific criteria to capture micro (up to 100 m), middle (~100–500 m), neighborhood (500 m–4 km), urban (4–50 km), and regional (tens to hundreds of kilometers) scales of air pollution. Among the considerations when using these types of data or other community-level monitoring data are how representative the data are of the true exposures of the population of interest, especially sensitive subgroups, given behavior and mobility patterns.

While community- or neighborhood-level sampling can accurately capture differences in exposure levels between communities or neighborhoods, it likely cannot capture differences in exposure within a community (neighborhood). It does not account for people's mobility, behavior, and time spent in microenvironments, which may contribute the most to their total exposure (e.g., being indoors ~85–90% of the time).

6.2.1.3 Personal Sampling

Personal monitoring is thought to be the gold standard in external exposure assessment, where physical, chemical, or biological contaminants of concern are measured at the point of contact using time-integrated or continuous methods. In the case of air pollution, monitors are typically worn by individuals as they go about their day, measuring concentrations in their breathing zone. Personal monitoring overcomes the challenges of community-wide or neighborhood-level monitoring described in the earlier section and is considered the most representative of true total exposure. As individuals move and conduct their daily activities, they often come in contact at close proximity to sources of air pollution (such as traffic exhaust or cooking), and as they transition from being indoors to outdoors to commuting and vice versa, they experience different levels of indoor-generated and outdoor-generated air pollution.

By combining personal, continuous monitoring with Global Positioning System (GPS) tracking, researchers can get an accurate depiction of the exposures experienced by study participants and their environments at high space and time resolutions. GPS-enabled personal exposure monitors to collect personal outdoor exposure readings are evolving and may offer an accurate means of characterizing pollution.

As with all monitoring applications, the choice of the method or approach needs to be driven by considerations of accuracy, precision, calibration needs, battery life, cost, participant burden, and complexity of execution or level of expertise needed to successfully carry out the sampling. Issues of contamination, chain of custody, and sample preservation for future analyses are also important.

6.2.2 Biomarkers of Exposure

A biomarker of exposure gives an assessment of an individual's absorbed dose of an environmental contaminant; that is, how much actually enters that body. Biomarkers of exposure can provide information not only about how much of a contaminant a person has inhaled or ingested, but also how rapidly the body can eliminate the toxicant and if there are individual differences in how one's body metabolizes a toxicant. In environmental health research, biomarkers of exposure are utilized to quantify levels in an individual with the goal of preventing or ameliorating adverse health effects. In some cases, such as health disparate populations, biomarkers are utilized in population-based research studies to gain a better understanding of the level of exposure and how it affects a population's health. Results from biomarkers of exposure measurements can guide efforts to prevent further exposures by informing researchers about populations that may be at highest risk or most vulnerable (i.e., children, pregnant women, the elderly, those living in close proximity to an exposure source) and guide targeted interventions to prioritize those populations to prevent occurrence of adverse health effects [5].

Environmental health research studies that utilize biomarkers [4] of exposure as a measurement tool generally involve collection of biospecimens from research participants, such as urine, blood, hair, and toenails. Biomarkers measured in biospecimens collected from research participants can help researchers reconstruct exposure histories to better understand how past exposures may be affecting current health conditions.

There are many different types of chemicals or metabolites of chemical exposures that can be measured in various biospecimens. Which biospecimen is chosen provides insights into different windows of exposure. Recent exposure to metals is measured in blood and urine samples. Metal concentrations measured in hair represent exposures occurring in the previous 1 to 4 months; toenails reflect exposures in the previous 6 to 9 months and are a reliable long-term biomarker of exposure. Adult's and children's toenails have been successfully used to measure metal concentrations in previous studies and have been associated with health outcomes, including cardiovascular disease mortality, skin cancer, and blood pressure in adults [6]. However, these noninvasive biomarkers have not been widely utilized in studies of health disparate populations. Though more invasive to collect, blood is often used as a biospecimen in which to measure biomarkers, in part because so many chemicals of interest can be measured in blood that reflect potential exposures of interest and may also be directly correlated with other biomarkers of disease.

The advantage of using biomarkers of exposure as a measurement tool in research studies is that it provides evidence-based data to demonstrate that the exposure of question has been absorbed into the body and provides a reflection of dose. Results would provide an indication of possible health effects and the chemicals present in the physical environment.

6.2.3 Qualitative Research

Qualitative research is of increasing interest in environmental health science and can provide complementary information to traditional exposure assessment and epidemiologic studies. Qualitative research, such as in-depth interviews, focus groups, or observation, can ascertain detailed information about perceptions of environmental exposures and of quality of health and life, as well as of the individual, local, and structural factors that contribute to exposure and disease. Qualitative approaches can engage residents on local environmental health problems and characterize the complexity of a community [7].

6.3 Importance of Physical Environment Determinants to Health Disparities Research

Certain characteristics of the built environment, such as industry, transportation, housing, and neighborhoods, have important implications for health disparities at the community and other levels. The built environments in which disadvantaged groups live, such as low-income communities and people of color who are segregated in neighborhoods with high levels of poverty and material deprivation, often adversely affect the health and well-being of residents [8]. Racial segregation is particularly important for understanding health disparities. Researchers have long argued that segregation is a crucial starting point for exploring the relationship between physical environment and health disparities [9]. Racial residential segregation is recognized as a structural factor that contributes to environmental exposure disparities and the health disparities experienced by racial and ethnic minorities. Environmental health research through the lens of segregation [8] clarifies and illuminates the experiences and conditions of health disparate populations and provides insight to a better understanding of why specific groups of people choose or are subjected to live in neighborhoods that are poorly built and maintained.

Research of the built environment explores health effects of asthma, obesity, lack of physical activity, cardiovascular disease, and cancer mortality, to name a few. Here, we discuss aspects of the built environment that assist in informing public policy to improve health and address health disparities.

6.3.1 Housing

Inequities in construction and maintenance of low-income housing have resulted in insufficient and poor-quality housing, overcrowding, and health problems [10]. People who live in such conditions are disproportionately exposed to lead, pests, air pollutants, and contamination. For example, metals exposure via contaminated drinking water, often due to lead pipes in older housing or contaminated water sources, is a common issue and of particular concern among health disparities communities. The recent lead water contamination crisis in Flint, Michigan and the chaotic response by government officials has highlighted the historical and structural conditions, as well as sociopolitical factors, that often contribute to a lack of access to safe water [11]. An analysis of the social and built environmental factors that exacerbated the health crisis in Flint found that children living in older homes in neighborhoods of generally poorer housing stock were disproportionately affected by higher rates of and greater increases in blood lead levels during this contamination event [12].

Disparities in housing are well established among low-socioeconomic status (SES) neighborhoods and racial and ethnic groups. Replacing leaded piping has succeeded in reducing the risk of exposure and improving health, especially for children. Such targeted interventions are critical examples of how housing disparities among disadvantaged groups can be reduced [13].

6.3.2 Food and Green Space Access

Societal and environmental inequalities influence access to basic human assets, including fresh food and green space. Limited access to these health resources can lead to poor behavioral choices, such as increased consumption of processed foods with high caloric density and decreased physical activity, which together contribute to negative health outcomes.

The most prominent health disparity stemming from both food and green space access is the obesity epidemic. Children and adults alike are more likely to engage in obesogenic behaviors based on their environment, leading to positive energy balance. Among children, the ethnic/racial groups most affected are Blacks (20.2%) and Latinos (22.4%) [14], many of whom live in inner-city, lower SES communities. This strongly suggests the need to examine the macro-level determinants of disease, such as the built environment.

With food access specifically, the accessibility of supermarkets, conveniences stores, and fast food shapes the eating environment and dictates food choices that vary in quality and caloric density. Methods used to directly analyze the environment and dietary intake with food choices are not consistent, making comparability between studies difficult. Body mass index (BMI) has been used as an alternative to associate food access to obesity [15]. Recent systematic reviews reveal that decreased access to healthy foods, such as limited supermarket access, and increased access to convenience stores, was not directly related to obesity [16]. Higher consumption of fast food has been associated with higher obesity rates, and fast food has been shown to be more available in lower SES neighborhoods [15].

Natural and green environments provide positive health benefits, such as promoting outdoor recreation and reducing stress. In a recent longitudinal study, a 10% increase in green space increased the likelihood of boys choosing outdoor pastimes compared to indoor, screen-based activities [17]. This suggests that restriction of green space access will in turn discourage and/or diminish physical activity.

We cannot yet assert if macro-level changes in the built environment will lead to improved food choices and physical activity. The roles of limited access to healthy food and green space warrant further study in the health disparities field.

6.3.3 Neighborhood Walkability

Understanding how neighborhood design could influence people's health-related behaviors and health outcomes is important for informing city planners and policy makers when they are designing or improving neighborhoods. Researchers have used neighborhood walkability, a measure of how conducive the built environment is to walking, to assess its impact on health disparities among disadvantaged groups. Some studies have found associations between highly walkable neighborhoods and increased physical activity among residents. However, disadvantaged groups tend to live in poorer neighborhoods with highly walkable areas, yet suffer from chronic diseases related to lack of physical activity [18].

To assess the level of walkability in neighborhoods, researchers can use GIS data, such as residential/population density, land use mix, street connectivity, and safety features, to develop a walkability index based on neighborhood characteristics. Field observations can also be used to assess neighborhood walkability in conjunction with GIS data [19]. For example, one study's walkability index components included population density, unique intersection density, distance along street network to nearest subway stop, a measure of balance among land use (e.g., residential, office, retail, education, and entertainment), and ratio of retail building floor area to retail land area. Their research showed that neighborhood conditions differed between poor and nonpoor neighborhoods that were equally walkable. This same study also examined safety measures (e.g., average speed limit of vehicles, vehicle collisions, felony complaints, narcotic arrests, and vacant housing units) and how they influence walking to help explain health disparities between advantaged and disadvantaged groups [18].

6.3.4 Air Pollution

There is mounting evidence that exposure to air pollution disproportionately affects individuals with lower income and education levels and some racial and ethnic groups. US counties with the highest levels of PM 2.5 and ozone had a higher fraction of African American residents and persons in poverty than did counties in the lowest quantile of air quality [20]. Disparities in long-term PM 2.5 exposures were observed across 215 US census tracts. Both African Americans and Latinos had significantly higher exposures than Whites. Moreover, persons with lower SES also generally had higher estimated exposures than those with higher SES [21]. Disparities in exposure to ambient air pollution during pregnancy was also documented in a nationwide study of over 4 million births in 264 US counties. Latino, African American, and Asian/Pacific Islander mothers were more likely to live in areas with higher air pollution exposures compared to White women [22].

Historically, studies investigating the disparate health effects of air pollution exposure have focused on premature mortality. The epidemiological evidence for premature mortality effects from long-term air pollution exposure is well established [23]. However, the evidence for differences of effects among disparate populations has been less clear, in part because low-SES and minority groups were often underrepresented in the early studies. A recent study investigating statewide mortality effects from long-term PM2.5 exposure over a 5-year period in New Jersey found a higher percentage increase in mortality associated with PM2.5 in census tracts with more African American residents, lower home value, or lower median income [24].

Increasing evidence suggests that socioeconomic factors may influence susceptibility to the health effects of air pollution, especially among children [21]. A recent study examined associations of emergency department visits for children with asthma in Atlanta, Georgia between 2002 and 2008 with daily zip code-level concentrations of ambient air pollutants and effect modification by neighborhood-level SES indicators and found stronger associations in areas of "deprivation" (defined as greater than 20% of the neighborhood unit living in poverty) compared with "nondeprivation areas" [25].

6.3.5 Toxic Metals Exposure

Toxic metals, such as lead and arsenic, rank among the top contaminants of public health concern. They are found in both water and soil and tend to disproportionately affect socially and economically disadvantaged communities. Lead, arsenic, and other metals such as cadmium, manganese, and mercury are considered systemic toxicants known to induce a multitude of adverse health effects, even at low levels of exposure. Children are particularly vulnerable to these contaminants. Lead is rapidly absorbed into the bloodstream and bioaccumulates in the bones and kidneys. Strong evidence indicates that even very low levels of lead exposure may lead to neurodevelopmental and cognitive impairment, particularly when exposure occurs in early life. Arsenic is a known carcinogen that also has been linked to various health effects in adults, including cardiovascular disease, neurological effects, and respiratory disease. Arsenic also has the potential to have damaging effects on children's health and development.

Metals are frequent and persistent soil contaminants. Contaminated soil near major roadways can contain lead, even decades after its removal from gasoline, as well as other metals released by combustion and vehicle brake and engine wear. Industrial activities can also emit high levels of these elements, which then settle in soil and dust, where they can remain for years. Disadvantaged communities tend to be disproportionately located near the sources of these pollutants, such as industrial operations, sites of legacy contamination, and dense roadway networks, and thus bear more of the health-related impacts. A study of low-income pregnant women in South Carolina found consistently higher soil arsenic and lead levels among households of Black mothers compared to White mothers, with a key factor being residential proximity to roadways [26]. In Los Angeles, a lead-acid battery smelter operated near predominantly low-income Latino communities has led to an estimated 100 000 individuals exposed to toxic metals emissions for decades. Screening of homes, parks, and schools in the facility's vicinity has found soil lead levels of >1000 ppm (>12× the 80 ppm California Human Health Residential Soil Screening Level for lead). Though remediation operations have been scheduled, government funds have only been allocated to cover the cleanup of about a quarter of the contaminated properties [27].

Examples of severe metal contamination incidents in disadvantaged communities abound, while many more individuals are chronically exposed to lower levels of these contaminants daily. It is also important to consider that people are likely exposed to combinations of toxic contaminants that may result in synergistic adverse effects or may exert effects as mixtures at much lower levels than predicted by toxicological models. While the individual toxicities of metals are relatively well known, the combined effects have not been studied in detail. Current work from the Maternal and Developmental Risks from Environmental and Social Stressors (MADRES) Center in Los Angeles has been working to fill the research gaps about the health effects of multiple exposures in an urban, low-income, Latino population of women and their children.

6.4 Case Study

6.4.1 Urban Built Environment: Park Equity in Los Angeles

A low-income, predominantly Latino neighborhood in South Los Angeles is situated between two freeways and lacks access to park space. The neighborhood is among the top 15% most disproportionately burdened communities in California, based on environmental pollution and social vulnerabilities. With the housing stock dominated by apartment buildings and multifamily housing, the children from these communities play in the streets or in abandoned junk yards. In South Los Angeles, with the support of a local community organization, Esperanza Community Housing, and the LA Neighborhood Land Trust, the neighbors transformed an abandoned lot into a playground and outdoor community space, Estrella Park.

Los Angeles County is one of the most park-deprived regions in the United States, with low-income neighborhoods and neighborhoods of color having the least access to nearby parks. For every 1000 children in Los Angeles County, White children have access to nearly 96 park acres compared to approximately 5 acres for children of color. Further, 14% of all county parks are within 200 m of a busy road or freeway in zones of elevated air pollution levels. With a high proportion of parks already near freeways, the higher concentration of near-roadway air pollutants could adversely affect the health of those playing in the park and walking or biking to the park. Parks near low-income neighborhoods and communities of color face higher exposure to near-roadway air pollutants compared to parks in predominantly White neighborhoods. The contrast is particularly high for Latino populations [28].

The urban built environment affects public health in key ways. Land use patterns, housing developments, transportation, and neighborhood amenities can be associated with either promoting or harming public health. Urban parks play an important role in encouraging physical activity among various subpopulations (i.e., different ages, ethnocultural groups, and socioeconomic groups). Beyond physical activity, a growing body of literature associates urban parks with multiple health benefits, including decreased stress levels, improved cognition, enhanced emotional well-being, and increased social cohesion [29].

The presence of persistent residential segregation suggests that differential exposure to the built environment may contribute to socioeconomic and racial and ethnic disparities in physical activity. Low-income communities of color have access to fewer parks, fewer park acres, and parks that are potentially more congested. The disparities in park acreage and quality likely exacerbate public health inequities (Figure 6.2).

6.5 Challenges and Opportunities

6.5.1 Challenges

Discovering links between the physical environment and health disparities is difficult because the problem is multifaceted and complicated by adverse environmental exposures, individual behaviors, and psychosocial stressors that can accelerate or protect against detrimental biological responses.

Nonetheless, a wealth of research has illustrated that negative attributes of the physical environment can be bad for human health, and that communities of color and low-income communities or indigenous populations are often disproportionately affected. Communities have recognized inequities and organized at the local level to decrease the hazards and improve health where people live, work, play, and study. With their experience, in-depth knowledge of place, cultural practices, and disease-coping strategies, such communities have become indispensable partners in the search for solutions. They are demanding a greater role in scientific research and the decision making that affects their health and well-being. The greatest challenge is in solving these known disparities, particularly since changing the physical environment requires long-term strategic planning that involves

Figure 6.2 This infographic is an example of how environmental health research is presented to communities in a way that can be easily understood. It helps to create awareness and engage communities to learn about their physical environment. *Source:* Gutschow, W. (2016) Children's health and urban parks [infographic]. University of Southern California Environmental Health Centers. (See color representation of this figure in color inserts section.)

many facets of society working together, including health professionals; urban planners; and local, state, and even federal officials.

6.5.2 Opportunities

Understanding the need to take a multidisciplinary approach to disentangling the disproportionate effects of the physical and social environment on disparate populations, the National Institute of Environmental Health Sciences, the National Institute on Minority Health and Health Disparities (NIMHD), and the EPA teamed up to create the Centers of Excellence on Environmental Health Disparities Research. The five centers are focused on addressing key questions related to health disparities, including understanding the cumulative effects of exposure to multiple environmental contaminants, how different socioeconomic groups are exposed differentially to environmental exposures, how different land uses and proximity to transportation infrastructure contribute to environmental health disparities, and how to build more sustainable communities and engage community stakeholders to build capacity.

Addressing disparities in the physical environment requires the coordination of multiple partners, experts, and community partners working collaboratively to address such complex public health issues. Community-engaged and multidisciplinary research approaches are needed that directly involve those communities disproportionately impacted by the physical environment. The collaborative approach responds to concerns of a local population and involves participatory research, education, and action. These approaches can be utilized to reduce, and ultimately eliminate, environmental health disparities in communities. The NIMHD Research Framework was developed to bring attention to the need for researchers to utilize a multilevel and multidomain approach for advancing our knowledge of how health, environmental health, and health disparities intersect [30].

6.6 Key Points

- The built environment and its environmental exposure effects, especially with regard to differential environmental exposures, are key aspects of the physical environment that may contribute to the development and exacerbation of health disparities among disadvantaged population groups.
- Utilizing the appropriate framework and the multiple methodologies available to environmental researchers is key to assessing and understanding health disparities in relation to the physical environment.
- A greater understanding of the complexities of the impact of the physical environment on disparities observed among multiple health outcomes requires a multifaceted research approach that engages the communities affected to make progress on eliminating these disparities among disadvantaged groups.

Disclaimer

The views and opinions expressed in this chapter are those of the authors only and do not necessarily represent the views, official policy, or position of the US Department of Health and Human Services or any of its affiliated institutions or agencies.

References

1 Office of Environmental Health Hazard Assessment (2017). CalEnviroScreen 3.0. Accessed June 19, 2020 https://oehha.ca.gov/calenviroscreen/report/calenviroscreen-30.

2 Oyana, T.J., Matthews-Juarez, P., Cormier, S.A. et al. (2015). Using an external exposome framework to examine pregnancy-related morbidities and mortalities: implications for health disparities research. *International Journal of Environmental Research and Public Health* 13, 1): 13, https://doi.org/10.3390/ijerph13010013-13.

3 Wilson, S., Burwell-Naney, K., Jiang, C. et al. (2015). Assessment of sociodemographic and geographic disparities in cancer risk from air toxics in South Carolina. *Environmental Research* 140: 562–568.

4 Mielke, H.W., Gonzales, C.R., and Powell, E.T. (2017). Soil lead and children's blood lead disparities in pre- and post-Hurricane Katrina New Orleans (USA). *International Journal of Environmental Research and Public Health* 14 (4): E407.

5 Harris, J.R., Gruenewald, T.L., and Seeman, T. (2008). An overview of biomarker research from community and population-based studies on aging. In: *National Research Council (US) Committee on Advances in Collecting and Utilizing Biological Indicators and Genetic Information in Social Science Surveys* (eds. M. Weinstein, J.W. Vaupel and K.W. Wachter). Washington, DC: National Academies Press, chap. 5. Available from: https://www.ncbi.nlm.nih.gov/books/NBK62427 (accessed June 26, 2020).

6 Paustenbach, D. and Galbraith, D. (2006). Biomonitoring and biomarkers: exposure assessment will never be the same. *Environmental Health Perspectives* 114 (8): 1143–1149.

7 Brown, P. (2003). Qualitative methods in environmental health research. *Environmental Health Perspectives* 111 (14): 1789–1798.

8 Morello-Frosch, R. and Lopez, R. (2006). The riskscape and the color line: examining the role of segregation in environmental health disparities. *Environmental Research* 102 (2): 181–196.

9 Gee, G.C. and Payne-Sturges, D.C. (2004). Environmental health disparities: a framework integrating psychosocial and environmental concepts. *Environmental Health Perspectives* 112 (17): 1645–1653.

10 Fullilove, M.T. and Fullilove, R.E. 3rd (2009). What's housing got to do with it? *American Journal of Public Health* 90 (2): 183–184.

11 Balazs, C.L. and Ray, I. (2014). The drinking water disparities framework: on the origins and persistence of inequities in exposure. *American Journal of Public Health* 104 (4): 603–611.

12 Sadler, R.C., LaChance, J., and Hanna-Attisha, M. (2017). Social and built environmental correlates of predicted blood lead levels in the Flint water crisis. *American Journal of Public Health* 107 (5): 763–769.

13 Levin, R., Brown, M.J., Kashtock, M.E. et al. (2008). Lead exposures in US children, 2008: implications for prevention. *Environmental Health Perspectives* 116 (10): 1285–1293.

14 Ogden, C.L., Carroll, M.D., Kit, B.K., and Flegal, K.M. (2014). Prevalence of childhood and adult obesity in the United States, 2011-2012. *The Journal of the American Medical Association* 311 (8): 806–814.

15 Hilmers, A., Hilmers, D.C., and Dave, J. (2012). Neighborhood disparities in access to healthy foods and their effects on environmental justice. *American Journal of Public Health* 102 (9): 1644–1654.

16 Cobb, L.K., Appel, L.J., Franco, M. et al. (2015). The relationship of the local food environment with obesity: a systematic review of methods, study quality, and results. *Obesity* 23 (7): 1331–1344.

17 Sanders, T., Feng, X., Fahey, P.P. et al. (2015). The influence of neighbourhood green space on children's physical activity and screen time: findings from the longitudinal study of Australian children. *International Journal of Behavior Nutrition and Physical Activity* 12: 126.

18 Neckerman, K.M., Lovasi, G.S., Davies, S. et al. (2009). Disparities in urban neighborhood conditions: evidence from GIS measures and field observation in New York City. *Journal of Public Health Policy* 30 (Suppl 1): S264–S285.

19 Frank, L.D., Sallis, J.F., Saelens, B.E. et al. (2010). The development of a walkability index: application to the Neighborhood Quality of Life Study. *British Journal of Sports Medicine* 44 (13): 924–933.

20 Miranda, M.L., Edwards, S.E., Keating, M.H., and Paul, C.J. (2011). Making the environmental justice grade: the relative burden of air pollution exposure in the United States. *International Journal of Environmental Research and Public Health* 8 (6): 1755–1771.

21 Bell, M.L. and Ebisu, K. (2012). Environmental inequality in exposures to airborne particulate matter components in the United States. *Environmental Health Perspectives* 120 (12): 1699–1704.

22 Woodruff, T.J., Parker, J.D., Kyle, A.D., and Schoendorf, K.C. (2003). Disparities in exposure to air pollution during pregnancy. *Environmental Health Perspectives* 111 (7): 942–946.

23 Pope, C.A. 3rd (2007). Mortality effects of longer term exposures to fine particulate air pollution: review of recent epidemiological evidence. *Inhalation Toxicology* 19 (Suppl 1): 33–38.

24 Wang, Y., Kloog, I., Coull, B.A. et al. (2016). Estimating causal effects of long-term PM2.5 exposure on mortality in New Jersey. *Environmental Health Perspectives* 124: 1182–1188, 1188.

25 O'Lenick, C.R., Winquist, A., Mulholland, J.A. et al. (2017). Assessment of neighbourhood-level socioeconomic status as a modifier of air pollution-asthma associations among children in Atlanta. *Journal of Epidemiology and Community Health* 71 (2): 129–136.

26 Davis, H.T., Aelion, C.M., Liu, J. et al. (2016). Potential sources and racial disparities in the residential distribution of soil arsenic and lead among pregnant women. *Science of the Total Environment* 551–552: 622–630.

27 Johnston, J.E. and Hricko, A. (2017). Industrial lead poisoning in Los Angeles: anatomy of a public health failure. *Environmental Justice* 10 (5): 162–167.

28 Su, J.G., Jerrett, M., de Nazelle, A., and Wolch, J. (2011). Does exposure to air pollution in urban parks have socioeconomic, racial or ethnic gradients? *Environmental Research* 111 (3): 319–328.

29 Seltenrich, N. (2015). Just what the doctor ordered: using parks to improve children's health. *Environmental Health Perspectives* 123 (10): A254–A259.

30 National Institute on Minority Health and Health Disparities (2017). NIMHD Research Framework. Accessed September 29, 2020 https://www.nimhd.nih.gov/about/overview/research-framework.html.

7

Genome-wide Genetic Approaches to Metabolic and Inflammatory Health Disparities

Cheryl A. Winkler

Frederick National Laboratory for Cancer Research, Frederick, MD, USA

7.1 Introduction

While it is widely recognized that poverty and other environmental factors are major determinants of health disparities, there is mounting evidence that health disparities may result from differences in the frequencies of genetic variants that predispose people to certain diseases. Understanding the sources of human genetic diversity and identifying causal genes are crucial to understanding the biological basis for the association of race and ethnicity with disparities in health and disease. Disentangling the relative contributions of genetic and nongenetic risk factors and identifying their interactions are essential steps to improving clinical outcomes, delivering personalized care, and eliminating health disparities.

In this chapter, we provide an overview of the sources of human diversity, the relationship between genetic ancestry and self-identified ancestry, and genome-wide strategies to identify genetic variants associated with health disparities.

7.2 Landscape of Genetic Variation

The human genome is composed of over three billion nucleotide bases and some 20 000–25 000 protein coding genes [1]. The 1000 Genome Project (1000 GP), completed in 2015, has cataloged over 88 million genetic variant sites, identified in 2504 individuals from 26 populations across five continents [2]. Single nucleotide polymorphisms (SNPs) occur on average once every 100–300 nucleotides and account for 95% of variant sites in our genomes. Over 84 million SNPs have now been identified. Only 5% of the variation is due to structural variants, the most frequent of which are small insertion deletions (3.6 million). Large structural variants are rarer (60 000) (www.1000genomes.org). Because of their abundance, SNPs are the primary source of phenotypic variation among individuals. SNPs are broadly classified by their derived allele frequency as rare (minor allele frequency [MAF] <1%), low

The Science of Health Disparities Research, First Edition. Edited by Irene Dankwa-Mullan, Eliseo J. Pérez-Stable, Kevin L. Gardner, Xinzhi Zhang, and Adelaida M. Rosario.

frequency (MAF 1–5%), and common (MAF > 5%). Individuals with African ancestry contain the greatest number of variant sites, consistent with the out-of-Africa model.

The architecture of human genetic variation has been shaped by mutation, random genetic drift, and selection acting in concert with demographic events throughout the 150 000 to 200 000 years of human history [3]. The migration of humans out of Africa 60 000–70 000 years ago and dispersal to continental and island land masses resulted in reproductive isolation of continental and island populations, and local adaption to changing environments [2]. As a consequence of the interplay between historical demographic events, genetic drift, and selection, human populations show geographic and population-specific differences in the relative proportion of common, low frequency, and rare SNPs. The majority of common variants with frequencies >10% are shared among all global populations, and represent variants present in humans prior to the out-of-Africa human expansion. Most common variants are found in noncoding regions and are predicted to be neutral or nearly neutral with little or no effect on health, since purifying selection will almost always prevent deleterious variants from becoming too frequent. Empirical evidence for this comes from numerous genome-wide association studies (GWAS) for common diseases, which have identified nearly 2000 common variants associated with complex traits and diseases, most of which occur in noncoding regions and have small effect sizes (OR < 1.5) [4, 5].

There are exceptions to this tendency of common alleles to manifest functional neutrality and global distribution. First, newly arising variants that confer a selective advantage to new or changing environmental conditions or protect against lethal pathogens may undergo selective sweeps and rapidly rise to high frequency or fixation in a given population. For example, the FY*O null allele for the chemokine receptor Duffy antigen/chemokine receptor (DARC) underwent a selective sweep in sub-Saharan Africa and is now near or at fixation, defined as carriage by population of a single allele. Homozygosity for the null allele provides protection against malaria caused by *Plasmodium vivax* and *Plasmodium knowlesi*, which use the DARC to infect erythrocytes [6]. Genes under balancing selection, i.e., showing heterozygous advantage and homozygous disadvantage, are rare but have the potential to have moderate to large effect sizes for diseases. A classic example is the hemoglobin HbS allele (sickle cell), which reduces the morbidity and mortality of *Plasmodium faciparum* malaria in HbS carriers but causes sickle cell disease in homozygotes [7]. HbS heterozygosity (sickle cell trait) has recently been shown to increase risk for chronic kidney disease [8]. The G1 and G2 APOL1 coding variants, which protect against strains of trypanosomes that cause human African trypanosomiasis (sleeping sickness), have undergone a recent selective sweep in West Africa. The combined APOL1 G1 and G2 frequencies approach ~50% in Yoruba and ~33% in African Americans [9, 10] who share substantial ancestry with West Africans [10]. Individuals who carry two risk alleles (25% and 13% of West Africans and African Americans, respectively) are at increased risk for a diverse spectrum of chronic kidney disease [11].

Population bottlenecks and founder effects may increase or decrease allele frequencies for rare or low-frequency pathogenic variants by random genetic drift occurring in isolated populations [12]. Founder effects occur when a small founder population occupies a new territory followed by rapid population expansion. Examples of founder effects inflating the risk of rare monogenic diseases include the increased prevalence of Tay-Sachs disease in Ashkenazi Jews [13] and the set of Finnish heritage diseases, comprising over 30 monogenic diseases observed in ethnic Finns [14]. Puerto Rican children are at twice the risk of asthma compared to other Latino subgroups and White non-Latinos, even after adjusting for traditional risk factors. Founder effects may account for this predisposition [15].

7.3 Pathogenic Potential of Low-frequency and Rare Variants

The 1000 Genome Project [2], the Simons Genome Diversity Project [16], and the Exome Aggregation Consortium [17] have provided valuable insights into the geographic distribution of rare and low-frequency variants and their potential contributions to disease and health disparities. In contrast to the cosmopolitan distribution of common variants, 17% of low-frequency variants occur in a single continental or ancestry group, while 53% of rare variants occur only in single populations [2]. The greatest degree of variant frequency differentiation is between Africa and the Americas; the least is between Eurasia and East Asia, reflecting sequential founder events and levels of gene flow among these continental populations [12]. Because rare variants tend to be younger than common variants, with less time to be eliminated by purifying selection, these variants have higher potential to be disease causing and to have larger effect sizes (penetrance) [16].

7.4 Admixture in the Americas

> "The classic mirror reflects skin color; but the DNA mirror reflects our common ancestors."
>
> *Dr. Beatriz Marcheco-Teruel [18].*

The European invasion of the Americas that began in the late fifteenth century, and the trans-Atlantic slave trade from the sixteenth to nineteenth centuries (Figure 7.1), have resulted in two large admixed populations in the United States: Latinos and African Americans [10, 20]. Populations with Native American, European, and African two- and three-way admixture are found throughout the Caribbean and the Americas as a result of these transformative events. Ancestry contributions to Latino populations are complicated and vary by geographic region in their admixture composition. Caribbean individuals may have three-way admixture with varying proportions of African, European, and indigenous ancestry contributing to their genomes, while Mexican Americans from the southwestern and western regions of the United States tend to have a greater proportion of Native American and European ancestry and little to no African ancestry, as shown in Figure 7.2 [21].

On average, African American genomes are comprised of approximately 15–20% European and 80–85% African ancestry, resulting from admixture about 8–20 generations ago. The proportion of European and Native American ancestry among African Americans varies greatly among individuals and in different regions of the United States [20]. Admixture between two or more differentiated populations results in chromosomes derived from the parental populations, which over subsequent generations, due to recombination, become a mosaic of chromosomal segments (ancestry blocks) derived from each parental ancestry group, as shown in Figure 7.3 [22, 23]. As will be shown below (and was recently reviewed by Shriner [22]), recent admixture can be exploited to map genes associated with common diseases, such as asthma [24] and kidney disease [25], that show disparate prevalence in the parental populations.

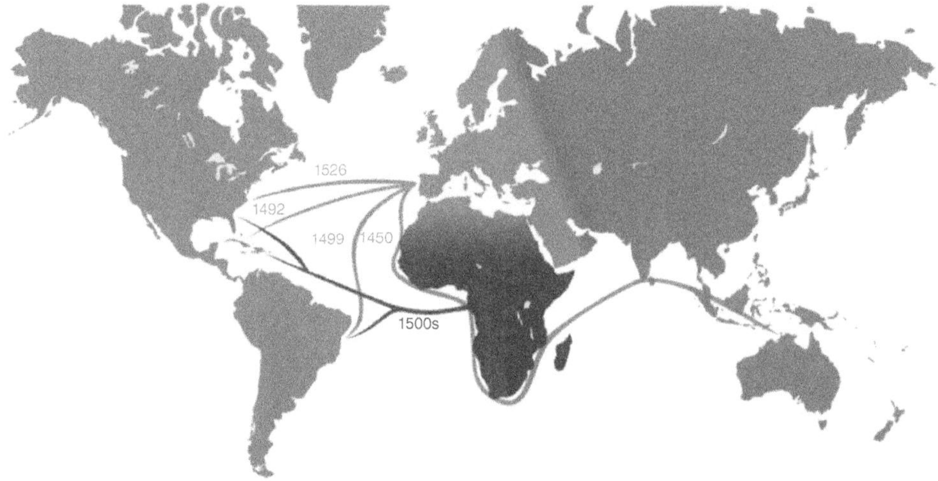

Figure 7.1 Major migrations and diasporas, 1400–1800, that are the sources of admixed populations in the Americas and elsewhere. The location of the ancient populations of Eurasia, Asia, sub-Saharan Africa, and the Americas are shown in distinct colors. *Source:* Adapted from Winkler et al. [19]. (See color representation of this figure in color inserts section.)

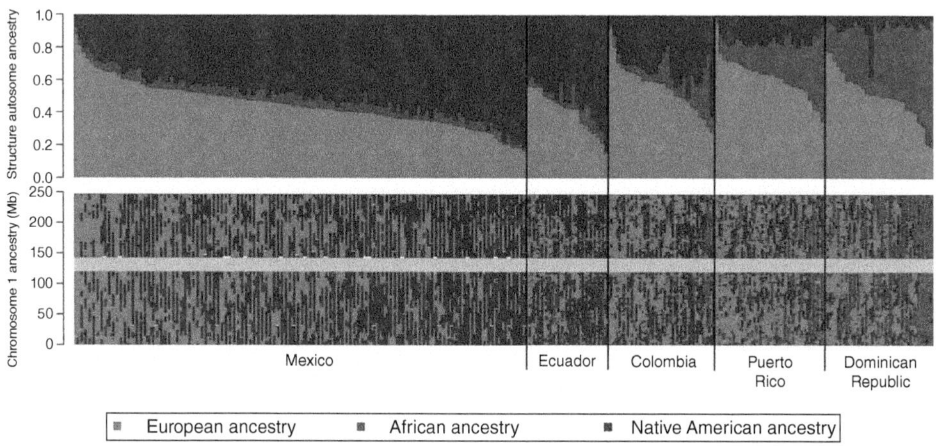

Figure 7.2 The top panel presents genome-wide ancestry of all autosomes for individuals from five Latino populations, showing the proportions of African, European, and Native American ancestry. The colors represent ancestry linkage disequilibrium blocks derived from the three ancestral populations contributing to the admixture. Each vertical slide represents a single individual, and the data for these individuals in each population are arranged from high to low European ancestry. The bottom panel presents the locus-specific (local) ancestry for chromosome 1, with each bar representing one individual. The two arms of the chromosomes are separated by a horizontal gray line representing the centromeric region. This figure makes the point that in an admixed population, each individual genome is a patchwork quilt of ancestral genomic segments. *Source:* Adapted from Bryc et al. [21] © 2010, Katarzyna Bryc. (See color representation of this figure in color inserts section.)

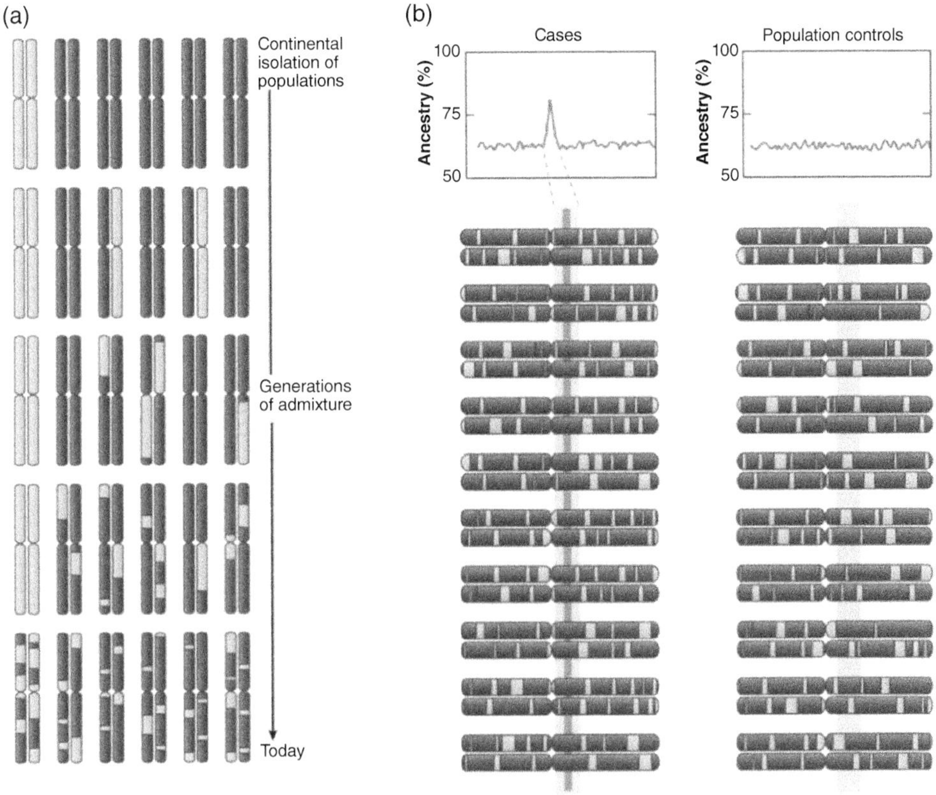

Figure 7.3 Shown is the process of admixture resulting from a moderate number of generations [4–18, 20, 21] since the initial admixture event. The boundary between light and dark green represents a recombination switch point. (a) The dark and light green color blocks represent chromosomal segments derived from matings between two parental (ancestral) populations. Starting with the second generation, meiotic cross-over (recombination) events produce chromosomal blocks of the two parental ancestries. Present-day admixed populations vary both in the extent of overall ancestry and block lengths due to differences in recombination events among individuals and because among individuals, original chromosomes have been subject to recombination for varying numbers of generations. (b) The graphs show the percentage of ancestry derived from the dark green segment of chromosome (~60%). In the 95% credible interval harboring the disease locus (yellow bar), there is an excess of majority ancestry blocks among cases, revealed as a spike in a graph of average ancestry for cases along the chromosome to ~80%. The orange bar indicates the location of the causal disease gene. Among population controls, the distribution of ancestry blocks is random across the ancestry. The spike of ancestry can be quantified either by comparing case ancestry versus control ancestry at the same locus (case-control statistic) or by comparing peak case local ancestry with case genome-wide average ancestry across all chromosomes (locus-genome statistic). *Source:* Adapted from Winkler et al. [19] © 2010, Elsevier. (See color representation of this figure in color inserts section.)

7.5 Identifying Disease Genes Associated with Health Disparities: Methods and Approaches

Here we provide a brief description of four commonly used genome-wide methods to map genes associated with common diseases and traits: (i) genome-wide association study (GWAS), (ii) mapping by admixture disequilibrium (admixture mapping), (iii) whole-exome sequencing (WES), and (iv) whole-genome sequencing (WGS).

With the large reference databases and browsers now available from sequencing an ever-increasing number of diverse human genomes and exomes, genetic studies use information from a variety of online resources and may employ multiple methods for detection of causal variants [26, 27]. For health disparities in Latinos and African Americans showing marked differences among ancestry groups, admixture mapping has a distinctive advantage over other genome-wide scans because of its ability to detect population-specific variants contributing to the disparity [19, 22]. The spectrum of human variation and haplotype structure derived from WGS and WES—of tens of thousands of diverse individuals from the 1000 Genomes Project [2], the Simons Diversity Project [17], the Exome Aggregation Consortium [16], and many other studies—is essential to admixture mapping and GWAS in identifying causal variants associated with diseases and traits.

7.5.1 Genome-wide Association Studies

GWAS are an unbiased method used to scan the entire genome for SNP association with specific phenotypes, including diseases, quantitative traits, and other measurable or observable phenotypes. GWAS have led to the identification of thousands of genetic variants associated with diseases and quantitative traits in over 35 million deoxyribonucleic acid (DNA) samples [5]. The approach involves genotyping a large number of SNPs (100 000 to over 2 million) using commercially available SNP arrays and DNA from a large number of individuals [28]. If an SNP allele is significantly more prevalent in cases compared to controls, the variant allele is considered to be associated with the phenotype [29]. Since even dense SNP arrays contain only a fraction of the variant sites in human genomes, the approach is indirect in that the causal SNP is usually not directly genotyped but is hitchhiking with nearby marker SNPs. The method depends on correlation between neighboring alleles that have not been separated by meiotic recombination due to their physical proximity on the same DNA strand. Neighboring SNP markers will tend to be inherited on the same haplotype block, resulting in linkage disequilibrium (LD) between marker SNPs and the causal SNP [29]. If the causal SNP has a strong enough effect on disease, other nearby SNPs in complete or incomplete LD with the causal SNP will also be associated with the disease. Results of a GWAS are typically displayed as a Manhattan scatter plot, so-named because it resembles a city skyline (Figure 7.4). Since even carefully conducted GWAS are prone to false positive associations, replication of implicated SNPs must be performed in a second independent study before embarking on follow-up fine mapping and functional assessment to identify causal variants responsible for the observed association [30].

GWAS are highly susceptible to a myriad of errors and biases that inflate false positive associations, including cryptic population substructure, cryptic relatedness among cases, and batch effects, as well as phenotyping and genotyping errors. Stringent quality control

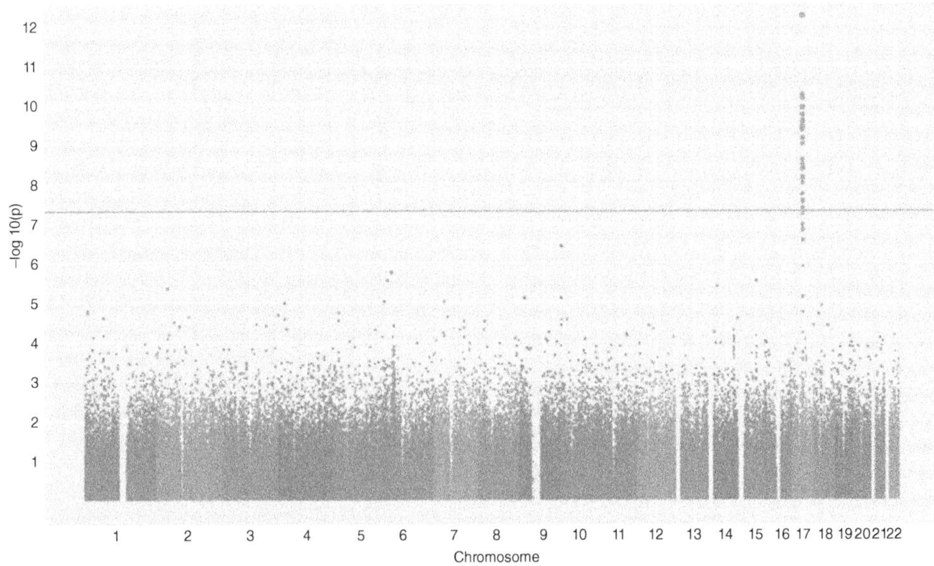

Figure 7.4 A Manhattan plot showing a peak of SNP association on chromosome 17q21 in a study of 1,893 Latinos with asthma and 1,881 Latino controls. The genomic coordinates, with chromosomes indicated, are displayed on the *X* axis, the negative log of the p values are displayed on the *Y* axis, and SNPs are represented by a point. The red line represents the genome-wide significance threshold, 5×10^{-8}. *Source:* Adapted from Galanter et al. [26]. (See color representation of this figure in color inserts section.)

procedures have been established and most journals require adherence to specific guidelines for publication [31]. It is especially critical to adjust for cryptic population substructure (differences in ancestral proportion) between cases and controls (a frequent source of type 1 errors) by adjusting for the proportion of global ancestry [32]. It is common practice in GWAS to use a conservative Bonferroni correction for genome-wide significance of $p < 5 \times 10^{-8}$, for $\alpha = 0.05$, accounting for a million independent SNPs tested. Other statistical methods may also be employed, including permutation tests and false discovery rate [33]. It is also standard practice to replicate statistically significant associations in a second independent group of cases and controls to securely identify associated loci. Meta-analyses combining results from several independent studies are frequently employed to securely identify associated loci and to narrow the confidence interval around point estimate for the odds ratio [34]. The avoidance of false positive (type 1) errors tends to inflate false negative (type 2) errors, which might account for a portion of the missing heritability for complex diseases revealed by GWAS studies [4]. Increasing sample size improves the power to detect associations; however, obtaining sufficient statistical power is a major hurdle for rare diseases, for diseases with polygenic inheritance where many variants with small effects contribute to the phenotype, for low-frequency and rare variants, and for variants with modest effect sizes. For minority populations experiencing health disparities, enrollment of large numbers of cases may be difficult to achieve.

As of 2016, over 80% of all GWAS subjects were of European ancestry, while less than 4% represented minority American racial/ethnic groups: African Americans (3%), Hispanic/Latin Americans (0.54%), and Native (indigenous) Americans (0.05%) [35]. Additionally, a disproportionate number of SNPs on genotyping arrays were originally identified in studies of modest size (resulting in a bias towards common variants) and in studies involving European individuals (resulting in a paucity of variants representative of non-European populations) [3]. Since African populations are older, they have a more complex haplotype structure and more diversity than any of the out-of-Africa founder populations that colonized Europe, Asia, Oceania, and the Americas [3]. This bias has prompted the development of population-specific and multiethnic SNP arrays with greater SNP density and the inclusion of both low-frequency and common variants informed by the HapMap Project, the 1000 Genomes Project, and other sequencing efforts [17].

A major advance in GWAS has been the development of statistical tools and population-specific haplotype reference panels to impute genotypes for missing SNPs not represented on a SNP array [36]. Imputation takes advantage of shared haplotypes between individuals over short stretches of chromosome sequence and reference haplotype panels to impute missing genotypes. Haplotypes are first estimated from the GWAS dataset using SHAPIT or other haplotype estimation programs [37]. Next, using an imputation program (e.g., IMPUTE2), genotypes from population-specific haplotype reference panels (e.g., 1000 Genome Project) are imputed into the pre-phased haplotypes [38]. Imputation of untyped SNPs increases the content of sparse arrays and is employed to harmonize genotypes across different genotyping arrays for association analysis of combined studies used in meta-analysis to increase statistical power. It allows for the imputation of untyped rare and low-frequency variants to identify potential causal variants. Imputation of genotypes in admixed populations is computationally more intensive because (i) each genome comprises a set of chromosomes originally derived from two or more ancestral populations and (ii) there is a shortage of sequenced genomes from some minority populations. The 1000 Genome Reference panels have greatly improved imputation for African American and Latino populations [39].

Population substructure resulting from ancestry differences between cases and controls is a frequent cause of type 1 errors. For this reason, most GWAS are performed on homogenous cases and controls matched on self-reported ancestry. Most GWAS studies adjust for global genetic ancestry using Bayesian (STRUCTURE) [40] or maximum likelihood (ADMIXTURE) [41] clustering models, or principal component analyses (EIGENSOFT) [42] to quantify the proportion of ancestry contributions. However, adding ancestry covariates to regression analyses may not be sufficient for admixed individuals, who vary greatly in proportion to parental ancestry contributions. As will be discussed, admixture mapping either alone or combined with GWAS increases the power to discover population-specific and shared genetic variants, respectively, associated with health disparities in African American, Latino, Native American, and other admixed populations.

7.5.2 Mapping by Admixture Linkage Disequilibrium

Admixture mapping provides a powerful alternative to GWAS to identify genetic variants associated with the subset of quantitative traits and diseases that differ in prevalence between the ancestral populations contributing to the admixture. The underlying hypothesis is that the causal variant(s) associated with a disease/trait will be more frequent on a chromosomal segment derived from the ancestral population with the higher disease/trait risk prevalence (Figures 7.3a,b and 7.5). It has particular application in the United States to

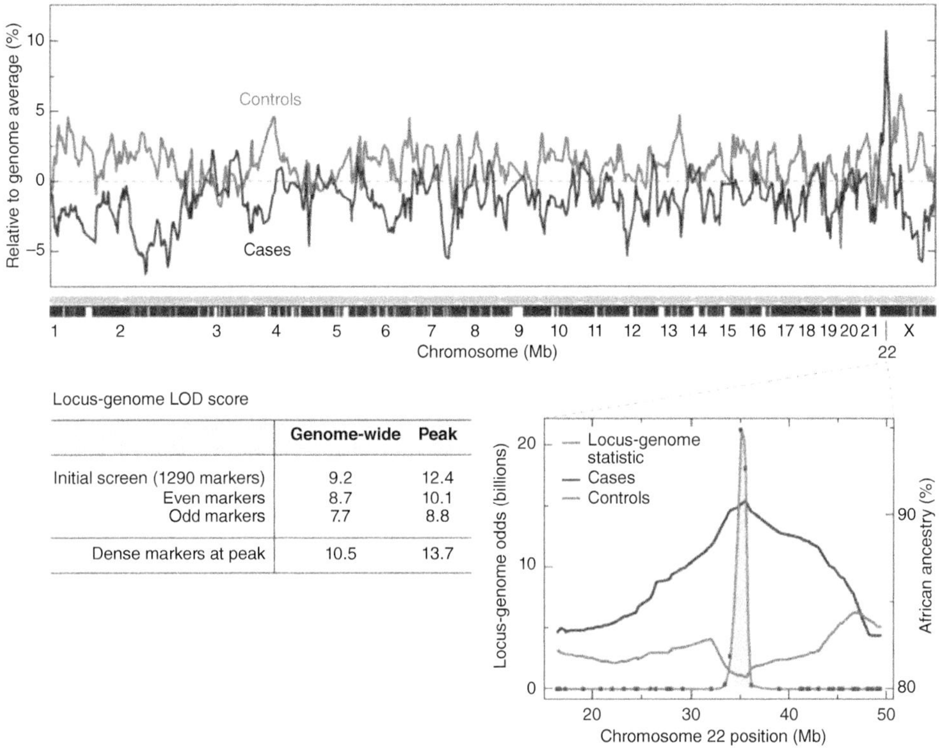

	Genome-wide	Peak
Initial screen (1290 markers)	9.2	12.4
Even markers	8.7	10.1
Odd markers	7.7	8.8
Dense markers at peak	10.5	13.7

Figure 7.5 An admixture scan for focal segmental glomerulosclerosis and HIV-associated nephropathy, which are, respectively, fourfold and 60-fold more frequent in African Americans compared to European Americans [43]. The sharp peak of African ancestry occurs on chromosome 22 centered on the plausible candidate gene MYH9 and 32 other genes. Subsequently, the ancestry signal was shown to be due to the presence of two African-specific coding variants in the APOL1 gene encoding apolipoprotein L1 [9]. *Source:* Adapted from Winkler et al. [19]. (See color representation of this figure in color inserts section.)

health disparities in Latinos and African Americans. Since far fewer markers are required to identify ancestry haplotype blocks derived from each population contributing to the admixture, statistical power is increased; to detect phenotype association with an ancestral haplotype block, approximately 500 cases are required for odds ratios of 2 or more and approximately 2500 cases for odds ratios of at least 1.2 [22]. Admixture mapping has been successfully used to identify genes associated with a number of traits and diseases, including ethnic neutropenia in African Americans [44], asthma, kidney disease, hypertension, and cardiovascular disease [19, 24].

Admixture mapping, similar to family studies, uses long-range LD introduced when two reproductively isolated (ancestral) populations interbreed to localize disease genes [23]. Unlike GWAS, which tests for SNP marker associations with a phenotype, admixture mapping tests for correlation between locus-specific (local) ancestry at each SNP and a phenotype. After the admixture event, meiotic recombination leads to a mosaic of haplotype blocks derived from the two or more populations contributing to the admixture (Figure 7.3a). As the number of generations increase after the beginning of the admixture process, the length of haplotypes will decay linearly due to recombination.

Although earlier admixture mapping algorithms worked best for recently admixed population (8–20 generations), it is now possible to use admixture mapping for admixed populations occurring 100 or more generations ago using improved statistical methods and haplotype panels from diverse global populations [22]. Resolution for localizing disease genes in intermediate family studies represents two to three generations of recombination, and GWAS studies of unrelated individuals represent the number of generations since the most recent common ancestor.

Admixture mapping involves genotyping SNPs spanning the genome and determining the ancestral origin of each of the two alleles at each genotyped locus. First-generation admixture scans employed sparse panels of ancestry informative markers (AIMs) (<2000 SNPs) that differed in allele frequencies between the two ancestry populations. Recently, software programs have been developed to analyze random SNPs obtained from dense SNP arrays, which increases resolution and power. Shriner et al. [22] estimate that 200 000 random SNPs are sufficient to detect nearly all ancestry switches in African Americans, which for the most part are two-way admixed. Local ancestry for each allele at each genotyped locus must be inferred indirectly using probabilistic methods based on the hidden Markov chain model [45]. The idea is that the frequency count of ancestry for each allele at each locus can be used to predict the ancestral origin of the alleles of the neighboring loci. At each locus, 2, 1, or 0 alleles will be inherited from one or the other contributing ancestry population. The accurate inference of the switch point or transition from one ancestry haplotype block to the other increases resolution and is an essential step for localizing causal genes (Figure 7.3b).

There are multiple statistical approaches that can be used to infer local ancestry. HAPMIX and PCADMIX use haplotype reference panels for each of the parental populations and only can be used for admixed populations with reference haplotypes [46]. RFMix assigns local ancestry and switch points by abstracting the ancestry information from the genotypes of the samples, which allows inference of local ancestry to populations with no or sparse reference data [47]. Once local ancestry for each locus is determined, global ancestry, the genome-wide average of all the local ancestries, is calculated for each individual.

The final step in admixture mapping is to test for local ancestry-phenotype correlations using regression models. Generalized linear models utilize different types of phenotypic data and permit the inclusion of covariates and interaction terms [22]. Two statistics are commonly generated.

First, the case-control statistic compares the extent of local ancestry in cases compared to controls. An excess or decrease of local ancestry in cases suggests the presence of a susceptibility or resistant disease locus (see Figures 7.3b and 7.5). Second, the case-only statistic compares local ancestry at each locus to global ancestry; a statistically significant difference in local ancestry compared to the genome-wide average suggests the presence of a causal disease locus (Figure 7.5). This is an important option if a control group is difficult to recruit, or for rare diseases where cases numbers may be small [48]. As with GWAS, replication of the most significant SNPs under the admixture peak is required. Fine mapping, conditioning on the most significant admixture SNP, is used to identify putative causal alleles responsible for the admixture signal [22].

7.6 Joint Admixture Mapping and Genome-wide Association Studies for Gene Discovery in Admixed Populations

Admixture mapping can be combined with GWAS to identify ancestry-specific variants and shared variants that contribute to disease disparities. The joint method utilizes dense genotyping to first perform admixture mapping to infer local ancestry and to test for correlation between local ancestry and the disease [26]. Genome-wide association and admixture mapping are dependent on the identification of causal variants by haplotype blocks and local ancestry blocks, respectively. Because the ancestry blocks are much longer than haplotype blocks, with admixture mapping there is a much lower statistical penalty for multiple comparisons with admixture mapping (5000–6000 comparisons, typically) compared to GWAS (one million comparisons, typically), but at the cost of yielding a much longer genomic region-of-interest (hundreds of kilobases) harboring the causal variant [22]. With joint analysis, admixture mapping is first used to identify causal disease variants that are ancestry-specific.

Next, GWAS is used to test for association between SNP and the disease phenotype using logistic regression models, stratifying each SNP for local ancestry and adjusting for global ancestry. This complementary approach has been applied to childhood asthma in Latinos [24, 26]. A joint admixture mapping study and GWAS using admixture mapping of an ancestry-specific, novel locus on chromosome 6, and by GWAS replicated in Latinos, identified an association with a previously identified chromosome 17q21 locus [26].

7.7 Whole-Genome and Whole-Exome Sequencing Approaches to Health Disparities

The development of high-throughput sequencing technologies has been transformative in advancing genomics in biomedical research and clinical practice. WGS provides a near complete DNA sequence (95%) of an individual's genome. WES employs capture methods to enrich for the subset of DNA that encodes proteins (1% of genome) and is now widely used to identify rare coding variants that cause monogenic diseases.

WGS and WES of diverse human populations have generated haplotype reference databases and provided a catalog of rare and low-frequency variants that have increased the resolution of GWAS to identify potential causal SNPs. Variant and haplotype data, from the 1000 Genomes Project and other sequencing studies, have tremendous potential to facilitate the discovery of causal variants that are in LD with disease-associated markers identified by GWAS and admixture scans. Sequence data from the 1000 Genomes Project, together with results from an admixture scan for chronic kidney disease, were used to identify African-specific coding variants in the APOL1 gene associated with a diverse spectrum of chronic kidney disease in African Americans [9]. WGS of diverse global populations will certainly identify other population-specific variants, like APOL1 coding variants, contributing to health disparities. It has been shown that rare variants are enriched in Puerto Ricans with asthma [49] and disease-causing variants are more abundant in endogamous populations [50]. Since WGS sequencing identifies most variant sites within the genomes of individuals, including common, low-frequency, and rare variants, traditional GWAS can be employed to test for association of low-frequency and common variants with locus-specific resolution. To identify disease-causing rare variants, which have greater potential to be more penetrant and are not shared across divergent populations, admixture mapping strategies to unravel hidden ancestry and fine structure of haplotypes embedded within ancestry blocks has been proposed, but not yet applied [24, 51]. More commonly, aggregation tests are employed to strengthen statistical signals by combining rare variants identified in a gene or pathway and comparing the proportion (burden) of variants in affected cases compared to controls.

7.8 Summary

> "Genetic background loads the gun, but the environment pulls the trigger."
>
> *Dr. George A. Bray* [52]

The discovery of genes contributing to health disparities is an important component in reducing the burden of health disparities. Admixture mapping, GWAS, and whole-genome/exome sequencing are complementary approaches that have been successfully used to identify shared and ancestry-specific variants spanning the allele frequency spectrum. The challenge will be to characterize the joint interactions between environmental and genetic risk factors that trigger pathophysiological processes leading to metabolic and inflammatory disease disparities. Accounting for joint gene X environmental interactions will lead to improvements in estimations of population-attributable risk; quantification of the relative impact of environmental exposures (e.g., poverty, toxins, diet) and genetic susceptibility variants on health outcomes; identification of relevant biological pathways for drug development and treatment; and identification of modifiable risk factors for genetically susceptible individuals [53].

7.9 Key Points

- Although poverty, socioeconomic factors, and systemic racism are major contributors to health disparities, for a subset of diseases, genetic variants may increase the risk.
- Geographic and population specific differences in allele frequencies and the relative proportion of common, low frequency, and rare SNPs contribute to differential disease risk among different ethnic and racial groups.
- Most common variants associated with common diseases have small effect sizes, but newly arising rare variants and pathogen-selected variants have the potential to be strongly pathogenic (e.g. APOL1 variants and chronic kidney disease).
- Admixture mapping is a powerful alternative to traditional GWAS to identify genetic variants associated with the subset of traits and diseases that differ in prevalence between the ancestral populations contributing to the admixture.
- The identification of genes contributing to health disparities is a critical component in reducing the burden of health disparities.

Definitions

Locus A fixed position on a chromosome harboring a gene or, more commonly, a genetic marker for a single nucleotide variant.

Allele One of two or more alternate forms of a gene. An individual inherits one paternal and one maternal allele of a gene. Generally, the common allele is referred to as the reference allele, and the alternative allele is the variant. Individuals with identical alleles at a specific locus are termed **homozygous** for that allele. Alternatively, an individual with the reference allele and the alternative allele would be termed **heterozygous**.

Genotype The combination of the two alleles inherited from each parent at a specific locus.

Phenotype Observable characteristics or traits resulting from genetic and environmental interactions. Although phenotypes can be assessed in many ways, it is most common in genetic studies to determine genotype-phenotype associations with binary phenotypes (e.g., presence or absence of a disease) or quantitative traits (e.g., height).

Single nucleotide variant (SNV) A nucleotide change at a specific position in the genome (e.g., A to G change).

Single nucleotide polymorphism (SNP) An SNV that is observed in more than 1 out of 100 chromosomes (minor allele frequency > 1% frequency) within a population.

Local ancestry As a result of meiotic recombination, an individual's chromosomes comprise recombined fragments from ancestral (parental) populations contributing to the admixture. Local ancestry is the locus-specific ancestry at each allele, where an individual can have 0, 1, or 2 copies of an allele from each ancestral population. Ancestry blocks are stretches of alleles derived from one or the other ancestral population.

Global ancestry Global ancestry is the average of the estimated local ancestries across the genome for each individual.

Ancestry informative markers Genetic markers, usually SNPs, that have different allele frequencies between two or more reproductively isolated populations.

Population structure or stratification Quantifiable differences in allele frequencies between subpopulations. Cryptic substructure in association studies may inflate false positive associations.

Linkage disequilibrium The nonrandom (correlated) association of alleles at different loci within a family or population, which tend to be inherited together from a common ancestry.

Haplotypes Segments of chromosomes sharing identity-by-descent from the most recent common ancestor are referred to as haplotypes. Haplotype length decays linearly as a function of time due to meiotic recombination at each generation (i.e., longest in families, intermediate in recently admixed populations, and shortest in random-mating populations).

Disclaimer

The views and opinions expressed in this chapter are those of the authors only and do not necessarily represent the views, official policy, or position of the U.S. Department of Health and Human Services or any of its affiliated institutions or agencies.

References

1 Ezkurdia, I., Juan, D., Rodriguez, J.M. et al. (2014). Multiple evidence strands suggest that there may be as few as 19,000 human protein-coding genes. *Human Molecular Genetics* 23 (22): 5866–5878.

2 Auton, A., Brooks, L.D., Durbin, R.M. et al. (Genomes Project Consortium) (2015). A global reference for human genetic variation. *Nature* 526 (7571): 68–74.

3 Lachance, J. and Tishkoff, S.A. (2013). Population genomics of human adaptation. *Annual Review of Ecological Evolution Systems* 44: 123–143.

4 Manolio, T.A., Collins, F.S., Cox, N.J. et al. (2009). Finding the missing heritability of complex diseases. *Nature* 461 (7265): 747–753.

5 MacArthur, J., Bowler, E., Cerezo, M. et al. (2017). The new NHGRI-EBI catalog of published genome-wide association studies (GWAS Catalog). *Nucleic Acids Research* 45 (D1): D896–D901.

6 Hamblin, M.T. and Di Rienzo, A. (2000). Detection of the signature of natural selection in humans: evidence from the Duffy blood group locus. *American Journal of Human Genetics* 66 (5): 1669–1679.

7 Aidoo, M., Terlouw, D.J., Kolczak, M.S. et al. (2002). Protective effects of the sickle cell gene against malaria morbidity and mortality. *Lancet* 359 (9314): 1311–1312.

8 Naik, R.P., Irvin, M.R., Judd, S. et al. (2017). Sickle cell trait and the risk of ESRD in Blacks. *Journal of the American Society of Nephrology* 28 (7): 2180–2187.

9 Genovese, G., Friedman, D.J., Ross, M.D. et al. (2010). Association of trypanolytic ApoL1 variants with kidney disease in African Americans. *Science* 329 (5993): 841–845.

10 Tishkoff, S.A., Reed, F.A., Friedlaender, F.R. et al. (2009). The genetic structure and history of Africans and African Americans. *Science* 324 (5930): 1035–1044.

11 Limou, S., Nelson, G.W., Kopp, J.B., and Winkler, C.A. (2014). ApoL1 kidney risk alleles: population genetics and disease associations. *Advanced Chronic Kidney Disease* 21 (5): 426–433.

12 Ramachandran, S., Deshpande, O., Roseman, C.C. et al. (2005). Support from the relationship of genetic and geographic distance in human populations for a serial founder effect originating in Africa. *Proceedings of the National Academy of Sciences* 102 (44): 15942–15947.

13 Kaback, M., Lim-Steele, J., Dabholkar, D. et al. (1993). Tay-Sachs disease-carrier screening, prenatal diagnosis, and the molecular era. An international perspective, 1970 to 1993. The International TSD Data Collection Network. *The Journal of the American Medical Association* 270 (19): 2307–2315.

14 de la Chapelle, A. (1993). Disease gene mapping in isolated human populations: the example of Finland. *Journal of Medical Genetics* 30 (10): 857–865.

15 Lara, M., Akinbami, L., Flores, G., and Morgenstern, H. (2006). Heterogeneity of childhood asthma among Hispanic children: Puerto Rican children bear a disproportionate burden. *Pediatrics* 117 (1): 43–53.

16 Lek, M., Karczewski, K.J., Minikel, E.V. et al. (2016). Analysis of protein-coding genetic variation in 60,706 humans. *Nature* 536 (7616): 285–291.

17 Mallick, S., Li, H., Lipson, M. et al. (2016). The Simons genome diversity project: 300 genomes from 142 diverse populations. *Nature* 538 (7624): 201–206.

18 Marcheco-Teruel, B., Parra, E.J., Fuentes-Smith, E. et al. (2014). Cuba: exploring the history of admixture and the genetic basis of pigmentation using autosomal and uniparental markers. *PLoS Genetics* 10 (7): e1004488.

19 Winkler, C.A., Nelson, G.W., and Smith, M.W. (2010). Admixture mapping comes of age. *Annual Review of Genomics and Human Genetics* 11: 65–89.

20 Salas, A., Carracedo, A., Richards, M., and Macaulay, V. (2005). Charting the ancestry of African Americans. *American Journal of Human Genetics* 77 (4): 676–680.

21 Bryc, K., Velez, C., Karafet, T. et al. (2010). Colloquium paper: Genome-wide patterns of population structure and admixture among Hispanic/Latino populations. *Proceedings of the National Academy of Sciences* 107 (Suppl): 28954–28961.

22 Shriner, D. (2017). Overview of admixture mapping. *Current Protocols in Human Genetics* 94: 1.23.1–1.23.8. https://doi.org/10.1002/cphg.44.

23 Chakraborty, R., Kamboh, M.I., and Ferrell, R.E. (1991). "Unique" alleles in admixed populations: a strategy for determining "hereditary" population differences of disease frequencies. *Ethnicity & Disease* 1 (3): 245–256.

24 Mersha, T.B. (2015). Mapping asthma-associated variants in admixed populations. *Frontier in Genetics* 6 (292): 1–21.

25 Kao, W.H., Klag, M.J., Meoni, L.A. et al. (2008). MYH9 is associated with nondiabetic end-stage renal disease in African Americans. *Nature Genetics* 40 (10): 1185–1192.

26 Galanter, J.M., Gignoux, C.R., Torgerson, D.G. et al. (2014). Genome-wide association study and admixture mapping identify different asthma-associated loci in Latinos: the genes-environments and admixture in Latino Americans study. *Journal of Allergy and Clinical Immunology* 134 (2): 295–305.

27 Shriner, D., Adeyemo, A., and Rotimi, C.N. (2011). Joint ancestry and association testing in admixed individuals. *PLoS Computational Biology* 7 (12): e1002325.

28 Pearson, T.A. and Manolio, T.A. (2008). How to interpret a genome-wide association study. *The Journal of the American Medical Association* 299 (11): 1335–1344.

29 Hirschhorn, J.N. and Daly, M.J. (2005). Genome-wide association studies for common diseases and complex traits. *Nature Review Genetics* 6 (2): 95–108.

30 MacArthur, D.G., Manolio, T.A., Dimmock, D.P. et al. (2014). Guidelines for investigating causality of sequence variants in human disease. *Nature* 508 (7497): 469–476.

31 Turner, S., Armstrong, L.L., Bradford, Y. et al. (2011). Quality control procedures for genome-wide association studies. *Current Protocols in Human Genetics* https://doi.org/10.1002/0471142905.hg0119s68.

32 Purcell, S., Neale, B., Todd-Brown, K. et al. (2007). Plink: a tool set for whole-genome association and population-based linkage analyses. *American Journal of Human Genetics* 81 (3): 559–575.

33 Johnson, R.C., Nelson, G.W., Troyer, J.L. et al. (2010). Accounting for multiple comparisons in a genome-wide association study (GWAS). *BMC Genomics* 11: 724.

34 de Bakker, P.I., Ferreira, M.A., Jia, X. et al. (2008). Practical aspects of imputation-driven meta-analysis of genome-wide association studies. *Human Molecular Genetics* 17 (R2): R122–R128.

35 Popejoy, A.B. and Fullerton, S.M. (2016). Genomics is failing on diversity. *Nature* 538 (7624): 161–164.

36 Li, Y., Willer, C., Sanna, S., and Abecasis, G. (2009). Genotype imputation. *Annual Review of Genomics and Human Genetics* 10: 387–406.

37 Delaneau, O., Marchini, J., and Zagury, J.F. (2011). A linear complexity phasing method for thousands of genomes. *Nature Methods* 9 (2): 179–181.

38 Howie, B., Fuchsberger, C., Stephens, M. et al. (2012). Fast and accurate genotype imputation in genome-wide association studies through pre-phasing. *Nature Genetics* 44 (8): 955–959.

39 Nelson, S.C., Stilp, A.M., Papanicolaou, G.J. et al. (2016). Improved imputation accuracy in Hispanic/Latino populations with larger and more diverse reference panels: applications in the Hispanic community health study/study of Latinos (HCHS/SOL). *Human Molecular Genetics* 25 (15): 3245–3254.

40 Pritchard, J.K., Stephens, M., and Donnelly, P. (2000). Inference of population structure using multilocus genotype data. *Genetics* 155 (2): 945–959.

41 Alexander, D.H., Novembre, J., and Lange, K. (2009). Fast model-based estimation of ancestry in unrelated individuals. *Genome Research* 19 (9): 1655–1664.

42 Price, A.L., Patterson, N.J., Plenge, R.M. et al. (2006). Principal components analysis corrects for stratification in genome-wide association studies. *Nature Genetics* 38 (8): 904–909.

43 Kopp, J.B., Smith, M.W., Nelson, G.W. et al. (2008). MYH9 is a major-effect risk gene for focal segmental glomerulosclerosis. *Nature Genetics* 40b (10): 1175–1184.

44 Reich, D., Nalls, M.A., Kao, W.H. et al. (2009). Reduced neutrophil count in people of African descent is due to a regulatory variant in the Duffy antigen receptor for chemokines gene. *PLoS Genetics* 5 (1): e1000360.

45 McKeigue, P.M., Carpenter, J.R., Parra, E.J., and Shriver, M.D. (2000). Estimation of admixture and detection of linkage in admixed populations by a Bayesian approach: application to African-American populations. *Annals of Human Genetics* 64 (Pt 2): 171–186.

46 Henn, B.M., Botigue, L.R., Gravel, S. et al. (2012). Genomic ancestry of North Africans supports back-to-Africa migrations. *PLoS Genetics* 8 (1): e1002397.

47 Maples, B.K., Gravel, S., Kenny, E.E., and Bustamante, C.D. (2013). RFMix: a discriminative modeling approach for rapid and robust local-ancestry inference. *American Journal of Human Genetics* 93 (2): 278–288.

48 Montana, G. and Pritchard, J.K. (2004). Statistical tests for admixture mapping with case-control and cases-only data. *American Journal of Human Genetics* 75 (5): 771–789.

49 Torgerson, D.G., Capurso, D., Mathias, R.A. et al. (2012). Resequencing candidate genes implicates rare variants in asthma susceptibility. *American Journal of Human Genetics* 90 (2): 273–281.

50 Nakatsuka, N., Moorjani, P., Rai, N. et al. (2017). The promise of discovering population-specific disease-associated genes in South Asia. *Nature Genetics* 49 (9): 1403–1407.

51 Montinaro, F., Busby, G.B., Pascali, V.L. et al. (2015). Unravelling the hidden ancestry of American admixed populations. *Nature Communications* 6: 6596.

52 Bray, G.A. (2004). The epidemic of obesity and changes in food intake: the fluoride hypothesis. *Physiology & Behavior* 82 (1): 115–121.

53 Hunter, D.J. (2005). Gene-environment interactions in human diseases. *Nature Reviews Genetics* 6 (4): 287–298.

8

Biologic Factors and Molecular Determinants in Inflammatory and Metabolic Diseases

Marquitta White[1], Esteban Burchard[1], Page Goddard[1], Anup K. Nair[2], Leslie J. Baier[2], Federico Fuentes[3], and Jeffrey B. Kopp[3]

[1] *University of California, San Francisco, CA, USA*
[2] *Phoenix Epidemiology and Clinical Research Branch, National Institute of Diabetes and Digestive and Kidney Diseases, National Institutes of Health, Phoenix, AZ, USA*
[3] *Kidney Diseases Branch, National Institute of Diabetes and Digestive and Kidney Diseases, National Institutes of Health, Bethesda, MD, USA*

8.1 Introduction and Approaches

The 2017 Global Burden of Disease [1, 2] study reported a decrease in the number of deaths from communicable, maternal, neonatal, and nutritional causes compared to the results in its 1990 study. There were also increased age-standardized death rates for several conditions. These included diabetes, other endocrine disorders, chronic kidney disease (CKD), cirrhosis, mental disorders, trauma, and musculoskeletal diseases. This chapter will address health disparities among some of these disorders and others. Figure 8.1 illustrates the disparity ratios, defined as disease prevalence in the most affected population group compared to the least affected population group, versus the estimated numbers of affected individuals, for several conditions that will be described in this chapter.

8.2 Asthma

Asthma is a pulmonary disorder characterized by chronic inflammation of the airways and is the most common chronic disease in children. It is a complex disease affected by genetic, environmental, and psychosocial factors. Asthma currently affects 334 million people worldwide, and recent prevalence estimates suggest that the global burden of asthma is increasing in the pediatric population [4]. Historically, asthma has been defined as a single condition; however, recent research suggests that asthma is a syndrome composed of distinct but overlapping symptom groups or endotypes [5].

Short-acting Beta2-agonist (SABA) drugs are considered first-line therapy for asthma, regardless of disease severity. The most commonly prescribed asthma medication in the

The Science of Health Disparities Research, First Edition. Edited by Irene Dankwa-Mullan, Eliseo J. Pérez-Stable, Kevin L. Gardner, Xinzhi Zhang, and Adelaida M. Rosario.

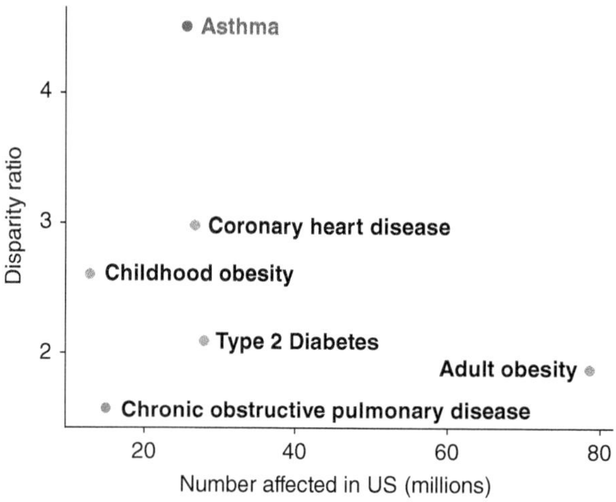

Figure 8.1 Common diseases characterized by health disparities, with the abscissa showing the number of individuals affected and the ordinate showing the ratio of incidence for the most affected group divided by the incidence for the least affected group. *Source:* Oh et al. [3] © 2016, American Thoracic Society.

world is albuterol (SABA) [6]. However, response to albuterol varies by racial/ethnic group. Puerto Ricans and African Americans display significantly lower bronchodilator response compared to individuals of Mexican and European descent [7]. Other medications commonly used to treat asthma include oral and inhaled corticosteroids, leukotrienes, and long-acting Beta2-agonists (LABAs). Due to the increased risk of asthma-related morbidity and mortality, LABAs are used only in conjunction with another asthma controller medication (such as inhaled corticosteroids) and only in instances where patient symptoms are not well controlled using other medications [6].

Genetic factors have a significant impact on variation in asthma, with asthma heritability estimates as high as 0.92 in certain populations [5]. Genome-wide association studies (GWAS) and candidate gene studies have implicated hundreds of genes in asthma susceptibility and/or morbidity, but the majority of this research has focused on populations of European origin [8].

Asthma-related genes that have emerged consistently across studies can be segregated into four groups: (i) innate immunity and immunoregulation (*CD14, TLR2, TLR4, TL6, TL10, IL10, TGFβ*, HLA class II molecules); (ii) Th2 cell activation and differentiation genes (*IL3)*; (iii) mucosal immunity genes (*DPP10)*; and (iv) genes involved in lung function and airway remodeling (*ADRB2, TNF*) [5]. Combined, the loci discovered through candidate gene studies and GWAS explain only a small portion of the estimated heritability of asthma, suggesting the existence of undiscovered genetic associations. It has been suggested that rare variation and/or gene by environment interaction effects may account for a portion of this "missing" heritability.

Clinical, epidemiological, and toxicological studies have established the strong impact of environmental factors on asthma susceptibility and disease progression. The most common environmental risk factors for asthma include tobacco smoke exposure (including from secondhand smoke); exposure to air pollutants (e.g., sulfur dioxide, nitrogen dioxide, particulate matter); exposure to allergens (e.g., cockroaches, dust mites, pet dander, dust); and respiratory viral infections [9]. In the United States, environmental risk factors tend to be more prominent in urban areas and/or areas occupied by populations with low socioeconomic status.

Psychosocial factors such as low socioeconomic status, emotional stress, and experiences of discrimination have all been associated with increased asthma risk and morbidity [10, 11]. Thakur et al. demonstrate that the perception of racial discrimination is significantly correlated with increased asthma risk and severity among African American children and adolescents [10]. Animal experiments indicate that stress may increase asthma exacerbation risk via modulations to the immune system and cytokine levels, leading to increased inflammation. It has been hypothesized that psychosocial stress may exacerbate the effects of environmental exposures, such as pollution, on asthma and inflammation, either through epigenetic modifications or alterations of biological function at the protein or pathway level [10].

Asthma prevalence, morbidity, and mortality vary significantly by population. In the United States, asthma is the most disparate common chronic disease among racial/ethnic groups (Figure 8.1). Asthma prevalence is highest in Puerto Rican (36.5%) and African American (13.4%) children, and lowest in White (7.6%) and Mexican American children (7.5%) [8]. Asthma mortality rates are four times higher in Puerto Rican children, and three times higher in African American children, compared to White children [8]. These disparities may be reflective of the increased disease severity and decreased asthma drug response displayed by Puerto Rican and African American children with asthma [7].

The populations that carry the highest disease burden, Puerto Ricans and African Americans, have been historically underrepresented in asthma studies. African Americans and Latinos constitute over 30% of the American population, but represent less than 4% of modern GWAS [12]. The few studies that have focused on minority populations have identified ethnic-specific loci associated with asthma [13]. In a 2012 meta-analysis, Torgerson et al. identified an asthma association unique to individuals with African ancestry at the PYHIN1 gene locus, which is part of a gene family linked to autoimmunity [13]. White et al. identified a novel variant in the patched domain containing 3 gene (PTCHD3) that is associated with increased asthma susceptibility in African Americans. Additionally, White et al. noted that only about 5% of asthma genetic associations previously reported in European populations are replicated in African American youth [12].

Genetic, environmental, and psychosocial risk factors coupled with the recent emergence of asthma endotypes contribute to the complexity of asthma. Presently, 47% of US children are members of racial/ethnic minority groups, the two largest being Latinos and African Americans. By 2020, it is estimated that the majority of children in the United States will be members of racial/ethnic minority groups [14]. The skew of asthma prevalence and mortality toward minority populations, coupled with the shifting demographics of the pediatric population in the United States, underscores the need to develop therapeutic interventions to improve asthma outcomes in children. With the bulk of asthma research and pharmacogenetic knowledge centered on populations of European ancestry, much work has yet to be done to improve asthma treatment and management for the populations carrying the majority of the disease burden (Latinos and African Americans). Advancements in asthma medications and pharmacogenetics must consider the biological impact of genetic, social, and environmental risk factors on patient outcomes. In an era of precision medicine, special attention should be given to the identification of population-specific risk factors and interventions, particularly for populations that carry a disproportionately high disease burden.

8.3 Metabolic Syndrome, Obesity, and Diabetes

Metabolic syndrome, obesity, and type 2 diabetes (T2D) are common diseases in developed and developing countries, but even among people living in a similar environment, their prevalence is ethnicity dependent. For example, data from the US National Health and Nutrition Examination Survey between 2007 and 2012 show that the prevalence of metabolic syndrome was highest in non-Hispanic White women (35.5%) and lowest in Mexican American men (27.8%) [15]. The prevalence of obesity among US adults between 2011 and 2014 was 36.5% with the highest prevalence among non-Hispanic black adults (48.1%) and lowest among non-Hispanic Asian adults (11.7%) [16]. According to the National Diabetes Statistics Report, the age-adjusted T2D prevalence among US adults was highest in American Indians from the Southwest region (22.2%) and lowest among the non-Hispanic White population [17].

Metabolic syndrome is defined as the presence of at least three of five medical conditions: abdominal obesity, hypertension, elevated fasting plasma glucose, high serum triglycerides, and low high density lipoprotein (HDL) levels. Most of these conditions are attributable to unhealthy weight gain; therefore, many risk factors for metabolic syndrome and obesity are similar. Both obesity and metabolic syndrome aggregate in families. Household members typically share lifestyle factors known to affect body weight, such as food choices and daily habits such as exercise. However, studies of twins have provided evidence that approximately two-thirds of the variability in obesity is attributable to genetics [18]. Obesity is also influenced by both energy intake and energy expenditure [19]. The largest determinant of energy expenditure is body composition (e.g., fat versus fat-free mass), particularly fat-free mass, which accounts for 80% of the variance in resting metabolic rate [20]. Body mass index (BMI), which is the most common index for obesity, does not describe body composition or body fat distribution (waist circumference, waist-hip ratio, or visceral fat) and studies have reported a discrepancy in the classification of obesity based on either body fat percentage or BMI [21]. The relationship between BMI, body composition, and body fat distribution also varies based on age, sex, and ethnic origin [22]. Therefore, ethnic-specific criteria are also used to define healthy weight. For example, a study found that women of Asian descent are at double the risk of developing T2D compared to White women with similar BMI [23]. This may be attributable to 3–5% higher total body fat in Asians compared to Europeans with the same BMI [24].

Metabolic syndrome and obesity are major risk factors for T2D, presumably due to the increased insulin resistance seen in these conditions. Insulin resistance corresponds to the inability of insulin-responsive tissues to respond to physiological levels of insulin, resulting in impaired glucose tolerance. Increased visceral adiposity is believed to be a major risk factor for the development of insulin resistance. To compensate for insulin resistance, pancreatic beta cells secrete more insulin, which leads to hyperinsulinemia. An inability of beta cells over time to compensate for the degree of insulin resistance leads to hyperglycemia and overt T2D.

However, there are population-specific differences in the contribution of insulin resistance and beta cell function to T2D pathogenesis. For example, African Americans have a high prevalence of T2D. It is assumed that obesity and dietary habits contribute to the high rates of T2D seen in this population. Studies have shown that early diabetes-related pathologies, such as hyperinsulinemia and insulin resistance, already exist in early childhood in this population and are independent of dietary habits [25]. According to the Randle hypothesis, a higher fatty acid content leads to hyperinsulinemia and insulin resistance. However, African American children have relatively lower plasma fatty acid and triacylglycerol concentration compared to White children. African American children also have higher HDL-C (HDL cholesterol) but are more insulin resistant and have higher prevalence of T2D, despite upregulated beta cell function [25, 26]. It has also been suggested that higher insulin secretion may be a cause rather than consequence of T2D in this population [26].

This leads to the question of what the primary cause of insulin resistance and upregulated beta cell function in African Americans is, and whether targeted intervention during early childhood may lead to a better outcome during adult life.

Genetic factors also play a key role in the manifestation of these conditions and the difference in their prevalence among different populations. The thrifty gene hypothesis proposed by geneticist James V. Neel remains an attractive but debatable explanation for the rise in conditions like obesity and T2D. A salient example to support this was the identification of a mutation in CREBRF (which encodes CREB3 Regulatory Factor) in the Samoan population that increases the risk for obesity [27]. This mutation, predominantly found in the Samoan population, is extremely rare in other populations. In cellular models, this mutation leads to the generation of an energy efficient adipocyte, providing carriers of this mutation an advantage during famine [27]. Intriguingly, this mutation is also associated with a 1.6-fold lower risk of T2D and shows evidence of positive selection, which might explain why this mutation has not been eliminated from the Samoan population [27].

GWAS have identified hundreds of loci that associate with T2D, obesity, and metabolic syndrome, and which differ in frequency and effect size in different populations [28]. Among these, the TCF7L2 and FTO loci are the most replicated loci for T2D and obesity, respectively. However, targeted studies have also identified population-specific mutations. Examples of ethnic-specific mutations include: (i) a mutation in ABCC8, which encodes the ATP sensitive potassium channel involved in insulin secretion, in American Indian populations from Arizona (twofold increase in T2D risk), (ii) SLC16A11 mutations in Mexican Americans and American Indians, which increase T2D risk, particularly in lean individuals, (iii) TBC1D4 mutation, which encodes AS160 and participates in insulin-mediated GLUT4 translocation in Greenlandic Inuit populations (10-fold increase in T2D risk), and (iv) HNF1A mutation, identified in the Canadian Oji-Cree population (fourfold increase in T2D risk) [29].

The manifestation of T2D and obesity in adult life is also influenced by epigenetic mechanisms. Epidemiological studies with longitudinal designs have shown a vital role of the intrauterine environment in adulthood, providing evidence that the offspring of women who had diabetes during pregnancy were more obese and had a higher prevalence at a younger age of the onset of T2D [30]. Changes in DNA methylation have been proposed as possible mechanisms by which the intrauterine environment affects disease pathogenesis during adulthood. Studies have now started to identify genes that are differentially methylated (e.g., LHX3) in offspring of mothers with diabetes during pregnancy compared to offspring of mothers without diabetes during pregnancy [31]. T2D risk is also modified by an interaction between epigenetic and genetic mechanisms. For example, the KCNQ1 locus is known to be imprinted, whereby only the maternal allele is expressed. Variation at this locus can increase risk for T2D only if carried on the maternal allele; no effect on T2D is observed if it is paternally inherited [32].

The prevalence of metabolic syndrome, obesity, and T2D are highest in minority populations, yet the clinical definitions of these diseases and studies to identify the genetic basis of these diseases have been done predominantly in individuals of European descent. However, evidence suggests that the molecular, genetic, and epigenetic determinants of these diseases differ by ethnicity. Therefore, research to determine the determinants of T2D and obesity in minority populations who suffer disproportionately from these conditions remains an area of high priority for future research.

8.4 Lupus and Other Rheumatologic Diseases

Systemic lupus erythematosus (SLE) is a chronic, multisystem autoimmune disorder characterized by great clinical heterogeneity. The broad nature of clinical manifestations and outcomes, from vague symptoms to life-threatening organ damage, makes diagnosis and treatment of this disease a challenge. The current diagnostic criteria have been validated by the Systemic Lupus International Collaborating Clinics (SLICC) group. They include 17 criteria, encompassing clinical, biochemical, and immunological features, which together achieve a diagnosis sensitivity of 94% and specificity of 93% [33]. The specific etiology has not been clarified yet, but there is solid evidence that both genetic and nongenetic factors act as triggers. The natural course of SLE typically shows alternating periods of activity or exacerbation and remission. Therefore, SLE outcomes can be unpredictable and may range from sustained remission to severe organ damage or death [34].

The pathogenesis of lupus remains only partially understood. The cornerstone is the loss of immunologic tolerance to self-components as the outcome of diverse abnormalities of the immune system. This leads to the production of autoantibodies and the formation of immune complexes against ubiquitous nuclear antigens, culminating in diffuse organ damage [35]. Apoptotic or necrotic cells, independently of the initiating event, may expose multiple intracellular antigens on their surface that can be recognized as pathogenic and presented to antigen-presenting cells, which can also show aberrant activity. These antigens are then introduced to T cells that promote the stimulation of hyperactive B cells, leading to the production of autoantibodies. Moreover, abnormalities in the clearance of apoptotic cells, seen, for example, with deficiencies in complement factors C1Q or C4, can be present in SLE as well as in other autoimmune processes, provoking the exacerbation of autoimmunity. Some of the most prevalent autoantibodies involved in SLE pathogenesis are anti-double-stranded DNA (dsDNA; 78% of cases), anti-Sm, anti-nuclear ribonucleoprotein (RNP), anti-Ro/SSA, and anti-La/SSB. Despite the fact that these autoantibodies have been described as reliable markers of the disease activity and are used for diagnosis and monitoring of SLE, they can be present even years before its clinical onset [36].

Lupus is notable for marked health disparities in three domains. First, lupus particularly affects women (female/male ratio 9:1) during childbearing age, although it can affect males and people of all ages [36, 37]. The prevalence of this disease in the general population ranges from 20 to 70 per 100 000 persons with an incidence of 1 to 10 per 100 000 [35, 37]. The incidence rate may be rising due to the diagnosis of milder forms of the disease based on the current accepted criteria and the increased likelihood of survival since the advent of new therapeutic agents [36]. The conspicuous difference in SLE prevalence in favor of female patients (9:1) might be partially explained by direct effects of sex chromosomes or by endogenous hormones indirectly affecting them. Female hormones interact at multiple steps with the immune system—for example, with T and B cells—which contributes to autoimmune reactivity. Supporting this, some cohorts have reported that women who had an early menarche (age ≤ 10 years) or exposure to estrogen-based therapies (hormonal replacement or oral contraceptives) manifest a twofold increased risk of developing SLE [35]. The latter statement also endorses the decrease in SLE incidence rate in prepubertal or postmenopausal women. Several authors have postulated a protective role of some androgens, like dehydroepiandrosterone (DHEA), which exacerbates the gender difference [38]. The clinical manifestations and outcomes can also be different between genders. Lower rates of serositis and higher rates of arthritis, oral ulcers, and skin involvement have been reported in females with SLE, while males have a worse overall prognosis [35, 38].

Second, the onset of lupus manifests a distinct age profile. Pediatric SLE (<16 years of age) is a very infrequent condition, with a reported annual incidence rate of 1 per 100 000 in North America [37]. Among European populations, this rate is approximately 8% according to the extensive Euro-Lupus cohort [39]. These patients usually show a higher incidence of hematologic, neurologic, renal, and psychiatric conditions [37]. On the other hand, those SLE cases with an age of onset over 50 years of age (9% according to the Euro-Lupus cohort) usually show a more benign course with lower degrees of activity and organ damage, perhaps due to the senescence of the immune system. However, these patients may have a poorer prognosis based on the accrued comorbidities associated with this population [37, 39]. It is of note that in both subsets gender predominance was less evident, with 4:3 for pediatric-onset SLE and 5:1 for the older group [39].

Third, people with Asian or African ancestry/background are at higher risk of developing SLE, with a two- or threefold increase in prevalence rates. The same fact was reported for persons of Latino background in the United States [35–37]. These findings support the notion that there is an intimate linkage between genetic and nongenetic factors in SLE pathogenesis, because despite the greater frequency among people of African descent, the disease is infrequent in West Africa. This situation may be explained by different environmental conditions, including solar exposure [35–37]. Various cohorts reported that Latino, Asian, and African-descent SLE patients are more prone to earlier age onset and greater disease activity with multiple organ system damage, including serosal, hematological, neurological, and renal involvement. Also, the rates of end-stage kidney disease (ESKD) due to lupus nephritis are higher among African-descent and Latino patients (RR [relative risk] of 3.12 and 1.18, respectively; $p < 0.05$) leading to higher rates of ESKD [34].

Genetic factors contribute to lupus pathogenesis, as first-degree relatives of SLE patients have an increased risk of developing lupus (RR of 315.94 for twins, 23.68 for siblings, 14.42 for offspring, and 11.44 for parents) [35]. Many genetic alterations have been associated with SLE. Several susceptibility genes have been identified via whole-genome scans of families. A specific interval on the long arm of chromosome 1 (1q23-24) and also some genes involved in T- and B-cell activation pathways, such as IRF5, STAT4, IRAK1, TREX1, TLR8, PTPN22, TNFSF4, PDCD1, BANK1, BLK, LYN, and PTPN22, have been linked to the development of SLE [36]. On the other hand, the rate of SLE among monozygotic twins is only 25%, reinforcing the idea that SLE may develop in genetically prone patients exposed to particular environmental conditions.

One the most widespread factors is exposure to ultraviolet radiation (UV), mainly UV-B and UV-A2. UV light induces apoptosis of skin cells, leading to the release of antigens that can trigger an inflammatory response, which can include B-cell activation and the production of several autoantibodies [40]. On the other hand, avoidance of sunlight exposure may lower vitamin D levels. Cross-sectional studies have reported a direct relationship between vitamin D deficiency and SLE activity, possibly due to the lack of an inhibitory modulation on T and B cells. Further, vitamin D levels are lower in African American SLE patients than in Caucasians, which may be explained by the reduced cutaneous conversion of vitamin D precursors among these patients [40].

Drug-induced lupus nephritis may follow exposure to hydralazine, procainamide, quinidine, isoniazid, and minocycline. The syndrome usually develops several months after drug initiation. Manifestations can remit once the drug is discontinued. In contrast to other forms of lupus, drug-induced lupus is more frequent among older patients and less frequent among females and African Americans. The clinical manifestations are usually mild, mostly cutaneous (5–40%) and in the joints (80–95%), and rarely constitute renal (0–5%) or neurological damage (0–2%). It is noteworthy that elevated titers of anti-histone antibodies are present in 95% of DILE (drug-induced lupus erythematosus) cases [35, 41]. Thus, it is clear that health disparities arising from genetic and nongenetic factors should be considered when analyzing SLE epidemiology, pathogenesis, clinical manifestations, and outcomes.

8.5 Kidney Disease

CKD is associated with substantial morbidity and mortality and is a rising global problem as populations age and experience the afflictions of abundance, including obesity, hypertension, and diabetes. In the Global Burden of Disease study cited in the introduction, for the period 1992 to 2013, the increase in CKD was 37%, compared to 9% for diabetes. Thus, CKD represents a major health problem, and this is true in both developed and least-developed countries.

The US Centers for Disease Control and Prevention (CDC) report that 2% of the adult population (5 million) have diagnosed CKD. The CDC further estimates that 15% of US adults, representing 15 million people, have CKD [41]. In developed countries, treatment of CKD and ESKD is a major social and economic burden. In the United States in 2014, 600 000 individuals were receiving treatment, at an aggregate annual cost of $30 billion. The total Medicare expenditure per person, which would include dialysis and hospitalizations and medications, was nearly $90 000 per year.

Diabetes is the major cause of kidney disease in the developed world and is an increasing cause of CKD in the developing world. Other causes of CKD include systemic diseases (e.g., diabetes, lupus, and vasculitis), primary glomerular diseases (e.g., glomerular diseases such as focal segmental glomerulosclerosis, membranous nephropathy, and IgA nephropathy), tubular diseases (e.g., medication toxicity and chronic tubular interstitial nephritis), and genetic disorders (e.g., polycystic kidney disease).

CKD is associated with increased risk for cardiovascular disease. This association is multifactorial, including the effects of hypertension, hyperlipidemia, and retained biomolecules (e.g., indoxyl sulfate), which are vascular toxins. When individuals have diabetes mellitus and CKD, the adverse cardiovascular effects act in a combinatorial fashion to drive vascular injury. In high- and middle-income countries, the peak incidence of CKD occurs after age 50. The combined effects of diabetes, hypertension, obesity, hyperlipidemia, and CKD cause death in many individuals with CKD before reaching ESKD, especially individuals with comorbidities and particularly from cardiovascular causes.

Therapies to slow the progression of CKD include treatment of the underlying disease, management of blood pressure and lipid disorders, and the use of progression-slowing medications, such as renin-angiotensin antagonist therapy. In least-developed countries (LDC), access to medical care is often limited, so diagnosis may be delayed and therapy may be too expensive for many citizens. Currently, the number of CKD deaths in LDCs include 24 in sub-Saharan Africa, nine in Asia, four in the Pacific Islands, and one in the Americas (Haiti). Thus, countries in sub-Saharan Africa are disproportionately represented among LDC.

Several US minority populations are at notable increased risk for CKD and ESKD, including African Americans, Latinos, and Native Americans. Increased rates of diabetes account for part of this disparity, but other factors contribute as well. Several approaches have been taken to identify root causes of these disparities.

First, genetic factors have been explored, as noted above. In one salient example, genetic variants in *APOL1*, encoding apolipoprotein L1, were identified in 2010 as an important cause of increased rates of nondiabetic kidney disease among populations of sub-Saharan African descent [42].

Second, toxins in the physical environment may contribute to CKD pathogenesis. One particularly salient example is the identification of aristolochic acid in the Aristolochia plant as the cause of Balkan nephropathy. Agricultural workers in tropical regions around the world are at increased risk for progressive CKD of undetermined etiology [43]. The available reports, particularly from Nicaragua (where it has been termed meso-American nephropathy) and Sri Lanka, suggest tubulointerstitial nephritis. While the causes remain elusive, reports have suggested a role for heat stress and for heavy metals, including arsenic. Other heavy metals identified with chronic exposure that have been implicated in CKD causation or progression include arsenic, cadmium, lead, and uranium. For all of these, the evidence remains short of convincing. Future studies might change this conclusion. The potential for nephrotoxicity may be of particular concern for exposed children [44].

Third, toxins and other factors in the social environment may contribute to CKD. Heroin was formerly strongly associated with CKD. Contributors, including possibly HIV, are known to include genetic variants in *APOL1*, as discussed above. Cocaine is a nephrotoxin, as acute ingestion has been associated with renal infarcts and interstitial nephritis.

Levamisole, sometimes added as an adulterant to cocaine, is associated with vasculitis. At the neighborhood level, many volatile organic compounds are present both outdoors and inside homes. This is particularly true of neighborhoods, often of low socioeconomic status, situated near industrial plants, including petrochemical operations. These compounds include many volatile organic chemicals, such as benzene, toluene, xylenes, styrene, chloroform, limonene, and pinenes [45]. They are found in gasoline vapors, diesel exhaust, moth repellants, and cleaning products. While some of these compounds accumulate in CKD, little is known about potential nephrotoxicity. At the personal level, cigarette smoking and exposure to household cigarette smoking is associated with CKD and with CKD progression.

Fourth, dietary factors are associated with CKD and with CKD progression. Excess calories promote excess weight and obesity, which are risk factors for CKD and for hypertension and diabetes, both risk factors for CKD. Similarly, excess dietary sodium promotes hypertension and limits the anti-fibrotic benefits of renin-angiotensin-aldosterone antagonist therapy in CKD.

Fifth, social and psychological factors may contribute to CKD incidence and, particularly, to its progression. Bruce and colleagues propose that the social environment is a contributor in the pathway from CKD risk to CKD. They discuss the role of depression, anxiety, anger/hostility, stress, and social relations [46]. They note that racial discrimination and institutionalized racism remain facts of life for many African Americans, despite the improvements brought about by the civil rights movement. Many of these stressors likely affect other populations, including undocumented immigrants and economically disadvantaged individuals of all backgrounds. Clarifying the role of these social and psychological factors, effective coping mechanisms, and biological pathways are high priorities for future investigations.

8.6 Key Points

- Asthma, arthritis, metabolic syndrome, diabetes, and chronic kidney disease manifest different prevalence rates among different racial and ethnic populations.
- These are complex diseases where genetics, environment, and social factors contribute to disease manifestations and clinical outcomes.
- The discovery of APOL1 genetic variants, common among African Americans, has provided insight into the excess risk for chronic kidney disease in this population.

Disclaimer

The views and opinions expressed in this chapter are those of the authors only and do not necessarily represent the views, official policy, or position of the U.S. Department of Health and Human Services or any of its affiliated institutions or agencies.

References

1 Naghavi, M., Wang, H., Lozano, R. et al. (2015). Global, regional, and national age-sex specific all-cause and cause-specific mortality for 240 causes of death, 1990–2013: a systematic analysis for the global burden of disease study 2013. *Lancet* 385 (9963): 117–171.

2 Naghavi, M., Abajobir, A.A., Abbafati, C. et al. (2017). Global, regional, and national age-sex specific mortality for 264 causes of death, 1980–2016: a systematic analysis for the global burden of disease study 2016. *Lancet* 390 (10100): 1151–1210.

3 Oh, S.S., White, M.J., Gignoux, C.R., and Burchard, E.G. (2016). Making precision medicine socially precise. Take a deep breath. *American Journal of Respiratory and Critical Care Medicine* 193 (4): 348–350.

4 Asher, I. and Pearce, N. (2014). Global burden of asthma among children. *International Journal of Tuberculosis and Lung Disease* 18 (11): 1269–1278.

5 Mathias, R.A. (2014). Introduction to genetics and genomics in asthma: genetics of asthma. In: *Advances in experimental medicine and biology*, vol. 79 (ed. A. Brasier), 125–155. Boston: Humana Press.

6 Shahidi, N. and Fitzgerald, J.M. (2010). Current recommendations for the treatment of mild asthma. *Journal of Asthma and Allergy* 3 (2010): 3169–3176.

7 Naqvi, M., Thyne, S., Choudhry, S. et al. (2007). Ethnic-specific differences in bronchodilator responsiveness among African Americans, Puerto Ricans, and Mexicans with asthma. *Journal of Asthma* 44 (8): 639–648.

8 Walker, B. Jr., Stokes, L.D., and Warren, R. (2003). Environmental factors associated with asthma. *Journal of the National Medical Association* 95 (2): 152–166.

9 Yonas, M.A., Lange, N.E., and Celedon, J.C. (2012). Psychosocial stress and asthma morbidity. *Current Opinion in Allergy and Clinical Immunology* 12 (2): 202–210.

10 Thakur, N., Barcelo, N.E., Borrell, L.N. et al. (2017). Perceived discrimination associated with asthma and related outcomes in minority youth: the GALA II and SAGE II studies. *Chest* 151 (4): 804–812.

11 Popejoy, A.B. and Fullerton, S.M. (2016). Genomics is failing on diversity. *Nature* 538 (7624): 161–164.

12 White, M.J., Risse-Adams, O., Goddard, P. et al. (2016). Novel genetic risk factors for asthma in African American children: precision medicine and the sage ii study. *Immunogenetics* 68 (6–7): 391–400.

13 Torgerson, D.G., Ampleford, E.J., Chiu, G.Y. et al. (2011). Meta-analysis of genome-wide association studies of asthma in ethnically diverse North American populations. *Nature Genetics* 43 (9): 887–892.

14 Colby, S.L. and Ortman, J.M. (2014). Projections of the size and composition of the US Population: 2014 to 2016. In: *Current Population Reports*, 25–1143. Washington: US Census Bureau.

15 Moore, J.X., Chaudhary, N., and Akinyemiju, T. (2017). Metabolic syndrome prevalence by race/ethnicity and sex in the United States, national health and nutrition examination survey, 1988–2012. *Preventing Chronic Disease* 14: E24.

16 Ogden, C.L., Carroll, M.D., Fryar, C.D., and Flegal, K.M. (2015). Prevalence of obesity among adults and youth: United States, 2011–2014. In: *NCHS Data Brief, no. 219*. Hyattsville: National Center for Health Statistics.

17 Centers for Disease Control and Prevention (2017). Estimates of Diabetes and its Burden in the United States, National Diabetes Statistics Report. (accessed July 1, 2020) https://www.cdc.gov/diabetes/library/features/diabetes-stat-report.html.

18 Stunkard, A.J., Foch, T.T., and Hrubec, Z. (1986). A twin study of human obesity. *The Journal of the American Medical Association* 256 (1): 51–54.

19 Piaggi, P., Vinales, K.L., Basolo, A. et al. (2018). Energy expenditure in the etiology of human obesity: spendthrift and thrifty metabolic phenotypes and energy-sensing mechanisms. *Journal of Endocrinological Investigation* 41 (1): 83–89.

20 Ravussin, E., Lillioja, S., Anderson, T.E. et al. (1986). Determinants of 24-hour energy expenditure in man. Methods and results using a respiratory chamber. *Journal of Clinical Investigation* 78 (6): 1568–1578.

21 Peltz, G., Aguirre, M.T., Sanderson, M., and Fadden, M.K. (2010). The role of fat mass index in determining obesity. *American Journal of Human Biology* 22 (5): 639–647.

22 Jackson, A.S., Stanforth, P.R., Gagnon, J. et al. (2002). The effect of sex, age and race on estimating percentage body fat from body mass index: the heritage family study. *International Journal of Obesity and Related Metabolic Disorder* 26 (6): 789–796.

23 Shai, I., Jiang, R., Manson, J.E. et al. (2006). Ethnicity, obesity, and risk of type 2 diabetes in women: a 20-year follow-up study. *Diabetes Care* 29 (7): 1585–1590.

24 Mehta, S., Mahajan, D., Steinbeck, K.S., and Bermingham, M.A. (2002). Relationship between measures of fatness, lipids and ethnicity in a cohort of adolescent boys. *Annals of Nutritional Metabolism* 46 (5): 192–199.

25 Kasim-Karakas, S.E. (2000). Ethnic differences in the insulin resistance syndrome. *The American Journal of Clinical Nutrition* 71 (3): 670–671.

26 Hasson, B.R., Apovian, C., and Istfan, N. (2015). Racial/ethnic differences in insulin resistance and beta cell function: relationship to racial disparities in type 2 diabetes among African Americans versus Caucasians. *Current Obesity Report* 4 (2): 241–249.

27 Minster, R.L., Hawley, N.L., Su, C.T. et al. (2016). A thrifty variant in CREBRF strongly influences body mass index in Samoans. *Nature Genetics* 48 (9): 1049–1054.

28 Nair, A.K. and Baier, L.J. (2015). Complex genetics of type 2 diabetes and effect size: what have we learned from isolated populations? *Review of Diabetes Studies* 12 (3–4): 299–319.

29 Baier, L.J., Muller, Y.L., Remedi, M.S. et al. (2015). ABCC8 R1420H loss-of-function variant in a Southwest American Indian community: association with increased birth weight and doubled risk of type 2 diabetes. *Diabetes* 64 (12): 4322–4332.

30 Pettitt, D.J. and Knowler, W.C. (1998). Long-term effects of the intrauterine environment, birth weight, and breast-feeding in Pima Indians. *Diabetes Care* 21 (Suppl 2B): 138–141.

31 Chen, P., Piaggi, P., Traurig, M. et al. (2017). Differential methylation of genes in individuals exposed to maternal diabetes in utero. *Diabetologia* 60 (4): 645–655.

32 Kong, A., Steinthorsdottir, V., Masson, G. et al. (2009). Parental origin of sequence variants associated with complex diseases. *Nature* 462 (7275): 868–874.

33 Petri, M., Orbai, A.M., and Alarcon, G.S. (2012). Derivation and validation of the systemic lupus international collaborating clinics classification criteria for systemic lupus erythematosus. *Arthritis Rheumatology* 64 (8): 2677–2686.

34 Fernandez, M., Alarcon, G.S., Calvo-Alen, J. et al. (2007). A multiethnic, multicenter cohort of patients with systemic lupus erythematosus (SLE) as a model for the study of ethnic disparities in SLE. *Arthritis Rheumatology* 57 (4): 576–584.

35 Maidhof, W. and Hilas, O. (2012). Lupus: an overview of the disease and management options. *Physical Therapy* 37 (4): 240–249.

36 D'Cruz, D.P., Khamashta, M.A., and Hughes, G.R. (2007). Systemic lupus erythematosus. *Lancet* 369 (9561): 587–596.

37 Pons-Estel, G.J., Alarcon, G.S., Scofield, L. et al. (2010). Understanding the epidemiology and progression of systemic lupus erythematosus. *Seminars in Arthritis and Rheumatism* 39 (4): 257–268.

38 Renau, A.I. and Isenberg, D.A. (2012). Male versus female lupus: a comparison of ethnicity, clinical features, serology and outcome over a 30 year period. *Lupus* 21 (10): 1041–1048.

39 Cervera, R., Khamashta, M.A., and Hughes, G.R. (2009). The euro-lupus project: epidemiology of systemic lupus erythematosus in Europe. *Lupus* 18 (10): 869–874.

40 Zandman-Goddard, G., Solomon, M., Rosman, Z. et al. (2012). Environment and lupus-related diseases. *Lupus* 21 (3): 241–250.

41 Centers for Disease Control and Prevention (2017). *National chronic kidney disease fact sheet*. Atlanta: CDC.

42 Genovese, G., Friedman, D.J., Ross, M.D. et al. (2010). Association of trypanolytic apol1 variants with kidney disease in African Americans. *Science* 329 (5993): 841–845.

43 Lunyera, J., Mohottige, D., Von Isenburg, M. et al. (2016). CKD of uncertain etiology: a systematic review. *Clinical Journal of American Society of Nephrology* 11 (3): 379–385.

44 Weidemann, D.K., Weaver, V.M., and Fadrowski, J.J. (2016). Toxic environmental exposures and kidney health in children. *Pediatric Nephrology* 31 (11): 2043–2054.

45 Batterman, S., Su, F.C., Li, S. et al. (2014). Personal exposure to mixtures of volatile organic compounds: modeling and further analysis of the RIOPA data. *Research Report on Health Effects Institute* 181: 3–63.

46 Bruce, M.A., Beech, B.M., Sims, M. et al. (2009). Social environmental stressors, psychological factors, and kidney disease. *Journal of Investigative Medicine* 57 (4): 583–589.

9

Insights into the Genomic Landscape of African Ancestry Populations

Implications for Health and Disease Disparities

Charles N. Rotimi, Amy R. Bentley, Ayo P. Doumatey, Guanjie Chen, Daniel Shriner, and Adebowale Adeyemo

Center for Research on Genomics and Global Health, National Human Genome Research Institute, National Institutes of Health, Bethesda, MD, USA

9.1 Introduction: Viewing the Complex Architecture of African Genomes from a Global Perspective

Describing the complex architecture of African genomes requires contextualizing Africa in human history. Historical, anthropological, and genomic evidence unequivocally supports Africa as the birthplace of anatomically modern humans (AMH) and the geographic origin of human expansion across the globe over the past 100 000 years [1]. AMH have lived longer on the African continent than anywhere else in the world [2]. This long evolutionary history has led to several important genetic characteristics in African ancestry populations, as revealed by the analyses of genomic data generated by such major consortia as the International HapMap, the 1000 Genomes, and the African Genome Variation Project (AGVP) [3, 4]. Compared to other global ancestry populations, these genetic characteristics include more haplotypes (Figure 9.1 [5]), lower levels of linkage disequilibrium (LD), more divergent patterns of LD, and more complex patterns of population substructure with implications for recent and ancient admixture. Equally diverse are the cultural and social characteristics of African populations, including linguistic diversity (representing approximately a third of the world's ~2000 languages), complex dietary patterns, varied farming practices, and uneven economic development [2] resulting in varying levels of urbanization and differences in the distribution of nongenetic disease risk factors.

Globally, African ancestry populations display a high level of population subdivision, with nearly as many ancestries within Africa as in the rest of the world combined (recently estimated as 11 vs. 12 [2]). Genetic differentiation (i.e., accumulation of differences in allelic frequencies due to evolutionary forces, such as selection or genetic drift) between African ancestries can exceed that between pairs of non-African ancestries. For example, differentiation measured by the fixation index (F_{ST}) is 0.054 between Khoisan and Omotic ancestries; 0.024 between Southern and Northern European ancestries, and 0.042 between

The Science of Health Disparities Research, First Edition. Edited by Irene Dankwa-Mullan, Eliseo J. Pérez-Stable, Kevin L. Gardner, Xinzhi Zhang, and Adelaida M. Rosario.

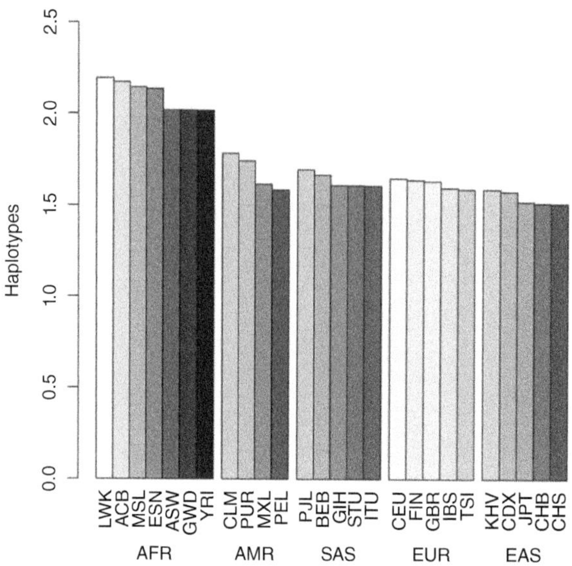

Figure 9.1 Haplotype diversity in the 1000 Genomes Project. Haplotypes were counted over windows of six consecutive diallelic SNPs, with a slide of one SNP. For each sample, all counted haplotypes have a frequency ⩾1%. See Supplementary Material, table S1 [3] for sample descriptors. *Source:* Rotimi et al. [5] © 2017, Oxford University Press. (See color representation of this figure in color inserts section.)

Arabian and Indian ancestries [1, 2]. As a function of physical genetic distance, LD decays faster in sub-Saharan African (SSA) ancestry populations than in non-African populations, resulting in shorter haplotypes among SSA ancestry populations.

These shorter haplotypes have significant implications for disease gene fine-mapping and more efficient causal disease genetic variant identification [3]. Recent admixture events—primarily, the trans-Atlantic slave trade—have resulted in genetic admixture among Africans, Europeans, and Native Americans. The average percentage of African ancestry in these populations varies considerably according to slave disembarkation points, as well as to sociocultural practices, such as ancestry-positive assortative mating. Estimates range from 87% in African Caribbeans from Barbados, 75% in African Americans in the southwestern United States, 19% in Puerto Ricans, 12% in Colombians in Medellín, 7% in people with Mexican ancestry in Los Angeles, and 4% in both Peruvians in Lima and Argentinians sampled from across Argentina [3] (Figure 9.2 [2]). In addition to these relatively recent demographic events, admixture has occurred between SSA ethnolinguistic groups and both Eurasian and hunter-gatherer populations in regionally distinctive ways [4].

Several adaptive forces, including exposure to fatal pathogens and climate, have shaped African genomes, with the resulting elevated frequencies of alleles conferring survival advantages [4]. Growing understanding of the evolution of common complex disorders is revealing how some of these alleles are maladaptive in modern-day environments within and outside the African continent. The following is a review of how African genomes and environments have shaped the distributions of diseases and variable drug response across global African ancestry populations, with implications for health disparities.

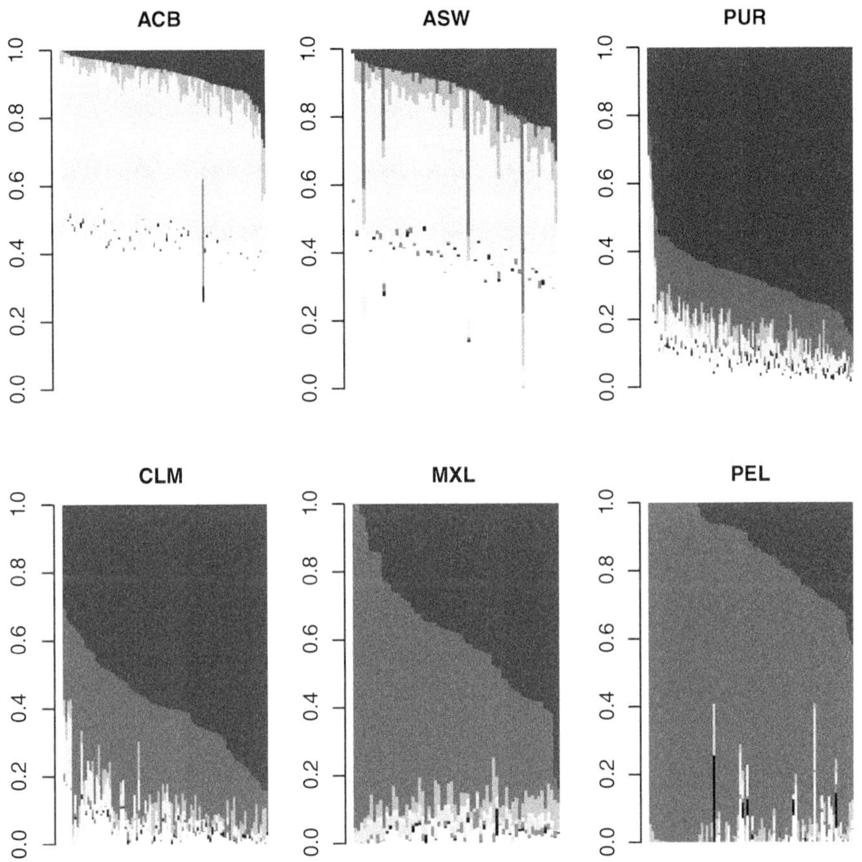

Figure 9.2 Individual admixture proportions from New World samples. The six samples are African Caribbean in Barbados (ACB), People with African Ancestry in Southwest USA (ASW), Puerto Ricans in Puerto Rico (PUR), Colombians in Medellín, Colombia (CLM), People with Mexican Ancestry in Los Angeles, California (MXL), and Peruvians in Lima, Peru (PEL) [3]. Conditional on eight ancestries, yellow corresponds to Western African ancestry, aquamarine corresponds to West-Central African ancestry, blue corresponds to Southern European ancestry, orange corresponds to Northern European ancestry, purple corresponds to Native American ancestry, red corresponds to South Asian ancestry, green corresponds to East Asian ancestry, and black corresponds to Southeastern Asian ancestry. Summing the Western and West-Central African ancestries, the range of individual admixture proportion across these six samples goes from 0 to 97.8%. Similarly, the range of individual admixture proportion for Native American ancestry goes from 0 to 100%. *Source:* Rotimi et al. [2] © 2016, Elsevier. (See color representation of this figure in color inserts section.)

9.2 Adaptive Forces that Shaped the Human Genome in Health and Disease among African Ancestry Populations

9.2.1 Protection Against Malaria

Years of continuous exposure to infectious agents on the African continent have resulted in increased frequency of alleles conferring protection against fatal pathogens [6]. Some of these genetic protections have occurred at the cost of increased susceptibility to noninfectious diseases [2]. Examples include six genes in the human genome associated with protection against malaria.

1) Mutations in the hemoglobin gene *HBB* confer protection against malaria. The HbS allele at rs334, corresponding to Glu6Val, causes sickle cell disease in homozygotes and appears on the Benin, Cameroon, Central African Republic/Bantu, Arabian/Indian, and Senegal haplotypes. HbS is absent among Native Americans and was introduced to the Americas during the trans-Atlantic slave trade. More generally, differences in the global distribution of HbS-containing haplotypes reflect differences in immigration patterns [7].
2) α and β thalassemias are the result of defects in the globin subunits HBA and HBB, respectively, and are prevalent in areas of epidemic malaria. Clinical presentation ranges from mild to severe, depending on the degree of imbalance in the subunit ratio [8].
3) Glucose-6-phosphate dehydrogenase (G6PD) deficiency, also known as favism, is an X-linked recessive inborn error of metabolism that predisposes to hemolysis and jaundice. It is the most common human enzyme defect [9]. Three G6PD deficiency alleles are associated with protection against cerebral malaria but also with increased risk for severe malaria anemia [10]. Primaquine is used to treat and prevent malaria due to *Plasmodium vivax* and *Plasmodium ovale* but can cause severe hemolytic anemia in G6PD-deficient individuals [11].
4) Positive selection in primates has been detected at *CD36*, a receptor that adheres to *Plasmodium*-infected erythrocytes to promote phagocytic clearance [12]. Mutations in *CD36* are associated with impaired fatty acid metabolism, glucose intolerance, atherosclerosis, arterial hypertension, diabetes, cardiomyopathy, and Alzheimer's disease [13].
5) The *ACKR1* null promoter mutation rs2814778 is nearly fixed in West and West-Central Africa [3]. rs2814778 is associated with protection against *P. vivax* but more severe disease in *Plasmodium falciparum* infection [14]. Local enrichment of African ancestry has been observed in Sudanese Arabs and Nubians, suggesting that the protective benefit of rs2814778 extends across the Sahel to eastern Africa [15] and to Madagascar [16]. rs2814778 is associated with benign ethnic neutropenia [17]. Benign ethnic neutropenia may be associated with increased risk of human immunodeficiency virus (HIV) infection and has been identified as a factor that may limit access to persons of African ancestry in clinical cancer trials due to the use of the absolute neutrophil count of at least 1.5×10^9 cells l^{-1} as an eligibility criterion [14].
6) Increased expression of γ-globin confers protection against malaria by increasing levels of fetal hemoglobin in individuals without HbS but not in individuals homozygous for HbS [18]. A mutation upstream of the γ-globin gene *HBG1* is associated with serum uric acid levels, which in turn are associated with increased risk of hypertension [19]. A seventh locus downstream of the cluster of glycophorin genes, which encode the MNSs blood group, contains a haplotype that confers resistance to severe malaria [20]. Although this locus shows genetic signatures of balancing selection, it may also result in ovalocytosis.

9.2.2 Protection Against African Sleeping Sickness (Trypanosomiasis)

Two genome-wide admixture mapping studies performed in African Americans identified a locus at chromosome 22q12.3 at which African ancestry conferred increased risk for idiopathic focal segmental glomerulosclerosis, HIV-associated nephropathy, nondiabetic end-stage kidney disease, and hypertensive end-stage kidney disease, but not for diabetic end-stage kidney disease [21]. Subsequent fine-mapping identified two haplotypes in *APOL1* called G1, defined by rs73885319 and rs60910145, and G2, defined by rs71785313 [22]. APOL1 carrying G2 affords protection against *Trypanosoma brucei rhodesiense* because the G2 deletion prevents binding of a trypanosomal virulence factor called serum resistance-associated protein [23]. Derived allele frequencies for rs73885319 are highest in West Africa, particularly in Nigeria and Ghana, and zero among non-African ancestry populations [6]. *APOL1* variants have been associated with increased risk of multiple nondiabetic nephropathies and also with cardiovascular disease [24]. Overcoming the fact that *APOL1* appears to exist only in a subset of primate lineages, transgenic mice with podocyte-specific inducible expression show that *APOL1* risk variants adversely affect endosomal trafficking and autophagy, leading to inflammatory-mediated podocyte death and glomerular scarring [25].

9.2.3 Protection Against Lassa Fever

Lassa virus, the cause of Lassa fever, has origins in the reservoir of the rodent *Mastomys natalensis* in Nigeria [26]. The main cellular receptor for Lassa virus is α-dystroglycan (*DAG1*) [27]. Glycosylation of α-dystroglycan by the enzyme-like acetylglucosaminyltransferase (*LARGE*) is required for binding of Old World arenaviruses, including Lassa virus [28]. Signals for natural selection have been detected at *LARGE* in populations with West African ancestry [29]. Mutations in *DAG1* and *LARGE* are associated with congenital muscular dystrophies [30], raising the possibility that an increased risk of congenital muscular dystrophies may result in an increased resistance to arenavirus infection.

9.2.4 Ecological Adaptations to Tropical Climate (Hypertension and Skin Pigmentation)

9.2.4.1 Hypertension

Over the past 50 years, a higher prevalence of hypertension has been reported in African Americans compared to European Americans. For 2011 and 2012, the Centers for Disease Control and Prevention estimated the prevalence of hypertension among all US adults to be 29.1%, with African Americans having a much higher prevalence (42.1%) compared to European Americans (28.0%) [31]. This risk for hypertension and associated comorbidities among African Americans, including stroke and heart and kidney diseases, has provoked premature ethnic-specific disease risk explanations and associated treatment approaches.

Using data from the African diaspora and other global ancestry populations, the essential nature of hypertension in African origin populations has been challenged. It is clear that the prevalence of hypertension is not uniform across African populations, with estimates ranging from a low of about 7% in rural West Africa to a high of 35% among African Americans. The Black nations of the Caribbean display intermediate rates of about 26% (Figure 9.3 [32]). Using data from 85 000 genetically related populations of African and European descent, Cooper *et al.* [33] demonstrated that the prevalence of hypertension across human populations did not follow racial/ethnic groupings. They showed that the hypertension prevalence ranged from 27 to 55% for Whites and 14 to 44% for Blacks (Figure 9.4 [33]).

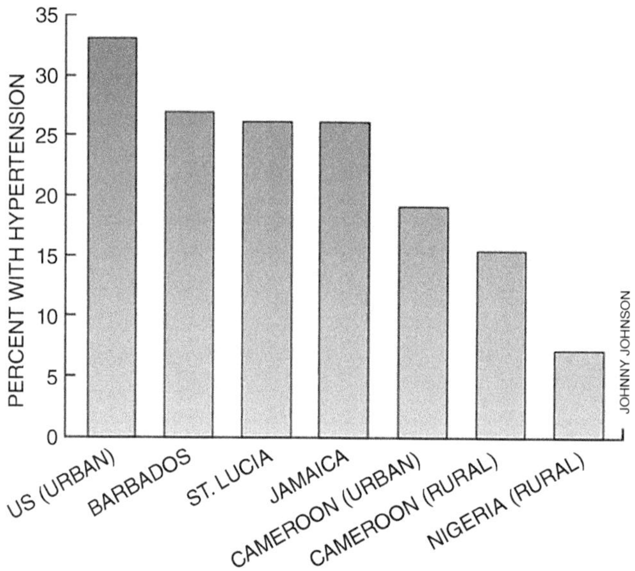

Figure 9.3 Proportion of individuals with hypertension among those of African descent. *Source:* Cooper et al. [32].

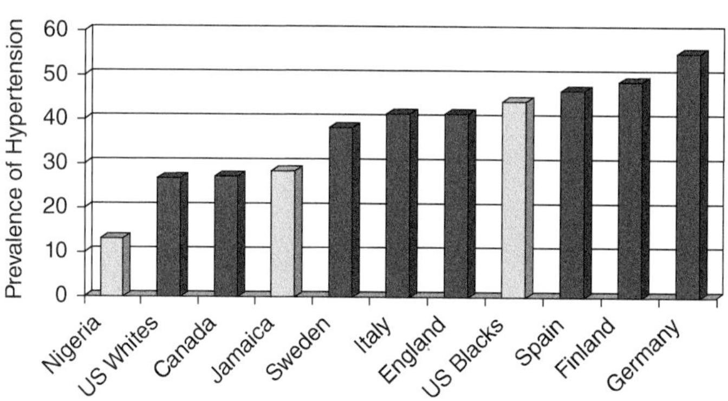

Figure 9.4 Hypertension prevalence across African and European descent populations. Hypertension is defined by blood pressure >140/90 mmHg or presence of hypertension treatment. Includes individuals aged 35–64. Results are age-adjusted. *Source:* Cooper et al. [33] © 2005, Springer Nature.

Signatures of adaptation to tropical climate are evident in the genomes of African ancestry populations. Local temperature-induced adaptive genetic changes may be one mechanism involved in hypertension and differences in salt sensitivity in humans [34]. Heat-adapted people, particularly members of the African diaspora, have greater risk of hypertension due to exposures of the modern world (e.g., increased salt intake) interacting with ancestral susceptibility [34]. Consistent with this hypothesis, signals of selection have been detected in *ATP1A1* (ATPase Na+/K+ transporting subunit alpha 1), *AQP2* (Aquaporin 2), and *CSK* (C-terminal Src kinase), genes previously implicated in hypertension and osmoregulation [4]. The ancestral allele of the *CSK* locus displaying high differentiation among African populations is strongly correlated with hypertension risk and its frequency is inversely correlated with latitude [4].

These findings support the hypothesis that adaptation to climate produced a latitudinal cline in hypertension susceptibility. If validated with more mechanistic studies, these observations could provide significant insight into the pathogenesis of hypertension. However, the ubiquity of exposure to environmental risk factors for hypertension at the global level is likely to make validating this hypothesis difficult. Of note: some of the highest rates of hypertension have been observed in non-African diaspora populations [33].

9.2.4.2 Dark Skin

Climatic adaptations in low-latitude and high-ultraviolet-radiation (UVR) regions of the tropics are thought to have maintained dark skin pigmentation to protect against UVR-induced deoxyribonucleic acid (DNA) damage and folate photolysis [35]. After modern humans migrated out of Africa, the challenges of producing vitamin D in the skin from the low UVR outside the tropics were met by natural selection acting on mutations, producing skin depigmentation more than 30 000 years ago [36]. Dark skin pigmentation is maladaptive in low-UVR environments, potentially contributing to health disparities in diseases associated with vitamin D deficiency, including several forms of cancer and cardiometabolic diseases [37]. Future mechanistic studies are needed to disentangle the effects of the ancestral alleles of genes selected for skin pigmentation from those of dietary and lifestyle changes leading to vitamin D deficiency and related comorbidities.

9.2.4.3 Evidence of Adaptation with Implications for Metabolic Processes

In a recent genome-wide analysis of individuals of Wolaita ethnicity from Ethiopia, we found evidence of positive selection in several genes involved in metabolic processes [38]. These included a novel selection signal in the *CDKAL1* gene (CDK5 regulatory subunit associated protein 1 like 1) that has been implicated in the etiology of type 2 diabetes, pancreatic β-cell function, and insulin secretion [39]. The inhibition of the CDK5 (cyclin dependent kinase 5) protein by *CDKAL1* leads to enhanced insulin secretion under conditions of high glucose levels [40]. Thus, reduced expression of *CDKAL1* inhibits insulin secretion, leading to an impaired response to glucotoxicity and increased risk for type 2 diabetes [41]. In our study, the frequency of the *CDKAL1* rs9348453 ancestral A allele in the haplotype favored by selection was over 70% in all population groups analyzed. Sequence data from the 1000 Genomes Project show that rs9348453 has a potential regulatory role in human skeletal muscle myoblast cell lines used to study diabetes and insulin resistance. Observed high correlation between rs9348453 and rs79915874 (chr6.hg19:g.21005146T>C) has potential functional relevance, given that rs79915874 has been predicted to disrupt binding motifs of the hepatocyte nuclear factor 4 transcription factor that is expressed primarily in the liver and pancreatic β cells. These findings suggest that *CDKAL1* is likely a key energy metabolism gene with the potential to provide novel insights into the pathobiology of diabetes and other metabolic traits [42].

We also found selection signals in the *PPARA* (peroxisome proliferator activated receptor alpha) gene involved in energy metabolism during prolonged food deficiency. This finding is consistent with the dietary use of enset, a crop with high levels of carbohydrate and low levels of fat and protein and with metabolic adaptation to high-altitude hypoxia, domesticated in Ethiopia subsequent to food deprivation 10 000 years ago [43]. A well-known feature of the enset plant is its drought tolerance. It was called the "tree against hunger" by European travelers to Ethiopia in the 1600s [44]. We postulate that the diet-induced positively selected haplotype we observed may inhibit *PPARA* expression, leading to a metabolic adaptation in which lipid oxidation is reduced and carbohydrate oxidation is enhanced even in times of energy deprivation. Consistent with our observation, others have reported signals of natural selection in *PPARA* in populations of diverse ancestry from widely ranging ecological regions [45].

9.2.4.4 Neglected Tropical Disease

Podoconiosis is nonfilarial elephantiasis characterized by an inflammatory reaction to mineral particles in irritant red clay soil derived from volcanic rock. HLA-DRB*0701 (major histocompatibility complex [MHC], class II, DR beta), HLA-DQA1*0201 (major histocompatibility complex, class II, DQ alpha), and HLA-DQB*0202 (major histocompatibility complex, class II, DQ beta) have been identified as risk variants, suggesting that the disease is associated with MHC class II antigen presentation [46]. Signal for recent positive selection suggests that podoconiosis-protective alleles have been favored [38]. HLA-DRB has been associated with complement C3 and C4 levels [47] and multiple sclerosis [48]. HLA-DQA has been associated with IgA nephropathy [49]. HLA-DQB has been associated with complement levels [47] and Parkinson's disease [50]. It is unknown if protection against podoconiosis occurs at the cost of increased risk for any of these diseases.

9.3 Pharmacogenomics

The vast genetic diversity in African ancestry individuals [3, 4] has pharmacogenomic consequences [6, 51]. Not only are there marked differences in allele frequencies at pharmacogenomic loci in African vs. other ancestry populations, but these variants also differ in significant ways among African ancestry populations [52]. Two examples illustrate this point:

1) The HLA-B*57:01 allele is strongly associated with serious adverse reactions to the HIV/AIDS drug abacavir. Among the Yoruba in Nigeria, the risk allele is absent, but it is carried by 14% of Kenyan Masai. Importantly, among the Luhya of Kenya, this variant is much less frequent (3%), showing that even the descriptor "Kenyan" is inadequate to describe potential pharmacogenomic risk [53].
2) Metabolism of codeine resulting from genetic variability in *CYP2D6* is another example. Because of multiple gene duplication events, some individuals carry several copies of *CYP2D6*, which leads to ultra-rapid metabolizing activity of codeine [54]. These individuals quickly convert codeine into morphine, a dangerous outcome that can lead to severe adverse reactions, including death. The global distribution of carriers varies widely, with very low rates reported in West Africa, 3% of Northern Europeans, 5–10% of Southern Europeans, and a high of about 30% of Ethiopians [54]. These data and the current lack of infrastructure to genotype individuals have led the Ethiopian Food, Medicine, and Health Care Administration and Control Authority to ban its use entirely [55].

Differences exist in pharmacogenomic variants affecting drug response in the treatment of hepatitis C virus with peginterferon and ribavirin to prevent progressive hepatic fibrosis, cirrhosis, and hepatocellular carcinoma. As has been noted by clinicians, response to treatment varies considerably, with about half of patients showing sustained virological response after the standard course of treatment. For reasons that remain unclear, African Americans have been known to be less responsive to treatment, independent of other host and viral factors. Genomic studies provide insight into observed ethnic disparities in treatment response by identifying a genetic variant in the *IL28B* gene that is an important predictor of treatment response and spontaneous clearance in patients infected with hepatitis C virus [56]. Considerable differences in allele frequencies between ancestry groups explain a substantial proportion of the observed differences in response rates between European Americans (68%), African Americans (36%), and Asians (95%). Indeed, the *IL28B* genetic variant may explain at least half of the difference in response rates observed between European Americans and African Americans receiving the same treatment.

Effective use of pharmacogenetics to increase the safety and efficacy of drug treatment will require research in diverse populations to ensure accurate identification of all relevant variants. Most genomic data currently available derive from European ancestry populations. Efforts to increase the diversity of populations participating in genomic research will help to prevent healthcare disparities in genomic medicine in the future.

9.4 Considerations for Future Studies

In the context of the growing success stories of the genomics of African ancestry populations, we highlight some issues for consideration for the design of future initiatives. First, a careful review of the literature shows that the sampling of African ancestry populations remains a patchwork process, with efforts ranging from collecting convenience samples (i.e., samples that are easy to get or already available) to the more systematic approaches implemented by the 1000 Genomes Project, AGVP, and the H3Africa project, among others. Recruitment efforts in African ancestry populations should be undertaken with an awareness of factors that may influence potential participants' trust. These factors include historical abuses inflicted upon African American research subjects and the tendency for researchers to simply collect samples on the African continent without engagement or follow-up. Effective community engagement is a key to inclusion of diverse populations through encouraging participants' trust. These approaches have been modeled in projects such as the HapMap Project [57] and MalariaGEN. While collecting the samples of diverse, engaged participants is key, care must also be taken to ensure that diversity is not missed due to selection of a genotyping platform that is not optimized for individuals of African descent. Growing awareness of this concern has led to the creation of genotyping chips designed using data from an increasing range of African samples. Notable is the H3Africa chip, based on sequence data from Africans from the largest number of ancestral groups [58].

9.5 Conclusions

Our genomes comprise an extraordinary record of the historical movements and signatures of adaptive capabilities to environmental challenges, including pathogens, diet, and climactic forces. These challenges that have led to unique characteristics in African genomes have significant implications for health and disease.

First, the notable genomic diversity in African ancestry populations means that it is not appropriate to evaluate genomic risk for African ancestry individuals based on analyses in populations of other ancestries. In fact, given the large degree of variation across African ancestry ethnic groups, extrapolations from one African ancestry group to another may be flawed. This issue is especially salient with respect to pharmacogenomics, where incorrect assumptions about genomic risk may lead to adverse reactions, including death.

Second, in some cases, adaptations to improve outcomes related to infectious disease and other environmental forces within Africa may prove maladaptive in the context of many African diasporan individuals, who may be living in environments distinct from that of their ancestors and for whom chronic disease is the predominant risk.

Both of these characteristics of African genomes highlight the importance of conducting genomic research directly in African ancestry individuals and, moreover, in a wide range of African ancestry individuals sampled in terms of geography, lifestyle, and ethnicity. The future of genomic research for African ancestry individuals depends on an adequate understanding of the history, culture, and systematic sampling of the ethnolinguistic groups across Africa and its diaspora peoples.

9.6 Key Points

- Evolutionary history has led to genetic characteristics among those of African ancestry that are relevant for genomic analyses, including lower levels of linkage disequilibrium, more divergent patterns of linkage disequilibrium, and more complex population substructure.
- Adaptive forces have shaped African genomes, leading to a rise in frequency of alleles conferring survival advantages, some of which have proven maladaptive in modern environments.
- Increases in modern complex disease risk have been observed among African populations due to: ecological adaptations to tropical climates; protection from malaria, trypanosomiasis, and Lassa fever; and selection for metabolic traits that confer an advantage in famine.
- Given the marked differences in allele frequencies at pharmacogenomic loci in African vs. non-African ancestry individuals as well as the substantial heterogeneity of allele frequencies within African ancestry populations, careful attention must be given to the study of pharmacogenomics in these populations.

Disclaimer

The views and opinions expressed in this chapter are those of the authors only and do not necessarily represent the views, official policy, or position of the U.S. Department of Health and Human Services or any of its affiliated institutions or agencies.

References

1 Shriner, D., Tekola-Ayele, F., Adeyemo, A., and Rotimi, C.N. (2014). Genome-wide genotype and sequence-based reconstruction of the 140,000 year history of modern human ancestry. *Scientific Reports* 4: 6055.

2 Rotimi, C.N., Tekola-Ayele, F., Baker, J.L., and Shriner, D. (2016). The African diaspora: history, adaptation and health. *Current Opinion in Genettics & Development* 41: 77–84.

3 The 1000 Genomes Project Consortium, Auton, A., Brooks, L.D. et al. (2015). A global reference for human genetic variation. *Nature* 526: 68–74.

4 Gurdasani, D., Carstensen, T., Tekola-Ayele, F. et al. (2015). The African genome variation project shapes medical genetics in Africa. *Nature* 517: 327–332.

5 Rotimi, C.N., Bentley, A.R., Doumatey, A.P. et al. (2017). The genomic landscape of African populations in health and disease. *Human Molecular Genetics* 26: R225–R236.

6 Baker, J.L., Shriner, D., Bentley, A.R., and Rotimi, C.N. (2016). Pharmacogenomic implications of the evolutionary history of infectious diseases in Africa. *Pharmacogenomics Journal* 17: 112–120.

7 Lindenau, J.D., Wagner, S.C., de Castro, S.M., and Hutz, M.H. (2016). The effects of old and recent migration waves in the distribution of HBB*S globin gene haplotypes. *Genetics and Molecular Biology* 39: 515–523.

8 Shang, X. and Xu, X. (2017). Update in the genetics of thalassemia: what clinicians need to know. *Best Practice & Research: Clinical Obstetrics & Gynaecology* 39: 3–15.

9 Cappellini, M.D. and Fiorelli, G. (2008). Glucose-6-phosphate dehydrogenase deficiency. *Lancet* 371: 64–74.

10 Shah, S.S., Rockett, K.A., Jallow, M. et al. (2016). Heterogeneous alleles comprising G6PD deficiency trait in West Africa exert contrasting effects on two major clinical presentations of severe malaria. *Malaria Journal* 15: 13.

11 Carson, P.E., Flanagan, C.L., Ickes, C.E., and Alving, A.S. (1956). Enzymatic deficiency in primaquine-sensitive erythrocytes. *Science* 124: 484–485.

12 Huang, B.H. and Liao, P.C. (2015). Tracing evolutionary relicts of positive selection on eight malaria-related immune genes in mammals. *Innate Immunity* 21: 463–476.

13 Rać, M.E., Safranow, K., and Poncyljusz, W. (2017). Molecular basis of human CD36 gene mutations. *Molecular Medicine* 13: 288–296.

14 Thobakgale, C.F. and Ndung'u, T. (2014). Neutrophil counts in persons of African origin. *Current Opinion in Hematology* 21: 50–57.

15 Triska, P., Soares, P., Patin, E. et al. (2015). Extensive admixture and selective pressure across the Sahel Belt. *Genome Biology and Evolution* 7: 3484–3495.

16 Pierron, D., Heiske, M., Razafindrazaka, H. et al. (2018). Strong selection during the last millennium for African ancestry in the admixed population of Madagascar. *Nature Communications* 9: 932.

17 Reich, D., Nalls, M.A., Kao, W.H. et al. (2009). Reduced neutrophil count in people of African descent is due to a regulatory variant in the Duffy antigen receptor for chemokines gene. *PLoS Genetics* 5: e1000360.

18 Mmbando, B.P., Mgaya, J., Cox, S.E. et al. (2015). Negative epistasis between sickle and foetal haemoglobin suggests a reduction in protection against malaria. *PLoS One* 10: e0125929.

19 Shriner, D., Kumkhaek, C., Doumatey, A.P. et al. (2015). Evolutionary context for the association of γ-globin, serum uric acid, and hypertension in African Americans. *BMC Medical Genetics* 16: 103.

20 Malaria Genomic Epidemiology Network (2015). A novel locus of resistance to severe malaria in a region of ancient balancing selection. *Nature* 526: 253–257.

21 Kao, W.H.L., Klag, M.J., Meoni, L.A. et al. (2008). MYH9 is associated with nondiabetic end-stage renal disease in African Americans. *Nature Genetics* 40: 1185–1192.

22 Genovese, G., Friedman, D.J., Ross, M.D. et al. (2010). Association of trypanolytic ApoL1 variants with kidney disease in African Americans. *Science* 329: 841–845.

23 Thomson, R., Genovese, G., Canon, C. et al. (2014). Evolution of the primate trypanolytic factor APOL1. *Proceedings of the National Academy of Sciences* 111: E2130–E2139.

24 Freedman, B.I., Langefeld, C.D., Andringa, K.K. et al. (2014). End-stage renal disease in African Americans with lupus nephritis is associated with APoL1. *Arthritis & Rheumatology* 66: 390–396.

25 Beckerman, P., Bi-Karchin, J., Park, A.S. et al. (2017). Transgenic expression of human *APoL1* risk variants in podocytes induces kidney disease in mice. *Nature Medicine* 23: 429–438.

26 Andersen, K.G., Shapiro, B.J., Matranga, C.B. et al. (2015). Clinical sequencing uncovers origins and evolution of Lassa virus. *Cell* 162: 738–750.

27 Cao, W., Henry, M.D., Borrow, P. et al. (1998). Identification of α-dystroglycan as a receptor for lymphocytic choriomeningitis virus and Lassa fever virus. *Science* 282: 2079–2081.

28 Hara, Y., Kanagawa, M., Kunz, S. et al. (2011). Like-acetylglucosaminyltransferase (*LARGE*)-dependent modification of dystroglycan at Thr-317/319 is required for laminin binding and arenavirus infection. *Proceedings of the National Academy of Sciences* 108: 17426–17431.

29 Andersen, K.G., Shylakhter, I., Tabrizi, S. et al. (2012). Genome-wide scans provide evidence for positive selection of genes implicated in Lassa fever. *Philosophical Transactions of the Royal Society of London Series B: Biological Sciences* 367: 868–877.

30 Clarke, N.F., Maugenre, S., Vandebrouck, A. et al. (2011). Congenital muscular dystrophy type 1D (MDC1D) due to a large intragenic insertion/deletion, involving intron 10 of the *LARGE* gene. *European Journal of Human Genetics* 19: 452–457.

31 Nwankwo, T. (2011–2012). Hypertension among adults in the United States: National Health and nutrition examination survey. *NCHS data brief*: 1–8.

32 Cooper, R.S., Rotimi, C.N., and Ward, R. (1999). The puzzle of hypertension in African-Americans. *Scientific American* 280: 56–63.

33 Cooper, R.S., Wolf-Maier, K., Luke, A. et al. (2005). An international comparative study of blood pressure in populations of European vs. African descent. *BMC Medicine* 3: 2.

34 Young, J.H., Chang, Y.P., Kim, J.D. et al. (2005). Differential susceptibility to hypertension is due to selection during the out-of-Africa expansion. *PLoS Genetics* 1: e82.

35 Jablonski, N.G. and Chaplin, G. (2010). Colloquium paper: human skin pigmentation as an adaptation to UV radiation. *Proceedings of the National Academy of Sciences* 107 (Suppl 2): 8962–8968.

36 Beleza, S., Santos, A.M., McEvoy, B. et al. (2013). The timing of pigmentation lightening in Europeans. *Molecular Biology and Evolution* 30: 24–35.

37 Batai, K., Murphy, A.B., Shah, E. et al. (2014). Common vitamin D pathway gene variants reveal contrasting effects on serum vitamin D levels in African Americans and European Americans. *Human Genetics* 133: 1395–1405.

38 Tekola-Ayele, F., Adeyemo, A., Chen, G. et al. (2014). Novel genomic signals of recent selection in an Ethiopian population. *European Journal of Human Genetics* 23: 1085.

39 Ng, M.C.Y., Saxena, R., Li, J. et al. (2013). Transferability and fine mapping of type 2 diabetes loci in African Americans. *Diabetes* 62: 965.

40 Wei, F.-Y., Nagashima, K., Ohshima, T. et al. (2005). Cdk5-dependent regulation of glucose-stimulated insulin secretion. *Nature Medicine* 11: 1104.

41 Steinthorsdottir, V., Thorleifsson, G., Reynisdottir, I. et al. (2007). A variant in CDKAL1 influences insulin response and risk of type 2 diabetes. *Nature Genetics* 39: 770.

42 Tishkoff, S.A., Reed, F.A., Ranciaro, A. et al. (2007). Convergent adaptation of human lactase persistence in Africa and Europe. *Nature Genetics* 39: 31–40.

43 Ehret, C. (2009). On the antiquity of agriculture in Ethiopia. *The Journal of African History* 20: 161–177.

44 Brandt, S.A., Spring, A., Hiebsch, C. et al. (1997). *Tree against Hunger: Enset-Based Agricultural Systems in Ethiopia*. Washington, DC: American Association for the Advancement of Science.

45 Pickrell, J.K., Coop, G., Novembre, J. et al. (2009). Signals of recent positive selection in a worldwide sample of human populations. *Genome Research* 19: 826–837.

46 Tekola Ayele, F., Adeyemo, A., Finan, C. et al. (2012). HLA class II locus and susceptibility to podoconiosis. *New England Journal of Medicine* 366: 1200–1208.

47 Yang, X., Sun, J., Gao, Y. et al. (2012). Genome-wide association study for serum complement C3 and C4 levels in healthy Chinese subjects. *PLoS Genetics* 8: e1002916.

48 Sanna, S., Pitzalis, M., Zoledziewska, M. et al. (2010). Variants within the immunoregulatory *CBLB* gene are associated with multiple sclerosis. *Nature Genetics* 42: 495–497.

49 Yu, X.-Q., Li, M., Zhang, H. et al. (2011). A genome-wide association study in Han Chinese identifies multiple susceptibility loci for IgA nephropathy. *Nature Genetics* 44: 178–182.

50 Nalls, M.A., Pankratz, N., Lill, C.M. et al. (2014). Large-scale meta-analysis of genome-wide association data identifies six new risk loci for Parkinson's disease. *Nature Genetics* 46: 989–993.

51 Li, J., Zhang, L., Zhou, H. et al. (2011). Global patterns of genetic diversity and signals of natural selection for human ADME genes. *Human Molecular Genetics* 20: 528–540.

52 Ramos, E., Doumatey, A., Elkahloun, A.G. et al. (2014). Pharmacogenomics, ancestry and clinical decision making for global populations. *The Pharmacogenomics Journal* 14: 217–222.

53 Rotimi, C.N. and Jorde, L.B. (2010). Ancestry and disease in the age of genomic medicine. *New England Journal of Medicine* 363: 1551–1558.

54 Aklillu, E., Persson, I., Bertilsson, L. et al. (1996). Frequent distribution of ultrarapid metabolizers of debrisoquine in an Ethiopian population carrying duplicated and multiduplicated functional CYP2D6 alleles. *Journal of Pharmacology and Experimental Therapeutics* 278: 441–446.

55 Anberbir, Y. (2015) Ethiopia: Authority Issues Red Alert On Codeine Drug. http://allafrica.com/stories/201511241318.html (accessed June 6, 2020).

56 Ge, D., Fellay, J., Thompson, A.J. et al. (2009). Genetic variation in IL28B predicts hepatitis C treatment-induced viral clearance. *Nature* 461: 399.

57 Rotimi, C., Leppert, M., Matsuda, I. et al. (2007). Community engagement and informed consent in the international HapMap project. *Journal of Community Genetics* 10: 186–198.

58 Ramsay, M. and Sankoh, O. (2016). African partnerships through the H3Africa Consortium bring a genomic dimension to longitudinal population studies on the continent. *International Journal of Epidemiology* 45: 305–308.

10

Applying Self-report Measures in Minority Health and Health Disparities Research

Mariana Sanchez[1]*, Bertha Hidalgo*[2]*, Adelaida Rosario*[5]*, Ligia Artiles*[3]*, Anita L. Stewart*[4]*, and Anna M. Nápoles*[3]

[1] *Robert Stempel College of Public Health and Social Work, Florida International University, Miami, FL, USA*
[2] *Ryals School of Public Health, The University of Alabama at Birmingham, Birmingham, AL, USA*
[3] *National Institute on Minority Health and Health Disparities, National Institutes of Health, Bethesda, MD, USA*
[4] *Institute for Health and Aging, Center for Aging in Diverse Communities, University of California, San Francisco, CA, USA*
[5] *Commissioned Corps Headquarters, Office of the Surgeon General, Rockville, MD, USA*

10.1 Introduction

Research on minority health and health disparities relies heavily on self-report measures that accurately assess the lived experiences of diverse, vulnerable groups. Some of these experiences represent mechanisms by which health disparities occur. By nature, many of the constructs representing these mechanisms are abstract sociodemographic, behavioral, social, and structural factors that are not readily observable [1, 2].

Self-report measures are collected through surveys, which is an efficient method to gather data from large numbers of people. Survey questions can use structured or open-ended responses. In this chapter, we focus on surveys that utilize structured response options.

Despite the important role of self-report measures in setting public health targets and assessing healthcare quality metrics, patients with low literacy, limited English proficiency, and functional limitations are frequently excluded from routine collection of these data. Often, this exclusion is due to the costs associated with paper-and-pencil format, interviewer administration, and language translation of surveys [3]. Researchers seeking to include vulnerable populations need to be familiar with pertinent measurement issues and methodological tools to ensure that self-report measures used in these groups are valid and reliable. Omitting vulnerable groups from health research and ignoring issues surrounding the use of self-report measures in these populations severely impede our ability to address health disparities.

The Science of Health Disparities Research, First Edition. Edited by Irene Dankwa-Mullan, Eliseo J. Pérez-Stable, Kevin L. Gardner, Xinzhi Zhang, and Adelaida M. Rosario.

10.2 Measurement Issues When Using Self-report Measures in Diverse Populations

To understand the determinants of health disparities and poor health among minority populations, researchers need to be familiar with measurement issues in these populations. Because most self-report health measures have been developed in mainstream populations, they need to be culturally adapted and translated into other languages before they can be applied in diverse populations. Measures may need to be simplified for those with lower levels of education or limited English proficiency. These issues can be grouped into two major categories: the conceptual and the psychometric properties of a specific measure [2].

10.2.1 Conceptual Adequacy and Equivalence

Because health-related concepts and behaviors are embedded within cultural contexts and vary across cultures, researchers need to concern themselves with the *conceptual adequacy* of a construct represented by the measure within a specific population group, and with the *conceptual equivalence* of a construct across groups [2]. Conceptual adequacy refers to evidence that the measure comprehensively, accurately, and sensitively assesses the dimensions of the construct that are relevant for a particular group. Conceptual equivalence pertains to the extent to which measures have similar meanings across groups.

For a measure to have conceptual adequacy, it must be relevant and meaningful for the population of interest. Assessing conceptual adequacy requires securing the input of the target population in evaluating the meaning and importance of specific concepts and measures. Measures may include items that are not relevant to the lived experiences of particular groups or may omit constructs that are important to them [4]. For example, a study of a multidimensional social support measure using qualitative interviews found that the study needed to add two dimensions that were important to older Chinese and Korean immigrants, language support and financial support [5].

Translations of survey questions should aim for equivalence of meaning across languages. Assessing conceptual equivalence reduces the risk of imposing a concept from one culture on another that may be irrelevant for that group. For example, a study of a measure of neighborhood quality among Whites and English- and Spanish-speaking Mexican Americans found that a subscale assessing physical environment lacked equivalence across language groups. This was most likely due to differences in the meaning of translated terms that might also reflect differences in meaning across social class (e.g., "run-down" home) [6].

10.2.2 Psychometric Adequacy and Equivalence

Researchers must also consider the *psychometric adequacy* of measures within specific groups and their *psychometric equivalence* across population groups if making such comparisons. Psychometric adequacy refers to evidence that the measure demonstrates good psychometric properties within a particular cultural group. Psychometric equivalence pertains to evidence that measures demonstrate invariance across groups and are not biased.

Otherwise, observed relationships between variables may reflect bias and not the true underlying relationships.

When comparing groups in health disparities research, the assumption is often made that the instrument is measuring the construct in the same manner across groups. In practice, this assumption is rarely tested [1]. Measurement invariance indicates that the same underlying construct is being measured across groups and is a prerequisite before valid comparisons can be made [1]. The growing demographic diversity in the United States underscores the importance of developing self-report measures that allow for valid comparisons across groups. Thus, testing for measurement invariance has become an increasingly important issue in recent years.

Cultural norms or language translation can introduce response bias or culturally mediated response styles that obscure true levels of study constructs [4, 7]. *Psychometric analyses* of self-report measures within a specific disparity or minority population help us know whether the measures demonstrate evidence of reliability and validity in those groups. In some cultures, use of the extreme choices in a Likert-type scale are preferred over the middle options, while in other cultures the opposite is true [7]. Such response styles will introduce measurement error. If surveys ask about sensitive information or behaviors, social desirability bias may result in the overreporting of socially acceptable behaviors and underreporting of unacceptable behaviors.

10.3 Methods for Evaluating Conceptual and Psychometric Properties of Self-report Measures

10.3.1 Methods for Assessing Conceptual Adequacy and Equivalence

Assessing the conceptual properties of self-report measures generally involves the use of qualitative research methods to pretest measures within the targeted groups. These methods consist of focus groups and cognitive pretest interviews that use probing, open-ended questions to assess whether survey items are comprehended as intended by survey developers. Testing existing measures in this way can provide evidence for the validity of the constructs or domains to be assessed in a structured survey for use in a targeted group. Such qualitative interviews can be stratified to assess the conceptual adequacy of measures within population strata (e.g., older African American men) or to compare conceptual equivalence across minority or language groups (e.g., compare Chinese and English language versions of the same instrument). To facilitate these types of comparisons, qualitative data analysis software can be used to sort and code responses by population strata.

Focus groups are used widely in health disparities and minority health research to develop and refine concepts and pretest measures. They allow research to reflect the concerns and issues of those being studied, making the constructs and measures meaningful and relevant [8]. In one study, focus groups were used to understand how being African American shaped experiences as employees and parents, resulting in the creation of new survey items designed to assess interpersonal racism and racial socialization [9]. Focus groups can be used as well to determine if existing measures are culturally appropriate and relevant.

Cognitive interviews are used widely during the pretesting phase of questionnaires to detect items that are not understood by respondents. Cognitive interview methods reflect a theoretical model of the survey response process involving four stages: comprehension or interpretation, information retrieval, judgment formation, and response editing [10]. In other words, the respondent must understand the question, recall information, decide on its relevance, and formulate an answer in the format provided by the interviewer. Cognitive interviews can help identify the types of errors made by respondents and understand how they interpret and answer questions. In studies of health disparities, cognitive interview pretesting can be critical to assessing the conceptual equivalence of self-report measures across groups varying in socioeconomic status (SES) or race/ethnicity. Results can be used to revise items for appropriateness to respondents' cultural context and lifestyle. These methods rely primarily on verbal probes about the interpretation of questions and recall strategies, which may be scripted or spontaneously created by the interviewer. Typically, interviews are audio-recorded and transcribed, and transcripts become the data to be analyzed using qualitative methods.

Cognitive pretest interviews are becoming standard practice in the translation and cultural adaptation of large-scale health and healthcare surveys, such as the Patient-Reported Outcomes Measurement Information System (PROMIS®) [11] and the Consumer Assessment of Healthcare Providers and Systems (CAHPS®) [12]. These interviews can help identify potentially costly and error-prone items that require modification prior to field-testing the instrument.

10.3.2 Methods for Assessing Psychometric Adequacy and Equivalence

Self-report measures must demonstrate the fundamental psychometric properties of reliability and validity to be methodologically sound and robust. Reliability refers to the ability of the measure to obtain consistent results if used again under similar circumstances (e.g., in the same population later in time) [13]. Validity refers to the ability of a measure and its items to accurately represent the particular theoretical or latent construct being assessed. That is, does it actually measure what it is intended to measure? [13]. The validity and reliability of self-report measures are required whether researchers seek to draw valid comparisons within one minority population group or make valid comparisons across groups.

There are three basic ways to judge the reliability of a measure: (i) consistency across items within a scale (internal consistency), (ii) over time (test–retest reliability), and (iii) across different raters (inter-rater reliability) [14]. The most commonly used measure of internal consistency is the Cronbach alpha coefficient (α), with values of 0.70–0.80 or greater indicating good internal consistency and higher scores indicating greater consistency. Test–retest reliability is relevant for measuring constructs that are assumed to be relatively stable over time. To assess test–retest reliability, one administers the measure to the same group of people at two time points and then examines the correlations between the scores. Inter-rater reliability is relevant when multiple raters or coders conduct the assessments and examine correlations across raters. It is applied generally when direct observations of behaviors are rated by observers [14].

Validity is particularly important in health disparity studies that aim to compare health determinants and outcomes across groups. A measure can be reliable even though it is not accurately assessing the intended underlying construct. Validity has many dimensions. Content validity refers to the extent to which a measure samples a representative range of the content of a concept. Construct validity refers to evidence that a measure is assessing what it intends to measure. Construct validity is often supported by demonstrating the extent to which a score on a measure correlates in the hypothesized direction with another measure with which it is expected to be related (e.g., Does a measure of depressive symptoms correlate inversely with a measure of emotional well-being?). In-depth discussions of these issues are available elsewhere [14, 15].

The reliability and validity of measures are not established in any one study; judgments are made based on cumulative evidence for a specific measure. *A measure that has demonstrated reliability and validity across many mainstream populations will not necessarily demonstrate adequate psychometric properties in diverse populations.* Health disparities researchers should examine and report on the properties of specific measures when used among diverse populations so that a body of evidence can inform subsequent use of these measures.

Modern psychometric approaches have been applied to assess the validity of self-report measures in general and in diverse populations. We briefly describe two of these approaches: confirmatory factor analysis (CFA) and item response theory (IRT). Both methods can be used to flag items for review and make decisions as to whether to drop or modify items.

A CFA framework allows researchers to test how well observed variables (i.e., items on a self-report instrument) represent the constructs being measured. CFA can be used to test for measurement equivalence (invariance) across groups and whether comparisons of sample estimates reflect true group differences and not group-specific factors that are unrelated to the construct of interest [1]. Such analyses are important because: (i) items may have different meaning across groups; (ii) cultural factors may lead a group to exhibit extreme response styles (e.g., use of "never" or "always" response options in a five-point Likert-type response set); or (iii) cultural norms unrelated to the construct of interest may cause one group to systematically demonstrate higher or lower scores [1].

IRT is an item-level statistical method used to analyze the relationship between a latent trait (construct that a scale is attempting to measure), the properties of the items in the scale, and respondents' answers to the individual items. The approach is called *item response theory* because it is based on assumptions about the mathematical relationships between abilities, attitudes or traits, and item responses. IRT can assess the degree to which differential item functioning (DIF) or bias occurs across subpopulations that vary by age, gender, race, language, and educational levels. In essence, an item is said to have DIF when people from different subgroups who have the same level of the trait have unequal probabilities of endorsing the item similarly [16]. For example, an analysis of the PROMIS pain interference short-form scale found DIF for Latinos relative to Whites on an item that assessed the degree to which pain interfered with engagement in social activities. Subsequently, the item was dropped [16].

10.4 Locating and Selecting Self-report Measures for Use in Diverse Populations

10.4.1 Resources for Locating Measures

Selecting self-report measures can begin with a literature review to identify instruments used by other researchers. Additionally, a wide array of instruments is available in the public domain. These resources were developed to guide researchers in locating well-established and high-quality measures intended for use in the health sciences. The use of standard measures across studies facilitates cross-study comparisons, thereby potentially increasing the scientific impact of individual studies. Table 10.1 provides a brief description of some widely used resources for locating measures.

While many instruments are available, not all are in the public domain. Some measures are copyrighted and copyright protections apply whether or not a copyright notice appears on the measure or is registered with the copyright office [17]. In some cases, fines have been levied for copyright violations related to use of survey measures. Investigators are responsible for contacting the instrument author or publisher to request permission to use it [17]. This issue is important for minority health and health disparities researchers to consider because measures often must be modified to be suitable for disparity populations.

10.4.2 Template for Reviewing Self-report Measures for Appropriateness in Your Study

Table 10.2 describes a comprehensive template or organizing framework that can be used to review characteristics of a self-report measure. This type of instrument review is important to understand how appropriate that measure is to address the study questions and for use in your targeted populations. In applying the template, researchers may find that the information specified by these guidelines for reviewing measures is not readily available. Clearinghouses for measures are making efforts to include this information, but it is also incumbent upon health disparities researchers to publish such data to inform future research efforts.

10.5 Adapting Measures for Diverse Populations

Applying the template in Table 10.2 often reveals evidence that specific items or measures being considered for application in a new population lack conceptual or psychometric adequacy in specific groups. Researchers then face the challenge of adapting measures. Furthermore, adaptations may be necessary whenever an instrument is translated into another language.

Social norms exert a powerful influence over many aspects of health and health behavior. These norms may include cultural attitudes, values, or stigma attached to certain health conditions, health interventions, or help-seeking behaviors. For measures to be culturally sensitive, it requires not only that they be translated into other languages, but that necessary adaptations be made to ensure that cultural nuances are considered [18]. Cultural

Table 10.1 Resources for locating self-report measures.

Electronic databases	
PhenX toolkit https://www.phenxtoolkit.org/	• Provides well-established, high priority, and broadly validated measures relevant to biomedical research • Includes focused collections in four areas (substance abuse/addiction/tobacco, mental health, blood sciences, and social determinants of health)
NIH toolbox http://www.healthmeasures.net/explore-measurement-systems/nih-toolbox	• Includes over 100 measures assessing cognitive, sensory, motor, and emotional function across the lifespan • Contains self-report and objective measures
PROMIS (Patient-Reported Outcomes Measurement Information System) http://www.healthmeasures.net/explore-measurement-systems/promis	• Provides over 300 self-report measures of physical, mental, and social health for use with the general population and those with chronic conditions
Neuro QoL Neurological Quality of Life Measures http://www.healthmeasures.net/explore-measurement-systems/neuro-qol	• Provides self-report measures assessing quality of life across physical, mental, and social health among adults and children with neurological disorders
Science of Behavior Change Measures Repository https://scienceofbehaviorchange.org/measures	• Contains over 100 measures of mechanisms of behavior change used by Science of Behavior Change Network investigators; includes evidence of validation in specific populations
National Institutes of Health Common Data Element (CDE) Resource Portal https://www.nlm.nih.gov/cde	• Includes lists of CDEs, surveys, questionnaires, instruments, instrument items, and other methods of data collection
Health and Psychosocial Instruments Database (HaPI) http://bmdshapi.com	• Contains reviews of measures used in nursing, physical therapy, psychology, and various medical specialties (requires OVID account)
Center for HIV Identification, Prevention and Treatment Services (CHIPTS) http://chipts.ucla.edu/resources	• Provides various health-related surveys/scales • Includes external links to other sites with measures • Provides technical assistance on choosing instruments for specific populations
Organizations and research centers	
Rand Health https://www.rand.org/health/surveys_tools.html	• Provides measures for use in health and healthcare research, including quality of care, patient satisfaction, and health-related quality of life
MacArthur Network https://www.macses.ucsf.edu	• Aims to further understand effects of SES on health of individuals/communities • Reviews measures of social and physical environment, development, psychosocial, and physical factors related to health

(Continued)

Table 10.1 (Continued)

Organizations and research centers	
Michigan Diabetes Research and Training Center http://diabetesresearch.med.umich.edu/Tools_SurveyInstruments.php	• Provides access to several diabetes survey instruments for diabetes patients and health professionals
Commonwealth Fund Surveys http://www.commonwealthfund.org/interactives-and-data/surveys#/sort=%40fdate63677%20descending	• Provides surveys on health insurance, Medicare, health system performance, healthcare quality, patient-centered care, child health, and underserved and aging populations

Note: URLs available at time of publication are subject to change.

Table 10.2 Template for reviewing self-report measures for appropriateness in your study.

1) **Overview of measure**
   - Original publication (full citation)
   - Any *subsequent* key studies that contributed substantially to the measure's development (include full citations)
2) **Definition of concepts**
   - This measure's "general" definition of the concept being measured
   - How similar is the measure's concept definition to what you intend to measure *based on your review of instructions*, *item stems*, and *response scales*?
3) **Methods of administration**
   - Self- or interviewer-administered via computer, telephone, interactive voice, etc.
   - Appropriateness of methods for your specific population
4) **Scale construction methods**
   - What scaling methods were used to develop the measure (e.g., factor analysis)?
5) **Description of measure content**
   - Structure of measure (measurement model)
   - Number of items, domains, subscales, and summary scores
   - Object of measurement (i.e., Are questions about respondent or someone else, such as a child or doctor?)
   - Context for the questions (usually stated in the instructions; e.g., doctors you have seen)
   - Timeframe of questions (e.g., past week, past month)
6) **Interpretability of scores**
   - Direction of scoring (What does a high score mean?)
   - Is the numeric value of the measure(s) intuitively meaningful?
7) **Nature of the samples in which it has been tested**
   - Has the measure been used or tested in a group *similar* to the group in which you are interested?
   - Descriptive sample characteristics of studies in which it has been tested

Table 10.2 (Continued)

8) **Variability and central tendency**
   - *Possible* score range for all subscales and for summary scale
   - *Observed score range* and *mean and standard deviation* in samples for all subscales and for summary scale
9) **Reliability**
   - Types of reliability reported, *including reliability coefficients* for all subscales and summary scale
10) **Validity**
    - Construct validity tests in any of the publications (e.g., known groups, convergent, convergent/discriminant)
11) **Responsiveness, sensitivity to change**
    - Evidence of responsiveness or sensitivity to change in any publication
12) **Scoring**
    - Is there a manual or guide on how to specifically create scores from the questionnaires? If not:
      - Is there enough information in any publication to determine the scoring rules?
13) **Translations** (if you will need them)
    - Other than English, is the measure available in the language you are interested in?
    - If you are using translated versions, what is the quality of the translation (adequacy of methods of translation)?
14) **Equivalence across diverse groups**
    - Conceptual equivalence
    - Psychometric equivalence, invariance
15) **Acceptability for your population**
    - "Real" burden—length of time needed to complete or to administer
    - Your estimate of perceived burden for your population
    - Any statistics or information on reading level? Is the literacy level appropriate?
16) **Can it be modified if necessary and costs**
    - Is there a statement on the published measure that modifications are permitted (with conditions)?
    - Are the measures copyrighted?
    - Is there any cost of using the measure?

Source: Adapted with permission from Anita L. Stewart, Ph.D.

differences, such as the willingness to discuss certain topics or interpersonal communication preferences, must be considered when conducting research with diverse populations. For example, among less acculturated Latinos, stigma associated with mental health issues may result in respondents being more likely to endorse somatic symptom measures than depression items [19].

Within-group heterogeneity must be considered also. Racial and ethnic groups are composed of subpopulations that vary by national origin, immigration history, social class, and context. These factors affect the manner in which determinants of health disparities are experienced and perceived within subgroups. Concha and colleagues conducted an exploratory factor analysis of a widely used measure of acculturative stress in a sample that included Cuban, Central American, and South American Latino immigrants in Miami [20]. The measure had previously been used primarily with Mexican immigrants. Their findings revealed that only 7 of 18 original items performed adequately in this sample, due most likely to differences in social and immigration status and the context of South Florida.

In addition to adaptations to accommodate cultural and language differences, measures and response options may need to be adapted due to literacy issues. Often, simplification of terms and sentence structure are needed. Such adaptations can facilitate translation into other languages as well. Data collection methods may also need to be adapted for low-literacy populations. For example, low-literacy populations may need alternatives to paper-and-pencil administration, such as audio-assisted or interviewer-administered modes.

Stewart and colleagues outlined several key reasons why measures developed in one group may not be suitable for use in another group, requiring that they be adapted [21]. These include that: a concept or dimension may be missing from a measure; the meaning of the concepts or items may differ by group; items/phrases may not be interpreted as intended; the process of responding to the questions may be difficult; and the context or mode of administration may differ from the original. Pertinent issues when adapting measures for health disparities research include examining the breadth (content validity) and meaning of constructs (construct validity), and language translation.

10.5.1 Examining the Breadth and Meaning of Constructs

When conducting research with disparity populations, language and cultural factors may influence respondents' ability to understand and accurately respond to self-report measures. Two broad issues apply. First, a measure may not contain items or constructs that are relevant to a specific population. Second, certain concepts or items may mean different things across groups or may exist in one cultural group but not the other.

For a measure to have content validity in a new group, researchers may need to add or change the wording of an item or even add items to fully capture the concept in that group. For example, Hays *et al.* suggest that, when applied to persons with disabilities, the RAND 36-item Short Form Survey (SF-36) related to "Does your health now limit you in these activities?" be adapted to read, "Does your health or disability now limit you in these activities?" [22]. In other cases, a whole new domain may need to be added. Based on focus group findings, researchers developed a new measure of interpersonal processes of care to address clinician cultural sensitivity, including subscales to assess discrimination experienced by diverse patient groups due to education, race/ethnicity, language, sensitivity to language needs, and immigrant status [23].

In other cases, concepts may not translate well across culture, language, or social class. For example, the term "feeling blue," found in some of the most widely used measures of depression in the United States, such as the SF-36 [24] and the Center for Epidemiological Studies-Depression scales (CES-D) [25], can be problematic when applied in nonmainstream groups. "Feeling blue" has different meanings in different languages. For instance, the color blue denotes a state of joy in some Latin American cultures [26].

10.5.2 Language Translation of Measures

Based on the American Community Survey conducted by the US Census Bureau, the nation's immigrant population was 43.7 million (13.5%) in July 2016, representing one out of eight residents [27]. Growth in the immigrant population in the United States is expected to continue [27]. Rigorously adapted and translated measures form the basis for comprehensive and accurate assessments of health outcomes in ethnic minority and immigrant populations. The past two decades have seen increased attention to identifying optimal methods for translating instruments into other languages. Based on the literature, the following illustrates a synthesis of recommended steps [4]. It should be noted that experimental and quasi-experimental studies comparing translation and adaptation methods suggest that back translation has limited value, and expert panels and cognitive pretesting with the target population are essential (Figure 10.1) [4].

Step 1: Selection of adequate measure. As previously described, the selection of measures should be guided by evidence of their conceptual and psychometric adequacy and equivalence among the target populations. Careful consideration of the content, not just the name of the instrument, should occur to ensure the appropriate domains are present to address the research question.

Step 2: Forward translation. This step entails translation of the measure to the target language by a bilingual translator. This person should preferably be a health professional familiar with the topic area and terminology used to describe the constructs being assessed. The translator should be instructed to emphasize conceptual rather than literal translations, and the use of natural, clear, and concise wording.

Step 3: Back translation. During this phase, translation of the instrument back to the source language is conducted by a second bilingual translator. Comparison of the original measure with the back-translated version is conducted by both bilingual translators. If needed, modifications are made to the translated measures to arrive at a reconciled version.

Step 4: Expert panel review. In this step, a panel of bicultural and bilingual experts is convened to review the translated measure. The goal of this panel review is to identify and resolve any conceptual or semantic issues between the translation and the original source measures. Experts can include target population members.

Step 5: Pretesting and field-testing. In this step, the instrument is pretested with representatives of the target population, followed by a field test with a sample from the targeted population. Qualitative (cognitive interviews) and quantitative (psychometric analyses) approaches are used to review the adequacy and equivalence of measures across language groups.

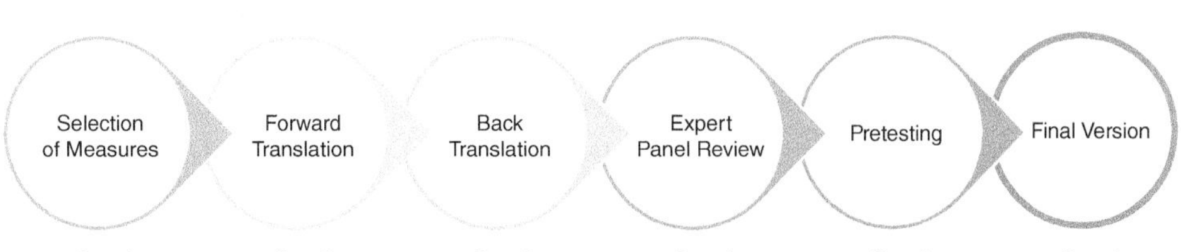

Figure 10.1 Steps for translation of self-report measures.

10.6 Future Directions

Healthy People 2030 recognizes the importance of the social determinants of health (SDOH) by including an overarching goal of creating social and physical environments that promote good health for all [28]. *Healthy People 2030* utilizes a place-based organizing framework of the SDOH that reflects five areas: economic stability, education, social and community context, health and healthcare, and neighborhood and built environment. New frameworks are emerging to guide the use of individual- and system-level data on the SDOH to address these factors within clinical settings [29]. Clearly, new self-report measures of many of these important constructs will need to be developed and validated among vulnerable populations if we are to make progress on reducing health disparities. Disentangling the complex and dynamic relationships among these factors will be essential to advance the science of health disparities. Two key challenges in this arena are examining interactions between SES and race/ethnicity across the life course and operationalizing the impact of cultural factors and acculturation processes within varying geographic and historical contexts.

10.6.1 Interactions of Socioeconomic Status and Race/Ethnicity Across the Life Course

SES is arguably one of the most widely used self-report measures in health disparities research. Extensive research indicates that individuals of lower SES experience higher morbidity and mortality across numerous health outcomes [30]. Low SES is associated with lack of access to healthcare, greater exposure to environmental toxins, riskier health behaviors, fewer resources, and higher mortality [30]. SES and race/ethnicity are highly confounded, with minority groups overrepresented in lower SES strata, making it a critical concept for understanding mechanisms by which racial/ethnic health disparities develop and are perpetuated.

While life-course models emphasize the cumulative effects of health determinants, most SES measures focus on current levels of SES [31]. Developing adequate life-course assessments to examine constructs, such as access to health resources at critical developmental time points, social class during childhood, and economic stress, could assist in disentangling the complex associations between cumulative socioeconomic disadvantage, race/ethnicity, and health [2]. Most SES measures have traditionally focused on individual-level variables, while studies are finding that community-level indicators, such as neighborhood concentrated disadvantage, contribute to disparities over and above individual factors [32].

Assessing a single aspect of identity (e.g., race) fails to take into account that many individuals claim multiple minority statuses (e.g., African American/female/lesbian). These multiple statuses, or intersectionalities, could potentially influence both discrimination experiences and subsequent health outcomes. Studies need to examine their potentially cumulative effects on outcomes.

For example, a study examining 33-year trajectories of SES among African Americans, Latinos, and Whites found that upwardly mobile African Americans and Latinos were more likely to experience acute and chronic discrimination than their socioeconomically stable counterparts [33]. Among Whites, upward mobility resulted in significantly less exposure to discrimination. Differential exposure to discrimination also explained a substantial number of Black/White, but not Latino/White, disparities in self-rated health. This study indicates that the nature of the SES/health gradient could vary by race. Another study, conducted among African American sexual minority women, found that negative patient-provider outcomes attributed to discrimination based on race/ethnicity, gender, and sexual orientation all contributed to reduced healthcare utilization [34].

10.6.2 Measuring Culture and Acculturation Processes

A recent *Lancet* commission defined culture as the "set of practices and behaviors defined by customs, habits, language, and geography that groups of individuals share" and added that "socioeconomic status produces its own cultures of security and insecurity that cut across nationality, ethnic background, gender orientation, age, and political persuasion" [35]. Culture forms the basis of a personal value system, which exerts significant influence on health and well-being and plays a key role in understanding health disparities [35]. Recognizing the personal and social contexts in which cultural mechanisms (i.e., practices, values, identity) influence health risk behaviors is critical in the development of effective and culturally suitable interventions [36].

Self-report measures need to be evaluated in the cultural context of the group being studied. For example, a study of Asian Indian immigrants in California found wide differences in the prevalence of ever using tobacco, depending on whether measures asked about conventional tobacco use (25%) versus traditional tobacco products (e.g., paan masala, gutka, bidis; 65%) [37]. Thus, population-based surveys may underestimate substantially the use of tobacco products in this and other immigrant populations.

Acculturation, the changes that occur when individuals come in contact with dissimilar groups, plays a key role in the health outcomes of immigrants [36]. Despite advancements in our conceptualizations of culture as a multidimensional construct, most research examining acculturation's influence on immigrant health has employed unidimensional proxy measures, such as language spoken at home, place of birth, and number of years in the United States [36]. Recent studies have revealed intricate and complex associations between cultural mechanisms and health risk behaviors among minorities [38]. An excellent review of the conceptual and methodological issues involved in assessing acculturation-related constructs can be found in a paper by Schwartz et al. [36]. Research has begun to address the effects of ethnic enclaves, controlling for individual-level factors, suggesting that ethnic density of neighborhoods may moderate the effects of ethnicity on healthcare access, health behaviors, and outcomes. These relationships may differ as well, based on ethnic group and outcome [39, 40]. Future research is needed to examine the direct and indirect pathways through which cultural beliefs, values, and identity exert their influence on health outcomes across diverse groups and locations.

10.7 Conclusions

The use of psychometrically sound and culturally relevant measures in health disparities research is essential. Failing to use adequate measures can lead to inaccurate conclusions with important ramifications for the provision of healthcare services and health policy that could be costly to vulnerable populations. Effectively addressing mechanisms by which health disparities are formed requires critical appraisal and informed use of self-report measures.

Health disparities researchers are charged with measuring complex health-related phenomena that function to produce and perpetuate disparities. Conversely, researchers must identify protective factors that operate in vulnerable populations in an effort to reinforce these. Data on many of these risk and protective factors are gathered via self-report. When utilizing self-report measures to assess health determinants and health outcomes, the quality, comparability, potential sources of bias, and validity and reliability of these instruments must be considered.

Self-report measures play a central role in research to better understand and address health inequities [41]. Health disparity and minority health researchers should consider trade-offs between using a measure as is, adapting a measure, or developing a new measure. Using a measure as is requires some evidence of the conceptual and psychometric adequacy and/or equivalence of the measure in the targeted group. There is no consensus on guidelines that should be applied when adapting measures, but researchers should continue to publish on the effectiveness of such methods to advance the field. Similarly, publishing on the development of new measures that are promising mechanisms of health and health disparities will continue to advance the field and contribute to our collective ability to reduce health inequities. Advancing the science of what and how we measure health determinants, mechanisms, and outcomes can lead to more effective and efficient interventions to reduce health disparities.

10.8 Key Points

- Essential measurement issues related to the use of self-report measures in disparity populations are presented.
- Self-report measures that assess determinants of health have been developed primarily with mainstream populations and may need to be culturally adapted or translated into other languages before they can be applied in diverse populations.
- This chapter provides a guide for locating and assessing the appropriateness of self-report measures for use with disparity groups.
- Opportunities and future directions for expanding the scope and availability of self-report measures in health disparities research are discussed.

Disclaimer

The views and opinions expressed in this chapter are those of the authors only and do not necessarily represent the views, official policy, or position of the U.S. Department of Health and Human Services or any of its affiliated institutions or agencies.

References

1 Gregorich, S.E. (2006). Do self-reported instruments allow meaningful comparisons across diverse population groups? Testing measurement invariance using the confirmatory factor analysis framework. *Medical Care* 44 (11 Suppl 3): S78–S94.

2 Stewart, A.L. and Napoles-Springer, A.M. (2003). Advancing health disparities research: can we afford to ignore measurement issues? *Medical Care* 41 (11): 1207–1220.

3 Cella, D., Hahn, E.A., Jensen, S.E. et al. (2015). *Patient-Reported Outcomes in Performance Measurement*. Research Triangle Park, NC: Research Triangle Institute.

4 Epstein, J., Santo, R.M., and Guillemin, F. (2015). A review of guidelines for cross-cultural adaptation of questionnaires could not bring out a consensus. *Journal of Clinical Epidemiology* 68 (4): 435–441.

5 Wong, S.T., Yoo, G.J., and Stewart, A.L. (2007). An empirical evaluation of social support and psychological well-being in older Chinese and Korean immigrants. *Ethnicity & Health* 12 (1): 43–67.

6 Kim, S.Y., Nair, R., Knight, G.P. et al. (2008). Measurement equivalence of neighborhood quality measures for European American and Mexican American families. *Journal of Community Psychology* 37 (1): 1–20.

7 Kemmelmeier, M. (2016). Cultural differences in survey responding: issues and insights in the study of response biases. *International Journal of Psychology* 51 (6): 439–444.

8 Napoles-Springer, A.M. and Stewart, A.L. (2006). Overview of qualitative methods in research with diverse populations. Making research reflect the population. *Medical Care* 44 (11 Suppl 3): S5–S9.

9 Hughes, D.L. and DuMont, K. (1993). Using focus groups to facilitate culturally anchored research. *American Journal of Community Psychology* 21 (6): 775–806.

10 Willis, G.B. (2005). *Cognitive Interviewing: A Tool for Improving Questionnaire Design*. Thousand Oaks, CA: Sage Publications.

11 Solorio, R., Ayala, N.C., Paez, E. et al. (2016). Use of cognitive interviews to adapt PROMIS measurement items for Spanish speakers living with HIV. *AIDS Research and Treatment* 2016: 8340863.

12 Harris-Kojetin, L.D., Fowler, F.J. Jr., Brown, J.A. et al. (1993). The use of cognitive testing to develop and evaluate CAHPS 1.0 core survey items. Consumer Assessment of Health Plans Study. *Medical Care* 37 (3 Suppl): MS10–MS21.

13 Hidalgo, B. and Goodman, M. (2012). Validation of self-reported measures in health disparities research. *Journal of Biometrics & Biostatistics* 3 (7): 1000e114.

14 Price, P.C., Jhangiani, R.S., and Chiang, I.A. (2015). *Research Methods in Psychology*. 2nd Canadian edition. British Columbia, Canada: B.C. Open Textbook Project.

15 Carmines, E.G. and Zeller, R.A. (eds.) (1979). *Reliability and Validity Assessment*, 1–70. Newbury Park, CA: SAGE University Press.

16 Teresi, J.A., Ocepek-Welikson, K., Cook, K.F. et al. (2016). Measurement equivalence of the Patient Reported Outcomes Measurement Information System® (PROMIS®) pain interference short form items: application to ethnically diverse cancer and palliative care populations. *Psychological Test and Assessment Modeling* 58 (2): 309–352.

17 Hays, R.D., Weech-Maldonado, R., Teresi, J.A. et al. (2018). Commentary: copyright restrictions versus open access to survey instruments. *Medical Care* 56 (2): 107–110.

18 Shrout, P.E., Alegria, M., Canino, G. et al. (2008). Testing language effects in psychiatric epidemiology surveys with randomized experiments: results from the National Latino and Asian American Study. *American Journal of Epidemiology* 168 (3): 345–352.

19 Interian, A., Allen, L.A., Gara, M.A. et al. (2006). Somatic complaints in primary care: further examining the validity of the Patient Health Questionnaire (PHQ-15). *Psychosomatics* 47 (5): 392–398.

20 Concha, M., Sanchez, M., de la Rosa, M., and Villar, M.E. (2013). A longitudinal study of social capital and acculturation-related stress among recent Latino immigrants in South Florida. *Hispanic Journal of Behavioral Sciences* 35 (4): 469–485.

21 Stewart, A.L., Thrasher, A.D., Goldberg, J., and Shea, J.A. (2012). A framework for understanding modifications to measures for diverse populations. *Journal of Aging and Health* 24 (6): 992–1017.

22 Hays, R.D., Hahn, H., and Marshall, G. (2002). Use of the SF-36 and other health-related quality of life measures to assess persons with disabilities. *Archives of Physical Medicine and Rehabilitation* 83: S4–S9.

23 Stewart, A.L., Napoles-Springer, A.M., Gregorich, S.E., and Santoyo-Olsson, J. (2007). Interpersonal processes of care survey: patient-reported measures for diverse groups. *Health Services Research* 42 (3 Pt 1): 1235–1256.

24 RAND Corporation (n.d.) 36-Item Short Form Survey (SF-36). https://www.rand.org/health/surveys_tools/mos/36-item-short-form/scoring.html. (accessed May 4, 2018).

25 Radloff, L.S. (1977). The CED-D scale: a self-reported depression scale for research in the general population. *Applied Psychological Measurement* 1: 385–401.

26 Streiner, D.L. and Norman, G.R. (2008). *Health Measurement Scales: A Practical Guide to Their Development and Use*, 4e. New York, NY: Oxford University.

27 Camarota, S.A. and Zeigler, K. (2017). *US Immigrant Population Hit Record 43.7 million in 2016*. Washington, DC: Center for Immigration Studies.

28 Office of Disease Prevention and Health Promotion, US Department of Health and Human Services, (n.d.) Healthy People 2030: Leading Health Indicators https://health.gov/healthypeople/objectives-and-data/browse-objectives#social-determinants-of-health. (accessed Sept. 29, 2020).

29 Gottlieb, L., Sandel, M., and Adler, N.E. (2013). Collecting and applying data on social determinants of health in health care settings. *JAMA Internal Medicine* 173 (11): 1017–1020.

30 Adler, N.E. (2009). Health disparities through a psychological lens. *American Psychologist* 64 (8): 663–673.

31. Shavers, V.L. (2007). Measurement of socioeconomic status in health disparities research. *Journal of the National Medical Association* 99 (9): 1013–1023.

32 Diez Roux, A.V. (2016). Neighborhoods and health: what do we know? What should we do? *American Journal of Public Health* 106 (3): 430–431.

33 Colen, C.G., Ramey, D.M., Cooksey, E.C., and Williams, D.R. (2018). Racial disparities in health among nonpoor African Americans and Hispanics: the role of acute and chronic discrimination. *Social Science & Medicine* 199: 167–180.

34 Li, C.C., Matthews, A.K., Aranda, F. et al. (2015). Predictors and consequences of negative patient–provider interactions among a sample of African American sexual minority women. *LGBT Health* 2 (2): 140–146.

35 Napier, A.D., Ancarno, C., Butler, B. et al. (2014). Culture and health. *Lancet* 384 (9954): 1607–1639.

36 Schwartz, S.J., Unger, J.B., Zamboanga, B.L., and Szapocznik, J. (2010). Rethinking the concept of acculturation: implications for theory and research. *American Psychologist* 65 (4): 237–251.

37 Patel, M., Mistry, R., Maxwell, A.E. et al. (2018). Contextual factors related to conventional and traditional tobacco use among California Asian Indian immigrants. *Journal of Community Health* 43 (2): 280–290.

38 Schwartz, S.J., Unger, J.B., Zamboanga, B.L. et al. (2015). Developmental trajectories of acculturation: links with family functioning and mental health in recent-immigrant Hispanic adolescents. *Child Development* 86 (3): 726–748.

39 Lim, S., Yi, S.S., De La , L., C.N., and Trinh-Shevrin, C. (2017). Defining ethnic enclave and its associations with self-reported health outcomes among Asian American adults in New York City. *Journal of Immigrant and Minority Health* 19 (1): 138–146.

40 Pruitt, S.L., Tiro, J.A., Xuan, L., and Lee, S.J. (2016). Hispanic and immigrant paradoxes in US breast cancer mortality: impact of neighborhood poverty and Hispanic density. *International Journal of Environmental Research and Public Health* 13 (12): 1238.

41 Burgard, S.A. and Chen, P.V. (2014). Challenges of health measurement in studies of health disparities. *Social Science & Medicine* 106: 143–150.

11

Conducting Community-based Participatory Research with Minority Communities to Reduce Health Disparities

Tung T. Nguyen[1], *Nina Wallerstein*[2], *Rina Das*[3], *Melanie D. Sabado-Liwag*[4], *Valarie Blue Bird Jernigan*[5], *Tvli Jacob*[5], *Tamela Cannady*[12], *Linda Sprague Martinez*[6], *Uchenna J. Ndulue*[14], *Abigail Ortiz*[7], *Andrea Williams Stubbs*[8], *Latrice C. Pichon*[9], *Sora Park Tanjasiri*[10], *Jane Pang*[13], *and Kent Woo*[11]

[1] *University of California, San Francisco, CA, USA*
[2] *College of Population Health, University of New Mexico, Albuquerque, NM, USA*
[3] *National Institute on Minority Health and Health Disparities, National Institutes of Health, Bethesda, MD, USA*
[4] *California State University, Los Angeles, CA, USA*
[5] *Center for Health Sciences, Oklahoma State University, Tulsa, OK, USA*
[6] *School of Social Work, Boston University, Boston, MA, USA*
[7] *Southern Jamaica Plain Health Center, Jamaica Plain, MA, USA*
[8] *St. Jude Children's Research Hospital, Memphis, TN, USA*
[9] *University of Memphis, Memphis, TN, USA*
[10] *University of California, Irvine, CA, USA*
[11] *NICOS Chinese Health Coalition, San Francisco, CA, USA*
[12] *Choctaw Nation Health Services, Talihina, OK, USA*
[13] *Pacific Islander Health Partnership, Huntington Beach, CA, USA*
[14] *Boston Public Health Commission, Child, Adolescent, and Family Health Bureau, Boston, MA, USA*

11.1 Introduction

Health disparities result from many factors interacting in complex ways, and research to explain and reduce such disparities necessarily involves stakeholders with diverse expertise and capacities. Community-based participatory research (CBPR), with its strong emphasis on stakeholder engagement and capacity building, is an ideal approach to describe health disparities and interventions to achieve health equity.

In this chapter, we will describe the rationale for using CBPR in health disparities research. A brief review of the evidence that CBPR is an effective orientation for health disparities research will follow, including results of a large cross-site National Institutes of Health (NIH) study to measure associations between promising practices and outcomes. This chapter will showcase examples of how CBPR can be successfully conducted with Black, Asian American, Latino, Native American, and Pacific Islander (PI) communities. Using emerging empirical evidence, we will describe how to design CBPR studies, with a focus on fundamental elements such as partnership capacity, principles alignment among partners, formal agreements, shared control and power, trust, mutual influence, and

The Science of Health Disparities Research, First Edition. Edited by Irene Dankwa-Mullan, Eliseo J. Pérez-Stable, Kevin L. Gardner, Xinzhi Zhang, and Adelaida M. Rosario.

leadership, as well as how to involve community voices and decision making in all phases of the research. This chapter will conclude with challenges that face CBPR practitioners and their potential solutions.

11.1.1 Learning Objectives

1) Understand the rationale for when and how to use CBPR.
2) Know about the key methodologies and measures utilized in CBPR.
3) Learn about the essential elements and steps needed to conduct successful CBPR projects.

11.2 Conducting Community-based Participatory Research with Minority Communities to Reduce Health Disparities

CBPR, the most established form of stakeholder and community engagement, has been defined by the W.K. Kellogg Foundation Community Health Scholars Program [1] as an overarching research orientation and methodology that equitably involves all partners, builds on community interests and strengths, and combines knowledge with social change action to improve community health and eliminate health disparities. Such research, which combines evidence and empowerment, engages communities, and involves stakeholders, has grown exponentially in the past 20 years as a valued approach to achieving health equity.

The ideal CBPR project involves community members and other nonacademic stakeholders in all phases of research, from defining the problem, through collecting, analyzing, and interpreting data, to codisseminating and translating findings to improve health [2]. Researchers have increasingly adopted CBPR as they recognize that the engagement of community partners can facilitate improved research processes and outcomes. Such research processes and outcomes are characterized by greater validity, community acceptance of research design and methods, culture-centered interventions that integrate community knowledge and practice-based evidence [3, 4], and strengthened attention to implementation contexts [5].

CBPR initiatives have specifically sought to challenge histories of research mistrust by building on core principles based on community priorities and strengths [6], and which support authentic partnerships among researchers, community members, and practitioners [7]. By addressing dissemination, implementation, and sustainability issues during the design phase, CBPR can decrease the time needed to translate research findings into sustainable programs, practices, and policies that can reduce health inequities [8, 9].

One common misconception about CBPR is that its methods differ from traditional research. CBPR can include *any* research method, from qualitative interviews, focus groups, and ethnographic observations, to quantitative data from epidemiologic surveys and large datasets. CBPR practitioners utilize research from disciplines such as anthropology, epidemiology, medicine, public health, and sociology. The key characteristic that separates CBPR from traditional research is its emphasis on community engagement throughout the research process.

While use of the term "community-engaged research" has grown, we have focused on CBPR for this chapter. Community-engagement strategies have been plotted along a continuum, from outreach, which can consist simply of educational sessions in community settings, to shared leadership of research decision making between academics and communities [10]. CBPR sits on the shared leadership end of the continuum and even beyond that, to community-directed research, where community agencies, for example, would invite academics to seek research funds under community-board oversight. Thus, CBPR, with its roots in community-based public health, insists on a rigorous, partnered approach, driven by community priorities and collaborative processes throughout the research process. While some projects may start with minimal engagement, we hope that this chapter provides insights that show how researchers can move toward stronger partnerships, collaborating "with" communities, rather than "in" community settings [2].

Planning such projects requires an understanding of how to partner well with community organizations and members; communicate with community leaders; develop and build trust (including budgeting in an equitable manner); develop research designs and timelines that take into account the reality of complex systems; and balance the different types of knowledge, evidence, and expertise between community stakeholders and researchers.

11.3 Evidence of CBPR Effectiveness and Advancement in CBPR Evaluation

In addition to positive changes in health behavior and outcomes, effective CBPR includes measures that are crucial in developing community capacity in areas such as trust; community involvement in the research process; and system, capacity, and policy changes to address health disparities. The first systematic review of CBPR, by the Agency of Healthcare Research and Quality, found that only 12 out of 60 studies had documented outcomes [11]. Since then, there has been a significant growth of systematic reviews and meta-analyses showing compelling evidence of positive impacts of projects using CBPR methods on health behaviors and health outcomes [12–15].

Impacts of CBPR methodology have also been found on the processes of engagement, such as increased community involvement in research and retention of minority participants [13, 16]. Intermediate outcomes have also been achieved that are distinct from traditional research models and that are crucial to changing systematic determinants of health disparities, such as community empowerment, partnership synergy and sustainability, and improved community care systems [14, 17]. Some studies have sought to correlate types of participatory engagement [15, 18] with evidence of the value of lay-delivered interventions [12]. Finally, a growing CBPR literature has documented substantial, improved health effects from policy changes [19, 20].

In 2006, the National Center on Minority Health and Health Disparities funded the Universities of New Mexico and Washington to assess the degree of value added by partnerships and community engagement to improved health outcomes and health equity. Through literature reviews and input from expert CBPR academic and community practitioners, a CBPR conceptual model was born (Figure 11.1) that proposed the importance of

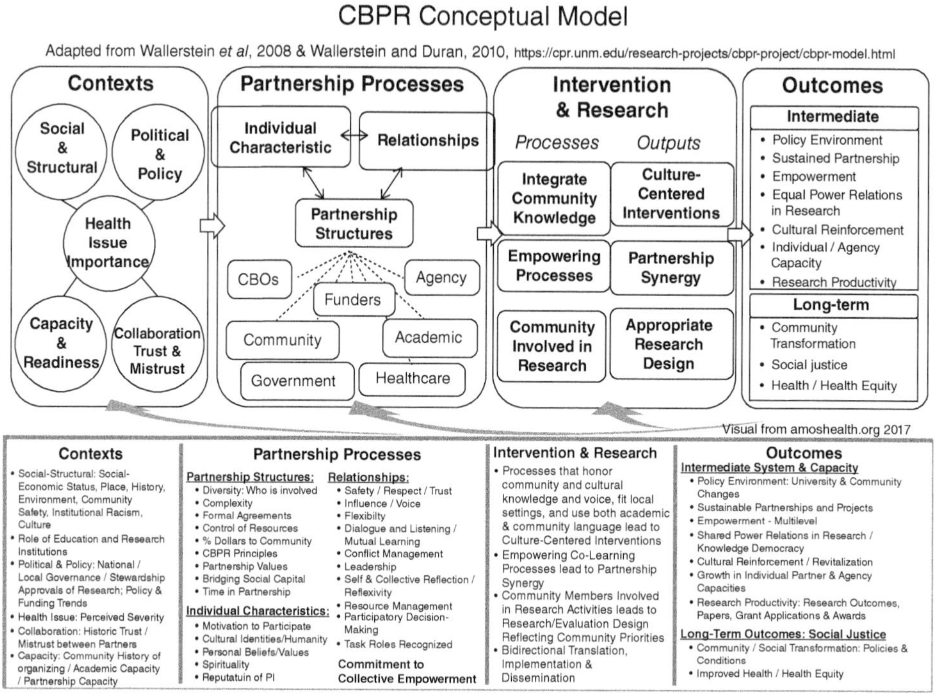

Figure 11.1 CBPR conceptual model. *Sources:* Adapted from Wallerstein et al, 2008; Wallerstein and Duran (2010) [2, 9].

context, partnering processes, the effect of these processes on the science of interventions and research design, and the contribution of partnering processes to outcomes [2].

The subsequent project, "Research for Improved Health," funded by NIH, was a mixed-methods study to test the CBPR model by identifying measures and metrics of partnering practices and outcomes, conducting internet surveys of 200 federally funded partnerships and validating measures [21], and examining the case studies of seven diverse partnerships [22]. Its aims were to describe partnership variability across the country, and assess associations between practices and outcomes [23]. This effort has been followed by the NIH-funded "Engage for Equity" study [24], with five university and community partners and a think tank of CBPR practitioners. Instruments have been refined (also translated into Spanish), internet surveys replicated (with another 179 partnerships responding), and tools have been developed to strengthen partnership evaluation and collective reflection.

From these studies of 379 partnerships, promising partnering and engagement practices associated with positive health outcomes have been identified, with two inter-related partnership pathways [25]. The relational pathway starts from partners having high capacity to leverage community strengths, which contributes to shared principles and values, and partnering practices of trust and respect. The structural pathway identifies formal partnering practices, such as community approval of research, and sharing of funds with the community. Both pathways interact to contribute to greater community involvement in the research, enhanced synergy, and to intermediate system and capacity changes, such as enhanced partner agency capacities, greater project sustainability, power sharing in research, and policy changes. These intermediate system and capacity changes contribute to longer-term community transformation and positive health outcomes.

One of the most promising practices identified in these studies is the involvement of community members in all stages of the research process. When a research project starts with community members identifying their concerns, it can be seen as meeting community needs and priorities and has from its inception a greater likelihood of community ownership and sustainability. Involving community members in the research or grant design, including the choice of methods, sampling, and development of instruments and interventions, fosters trust that the research and programs will be implemented appropriately for the culture and community setting. Content and linguistic validity of the sampling design, instruments, and interventions will be maximized. Community members may be hired to collect data, which is a good way to bring funds to the community, in addition to which community input and direction are essential for data interpretation. Even if analysis is conducted by university team members, collaborating on data interpretation through reviewing initial analyses can revise analytic questions, deepening understanding of the importance of particular results to the community, and encourage the translation of the data into actionable strategies for community benefit. Dissemination requires community involvement in order to know the best channels of communication and to identify levers for affecting policy change or program impact. The involvement of community members in implementation, evaluation, and dissemination of results facilitates project sustainability for long-term health outcomes, especially when grant funding ends.

In addition to the involvement of community members in every step of the research process, the above-mentioned studies with 379 partnerships have identified over 20 promising practices associated with or which contribute to a range of positive health outcomes [23]. Some of these practices, shown in Table 11.1 and demonstrated through the case studies discussed below, are also articulated in systematic reviews. While individual partnerships may choose to adopt the practices that best fit their own settings and populations, we recommend that all partnerships engage in periodic reflection about what they are doing well and what they would like to strengthen in order to achieve multiple outcomes [2, 26].

Table 11.1 Promising partnership practices.

Building from partnership capacity
Fostering trust and respect
Ensuring shared influence and community voice
Creating participatory decision making processes
Sharing power with community partners
Sharing financial resources with community partners
Having community advisory boards
Developing memoranda of agreement
Involving community members in all steps of research

Sources: Wallerstein et al. (2018) [2]; Wallerstein (2018) [26].

11.4 Case Studies

In this section, we describe CBPR projects conducted with African Americans, Asian Americans, Latinos, Native Americans, and PIs to illustrate how key CBPR principles and practices can be implemented among communities of color.

11.4.1 Connecting to Testing and Prevention

In 2012, pre-exposure prophylaxis (PrEP) became available for high risk human immunodeficiency virus (HIV)-negative populations who experience health disparities, such as young Black men who have sex with men (YBMSM) and transgender youth and young adults. PrEP usage in Memphis, Tennessee was low, an example of health disparity in access to life-saving prevention in the South. The historical underpinnings of racism, contextual factors of HIV stigma, and religious influences make it particularly difficult to reach and provide culturally meaningful prevention services to YBMSM and trans youth.

In order to increase testing and linkage to HIV care and PrEP for these marginalized communities, a CBPR partnership was formed between St. Jude Children's Research Hospital's Connect to Protect Memphis, the Shelby County Health Department, the Tennessee Department of Health, the University of Memphis School of Public Health, and YBMSM and transgender community gatekeepers. The aim was to increase and enhance HIV education and testing and streamline linkage to care, including access to PrEP.

The CBPR principles and practices underlying the work included: (i) recognizing YBMSM and trans youth as separate units of identity, (ii) facilitating collaborative and equitable involvement of YBMSM and trans youth, HIV service providers, and researchers in all phases of the research; and (iii) establishing long-term commitments with health departments for HIV testing resources and supportive services.

Community gatekeepers included professionals from HIV service organizations or PrEP clinics and self-identified YBMSM or trans youth. They met monthly to develop action plans and messaging for HIV education, draft guidelines for outreach activities, and establish social network strategies to increase testing and linkages to care or prevention services. Gatekeepers ensured cultural acceptability of recruitment strategies and project activities. Food was provided at meetings, and gift cards given to YBMSM or trans youth gatekeepers.

During the pilot phase of Connecting to Testing and Prevention, a project funded by the National Institute on Minority Health and Health Disparities (NIMHD), innovative testing strategies included the organization of a prevention-themed house and ball event employed in nontraditional settings (i.e., nightclub venues). Specially trained YBMSM and trans youth planned these events and delivered the PrEP education during them. Testing kits were supplied and PrEP materials were tailored for YBMSM and trans youth's culturally specific needs, which demonstrated promising resource management and partnership capacity practices. Through effective and shared leadership, 645 youths were tested for HIV, and 33.9% were linked to PrEP [27]. The 4.5% who were positive were all linked to and engaged in care. This case study identified the value of community involvement as a partnership best practice for engaging and gaining support within networks of YBMSM and trans youth living in the Southern Bible Belt and for carrying out interventions to improve their health.

11.4.2 Healthy Living: The Chinese Lay Health Workers Project

The 3.6 million Chinese Americans are the largest Asian American population, 70% of whom are immigrants and 46% of whom have limited English proficiency. Colorectal cancer (CRC) screening reduces mortality and is cost-effective, but compared to non-Hispanic whites, Chinese Americans are less likely to have had CRC screening.

A national meeting of Asian American community leaders and researchers in 1998 led to the creation of the Asian American Network for Cancer Awareness, Research, and Training (AANCART), a Special Populations Network (SPN) funded by the National Cancer Institute (NCI). A CBPR approach was adopted by the AANCART site in San Francisco, AANCART-SF, to overcome a history of distrust between the Chinese American community and University of California, San Francisco (UCSF) researchers. In 2001, AANCART-SF created the Chinese Council, a network of organizations interested in addressing cancer within the community. Members included UCSF, the American Cancer Society, the Chinatown Public Health Center, clinicians, and community-based organizations, including NICOS Chinese Health Coalition.

In 2004, the Chinese Council identified CRC screening as a priority and chose to study lay health worker (LHW) outreach as an intervention. UCSF and NICOS, which had met quarterly at Council meetings and had conducted outreach activities together from 2001 to 2004, were appointed to write a pilot grant, which received funding in 2005. Based on the results, UCSF and NICOS obtained funding from NCI to conduct a randomized controlled trial (RCT) comparing the efficacy of LHW plus print materials to print materials alone in promoting CRC screening among Chinese Americans.

From 2009 to 2015, UCSF and NICOS met monthly as a team and in subcommittees to design and implement the intervention and measure its outcomes. Decisions were made by consensus. NICOS staff worked with UCSF to develop intervention materials, such as a brochure, a flipchart, and the LHW training program. The control group received lectures focusing on healthy nutrition and physical activity. Through a subcontract from UCSF, NICOS recruited 58 LHWs and helped them to recruit participants for the study. UCSF and NICOS staff jointly trained the LHWs, who carried out their work under the supervision of NICOS staff.

In addition to building NICOS's capacity for conducting research and LHW interventions and UCSF's capacity to work with Chinese Americans, the outcomes of the project included identifying LHW effectiveness in promoting CRC screening [28] and a second randomized controlled trial that showed Chinese language lectures were effective in promoting healthy nutrition and physical activity [29]. This case study illustrates the process of establishing trust through incremental activities, such as coalition building, resource sharing, outreach activities, and pilot studies, and the value of community involvement and participatory decision making in codesigning and conducting effective intervention research to reduce health disparities among Asian Americans.

11.4.3 Nuestro Futuro Saludable [Our Healthy Future]

The Nuestro Futuro Saludable (NFS) partnership, funded by NIMHD, developed and pilot tested an after-school learning intervention for middle school students focused on health equity and the social determinants of health. This partnership began with a steering committee of researchers from Tufts University Schools of Public Health and Community Medicine, Engineering, and Nutrition, along with community investigators from the Center for Community Health Education Research and Service, the Southern Jamaica Plain Community Health Center, the Boston Public Health Commission Center for Health Equity and Social Justice, and the Latin American Health Institute. The steering committee convened a community advisory board comprised of four youths, two parents, two youth workers, two community health center-based social workers, and three researchers. An organizational structure composed of the interlocking membership of the steering committee and the community advisory board, and a memorandum of agreement (MOA) were used to guarantee that the steering committee was accountable to the recommendations of the community advisory board and the broader community [30].

The community advisory board, chaired permanently by a youth activist and artist, initially met biweekly and then monthly. Decisions were made by consensus. While the initial intent was to develop an intervention to address health inequities experienced by Caribbean Latino youth and young adults aged 16 to 24, the community advisory board was instrumental in shifting the focus to the critical age of middle school youth, and adopting pedagogical principles of racial justice and youth empowerment rather than traditional top-down strategies to change behavior. The community advisory board also insisted on hiring successful young adults of color to recruit participants and facilitate the intervention, ensuring ties to the local community.

The intervention was designed to increase the number of engaged citizens; promote awareness of the social determinants of nutrition, physical activity, and stress; and decrease stress. It employed an assets-based, youth-engaged approach, which included applied inquiry, critical pedagogy, and organizing. Sixty-eight enrolled middle school youths learned about social determinants and conducted exploratory research in their communities [31]. Post-intervention surveys showed that participants had significant improvement in knowledge about nutrition, physical activity, and community-level factors that influence those behaviors. Involved youth felt they gained skills to change their community for the better [32]. The intervention had additional community-level successes: the program at the school was sustained through the local public health department, which provided employment to facilitators and participants, and the NIMHD funded a new science education program at the school.

This case study illustrates the value of power-sharing through the use of a community advisory board with decision-making power. By acknowledging inequitable power dynamics and sharing power through an MOA, an environment was created that enabled the team to revise research eligibility criteria. In addition, community empowerment led to explicit discussions about exclusionary language, such as jargon, and acknowledgment of roles with respect to race, gender, socioeconomic status, positionality, and power in the academy or community [33]. Trusting the process and the expertise of the community advisory board allowed the NFS to develop a meaningful intervention oriented toward empowerment.

11.4.4 Participatory Health Impact Assessment to Inform Tribal Policy

Tribal sovereignty uniquely positions Native American nations to conduct CBPR to implement policy and environmental interventions. However, few such interventions exist, for reasons that include lack of familiarity by academic researchers with tribal policymaking processes as well as the lack of a CBPR orientation to engage tribal leaders. The Tribal Health and Resilience In Vulnerable Environments (THRIVE) study is a partnership formed in 2010 between University of Oklahoma researchers and Chickasaw and Choctaw Nations tribal health leaders [34]. The University's Principal Investigator, a Choctaw citizen, aimed to improve tribal food environments through policy, an approach that interested tribal health planners because this complemented individually focused clinical and health education programs.

The first year focused on partnership development. Academic and tribal health partners hoped to integrate tribal commerce and health resources to improve tribal food environments; however, health planners had yet to work with tribal commerce. The partners proposed implementing "healthy makeovers" in eight tribal convenience stores and shared the proposal with tribal commerce leaders, securing letters of support from new partners who endorsed the study but did not commit to participating. When NIH funding was obtained, it became critical to strengthen these new and somewhat tenuous relationships.

Using a Health Impact Assessment planning framework and a participatory orientation, partners engaged tribal government, commerce, and health leaders to identify common language and areas of overlap and mutual benefit. Twenty-four partners (12 per nation) met every three weeks, with decisions made by consensus. Early on, both nations established explicit financial and research memoranda of understanding with the University. In the first 2 years, the tribal Institutional Review Boards (IRBs) oversaw the partnership and the IRB chairs participated in most meetings to provide technical assistance and ensure community benefit. While commerce leaders were not focused on improving health, they were interested in offering new fresh foods to boost sales. In exchange for data on food preferences, marketing, and cost, they procured fresh vegetable and fruit quick packs, bottled water, salads, wraps, and other healthy snacks and displayed them prominently in stores. Tribal leaders were less concerned than expected about possible short-term revenue losses should healthy foods not sell, and highly interested in potential cost savings that healthy foods might have on tribal health. They also welcomed sending tribal citizens the message that they cared about their health.

While data have not been fully analyzed for the intervention's effect on primary outcomes (i.e., vegetable and fruit intake and body mass index), important changes to tribal food environments have been made. Significant increases in fruit and vegetable availability have been found in stores participating in the intervention. Initial sales data show that some items, including bottled water, nearly doubled in sales when compared to the same time period in the previous year [35]. The Choctaw Nation has since expanded the intervention to all 16 of its stores. Within both nations, study findings are providing solid data of a positive health impact as a foundation for evidence-based policy formulation to improve health.

This case study illustrates the importance of trust, which was promoted by having an academic research leader who was from the community and a process developed through participatory decision making. With tribal nations wielding significant political and economic power, power sharing during the research led to better implementation and sustainable system change, outcomes that are at least as important as individual health and behavior changes.

11.4.5 Pacific Islander "Let's Move"

Weaving an Islander Network for Cancer Awareness, Research, and Training (WINCART) was an NCI-funded Community Network Program Center (CNPC) to address cancer health disparities among Native Hawaiians and Pacific Islanders in Southern California [36]. In 2011, WINCART facilitated the development of the "Pacific Islander Let's Move" campaign, led by a steering committee of community leaders and university researchers. The committee applied the NCI Research-Tested Intervention Programs (RTIPs) adaptation guidelines to tailor an evidence-based program called "Instant Recess" [37] to be culturally and community appropriate for Pacific Islanders. "Instant Recess" developed educational materials, recruited social organizations to host the program, trained an organizational champion to facilitate activities, and conducted pre- and post-assessments of organizational and individual change.

Over 6 months, the program goals and activities were planned through shared control and leadership of the steering committee. CBPR was used throughout the planning, implementation, and evaluation processes. Power-sharing is a key aspect of CBPR, and hence, each meeting was cochaired by a community representative and a WINCART health educator, with remaining members comprising of 11 community organization representatives from Pacific Islander communities and researchers from three universities. Through participatory decision making, community representatives identified the goal of promoting physical activity targeting adults and their social environments. After reviewing research literature, it was the WINCART health educator who proposed the use of "Instant Recess" and the RTIPs guidelines.

The final program included four components: identification of an organizational program champion, establishment of a verbal policy to promote physical activity at every organizational meeting, creation of an environment conducive to physical activity through prominent display of an in-language poster and provision of space, and showing of a 10-minute video at every organizational meeting. Eight organizations, including churches, a support group, and neighborhood and youth groups, were recruited to pilot test the program. At baseline, only one organization had a physical activity program, and none specifically for Pacific Islanders. After 2 months, seven organizations had fully implemented all components of the "Let's Move" program. At baseline, over half of the 106 adult participants reported not being physically active. At the 2-month follow-up, participants increased the number of days they engaged in medium-intensity activity from 2.09 to 2.90 days per week.

Because of these positive results, the steering committee expressed interest in disseminating their materials through http://CES4Health.info, a peer-reviewed website [38]. This case study is an example of the importance of early involvement of community organizations in the development of research ideas and interventions, as well as of the value of power sharing and participatory decision making.

These five case studies showcase the principles and promising practices within CBPR and community engagement that have contributed to positive outcomes, many of which were identified within the Research for Improved Health study. These examples also illustrate the range of outcomes achievable through equitable and respectful partnering practices, including intermediate outcomes of increased sustainability, empowerment, capacity development, and policy changes that promote health and health equity outcomes.

11.5 Anticipating Challenges and Opportunities in CBPR

A key challenge in initiating CBPR is the preliminary, often unfunded, work necessary to establish the relationships and trust needed to identify a common research question and approach. On the funders' side, it is important to fund early partnerships through development grants, such as institutional pilot grants, the NCI Special Populations Network, or tiered Patient-Centered Outcomes Research Institute (PCORI) grants, even if they do not lead directly to fully realized research projects. Investigators can nurture connections with community leaders, either through personal relationships (easier if the researcher or other academic staff are from the community) or by gaining trust by proxy through researchers who have established trust with the community.

Time is needed to conduct CBPR effectively. Establishing trust and good working relationships with communities of color requires showing up and responding to community needs first, before asking research questions. This can be challenging for junior researchers, who face deadlines for productivity in obtaining grants and publishing results. Yet junior researchers, especially scholars of color, are often committed to the dual responsibilities of their communities and the academy. Encouraging them to join existing teams and supporting them to publish about partnership processes rather than waiting for outcomes can jumpstart publications.

A third challenge and opportunity is that of power sharing between academic investigators and community leaders. Aspects of this challenge can include who sets the agenda and chairs meetings to the details of decision-making processes. In addition, budgeting for a CBPR project is often a good indication of the power dynamics of the partnership, and researchers need to understand that sharing financial resources is a promising practice that is beneficial to both community and academic partners. Multiple Principal Investigator mechanisms, often with community Principal Investigators, enable CBPR partnerships to clearly delineate roles, which point out the need for a clear and mutually developed MOA or subcontracts. Subcontracting can bring mutual advantages. The community partner with the most culturally grounded staff brings deep knowledge of the community. Researchers bring specific research knowledge. Bringing this experience together enhances the local fit of the research design and the depth and sustainability of interventions. It can also promote other outcomes, such as increased community partner capacity to apply research skills to program evaluation.

Because so much of health inequities is the result of upstream determinants, such as educational disparities, poverty, and injustice, CBPR practitioners constantly have to deal with ever-changing policy and political environments. A program that may seem sustainable at the time of grant writing may not have the same support from policymakers at the completion of the project. However, CBPR practitioners, through their connections with stakeholders, often have the mechanisms in place to influence policy changes. Interventions that incorporate policy change add substantially to the possibility of long-term health equity outcomes.

11.6 Research Support for CBPR

The NIH, Centers for Disease Control and Prevention (CDC), and other research funders have led the way by providing specific funding initiatives for building community partnerships, training researchers and community members, conducting research studies to address health disparities, and establishing peer review panels with appropriate expertise [39]. The CDC began to fund community participation in research in 1986 through the Prevention Research Centers, and in 1999 through Racial Ethnic Approaches to Community Health capacity grants. The NIH began funding CBPR with a National Institute of Environmental Health Sciences 1995 initiative to address environmental injustices. Other initiatives followed, including the National Institute of Mental Health's Interventions and Practice Research Infrastructure Program, the NCI's SPNs, Community Networks Programs (CNPs), and, subsequently, the CNPCs [40]. Similar NIH CBPR efforts included Native American Research Centers for Health, a partnership between the Indian Health Service and NIH, and transdisciplinary health disparities research centers [41]. Recognizing the importance of a long-term commitment, NIMHD [8] created a unique CBPR funding stream in 2005, with 11 years of phased funding possible for planning, intervention, and dissemination. Since the mid-2000s, additional funding has included community engagement components within Clinical Translational Science Awards. The creation of PCORI extended the principles of CBPR to patient-engaged research.

Federal and health professional support has also been recently reported, such as the Institute of Medicine's call for educating health professionals to work with community partnerships to address social determinants; the Office of Behavioral and Social Science Research report on understanding cultural complexities in working with communities; and the North American Primary Care Group's recent policy on greater patient and community involvement in research. In sum, for close to three decades, NIH and other institutions have recognized the importance of CBPR and community engagement in research, particularly to reduce health disparities among underserved health disparities populations.

11.7 Conclusions

CBPR has evolved from a principled approach to conducting research through a social justice lens to an established research orientation that can provide a strong evidence-based and sustainable approach to reducing health disparities. CBPR principles and stages can be applied to similar orientations, such as patient-centered or practice-based research, as long as the appropriate stakeholders are identified. Conducting CBPR effectively requires an understanding of how to be an equitable partner, how to build capacity, and how to create sustainable changes, in addition to an understanding of traditional research skills. These are fundamental skills and values needed for any type of transdisciplinary research with the goal of reducing health disparities and achieving health equity.

11.8 Key Points

- CBPR is a research approach that can provide a strong evidence-based and sustainable approach to reducing health disparities.
- Key promising CBPR practices include building on existing partnership capacity, involving community members in all steps of research, fostering trust, and sharing power and resources with partners.

- Each CBPR project is unique depending on local needs and capacities as well as prior history of collaborations.
- CBPR outcomes include increased sustainability, empowerment, capacity development, and policy changes as well as better health and health equity.

Disclaimer

The views and opinions expressed in this chapter are those of the authors only and do not necessarily represent the views, official policy, or position of the U.S. Department of Health and Human Services or any of its affiliated institutions or agencies.

References

1 W. K. Kellogg Foundation Community Health Scholars Program (2001). *Stories of impact [brochure]*. Ann Arbor, MI: University of Michigan, School of Public Health, Community Health Scholars Program, National Program Office.

2 Wallerstein, N., Duran, B., Oetzel, J., and Minkler, M. (eds.) (2018). *Community-Based Participatory Research for Health: Advancing Social and Health Equity*, 3e. Jossey-Bass: San Francisco, CA.

3 Dutta, M.J., Anaele, A., and Jones, C. (2013). Voices of hunger: addressing health disparities through the culture-centered approach. *Journal of Communication* 63: 159–180.

4 Green, L.W. (2006). Public health asks of systems science: to advance our evidence-based practice, can you help us get more practice-based evidence? *American Journal of Public Health* 96: 406–409.

5 Yano, E.M., Green, L.W., Glanz, K. et al. (2012). Implementation and spread of interventions into the multilevel context of routine practice and policy: implications for the cancer care continuum. *Monographs-Journal of the National Cancer Institute* 44: 86.

6 Israel, B.A., Eng, E., Schulz, A.J., and Parker, E.A. (eds.) (2013). *Methods in Community-Based Participatory Research for Health*, 2e. Jossey-Bass: San Francisco, CA.

7 Muhammad, M., Wallerstein, N., Sussman, A.L. et al. (2015). Reflections on researcher identity and power: the impact of positionality on Community Based Participatory Research (CBPR) processes and outcomes. *Critical Sociology* 41 (7–8): 1045–1063.

8 Dankwa-Mullan, I., Rhee, K.B., Williams, K. et al. (2010). The science of eliminating health disparities: summary and analysis of the NIH summit recommendations. *American Journal of Public Health* 1 (100 Suppl 1): S12–S18. https://doi.org/10.2105/AJPH.2010.191619. Epub 2010 Feb 10.

9 Wallerstein, N. and Duran, B. (2010). Community-based participatory research contributions to intervention research: the intersection of science and practice to improve health equity. *American Journal of Public Health* 100 (Suppl 1): S40.

10 McCloskey, D.J., Aguilar-Gaxiola, S., Michener, J.L. et al. (2011). *Principles of Community Engagement*, 2e. National Institutes of Health Publication No. 11-7782. Bethesda, MD: CTSA Community Engagement Key Function Committee Task Force on the Principles of

Community Engagement. https://www.atsdr.cdc.gov/communityengagement/pdf/PCE_Report_508_FINAL.pdf (accessed December 21, 2017).

11 Viswanathan, M., Ammerman, A., Eng, E. et al. (2004). *Community-Based Participatory Research: Assessing the Evidence*. Summary, Evidence Report/Technology Assessment No. 99. Publication 04-E022-1. Rockville, MD: Agency for Healthcare Research and Quality.

12 Anderson, L.M., Adeney, K.L., Shinn, C. et al. (2015). Community coalition-driven interventions to reduce health disparities among racial and ethnic minority populations. *Cochrane Database of Systematic Reviews* 6: CD009905. https://doi.org/10.1002/14651858.CD009905.pub2.

13 de las Nueces, D., Hacker, K., DiGirolamo, A., and Hicks, L.S. (2012). A systematic review of community-based participatory research to enhance clinical trials in racial and ethnic minority groups. *Health Services Research* 47: 1363–1386.

14 Drahota, A., Meza, R., Brikho, B. et al. (2016). Community-academic partnerships: a systematic review of the state of the literature and recommendations for future research. *Milbank Quarterly* 94 (1): 163–214.

15 O'Mara-Eves, A., Brunton, G., Oliver, S. et al. (2015). The effectiveness of community engagement in public health interventions for disadvantaged groups: a meta-analysis. *BMC Public Health* 15 (1): 1352.

16 Cyril, S., Smith, B.J., Possamai-Inesedy, A., and Andre, M.N.R. (2015). Exploring the role of community engagement in improving the health of disadvantaged populations: a systematic review. *Global Health Action* 8: 1–12.

17 Salimi, Y., Shahandeh, K., Malekafzali, H. et al. (2012). Is Community-Based Participatory Research (CBPR) useful? A systematic review on papers in a decade. *International Journal of Preventive Medicine* 3 (6): 386–393.

18 Brunton, G., O'Mara-Eves, A., and Thomas, J. (2014). The "active ingredients" for successful community engagement with disadvantaged expectant and new mothers: a qualitative comparative analysis. *Journal of Advanced Nursing* 70: 2847–2860.

19 Minkler, M., Garcia, A., Rubin, V., and Wallerstein, N. (2012). *Community-Based Participatory Research: A Strategy For Building Healthy Communities And Promoting Health Through Policy Change*. Berkeley, CA: *PolicyLink*, U.C. Berkeley School of Public Health (accessed June 16, 2020). http://www.policylink.org/sites/default/files/CBPR.pdf.

20 Cacari Stone, L., Minkler, M., Freudenberg, N., and Themba, M.N. (2018). Community-based participatory research for health equity policymaking: advancing social and health equity. In: *CBPR for Health: Advancing Social and Health Equity*, 3e (eds. N. Wallerstein, B. Duran, J. Oetzel and M. Minkler). San Francisco, CA: Jossey-Bass.

21 Oetzel, J.G., Zhou, C., Duran, B. et al. (2015). Establishing the psychometric properties of constructs in a community-based participatory research conceptual model. *American Journal of Health Promotion* 29 (5): e188–e202.

22 Lucero, J., Wallerstein, N., Duran, B. et al. (2016). Development of a mixed methods investigation of process and outcomes of community based participatory research. *Journal of Mixed Methods* 12: 1. https://doi.org/10.1177/1558689816633309.

23 Duran, B., Oetzel, J., Magarati, M., Parker, M., Zhou, C.,Roubideaux, Y, Muhammad, M., Pearson, C., Belone, L., Kastelic, S.H., Wallerstein, N. Promising Practices and Outcomes: Learnings from a CBPR cross-site national study, Progress in CommunityHealth Partnerships: Research, Education, and Action, 2019, 13 (4), 337–352.

24 Wallerstein, N., Oetzel, J., Sanchez-Youngman,S., Boursaw, B., Dickson, E., Kastelic, S., Koegel, P., Lucero, J., Magarati, M., Ortiz, K., Parker, M., Peña, J., Richmond, A., Duran, B., (2020). Engage for Equity: A Long-Term Study of Community Based ParticipatoryResearch and Community Engaged Research Practices and Outcomes, Health Education and Behavior, 47(3): 380–390.

25 Oetzel JG, Wallerstein N, Duran B, Sanchez-Youngman S, Nguyen T, Woo K, Wang J, Schulz A, Keawe'aimoku Kaholokula J, Israel B, Alegria M. Impact of Participatory Health Research: A Test of the Community-Based Participatory Research Conceptual Model. Biomed Res Int. 2018 Apr 24;2018:7281405.

26 Wallerstein, N. (2018). Appendix 10: Instruments and measures for evaluating community engagement and partnerships. In: *Community-Based Participatory Research for Health: Advancing Social and Health Equity*, 3e (eds. N. Wallerstein, B. Duran, J. Oetzel and M. Minkler). San Francisco, CA: Jossey-Bass.

27 Williams, A., Pichon, L.C., Wiley, E., and Obong'o, C. (2017). Forecast of HIV Prevention: Using Pre-Exposure Prophylaxis (PrEP) to Change Community Environments. *Poster presented at the 145th annual meeting of the American Public Health Association*, Atlanta, GA.

28 Nguyen, T.T., Tsoh, J.Y., Woo, K. et al. (2017). Colorectal cancer screening and Chinese Americans: efficacy of lay health worker outreach and print materials. *American Journal of Preventive Medicine* 52: e67–e76.

29 Jih, J., Le, G., Woo, K. et al. (2016). Educational interventions to promote healthy nutrition and physical activity among older Chinese Americans: a cluster-randomized trial. *American Journal of Public Health* 106: 1092–1098.

30 Martinez, L.S., Ndulue, U., and Peréa, F.C. (2011). Nuestro Futuro Saludable: connecting public health research and community development in partnership to build a healthy environment. *Community Development* 42: 255–267.

31 Sprague Martinez, L.S., Gute, D.M., Ndulue, U. et al. (2012). All public health is local revisiting the importance of local sanitation through the eyes of youth. *American Journal of Public Health* 102: 1058–1060.

32 Sprague Martinez, L., Reich, A.J., Flores, C.A. et al. (2017). Critical discourse, applied inquiry and public health action with urban middle school students: lessons learned engaging youth in critical service-learning. *Journal of Community Practice* 25: 68–89.

33 Ndulue, U., Peréa, F.C., Kayou, B., and Martinez, L.S. (2012). Team-building activities as strategies for improving community-university partnerships: lessons learned from Nuestro Futuro Saludable. *Progress in Community Health Partnerships* 6: 213–218.

34 Jernigan, V.B.B., Wetherill, M., Hearod, J. et al. (2016). Cardiovascular disease risk factors and health outcomes among American Indians in Oklahoma: the THRIVE study. *Journal of Racial and Ethnic Health Disparities* 4: 1061–1068.

35 Jernigan, V.B.B., Wetherill, M., Hearod, J. et al. (2017). Food insecurity and chronic diseases among American Indians in rural Oklahoma: the THRIVE Study. *American Journal of Public Health* 107 (3): 441–446.

36 Tanjasiri, S.P. and Tran, J. (2008). Community capacity for cancer control collaboration: weaving an islander network for cancer awareness, research and training. *Cancer Detection and Prevention* 32S: S37–S40.

37 Yancey, A.K. (2010). *Instant Recess: Building a Fit Nation 10 Minutes at a Time*. Berkeley and Los Angeles, CA: University of California Press.

38 LaBreche, M., Cheri, A., Custodio, H. et al. (2016). Let's move for Pacific islander communities: an evidence-based intervention to increase physical activity. *Journal of Cancer Education* 31 (2): 261–267.

39 Leviton, L. and Green, L.W. (2018). Funding for CBPR re-considered. In: *CBPR for Health: Advancing Social and Health Equity*, 3e (eds. N. Wallerstein, B. Duran, J. Oetzel and M. Minkler). San Francisco, CA: Jossey-Bass.

40 Braun, K.L., Nguyen, T.T., Tanjasiri, S.P. et al. (2012). Operationalization of community-based participatory research principles: assessment of the National Cancer Institute's Community Network Programs. *American Journal of Public Health* 102 (6): 1195–1203.

41 Cooper, L.A., Ortega, A.N., Ammerman, A.S. et al. (2015). Calling for a bold new vision of health disparities intervention research. *American Journal of Public Health* 105 (Suppl 3): S374–S376.

12

Racial/Ethnic Health and Healthcare Disparities Measurement

The Application of the Principles and Methods of Causal Inference

Benjamin Cook[1, 3], *Sarah Forrester*[2], *Timothy Creedon*[3], *Joan Wasserman*[4], *Meryl Sufian*[5], *and Jeroan Allison*[2]

[1] *Harvard Medical School, Boston, MA, USA*
[2] *University of Massachusetts Medical School, Worcester, MA, USA*
[3] *Cambridge Health Alliance, Cambridge, MA, USA*
[4] *Uniformed Services University of the Health Sciences, Bethesda, MD, USA*
[5] *National Institute on Minority Health and Health Disparities, National Institutes of Health, Bethesda, MD, USA*

12.1 Introduction

This chapter describes methods of measuring health disparities and of identifying potential mechanisms underlying those disparities using statistical tools that support causal inference. Historically, disparities are often described as a difference between racial/ethnic minorities and Whites, in which minorities are disadvantaged. In this chapter, we provide examples related to racial/ethnic disparities but recognize that disparities exist between groups of any kind that are treated differently by people and institutions with greater social, economic, and political power. We also want to strongly emphasize that race is a social and political construct. We refer to race in this chapter as a variable used in statistical models that is sometimes useful in demonstrating important differences in healthcare access, process, and outcomes. This problem is only partially attenuated by our use of the expanded term, race/ethnicity, because ethnicity refers to a broader social, political, and cultural context.

First, we provide a background on how disparities have been defined, encouraging researchers to be explicit about their definition of disparities and to be mindful that any analytical method used to measure disparities carries an implicit definition of disparity, which may or may not fit with principles of fairness and equity. Second, we borrow ideas underlying the principle of the counterfactual, common to models of causal inference, to explore analytical methods that measure racial/ethnic disparities as due to unfair or unjust differences experienced by racial/ethnic minority groups but not due to other fair or just differences. Third, we extend these ideas to show how racial/ethnic differences can be broken down into more manageable parts, or "decomposed." Fourth, we investigate methods that identify points of intervention along the causal chain from race/ethnicity to health,

The Science of Health Disparities Research, First Edition. Edited by Irene Dankwa-Mullan, Eliseo J. Pérez-Stable, Kevin L. Gardner, Xinzhi Zhang, and Adelaida M. Rosario.

and discuss the larger question of how to understand race/ethnicity as a variable that can "cause" negative health outcomes.

In the second segment of the chapter, we assess statistical methods used to evaluate causal effects in intervention studies, including experimental and quasi-experimental methods (propensity score, instrumental variable [IV], and difference-in-difference). Although many of these techniques are frequently applied to experimental studies, the emphasis in this chapter will be on observational designs.

This chapter also discusses how analytical methods can be used to understand and measure disparities, and to test interventions aimed at reducing those disparities, with an emphasis on observational data.

12.2 Aligning Definitions of Disparity with Appropriate Statistical Methods

Progress in the measurement and tracking of racial/ethnic health and healthcare disparities requires agreement on definitions and methods of measuring disparities. Race/ethnicity is used throughout this chapter as a shorthand representation of the combined categories of race (e.g., Black, White, Asian, Native American) and ethnicity (Latino and non-Latino). These categories have been socially constructed to distinguish individuals based on physical features that groups and cultures consider to be socially significant (race) and to distinguish individuals based on shared culture, such as language, ancestry, and beliefs (ethnicity). We recognize the problematic use of these socially constructed categories but also their importance, given their strong linkages to social position, policy, politics, and interpersonal relationships, and their wide use in the medical and health services research literature.

Because the term "disparity" implies not only difference but also unfairness and inequality, reaching consensus on the definition of disparity is difficult and inherently controversial. More broadly, the measurement and tracking of health and healthcare disparities requires a conceptual framework about how disparities arise and what we should hold the healthcare system and other stakeholders "accountable for" as we assess progress in reducing disparities over time.

To implement definitions of health and healthcare disparities, statistical tools are needed that match the concepts underlying the definition. Statisticians and methodologists have long been interested in causal inference, or the ability to draw conclusions related to cause and effect.

The Rubin Causal Model (RCM) has been especially influential as a model to test and assess causal hypotheses [1–3]. However, the use of causal modeling in racial/ethnic comparisons is controversial because it relies on the notion that race and ethnicity can be assigned or manipulated. However, we intend to show how *certain principles* of causal models are useful for measuring and tracking racial/ethnic health and healthcare disparities and in identifying the mechanisms of disparities that will lead to interventions that reduce these disparities.

We recognize that the frequent use of the White population as the referent group in health disparities research is problematic in that it may engender a focus on the weaknesses instead of the strengths of already marginalized communities. Imposing standards from a dominant social group upon a marginalized group can be considered a form of "cultural imperialism" and may lead to harmful stereotyping [4].

To counterbalance these concerns, we are enthusiastic about work that looks within racial/ethnic subgroups to identify strengths that may be leveraged for positive change. An example of this approach lies in community-informed research, which, although relevant, is beyond the scope of this chapter. Further, in some instances, there are not enough data about particular racial/ethnic subgroups to use as a basis of comparison. Consequently, researchers are encouraged to provide sufficient rationale for choosing the referent group being used, whether it is White or otherwise.

12.2.1 Definitions of Racial/Ethnic Healthcare Disparity

We specifically focus in this section on *healthcare* disparities, recognizing healthcare as just one of the domains of influence on health disparities. Methods applicable to the measurement and understanding of health disparities are discussed below. Researchers are encouraged to use the National Institute on Minority Health and Health Disparities Research Framework as a conceptual model to guide methods for understanding and addressing the complex and multifaceted nature of minority health and health disparities.

Three commonly used healthcare disparity definitions and accompanying methods have been identified and compared [5, 6] (see Table 12.1). The first focuses on the race/ethnicity coefficient in a multivariate regression model, implicitly defining a racial/ethnic disparity as "residual direct effect" of race/ethnicity, after adjustment for all other available covariates. The second definition considers disparities to be *all differences* among populations in measures of health and healthcare. A third definition is the Institute of Medicine (IOM) definition from *Unequal Treatment* [10], which posits that disparities are all differences except those due to clinical appropriateness, patient needs, and patient preferences. This definition is distinct from the other definitions in that it requires the analyst to carefully distinguish between confounding variables that should be accounted for via statistical adjustment (e.g., clinical appropriateness, patient need, preferences) and those variables considered to be underlying components of the overall disparity that should not be adjusted for (e.g., institutional factors, interpersonal discrimination). Implementation of the IOM definition requires the analyst to fully specify a regression model of healthcare utilization, and then in a subsequent step, to identify, or "partial out" those differences that are amenable to healthcare system change. Adjusting for some variables but not others can also be thought of as the creation of counterfactual scenarios, a concept illustrated by the conceptual framework established in Rubin's Causal Model [2].

12.3 A Method of Measuring Healthcare Disparities Incorporating a "Counterfactual" Scenario

The RCM provides a useful framework for creating counterfactual scenarios and using them in statistical analysis. According to the potential outcomes framework [2], causal impact is measured as the difference between an individual being exposed to a treatment and the same individual not being exposed to a treatment. Obviously, only one of these

Table 12.1 Summary of three main definitions and methods of disparities measurement with examples, pros, and cons [6–9].

Definition	Method	Example	Pros	Cons
Residual direct effect	Compare mean White healthcare use with mean healthcare use in a counterfactual group that is similar to the White group in all aspects except having racial/ethnic minority identity	Logit results for models of any reported ambulatory visits by year 1997 (reported as odds ratios with Whites as comparison group). Black OR = 0.46 (0.35, 0.59); Latino OR = 0.7 (0.56, 0.89) (Guevara et al. 2006)	Readily interpreted from standard regression output. Identifies independent effect of race	Adjusts for racial/ethnic differences in care that are due to SES variables such as income, insurance, and education
Disparities equal to unadjusted means	Compare mean White healthcare use with mean healthcare use of racial/ethnic minority group	Percentage of people who had an office-based or outpatient department visit in the calendar year, by race and ethnicity, 2007. White 75.3% versus black 65.7% versus Hispanic 59.1% versus Asian 62.7% (AHRQ **2010**)	Easily computable and understandable comparison	Does not account for health status and age differences between populations
Institute of Medicine (IOM) definition	Compare mean White healthcare use with mean healthcare use of a counterfactual group that has the SES characteristics and racial/ethnic identity of the racial/ethnic group but the health status of the White group	Predicted probability of having any mental healthcare use in the last year, 2007: 15.7% Whites versus 7.0% Blacks versus 8.3% Latinos (Cook et al. 2010)	Accounts for racial/ethnic differences by health status, age, and gender	Uses a rank and replace prediction method that is more computationally intensive than other methods

Source: Based on Cook et al. (2009) [6].

scenarios can be true (an individual is either exposed or not exposed to a treatment). Thus, one of the two scenarios is "counterfactual," or something posited for the sake of comparison. Every unit of analysis (i.e., persons with healthcare needs) has different potential outcomes that depend on the conditions or interventions they are "assigned" to. With the potential outcomes framework in mind, health services researchers can design studies that test counterfactuals: What would happen to a person's health and the healthcare they receive if their race or ethnicity were "assigned" differently, holding all else constant?

The generation of a counterfactual scenario is relevant for implementing the IOM definition or other definitions that recognize that some factors (unjust or unfair differences) contribute to the disparity while other factors (just or fair differences) should be accounted for and not considered part of the disparity. As an example, a researcher implementing the IOM definition might compare the healthcare utilization of two racial/ethnic groups (e.g., "White" and "Black") by comparing Whites to a counterfactual group of Blacks that have the health status of Whites but maintain their own socioeconomic (SES) characteristics and experiences with discrimination. McGuire et al. [11] and Cook et al. [5, 6] describe a number of methods (e.g., rank and replace or propensity score-based methods) of implementing this type of counterfactual analysis. This framework borrows the idea of the counterfactual from the potential outcomes framework of the RCM but the intention is not to make causal claims. Instead, the counterfactual scenario is used to allow the researcher to carefully consider how multiple variables do, or do not, contribute to a disparity and how they should enter into a healthcare disparity calculation.

12.4 Extending the Use of "Partial Differencing" to Identifying Targets for Health and Healthcare Disparities Reduction Using Decomposition Methods

One methodological strategy for identifying potential candidates for interventions that reduce health and healthcare disparities is the decomposition of the differences between racial/ethnic groups using Oaxaca-Blinder methods [12, 13] or extensions for nonlinear models [14]. This method allows the analyst to identify exactly how much of the overall racial/ethnic difference can be explained by each observable covariate.

To implement decomposition in linear regression models, an ordinary least squares regression model is estimated for one of the subgroups (e.g., a group of White individuals) to describe the associations between the dependent variable (e.g., access to treatment) and the independent variables of interest (e.g., income, insurance status, health status, age, sex). The same model is estimated for the comparison group (e.g., racial/ethnic minority group). Next, these models are used to generate predictions of the dependent variable for the two groups. To assess the contribution of racial/ethnic differences in the distribution of each variable to the overall difference, we subtract the actual predicted mean from a counterfactual predicted mean calculated by replacing the minority mean with the White mean of that variable while holding the distributions of the other variables constant. To assess the contribution of racial/ethnic differences in the coefficient, or "return," of each variable to the overall difference, we subtract the actual predicted mean from a counterfactual predicted mean calculated by replacing the minority coefficient with the White coefficient of that variable without changing the other coefficients. Doing this for each variable in the model, one can assess how much of the overall difference between groups is explained by the distribution and coefficient of each variable. In doing so, we can identify variables (e.g., racial/ethnic differences in health insurance) that are amenable to intervention. More detail about these decomposition methods, adaptation of the methods to nonlinear models [14] and applications to disparities studies can be seen in prior studies [15–17].

12.5 Adapting Causal Models to Identify Interventions to Reduce Racial/Ethnic Health Disparities

12.5.1 Mediators of the Relationship Between Race/Ethnicity and Health Outcomes

Another application of the causal inference framework to understanding health and healthcare disparities is to assess how the relationship between race/ethnicity and health outcomes is mediated by other variables that are more amenable to manipulation and that lie upon the causal pathway from race to health outcomes. For example, if a measure of current income or other SES variable is in the regression model, then an assessment of how race is related to the health outcome at different levels of SES can be inferred. This can be considered to be a "weak interpretation" of the race effect [18]. The degree to which the health inequality could be reduced by intervening on inequalities in SES can be assessed by subtracting the difference in race coefficients in models with and without the SES covariate (the mediating effect of SES or "mediation racial inequality measure").

Assessing the race/ethnicity coefficient with and without adjustment for the SES covariate provides information about how much change in the health outcome would occur if the racial/ethnic minority population exchanged their SES distribution so that it equaled the SES distribution of the White group. The ideas underlying the estimation of this mediated effect of race is part of a vast literature on mediation and the complexities of estimation and interpretation of mediation analyses [19–23].

A true understanding of the causal pathway linking an individual's or group's racial/ethnic identity to health and healthcare outcomes requires disentangling or somehow manipulating the multiple phenotypic, genetic, and cultural constructs (typically unavailable in data used in health disparities research) that comprise the race/ethnicity variable. This requires the difficult, if not impossible, task of manipulating race/ethnicity or somehow identifying counterfactual scenarios that alter an individual's racial/ethnic category. As an alternative, methods that assess how observable variables mediate the relationship between race/ethnicity and health may be more feasible, and can help to preliminarily identify pathways for disparities reduction. As with other mediation analyses, the identification of plausible pathways for intervention requires that the mediator (SES, in the examples above) be unconfounded and temporally located after the "assignment" of race. Mediation analysis without experimental data provides only tentative causal results. Interactions between race/ethnicity and mediators can also be added to these models to gain further insights into the sources and pathways of disparities.

References to the rich literature in indirect and direct effects, and the requirements for assessing mediation, are provided in VanderWeele and Robinson [18]. In the presence of data on variables involved in the multiple underlying pathways between race and health, a more complex understanding of multiple mediating pathways can be achieved with the use of structural equation modeling, an analytical method deserving of more space than the current chapter allows and that is well described in numerous other articles [24, 25].

12.6 Investigating the Meaning of the Race/Ethnicity Coefficient in Regression Models

The principles of causal inference can be used to question and explore the use of race/ethnicity coefficients in regression equations in studies of health disparities. In regression equations, the interpretation of the coefficient on the race/ethnicity variable is straightforward (descriptive), but may not be causal, while the utility of the magnitude and significance of that coefficient for developing interventions that reduce disparities may not be clear. In these models, if there is an association with race, it remains unknown without proper controls in the model if the race/ethnicity difference is due to discrimination, physical or genetic characteristics, unequal educational or economic opportunities, or a common covariate that is both correlated with race/ethnicity and the outcome [18].

By parsing the relationship between the race/ethnicity variable and health and healthcare outcomes into differences due to the influence of discrimination, socioeconomic status, past experiences that inform current treatment, etc., one starts to uncover the variables that have been reduced into a single race/ethnicity variable. The identification of a causal effect of race/ethnicity is reductionist and inherently difficult because it requires that all confounders are accounted for by statistical adjustment.

12.6.1 Can One Identify the Causal Effect of Race?

According to the principles of causal inference, the understanding of the causal effect of race/ethnicity requires the manipulation of race/ethnicity [26]. This means that the physical phenotype, parental physical phenotype, genetic background, and social and cultural context would have to be manipulable if causal effects of race/ethnicity are to be identified. In order to determine the causal effect of race, racial categories, differential perceptions and treatment of racial groups, and/or associations between race/ethnicity and health outcomes need to be modifiable [26, 27].

The question of whether race and ethnicity are conditions that can be changed hypothetically in a potential outcomes framework remains controversial. Less controversial is the idea that race and ethnicity are social constructs and that people and institutions respond to those of different racial/ethnic backgrounds in different ways. To the extent that interpersonal and societal responses to race/ethnicity frequently have been and remain discriminatory, resulting in disparities, we see that these responses to race/ethnicity are conditions that can be changed or manipulated. For example, there are a number of controlled experiments (called audit studies or "kindred" field experiments) that manipulate the perception of someone's race/ethnicity in order to assess discrimination. This type of study has been used to test for discrimination in employment [28] and housing [29] applications. Another example is the use of actors presenting identical symptoms with the race/ethnicity of the actor randomly selected [30].

In sum, causal inference methods present a powerful set of scientific tools for identifying and measuring health and healthcare disparities. Extending these methods to identify the causal pathway linking an individual's or group's racial/ethnic identity to health and healthcare is controversial and can only begin to be realized in tightly controlled situations that manipulate interpersonal and societal responses to an individual's race/ethnicity.

12.7 Statistical Methods Used to Evaluate Causal Effects in Intervention Studies

12.7.1 Experimental Methods: Randomized Controlled Trials and Field Experiments

When seeking to infer causation in research, the goal is to be able to state that some independent variable causes change in a dependent variable. In order to unequivocally state that one unit causes change in another unit, we would need to be able to observe the counterfactual, as described above. However, it is impossible to observe the counterfactual. Randomized controlled trials (RCTs), often used in clinical research, are considered the gold standard for causal inference. Although the counterfactual cannot be observed in RCTs, it is the closest one can get because random assignment ensures that the control and treatment groups are similar. Random assignment creates two or more treatment groups that are balanced on both measured and unmeasured confounders, generating a higher level of evidence than would be found in observational studies.

Slightly less rigorous experimentally are field experiments, which are conducted in real-life environments. The independent variable (e.g., the treatment) is still manipulated by the researcher. The main differences between RCTs and field experiments are that field experiments eliminate the laboratory setting and there are fewer mechanisms to control for extraneous variables. The elimination of the laboratory setting is desirable because the inferences are more applicable to real life. RCTs and field experiments in social science research are often neither feasible nor ethical. Due to these limitations, it can be difficult to generate true casual inference in the absence of randomization. Another alternative to the RCT is a natural or social experiment where naturally occurring events or new social policies or programs are introduced in which people have different exposures that researchers can document by collecting and analyzing data. Planned and natural experiments face some of the same issues with confounding and other threats to validity, such as loss to follow-up and inaccurate assessment of exposure and outcomes.

Over the past few decades, quasi-experimental methods have been utilized to infer causation from observational studies and data associated with natural or social experiments. Three such methods are: propensity scores, IVs, and difference-in-differences. To supplement the discussion provided below and to provide specific illustrations of how these methods are applied to address health disparities, we have selected a set of peer-reviewed manuscripts for further reading (Table 12.2).

12.7.2 Quasi-Experimental Methods: Propensity Scores

A propensity score simulates randomized studies by attempting to balance participant characteristics that would be randomly distributed among the groups being studied in an RCT, but are not in an observational study. Rosenbaum and Rubin [36] define the propensity score as the conditional probability of assignment to a treatment given a vector of observed covariates; or the propensity to a particular treatment given the vector of observed covariates. It is important to note that propensity score methods may create balance only on observed characteristics. Therefore, if unobserved characteristics not related to the

Table 12.2 Examples of Quasi-experimental methods in peer-reviewed literature.

Quasi-experimental method	Use	Literature
Propensity scores	Matching	Massoglia [31]
Propensity scores	Adjusting for demographic differences	Trivedi and Ayanian [32]
Propensity scores	Balancing	Landrum and Ayanian [33]
Instrumental variables	Utilized density of cardiologists to represent access to specialty care	Landrum and Ayanian [33]
Difference-in-differences	Estimated effect of health reform on economic disparities in joint replacement	Hanchante et al. [34] and Ryan et al. [35]

observed characteristics are present, they will not be balanced using this method [37]. Most often, propensity scores are determined through logistic regression, with the treatment assignment as the outcome and a number of covariates as predictors. This probability, or propensity score, is then used in analyses in three main ways: matching, adjustment, and stratification [38].

Propensity score matching involves developing propensity scores and then matching participants from the control and treatment groups based on similar scores. Although matching sounds basic, there are a number of schemes that can be used and there are no generally accepted criteria for when to use a particular matching type. A second way to use propensity scores is to adjust for the score in further analysis. Propensity score adjustment is similar to multivariable regression in that observed covariates are included in the model and adjusted for in order to create a level comparison between the treatment and control groups. The difference lies in the way in which the variables are included in the model. In propensity score adjustment, the covariates are summarized into a single score that is included in the model for adjustment. In multivariable regression, all the covariates are included in the model. Although the methods are similar, there is a tradeoff between variance, bias, and empirical power that is dependent on the number of covariates, the strength of the association between the exposure and the outcome, and the number of events per covariate. See Cepeda et al. [39] for a comparison of logistic regression and propensity score analysis.

The third use of propensity scores is stratification, in which patients are stratified based on their observed propensity score. Five strata, based on quintiles of propensity scores, are most common in the literature. Observed covariates are then checked within strata to ensure that there are no significant differences between the treatment and control groups. Stratification can then be used to adjust in further analysis, or the main analysis can include comparing the outcome by strata.

Although propensity-score methods offer a powerful approach with the potential to allow for casual inference in studies that would otherwise not be amenable to such inference, there are certain assumptions that must be met in order for propensity score analysis to lead to an unbiased estimate of effects; specifically, ignorable treatment assignment (conditional independence) and common support (all individuals have a positive predicted probability of being both participants and nonparticipants). Propensity-score methods can be conducted when these assumptions are violated, but the inference from any analysis would be suspect.

Because the covariates included in the propensity score are assumed to be the only variables that affect both the treatment and the outcome, it follows that a propensity score including these covariates could be used to create an analysis in which the likelihood of being in the treatment group is random [31], or ignorable. Treatment assignment and outcome become conditionally independent given the vector of covariates [33]. Thus, inferences made from such analyses become similar to inferences made if the treatment were randomly assigned. Though true and successful randomization is required to experimentally balance both known and unknown confounders, this assumption can only simulate random assignment. Since the purpose of propensity score analysis is to create balanced groups, participants need to have similar observed characteristics in order to be included in the analysis.

12.7.3 Quasi-Experimental Methods: Instrumental Variables

Another method for drawing causal inferences from observational data is known as instrumental variables (IVs) method. IVs are variables, or instruments, that are associated with treatment allocation but not with the outcome, except through the treatment [40]. Thus, they can be used in place of the treatment variable to predict the outcome. IV methods are intended to estimate causal effects when unobserved differences between treatment groups are present, as opposed to propensity score methods, which balance only observed confounders. Tying into the above discussion on RCTs, successful randomization may be considered as a strong instrument. Typically, the instrument is used to predict treatment independently of unobserved confounders. Patient outcomes are then compared based on the IV rather than the actual treatment.

As with propensity scores, the use of IV is subject to a number of assumptions, most of which are not directly verifiable, in order to draw valid inferences [41]. One assumption, ignorable treatment assignment, is shared by propensity score analysis and IV, but the following assumptions are unique to IV:

1) The stable unit treatment assumption (SUTVA) assumes that the treatment of interest when applied to the "intervention group" does not affect the outcome for those in the comparison group.
2) The IV must be related to the treatment or have a nonzero causal effect on the treatment. This is the most basic assumption for an IV, since the variable cannot be used in place of treatment assignment if it is not associated with it.
3) The exclusion restriction states that the IV cannot have an effect on the outcome except through the treatment variable. Although this assumption is not directly verifiable, analyzing available variables that have a theoretical association with the outcome may give the researcher an idea of whether or not this assumption is violated. If it is discovered that the IV does have an observable effect on the outcome aside from through treatment, the researcher will have to control for those effects in analysis.
4) Monotonicity implies that although the treatment may not affect everyone, anyone who is affected is affected in the same way.

12.7.4 Quasi-Experimental Methods: Difference-in-Differences

The DiD method is most commonly used in economic research to look at before and after policy changes and other types of natural or social experiments. Although DiD sounds similar to a simple pre-post design, it is different in that it accounts for any time-dependent trends that could confound a simple pre-post design by using a comparison group that is exposed to the same time trends. DiD is similar to the IV method in that it is specifically designed to account for unobservable differences, but it only accounts for time-fixed unobservable differences. It does, however, differ from IV in a couple of key ways: (i) DiD is a useful method when longitudinal or panel data are available, as it is built specifically to take advantage of these characteristics and does not rely on an IV that, as previously mentioned, can be hard to find. (ii) DiD can be estimated in a regression framework, so the key to this analysis is the assumptions that must be met in order for the DiD estimate to accurately reflect the difference in change over time or between groups. One should note, however, that errors are not independent in DiD models, so, at the very least, standard errors that account for the clustering of repeated outcomes with the observation unit should be used.

The first and most important assumption in any DiD analysis is the parallel or common trend assumption, which states that the outcome for groups under comparison would follow the same time trend in the absence of treatment [42, 43]. Although this assumption is hard to verify, the assumption is tested through the use of leads and lags. A lead is a period of time prior to intervention that can be tested to ensure that trends would be similar without introduction of the exposure. Often, an interaction between exposure and time is created, and if there is no significant difference in trends between the groups, then it can be inferred that the pretreatment trends are not significantly different. If there is a significant difference between the trends, then the researcher would need to find a different comparison group in order to produce unbiased results. Lags follow a similar concept but they test for effects for a period of time after the treatment is over. It is crucial to note that the outcome does not have to be equal for each group prior to the intervention. The length of leads and lags can vary and is dependent upon available data.

The second assumption, common shocks, states that any event that occurs during or after treatment will affect the control and treatment groups equally. The researcher assumes that nothing obvious happened during the study that affected the groups differently. Ryan et al. [35] have created a checklist for researchers to ensure that the elements needed for DiD are present and that the assumptions are met. The checklist also includes how to test and what to do if the assumption is violated.

One of the most important issues in any DiD analysis is choosing a comparison group, as the incorrect group can result in a biased estimation of effects. There are many options for selecting a comparison group, from the general method of using all untreated units or untreated units selected to be similar to the treated persons, to the specific method of matching based on observable characteristic or level of outcome at preintervention.

12.8 Conclusion and Limitations

In the first part of this chapter, we encouraged researchers to be explicit about the definition of disparities used in their research and understand how the definition aligns with principles of fairness and equity. Methods to implement conceptually grounded definitions of health and healthcare disparities are not always readily available in standard statistical packages.

We demonstrate how the implementation of the IOM definition of healthcare disparities requires the development of counterfactual comparison groups using novel techniques. Techniques based in the idea of the counterfactual scenario from the RCM have been applied in international health equity literature to measure the independent influence of selected individual characteristics (SES) on health [44, 45]. These studies use counterfactual methods to identify which partial inequalities are causes for concern for policymakers. Factors that lead to disparities that are beyond the control of the individual or that are ethically questionable may rise in priority and, thus, be differenced out. The chapter also introduces a number of methods (decomposition techniques, mediation analyses, structural equation modeling) that are recommended for understanding pathways underlying disparities (racial/ethnic differences) and identifying intervention targets.

Both experimental and quasi-experimental models have strengths and limitations for analyses used to inform health equity. The utility of these models has been the focus of the second part of this chapter. At this point, it is worthwhile to briefly reiterate previously mentioned limitations and add a few more. RCTs and clustered randomized trials are not always feasible or ethical. They are often prohibitively expensive, and no internal review board will approve a study that wants to subject participants to health disparities or discrimination. In addition, even when RCTs are warranted, noncompliers can muddy the results.

The commonly used methods to simulate randomization also have their limitations. Propensity scores are susceptible to unmeasured confounding, since propensity scores only account for observed confounders and the results are often not generalizable beyond the region of common support. This means that the results may not be generalizable to those that do not have the same vector of covariates. IVs, though better at accounting for unmeasured confounders, are subject to limitations as well. IVs that strongly correlate with the exposure but do not associate with the outcome can be very hard to find, and using instruments with a weak correlation with the exposure risks loss of power and biased estimates. Finally, DiD is a good strategy for evaluating natural or social experiments in the presence of longitudinal or panel data, but it cannot be used otherwise. However, it can be difficult to find a control that does not violate the common trends assumption, and without lead data it is impossible to provide evidence that this assumption is likely met.

This chapter is designed to introduce selected techniques that may be useful for assessing disparities in healthcare and making causal inference related to health equity, with an emphasis on observational data. We recognize that a single chapter cannot transmit the necessary expertise for the proper use of these techniques, and we want to encourage collaboration with statisticians and methodologists at a point early enough that study design can be adapted. We encourage the pursuit of more advanced topics for understanding and addressing health disparities, such as: (i) structural equation modeling as an extension of mediation analyses [24, 25], (ii) doubly robust [46] regression as an extension of propensity-score methodology, (iii) complier average causal effects [47] as an emerging new use of IVs, and (iv) interrupted time series methodology as an extension of DiD analysis [48].

12.9 Key Points

- Researchers should be explicit about their definition of health and healthcare disparities, and be mindful that the analytical method used to measure disparities can implicitly determine how a disparity is defined.
- While it is debatable whether race/ethnicity as a variable can "cause" negative health outcomes, it is possible to identify points of intervention by carefully assessing the causal chain linking race/ethnicity to health.
- When measuring healthcare disparities, analytical methods based on the counterfactual framework can help to distinguish between racial/ethnic healthcare differences that are due to unfair or unjust differences (disparities), and racial/ethnic differences that are due to fair or just differences (not considered to be disparities).
- Propensity score, instrumental variable, and difference-in-difference methods are well-established statistical methods of evaluating causal effects in quasi-experimental studies. These and other well-established statistical methods can be applied to observational data in order to better understand and measure disparities, and to test how changes in policies and other events impact disparities.

Disclaimer

The views and opinions expressed in this chapter are those of the authors only and do not necessarily represent the views, official policy, or position of the U.S. Department of Health and Human Services or any of its affiliated institutions or agencies.

References

1 Holland, P. (1986). Statistics and causal inference. *Journal of the American Statistical Association* 81 (396): 945–960.
2 Rubin, D. (2005). Causal inference using potential outcomes. *Journal of the American Statististical Association* 100 (469): 322–331.
3 Rubin, D.B. (1991). Practical implications of modes of statistical inference for causal effects and the critical role of the assignment mechanism. *Biometrics* 47 (4): 1213–1234.
4 Young, I. (2004). Five faces of oppression. In: *Oppression, Privilege, & Resistance* (eds. L. Heldke and P. O'Connor). Boston: McGraw Hill.
5 Cook, B., McGuire, T.G., and Zaslavsky, A.M. (2012). Measuring racial/ethnic disparities in health care: methods and practical issues. *Health Services Research* 47 (3 Pt 2): 1232–1254.
6 Cook, B.L., McGuire, T.G., Zaslavsky, A.M., and Meara, E. (2009). Adjusting for health status in non-linear models of health care disparities. *Health Services Outcomes and Research Methodology* 9 (1): 1–21.
7 Guevara, J.P., Mandell, D.S., Rostain, A.L. et al. (2006). Disparities in the reporting and treatment of health conditions in children: an analysis of the Medical Expenditure Panel Survey. *Health Services Research* 41 (2): 532–549.

8 AHRQ (2010). *National Healthcare Disparities Report, 2009*. Agency for Healthcare Research and Quality: Rockville, MD.
9 Cook, B.L., McGuire, T.G., Lock, K., and Zaslavsky, A.M. (2010). Comparing methods of racial and ethnic disparities measurement across different settings of mental health care. *Health Services Research* 45 (3): 825–847.
10 IOM (2003). *Unequal Treatment: Confronting Racial and Ethnic Disparities in Health Care*. Washington, DC.: The National Academies Press.
11 McGuire, T.G., Alegria, M., Cook, B.L. et al. (2006). Implementing the Institute of Medicine definition of disparities: an application to mental health care. *Health Services Research* 5: 1979–2005.
12 Blinder, A. (1973). Wage discrimination: reduced form and structural elements. *Journal of Human Resources* 8 (4): 436–455, 693-709.
13 Oaxaca, R. (1973). Male-female wage differentials in urban labor markets. *International Economic Review* 14 (October): –693, 709.
14 Fairlie, R. (2006). An extension of the Blinder-Oaxaca decomposition technique to logit and probit models. In: *IZA Discussion Papers*. Bonn, Germany: Institute of Labor Economics (IZA).
15 Cook, B.L., Zuvekas, S.H., Chen, J. et al. (2017). Assessing the individual, neighborhood, and policy predictors of disparities in mental health care. *Medical Care Research and Review* 4: 404–430.
16 Saloner, B., Carson, N., and Cook, B. (2014). Explaining racial/ethnic differences in adolescent substance abuse treatment completion in the United States: a decomposition analysis. *Journal of Adolescent Health* 54 (6): 646–653.
17 Zuvekas, S. and Taliaferro, G. (2003). Pathways to access: health insurance, the health care delivery system, and racial/ethnic disparities, 1996–1999. *Health Affairs* 22 (2): 139–153.
18 VanderWeele, T.J. and Robinson, W.R. (2014). On the causal interpretation of race in regressions adjusting for confounding and mediating variables. *Epidemiology* 25 (4): 473–484.
19 Hafeman, D.M. (2009). Proportion explained: a causal interpretation for standard measures of indirect effect? *American Journal of Epidemiology* 170 (11): 1443–1448.
20 Hafeman, D.M. and Schwartz, S. (2009). Opening the black box: a motivation for the assessment of mediation. *International Journal of Epidemiology* 38 (3): 838–845.
21 Naimi, A.I., Schnitzer, M.E., Moodie, E.E., and Bodnar, L.M. (2016). Mediation analysis for health disparities research. *American Journal of Epidemiology* 184 (4): 315–324.
22 Richiardi, L., Bellocco, R., and Zugna, D. (2013). Mediation analysis in epidemiology: methods, interpretation and bias. *International Journal of Epidemiology* 42 (5): 1511–1519.
23 VanderWeele, T.J. (2009). Marginal structural models for the estimation of direct and indirect effects. *Epidemiology* 20 (1): 18–26.
24 Adler, N.E. and Rehkopf, D.H. (2008). US disparities in health: descriptions, causes, and mechanisms. *Annual Review of Public Health* 29: 235–252.
25 Artinian, N.T., Warnecke, R.B., Kelly, K.M. et al. (2007). Advancing the science of health disparities research. *Ethnicity and Disease* 3: 427–433.
26 Holland, P.W. (2003). *Causation and Race: Research Report 03–03*. Princeton, NJ: E.T. Service.
27 National Research Council (2004). *Measuring racial discrimination*. In: *Panel on Methods for Assessing Discrimination* (eds. R.M. Blank, M. Dabady and C.F. Citro), 121–128. Washington, DC.: National Academy Press.

28 Bertrand, M. and Mullainathan, S. (2004). Are emily and brendan more employable than Lakisha and Jamal? A field experiment on labor market discrimination. *American Economic Review* 4: 991–1013.

29 Turner, M., Santos, R., Levy, D. et al. (2013). *HousingDiscrimination Against Racial and Ethnic Minorities 2012*. Washington, DC: Policy Development and Research, U.S, Department of Housing and Urban Development.

30 Schulman, K.A., Berlin, J.A., Harless, W. et al. (1999). The effect of race and sex on physicians' recommendations for cardiac catheterization. *New England Journal of Medicine* 340 (8): 618–626.

31 Massoglia, M. (2008). Incarceration as exposure: the prison, infectious disease, and other stress-related illnesses. *Journal of Health Sciences Behavavior* 49 (1): 56–71.

32 Trivedi, A.N. and Ayanian, J.Z. (2006). Perceived discrimination and use of preventive health services. *Journal of General Internal Medicine* 21 (6): 553–558.

33 Landrum, M.B. and Ayanian, J.Z. (2001). Causal effect of ambulatory special care on mortality following myocardial infarction: a comparison of propensity score and instrumental variable analysis. *Health Services & Outcomes Research Methodology* 2: 221–245.

34 Hanchate, A.D., Kapoor, A., Katz, J.N. et al. (2015). Massachusetts health reform and disparities in joint replacement use: difference in differences study. *British Medical Journal* 350: h440.

35 Ryan, A.M., Burgess, J.F. Jr., and Dimick, J.B. (2015). Why we should not be indifferent to specification choices for difference-in-differences. *Health Services Research* 50 (4): 1211–1235.

36 Rosenbaum, P.R. and Rubin, D.B. (1983). The central role of the propensity score in observatinoal studies for causal effects. *Biometrika* 70 (1): 41–55.

37 Rubin, D.B. (1997). Estimating causal effects from large data sets using propensity scores. *Annals of Internal Medicine* 127 (8 Pt 2): 757–763.

38 Brookhart, M.A., Schneeweiss, S., Rothman, K.J. et al. (2006). Variable selection for propensity score models. *American Journal of Epidemiology* 12: 1149–1156.

39 Cepeda, M.S., Boston, R., Farrar, J.T., and Strom, B.L. (2003). Comparison of logistic regression versus propensity score when the number of events is low and there are multiple confounders. *American Journal of Epidemiology* 158 (3): 280–287.

40 Mays, G.P. and Smith, S.A. (2011). Evidence links increases in public health spending to declines in preventable deaths. *Health Affairs (Millwood)* 30 (8): 1585–1593.

41 Angrist, J.D., Imbens, G.W., and Rubin, D.B. (1996). Indentification of causal effects using instrumental variables. *Journal of the American Statistical Association* 91 (434): 444–455.

42 Angrist, J.D. and Pischke, J.S. (eds.) (2009). *Mostly Harmless Econometrics: an Empiricist's Companion*. Princeton, N.J.: Princeton University Press.

43 Dimick, J.B. and Ryan, A.M. (2014). Methods for evaluating changes in health care policy: the difference-in-differences approach. *The Journal of the American Medical Association* 312 (22): 2401–2402.

44 Fleurbaey, M. and Schokkaert, E. (2012). Equity in health and health care. In: *Handbook of Health Economics*, vol. 2 (eds. M.V. Pauly, T.G. McGuire and P.P. Barros). Amsterdam: Elsevier.

45 Roemer, J. (2002). Equality of opportunity: a progress report. *Social Choice and Welfare* 19 (2): 455–471.

46 Funk, M.J., Westreich, D., Wiesen, C. et al. (2011). Doubly robust estimation of causal effects. *American Journal of Epidemiology* 173 (7): 761–767.

47 Connell, A.M. (2009). Employing complier average causal effect analytic methods to examine effects of randomized encouragement trials. *American Journal of Drug and Alcohol Abuse* 35 (4): 253–259.

48 Wagner, A.K., Soumerai, S.B., Zhang, F., and Ross-Degnan, D. (2002). Segmented regression analysis of interrupted time series studies in medication use research. *Journal of Clinical Pharmacy and Therapeutics* 27 (4): 299–309.

13

Small Area Estimation and Bayesian Disease Mapping for Minority Health and Health Disparities

Xingyou Zhang[1], Benmei Liu[2], Shumei Yun[3], and Robert L. Phillips[4]

[1] *U.S. Bureau of Labor Statistics, Washington, DC, USA*
[2] *National Cancer Institute, National Institutes of Health, Bethesda, MD, USA*
[3] *Resolve to Save Lives and Vital Strategies, New York, NY, USA*
[4] *American Board of Family Medicine, Washington, DC, USA*

13.1 Introduction

Small area health data have been in great demand to characterize local health disparities, inform public health practice, and guide health policy development and service delivery because of their capacity for improved understanding of social, economic, and geographic disparities in population health outcomes [1]. The term “small area” usually refers to small geographic areas such as counties, cities, school districts, tribal areas, and census tracts or other locally recognized or defined neighborhoods in the United States. However, reliable small area health data are challenging to obtain, especially for small minority populations facing some of the worst health disparities and bearing a disproportionate disease burden [2]. Population-based health databases, collected mainly through large-scale probability sample surveys and administrative healthcare system or medical record reviews, are often used to produce health statistics and meet local health data needs for small geographic areas.

Population-based, large scale health surveys, such as the National Health Interview Survey (NHIS), the National Health and Nutrition Examination Survey (NHANES), and the Behavioral Risk Factor Surveillance System (BRFSS), are designed to produce national or state health statistics. Standard survey estimates for the national and major subgroups are termed “design-based” or “direct” estimates because they are based only on the survey data and the selection probabilities for the sample in the subgroup of interest. The direct estimates, based on observed survey data, become problematic when the sample sizes in the geographies of interest (e.g., counties or subcounty areas) are small or where those small areas have no sample at all. In this situation, small area estimation (SAE) techniques are commonly used.

SAE has a long history of development and application [3]. The essential idea is to utilize mathematical and/or statistical models that combine survey data with information from other sources, such as administrative or census records, to produce reliable estimates for small areas.

The Science of Health Disparities Research, First Edition. Edited by Irene Dankwa-Mullan, Eliseo J. Pérez-Stable, Kevin L. Gardner, Xinzhi Zhang, and Adelaida M. Rosario.

Several federal government agencies have developed long-term SAE programs to produce estimates for all counties and/or states for a range of health indicators using survey data, such as the National Cancer Institute (NCI) small area estimates (SAEs) of cancer-related measures,[1] the US Census Bureau's Small Area Health Insurance Estimates (SAHIE) program,[2] and the US Substance Abuse and Mental Health Services Administration (SAMHSA) state and substate estimates of substance use and mental disorders.[3] More SAE methods and applications for public health have been published in recent years [4–10]. The National Institutes of Health (NIH) has also funded several SAE studies [11, 12].

Population-based administrative health records or medical record reviews, such as vital records (e.g., birth and deaths) and disease registries (e.g., cancer registry) offer the most complete documentation of health events and disease cases but still might not be sufficient to produce reliable data for geographic areas with small populations. Population-based administrative healthcare records, such as medical claims data (e.g., Medicaid and Medicare data) are also important data sources for assessing population health. Several states now have All Payer Claims databases, but these are not yet standardized and many are incomplete. Health statistics computations using these data sources can still become very unstable for geographic areas with very small populations. For example, for a county with a population of 1000 or less, one additional or missing death record could substantially increase or decrease the county mortality rate. Even for those counties with large populations, this could also be true for mortality rates for small subpopulations. Thus, spatial, temporal, and spatiotemporal smoothing techniques have been commonly applied to produce reliable SAEs. Small area analysis, especially Bayesian disease mapping, has become a popular statistical tool to borrow information across space and/or time to estimate local disease burden and examine the causes of local health disparities [13].

In this chapter, we review two common statistical approaches for small area health data analysis: SAE for sample survey data and Bayesian disease mapping for population-based data. SAE generally refers to statistical methods that produce reliable estimates for small areas using sample survey data and other auxiliary data sources. Bayesian disease mapping generally refers to Bayesian statistical methods that produce reliable estimates for small areas using population-based administrative databases. Two SAE public health projects (the National Cancer Institute's Small Area Estimates for Cancer-Related Measures Project and the 500 Cities Project, led by the Centers for Disease Control and Prevention (CDC), the CDC Foundation, and the Robert Wood Johnson Foundation) are summarized. A practical guide to SAE is provided, followed by an example of how to apply SAE to quantifying small minority population health and health disparities: the model-based SAEs of current cigarette smoking prevalence among adult non-Hispanic American Indians and Alaska Natives at county and census tract levels. We also summarize the main spatial and spatiotemporal models for Bayesian disease mapping and provide applications of Bayesian hierarchical spatial models for small area health disparity studies related to public health practice and research. We conclude this chapter with thoughts on SAE and Bayesian disease mapping.

---

1 https://sae.cancer.gov
2 https://www.census.gov/programs-surveys/sahie.html
3 www.samhsa.gov/data

13.2 Basic Statistical Models for Small Area Estimation

Generalized Linear Mixed Models (GLMMs) that include both fixed effects and random area-specific effects have been widely used in SAE in recent decades [3]. The popularity of GLMMs in SAE is due to their flexibility in combining information from different sources and accounting for different sources of error. Please refer to McCulloch et al. for details on GLMM theory and Stroup for practical guidance on GLMM applications [14, 15].

There are two primary types of GLMMs for SAE: (i) area-level models and (ii) unit-level models. Area-level small area models model aggregated outcomes from original survey data at the small area level of interest. Direct survey estimates at the small area level are used to construct regression models that predict population health outcomes for those small areas with small sample sizes or without any sample at all. Unit-level small area models model the individual-level health outcomes from the original survey. These unit-level models predict the expected responses for all the individuals within a small area and summarize these predicted values for all individuals to obtain the values of population health outcomes for small areas.

13.2.1 Area-level Models

Various area-level models and methods of estimation have been developed. The most popular approach is the Fay-Herriot model, originally developed to estimate per-capita income for US areas with populations of less than 1000 [16]. Let y_{ij} denote the individual survey response for a health outcome of interest for respondent j ($j = 1, \ldots, n_i$) in area i ($i = 1, \ldots, m$), where n_i is the area-level sample size and m is the number of small areas. The parameters to be estimated are the small area means (or proportions) θ_i. To estimate θ_i, the Fay-Herriot type of model generally assumes the following regression form:

$$Y_i = \theta_i + e_i \tag{13.1}$$

$$\theta_i = X_i\beta + u_i, i = 1, \ldots, m \tag{13.2}$$

where Y_i is the direct survey estimate of the small area mean θ_i for each of the small areas of interest, and e_i is the area-specific sampling error. X_i denotes a set of area-specific predictors associated with the health outcome; for example, obtained from the US Census's ongoing American Community Survey and/or administrative records, and β denotes the corresponding regression coefficients. u_i are area-specific random effects that represent local variations unexplained by those explicit area-specific predictors (X_i), and m is the number of small areas. The random effect u_i is further assumed to be independent and identical normally distributed with a mean of zero and an unknown variance (denoted as ψ). The sampling error e_i is assumed to be independent, normally distributed with a mean of zero and known sampling variance (denoted as φ_i). In practice, the sampling variances φ_i are unknown and have to be estimated before fitting the model.

The generalized variance function (GVF) technique is commonly used to smooth the sampling variance for SAE problems. An alternative is to smooth the survey design effect [17]. The parameters β and σ_u^2 need to be estimated from the available data by fitting the model. The empirical best prediction (EBP) approach and the hierarchical Bayesian (HB) method have been used to estimate the unknown model parameters and infer for the small area means using mixed-effect models. The final estimate for area i derived from the Fay-Herriot type of model has the general form:

$$\hat{\theta}_i^{EBP} = (1 - p_i)\hat{\theta}_i + p_i X_i \hat{\beta} \tag{13.3}$$

$$p_i = \frac{\varphi_i}{\psi + \varphi_i} \tag{13.4}$$

where p_i is the sliding scale between the direct estimates $\hat{\theta}_i$ and the regression estimates $X_i\hat{\beta}, 0 \le p_i \le 1$. When there are sufficient data, p_i goes to zero; when there are little or no data, p_i goes to 1. For more details about the various mixed-effect models and how to infer using EBP or HB, we refer to Rao and Molina [3].

Once the area-level model is fitted and SAEs are conveniently obtained, area-level models require only area-level auxiliary data in model prediction. Area-level small area estimators tend to be design-consistent for small areas with larger sample sizes, which means model-based SAEs are expected to be consistent with their corresponding direct survey estimates when the sample sizes grow larger. However, there are at least two practical limitations [18]:

- Area-level models lack flexibility to produce estimates for health outcomes for subdomains at the level of small areas and cannot produce estimates for the domains of subarea levels.
- The sampling variances associated with the sampling errors are assumed to be known, but, in practice, direct variance estimates are not stable for many small areas. Additional smoothing techniques are used to produce stable sampling variance estimates.

13.2.2 Unit-level Models

A unit- (or respondent-)level mixed model can be used when unit-specific variables are available for populations in small areas. This class of models can incorporate auxiliary information at both the unit and area levels. Both unit-level linear mixed models and unit-level GLMMs have been used for SAE purposes in finite population frameworks. A simple example of a unit-level mixed model [19] takes the following regression model format:

$$y_{ij} = x_{ij}\beta + u_i + e_{ij}, j = 1,\ldots,N_i, i = 1,\ldots,m \tag{13.5}$$

where y_{ij} is the response of a health outcome of interest for unit j within small area i; x_{ij} are a set of individual-level and/or area-specific factors associated with health outcomes, and β are the corresponding regression coefficients; u_i are area-specific random effects that represent local variations unexplained by explicit individual-level and/or area-specific factors (x_{ij}); e_{ij} is the random error associated with individual health outcome responses; u_i and e_{ij} are assumed to be independent and identically distributed normal random variables $N(0, \sigma_u^2)$ and $N(0, \varphi)$; N_i is the population size in area i; and m is the number of small areas. The model parameters σ_u^2, φ will be estimated using the survey responses y_{ij},

$j = 1, \ldots, n_i$, by fitting the model. Predictions $(\hat{y}_{ij})$ for the nonsampled units $y_{ij}, j = n_{i+1}, \ldots, N_i$ will be derived using the same model. Finally, the SAEs can be computed using the following form:

$$\hat{Y}_i = \left(\sum_{j=1}^{n_i} y_{ij} + \sum_{j=n_{i+1}}^{N_i} \hat{Y}_{ij} \right) / N_i \tag{13.6}$$

For most health surveys, the ratio of sample and population is quite small. BRFSS, the largest health survey in the world, samples less than 0.2% of the target population. Thus, the Eq. (13.6) could be reduced to a model-based synthetic estimator:

$$\hat{Y}_i = \left(\sum_{j=1}^{N_i} \hat{Y}_{ij} \right) / N_i \tag{13.7}$$

Unit-level models do not use design-based estimators directly. One can build a unit-level model that incorporates all the design information to produce design-consistent estimators if the detailed design information (e.g., stratification and clustering) is available at the individual level. The modeling may become challenging if a very complex design is used. If the design information is not available at the individual level, one can incorporate survey weights following the approaches outlined in the literature. The unit-level model estimators could make use of survey weights to preserve the design consistency for small areas with large sample sizes [20].

Unit-level models are more computationally intensive than area-level models because they use all-individual survey sample records in model fitting and involve large population units for every small area in model prediction. Sometimes, the required computing time for unit-level model fitting and prediction is prohibitive for computationally intensive statistical estimation methods, such as Bayesian inference via Markov chain Monte Carlo (MCMC) techniques and bootstrapping. The ever-easier access to powerful computing resources makes these robust statistical methods more practical. Despite these limitations and challenges, unit-level models have great flexibility. Once the individual prediction is obtained, one could derive SAEs at any census geographic levels, from census block (the smallest census geographic unit) to the entire nation. SAEs are automatically consistent across geographic levels and between demographic domains.

13.2.3 Auxiliary Data and Model Selection for Small Area Estimation

Auxiliary data (or predictor variables) play an important role in SAE. The choice of small area models depends on the availability of auxiliary data and the relationship between these data and the variables of interest. Auxiliary data are often obtained from various administrative and census records. In essence, we want to "borrow strength" from these auxiliary data to increase the accuracy of estimates for small areas. Model selection techniques may be applied to select the best set of auxiliary variables. With the classical modeling approach, the Akaike information criterion (AIC) and Bayesian information criterion (BIC, also known as the Schwarz criterion) are commonly used criteria for model selection purposes. For detailed information on these criteria and how to use them in the SAE context, we refer to Rao and Molina (pp. 111–112) [3]. With the HB modeling approach, a commonly used model selection criterion is the deviance information criterion (DIC) proposed by Spiegelhalter et al. [21].

13.2.4 Applications of Small Area Estimation in Public Health

Both area-level and unit-level small area models are used to produce SAEs of a wide range of health indicators from various health surveys. In this subsection, we highlight two SAE projects: SAEs for Cancer-Related Measures led by NCI and the 500 Cities Project, jointly conducted by the CDC, the CDC Foundation, and the Robert Wood Johnson Foundation.[4]

NCI, collaborating with other government agencies and universities, used novel area-level statistical models to develop SAEs for cancer risk factors (e.g., smoking), screening, policies (e.g., smoke-free laws), and knowledge at the county, health service area, and state levels. Employing a unit-level multilevel regression and poststratification (MRP) approach [8, 9, 22, 23], the 500 Cities Project produces city- and census-tract-level SAEs for chronic disease risk factors, health conditions, and clinical preventive service use for the 500 largest US cities.

13.3 Small Area Estimation: A Brief Practical Guide

We reviewed two basic types of statistical models for SAE in Section 13.2. To conduct a practical SAE project using one or more selected surveys, several major steps are needed. When an area-level model is employed, the following major steps are involved:

- Produce direct estimates at small area level for the outcome of interest using standard survey software, such as SUDAAN[5] or SAS PROC SURVEY packages.[6]
- Smooth the sampling variance or design effects associated with the direct estimates.
- Prepare the area-level covariates from outside sources, such as the decennial US Census figures or its ongoing American Community Survey.
- If a big pool of covariates is available, run classical model selection to choose initial covariates.
- Fit the area-level model and make inferences using either the frequentist (EBP) or the HB approaches.
- Run model diagnosis and evaluation to validate SAEs.

When a unit-level model is employed, the following four basic steps are required, following Zhang et al. [8, 9]:

- Data linkage: This step is time-consuming but fundamental. An SAE project often requires three main data sources and their linkages at various stages: (i) health survey data, (ii) auxiliary data, and (iii) population data. Health survey data with a health

4 www.cdc.gov/500cities

5 RTI international, Research Triangle Park, NC.

6 SAS Institute, Cary, NC.

outcome of interest are essential for SAE. Each health survey has its own strengths and limitations for SAE. Sometimes, an SAE project might use multiple health survey data to derive its estimates. Auxiliary data could come from a wide variety of data sources. In the United States, two detailed population datasets are available from the US Census Bureau: decennial census population counts; and postcensal population estimates. The decennial census has population counts at census block level by age, sex, and race/ethnicity but is only available every 10 years. Census population estimates provide county-level population counts by age, sex, and race/ethnicity and are available annually.[7]

- Model formulation and fitting: construct a statistical model linking health outcome and relevant factors and obtain the model parameter estimates via statistical estimation methods. Model formulation, by nature, is a process that translates a subject matter question into a mathematical or statistical model [24]. How to formulate an appropriate model for a health outcome is challenging. Statistically, GLMM theory provides a basic framework to construct a statistical model for continuous or categorical health indicators. Model fitting estimates the unknown model parameters given the observed data via two common approaches: (i) likelihood-based methods and (ii) Bayesian estimation via MCMC. In the case of nonlinear mixed models, likelihood functions become analytically intractable, so linearization methods, such as pseudo-likelihood (PL), or numerical integral approximation methods, such as Laplace and adaptive quadrature, have been used for model parameter estimation. These estimation methods are usually available from standard statistical software that can fit GLMM, such as SAS, R, and Stata.
- Model prediction: use the fitted statistical models to make predictions based on census population data and generate SAEs of health indicators for both the point estimates and their predicted standard errors. Monte Carlo simulation and parametric bootstrapping are very practical choices to estimate the predicted standard errors.
- Model validation: validate SAEs internally and externally if possible. Internal validation compares model-based SAEs with those reliable direct survey estimates. External validation compares model-based SAEs with reliable estimates from data sources other than the health survey for model fitting.

13.4 Small Area Estimation for Quantifying Health and Health Disparities of Small Populations

SAE can be a very powerful and practical tool to explore and examine health and health disparities at local geographic areas, especially for small minority populations. According to the 2010 US Census, the adult population (18 and older) of non-Hispanic American Indians/Alaska Natives (NH-AIAN) is 1,599,777, accounting for only 0.68% of the total adult population of the United States. However, this minority population has the highest prevalence of cigarette smoking compared to all other racial/ethnic groups in the United States and experiences substantial disparities in tobacco-related diseases and deaths due to this prevalence [25]. To generate model-based estimates of current cigarette smoking

7 www.census.gov/programs-surveys/popest.html

prevalence among adults aged 18 and over in the United States [8], we used public-available BRFSS 2011 and Census 2010 figures and applied the unit-level MRP approach. The model-based prevalence estimates of current cigarette smoking are 18.7% for the entire US adult population and 34.8% for the NH-AIAN adult population. Table 13.1 presents the basic summaries of model-based SAEs of current cigarette smoking prevalence. Table 13.2 summarizes the distributions of predicted standard errors. Geographic disparities in model-based estimates of current cigarette smoking prevalence for NH-AIAN at state, county, and census tract levels are larger than the corresponding disparities for the total population. Similar patterns are also observed for the predicted standard errors. Figure 13.1 shows the model-based prevalence estimates of current cigarette smoking among adults aged 18 and over for all populated census tracts. Figure 13.2 shows the model-based prevalence estimates of current cigarette smoking among NH-AIAN adults aged 18 and over for census tracts with at least 100 NH-AIAN adults. This map can be used to identify local communities with a high prevalence of smokers and a high concentration of NH-AIAN adults to help inform public health program development and planning that targets NH-AIAN populations.

Using the MRP approach for SAE with BRFSS, we generated model-based estimates and produced similar tables and figures for other small minority populations of interest (e.g., Hawaii natives and Pacific Islanders) as well as for other health indicators at various geographic levels [8].

13.5 Bayesian Disease Mapping

Population sizes in small areas are too small to produce statistically reliable estimates. Bayesian disease mapping involves developing Bayesian hierarchical spatial models that borrow information from adjacent areas to produce reliable SAEs and examine the associations between local disparities in health outcomes and their predictors at small area levels. Small area data often exhibit substantial spatial cluster patterns: a small area is often surrounded by other small areas with similar high or low values in health outcomes. These clusters are associated with positive spatial correlations in health outcomes among small areas. Bayesian disease mapping could account for spatial correlations (also referred to as spatial autocorrelation, spatial homogeneity, or spatial continuity) in health outcomes for small areas via a flexible and powerful Bayesian hierarchical modeling framework. Unlike single-dimension and single-direction temporal correlations, spatial correlations have more than one dimension, following all possible directions. Thus, spatial correlations among small areas result in increased complexity in model specification. Bayesian hierarchical modeling has become a standard approach for spatial and spatiotemporal models for disease mapping in recent decades. In this section, we review basic statistical models often used in Bayesian disease mapping from a Bayesian hierarchical modeling perspective: mapping local geographic variations of disease and health outcomes and quantifying their predictors at small area levels. We first review the main statistical models for disease mapping and then provide a Bayesian hierarchical spatial modeling framework for model specification and fitting. We conclude this section with a few health applications of Bayesian hierarchical spatial models for disease mapping.

Table 13.1 BRFSS model-based prevalence estimates (percent) of current cigarette smoking among adults aged 18 and over in the United States, 2011.

Race	Geography	N	Minimum	1st quartile	Median	3rd quartile	Maximum	Range
All races	State	51	11.3	17.4	19.2	22.3	26.2	14.8
	County	3143	7.6	18.2	20.3	23.1	38.1	30.5
	Census tract	72 531	2.8	16.0	18.8	21.9	45.2	42.5
NH-AIAN	State	51	22.6	32.4	35.0	38.2	43.7	21.0
	County	3133	6.2	33.3	36.3	39.6	54.0	47.8
	Census tract	69 225	3.5	29.4	34.0	38.3	59.0	55.5

Table 13.2 Predicted Standard Errors (percent) of BRFSS model-based prevalence estimates of current cigarette smoking among adults aged 18 and over in the United States, 2011.

Race	Geography	N	Minimum	1st quartile	Median	3rd quartile	Maximum	Range
All races	State	51	0.63	0.87	1.02	1.21	3.48	2.85
	County	3143	0.91	1.71	2.53	3.18	4.56	3.65
	Census tract	72 531	0.34	1.53	1.94	2.58	5.18	4.84
NH-AIAN	State	51	1.09	1.31	1.49	1.81	4.58	3.49
	County	3133	0.58	2.52	3.59	4.42	5.64	5.06
	Census tract	69 225	0.44	2.39	2.93	3.70	6.03	5.60

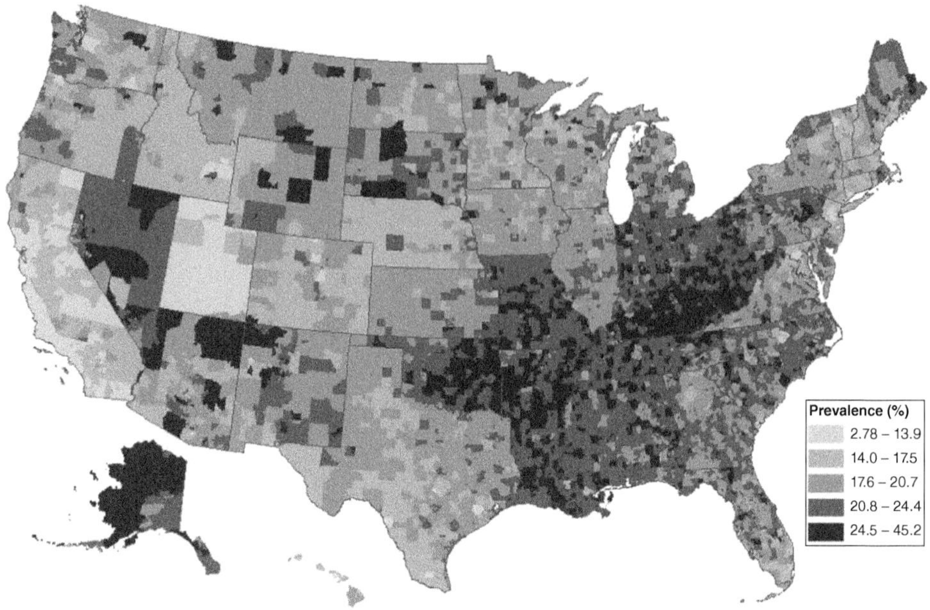

Figure 13.1 BRFSS model-prevalence estimates of current cigarette smoking among adults aged 18 and over in the United States, 2011 by 2010 census tract. (See color representation of this figure in color inserts section.)

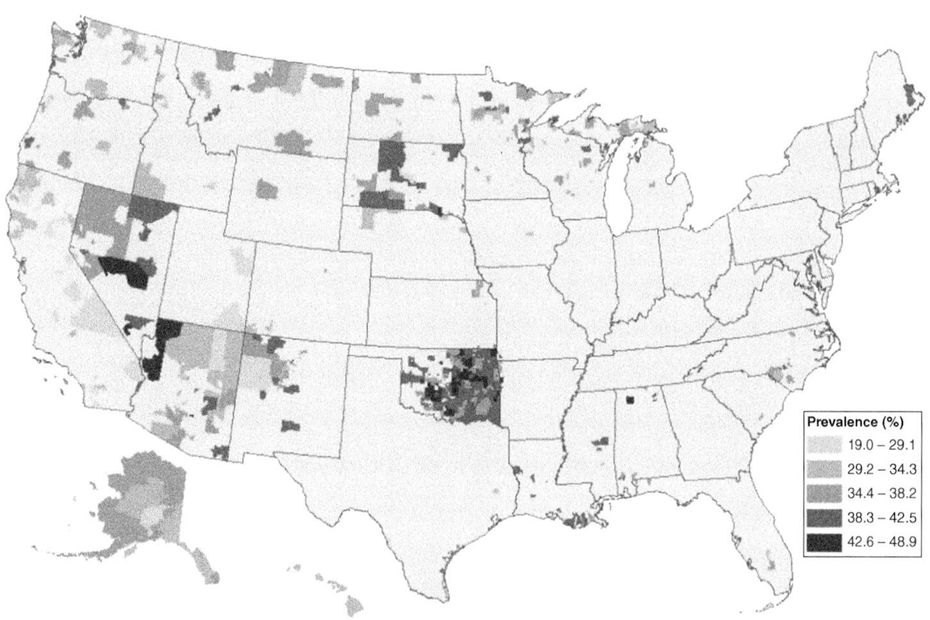

Figure 13.2 BRFSS model-prevalence estimates of current cigarette smoking among non-Hispanic American Indian and Alaska Native (NH-AIAN) adults aged 18 and over in the United States, 2011 by 2010 census tract with at least 100 NH-AIAN adults. (See color representation of this figure in color inserts section.)

13.5.1 Basic Statistical Models for Bayesian disease mapping

A large variety of statistical models, with varying levels of complexity, have been developed for Bayesian disease mapping. They can be grouped into six major model categories: (i) nonspatial model, (ii) spatial model, (iii) spatial convolution model, (iv) spatial mixture model, (v) spatiotemporal model, and (vi) spatial random coefficient model. We use the common Poisson model for disease risk to illustrate the similarity and differences among these spatial models.

A nonspatial Poisson model takes the following general format:

$$y_i \mid \mu_i \sim \text{Poisson}\left(E_i \mu_i\right), i = 1,2,3,\ldots,m$$

$$\log\left(\mu_i\right) = x_i \beta + u_i \tag{13.8}$$

where y_i is the observed disease count (e.g., the number of deaths or disease events) for small area (i) that follows Poisson distribution; μ_i is the disease rate (risk); E_i is the known population size; and m is the number of small areas. If E_i is the known expected disease count, then μ_i becomes the relative disease risk.

Following standard GLMM model specification, the disease rate is linearly associated with other small area covariates via a log link function (Eq. (13.8)), where x_i is a matrix of known covariates (also called explanatory variables); β is a vector of the corresponding regression coefficients associated with x_i; and u_i is the area-specific random effect, which is often assumed to be normally distributed with a mean of zero and a variance of σ_u^2 ($u_i \sim N\left(0, \sigma_u^2\right)$). The area-specific random effect (u_i) is used to capture local spatial heterogeneity effect. Epidemiologically, u_i represents the local unobserved contextual effect. The nonspatial Poisson model assumes spatial independence (no spatial correlation) among small areas. This is a very strong assumption for small area data.

If we assume spatial dependence (spatial correlations) among small areas, then the linking model in the Eq. (13.8) becomes:

$$\log\left(\mu_i\right) = x_i \beta + s_i \tag{13.9}$$

where s_i is a spatially structured random effect. There are various ways to specify s_i to account for spatial correlations among small areas. Conditionally autoregressive (CAR) structure is most commonly assumed for s_i:

$$s_i \mid s_{j \neq i} \sim N\left(\overline{s}_i, \frac{\sigma_s^2}{m_i}\right), \text{where } \overline{s}_i = \frac{1}{m_i} \sum_{j=1}^{m_i} s_j \tag{13.10}$$

where σ_s^2 is the variance associated with spatial random effect (s_i), and m_i is the number of adjacent small areas or spatial neighbors of small areas (i). The spatial random effect (s_i) is used to capture the regional spatial clustering effects. Statistically, s_i could account for the spatial correlations between small areas; epidemiologically, s_i represents the regional unobserved contextual effect.

For actual small area data, both spatial heterogeneity and spatial clustering effects could coexist; thus, the spatial convolution model that includes both spatially unstructured (u) and structured random effects (s_i) may be preferred:

$$\log(\mu_i) = x_i\beta + u_i + s_i \tag{13.11}$$

The model (Eq. (13.11)) is also referred to as the BYM model and has been widely applied for Bayesian disease mapping [26]. However, the spatial convolution model assumes universal spatial correlation among all small areas, which is hardly true. Unlike geophysical processes, where spatial continuity is usually assumed, the spatial process of disease outcomes and health behaviors could experience abrupt spatial discontinuity. Human behaviors and socioeconomic processes could be very different among adjacent small areas. We have observed spatial spillover effects of social and economic processes, but we also have observed abrupt changes in neighborhood social and built environments. The complexity of spatial patterns of socioeconomic processes also reflects the coexistence of spatial continuity and discontinuity for disease and health behaviors. Some parts of small areas could have strong spatial clustering effects while other parts might not. The simple incorporation of spatially structured random effects in the spatial convolution model could cause potential over-smoothing of local disease risks [27]. Thus, various spatial mixture models have been developed to deal with the complexity of spatial processes of diseases and health behaviors. One solution is to introduce a weighting factor to balance spatially unstructured (or uncorrelated) and structured (correlated) random effects. The spatial Poisson convolution model then becomes the spatial Poisson mixture model, as follows:

$$\log(\mu_i) = x_i\beta + r * u_i + (1-r) * s_i \tag{13.12}$$

where r is the weighting factor ($0<r<1$). This spatial Poisson mixture model has greater flexibility in accounting for both spatial clustering effects and potential spatial outliers [28].

The above spatial models only consider the small area data at one time point. But these spatial models could be conveniently extended to deal with small area data across both space and time. A typical spatiotemporal model can be formulated to model space-time disease risk as follows [29]:

$$\log(\mu_{it}) = x_i\beta + u_i + s_i + \tau_t \tag{13.13}$$

where τ_t is a temporal random effect that statistically specifies correlations in disease rate over time. This spatiotemporal model assumes the independence between space and time for disease rate, which means spatial and temporal correlations for small area data are separable. For most small area data collected over time, a space and time interaction is common. The following spatiotemporal model could account for such potential space-time interactions:

$$\log(\mu_{it}) = x_i\beta + u_i + s_i + \tau_t + \psi_{it} \tag{13.14}$$

where ψ_{it} is a space-time interaction random effect that specifies the spatiotemporal correlations among small area data. Thus, the disease rate of a small area is influenced by its spatial neighbors as well as by its spatiotemporal neighbors. Knorr-Held has provided a unified framework to specify such space-time interaction terms under different spatiotemporal correlation scenarios [30].

In all the above spatial models, the regression coefficients (β) are assumed to be constant across small areas. However, the relationships between diseases or health behaviors could change over space. For example, the relationships between mortality and income could vary across small areas or vary between urban and rural areas. A constant regression coefficient for income in the model could not account for such spatially varying processes. Spatial random coefficient models provide a natural and flexible framework to represent such complex spatial processes [31]:

$$\log\left(\mu_{it}\right) = x_i\beta_i + u_i + s_i + \tau_t + \psi_{it} \tag{13.15}$$

where β_i is a vector of spatially varying coefficients. A more complex spatial model could be formulated to account for spatiotemporal varying coefficients:

$$\log\left(\mu_{it}\right) = x_i\beta_{it} + u_i + s_i + \tau_t + \psi_{it} \tag{13.16}$$

where β_{it} is a vector of spatiotemporally varying coefficients. Gelfand et al. [32] have provided a theoretical summary of spatially random coefficient models and their applications. These spatial models could be further extended to model multiple correlated disease outcomes simultaneously [33].

13.5.2 Bayesian Hierarchical Spatial Model Specification and Fitting

With the advances in Bayesian computation, Bayesian spatial and spatiotemporal models have become more popular for disease mapping. Bayesian hierarchical spatial modeling provides a flexible framework to formulate spatial and spatiotemporal models for disease mapping. Despite the complexity of spatial and spatiotemporal models, it takes three model stages to specify a Bayesian hierarchical spatial or spatiotemporal model: data, process, and parameter [34]. We use the spatial Poisson mixture model as an example to illustrate these three stages.

- Stage one: data model. We have observed disease counts for small areas. These small area count data are not statistically independent. However, we can assume these data are independently following Poisson distribution, given the model relationships and model parameters (μ_i, β, u_i, s_i, r): $y_i \mid \mu_i, \beta, u_i, s_i, r \sim$ independent Poisson($E_i\mu_i$).
- Stage two: process model. At this stage, we specify the structural relationships between the mean parameter (disease rate) and model covariates and parameters as Eq. (13.12).
- Stage three: parameter model. This stage is to specify the probability models for all model parameters (β, u_i, s_i, r) to account for their statistical uncertainties. We assume normal distribution priors for β, u_i, and s_i, and beta distribution prior for r. We further specify inverse Gamma distribution hyperpriors for variance parameters (σ_u^2 and σ_s^2) of u_i and s_i.

Bayesian model fitting is based on the production of data, process, and parameter models. We estimate all model parameters and their uncertainties, and can make statistical inferences based on model parameters' posterior distributions. In other words, we can estimate the means, standard errors, and credible intervals for all model parameters based on their posterior distributions. Bayesian fitting via MCMC simulation for spatial and spatiotemporal models can be implemented in various statistical software packages, such as

WinBUGS/OpenBUGS[8,9] JAGS,[10] R-INLA (Bayesian computing with INLA),[11] and STAN.[12] For advanced statistical programming, R, C++, Fortran, and MatLab could be used to implement spatial and spatiotemporal models via a fully Bayesian approach.

Many model diagnostics are available to check Bayesian model fitting of spatial and spatiotemporal models for small area data. The DIC is particularly useful for Bayesian model selection [21]. DIC is a hierarchical modeling generalization of the AIC and the BIC. DIC balances model fitting for data and model complexity. DIC is defined as the average of model deviance for model parameters ($\overline{D(\theta)}$) minus the model deviance for mean posterior model parameters ($D(\bar{\theta})$): $DIC = 2\overline{D(\theta)} - D(\bar{\theta})$. The smaller the DIC, the better the model is supported by the data.

13.5.3 Applications of Bayesian Hierarchical Spatial Models for Health Disparities Research

Bayesian hierarchical spatial and spatial temporal models have become the most popular statistical tools for disease mapping and for exploring geographic and/or temporal health disparities. In general, these small area analyses have two main focuses: (i) quantifying disease burden and spatial patterns, and (ii) identifying and evaluating disease risk factors. Holt et al. [35] constructed a spatial mixture model to study geographic disparities in chronic obstructive pulmonary disease hospitalization among Medicare beneficiaries for 949 health service areas in the United States. They identified both spatial clusters and spatial outliers with substantial excessive risks [35]. Zhu et al. applied Bayesian hierarchical spatial models to investigate the links between alcohol availability, drug "hot spots," and violent crime [36]. Hao et al. applied a series of spatial models to evaluate the associations between ozone, fine particulate matter, and chronic lower respiratory disease mortality among 3,109 contiguous counties in the United States [37]. Recently, spatiotemporal models have been applied to examine the trends and patterns of geographic disparities in 1980–2014 mortality rates for major causes of death in the United States [38].

13.6 Conclusions

SAE and Bayesian disease mapping share numerous similarities in statistical model formats and estimation. In this chapter, we define SAE as statistical models used to produce reliable estimates for small areas with small or no survey samples using survey sample-based data. We define Bayesian disease mapping as statistical models used to produce reliable estimates for small areas with small populations using population-based data. We provide methodological guidance but do not cover all the statistical issues associated with these two topics. At the end of this chapter, we provide a list of further reading.

---

8 www.mrc-bsu.cam.ac.uk/software/bugs/the-bugs-project-winbugs
9 www.openbugs.net
10 http://mcmc-jags.sourceforge.net
11 www.r-inla.org
12 http://mc-stan.org

For SAE, there are two basic types of small area models: area level and unit level. The model selected depends on many factors, but, in general, if unit-level information is available, a unit-level model approach is preferred [39, 18]. In many situations, SAE is used to make predictions for unobserved data, while small area models fitting the observed data best may not result in the best predictions [40]. Traditional model diagnostics focus on data fitting; thus, validation of model-based SAEs is always needed. How to estimate predicted standard errors for SAEs is straightforward when a Bayesian approach is applied, but it could become complex under classical likelihood-based approaches, especially for unit-level small area models. This important issue is only briefly mentioned in this chapter.

For Bayesian disease mapping, the specification of appropriate statistical models for health conditions and/or outcomes under study is critical, and Bayesian hierarchical modeling provides a flexible framework for model specification and fitting. In most cases, Poisson modeling is an appropriate choice for disease count data. If the disease rate is not low, a binomial model may be a better choice for disease risk modeling. Without prior knowledge of the underlying true model, comparison of a few models with different spatially structured random effects may be necessary to choose the most appropriate model that fits the data.

Health policymakers, academic researchers, and public health practitioners are demanding reliable health statistics with more geographic granularity as well as more information about subpopulations, especially in the field of minority health and health disparities research and practice. SAE and Bayesian disease mapping could and should become more practical statistical tools that link multiple data sources and produce statistically reliable health data for small populations in local geographic areas. We have demonstrated that multilevel SAE could produce more reliable small area health statistics for small populations and we expect more applications in the near future. In the era of big data, additional research is also needed to explore how big data sources and big data analytics could be used to further improve SAE and meet small area health data needs for small populations.

13.7 Key Points

- Small area estimation utilizes statistical models that can integrate health survey data with information from other auxiliary data sources to produce reliable local population health indicators.
- Multilevel Regression and Poststratification (MRP) for small area estimation is a powerful statistical tool that can quantify minority population health and health disparities and establish local population health surveillance systems.
- Bayesian disease mapping is a flexible and robust statistical modeling framework that can borrow information across space and/or time to estimate local disease burden and examine the causes of local health disparities.

Disclaimer

The views and opinions expressed in this chapter are those of the authors only and do not necessarily represent the views, official policy, or position of the U.S. Bureau of Labor Statistics or the U.S. Department of Health and Human Services or any of its affiliated institutions or agencies.

Acknowledgments

The authors would like to thank Eric J. Feuer and Li Zhu at the National Cancer Institute and James B. Holt, Hua Lu, and Yan Wang at the Centers for Disease Control and Prevention for their helpful comments and suggestions.

References

1 National Academies of Sciences, Engineering, and Medicine (2017). *Federal Statistics, Multiple Data Sources, and Privacy Protection: Next Steps*. Washington, DC: The National Academies Press.

2 Korngiebel, D.M., Taualii, M., Forquera, R. et al. (2015). Addressing the challenges of research with small populations. *American Journal of Public Health* 105 (9): 1744–1747.

3 Rao, J.N.K. and Molina, I. (2015). *Small Area Estimation*, 2e. Hoboken, NJ: Wiley.

4 Congdon, P. (2009). A multilevel model for cardiovascular disease prevalence in the US and its application to micro area prevalence estimates. *International Journal of Health Geographics* 8: 6.

5 Lin, M., Zhang, X., Holt, J.B. et al. (2018). Multilevel Model to Estimate County-level untreated dental caries among US children aged 6–9 years using the national health and nutrition examination survey. *Preventive Medicine* 111: 291–298.

6 Malec, D., Sedransk, J., Moriarity, C.L., and LeClere, F.B. (1997). Small area inference for binary variables in the national health interview survey. *Journal of the American Statistical Association* 92 (439): 815–826.

7 Twigg, L., Moon, G., and Jones, K. (2000). Predicting small-area health-related behaviour: a comparison of smoking and drinking indicators. *Social Science & Medicine* 50 (7–8): 1109–1120.

8 Zhang, X., Holt, J.B., Lu, H. et al. (2014). Multilevel regression and poststratification for small-area estimation of population health outcomes: a case study of chronic obstructive pulmonary disease prevalence using the behavioral risk factor surveillance system. *American Journal of Epidemiology* 179 (8): 1025–1033.

9 Zhang, X., Holt, J.B., Yun, S. et al. (2015). Validation of multilevel regression and poststratification methodology for small area estimation of health indicators from the behavioral risk factor surveillance system. *American Journal of Epidemiology* 182 (2): 127–137.

10 Berkowitz, Z., Zhang, X., Richards, T.B. et al. (2016). Multilevel small-area estimation of multiple cigarette smoking status categories using the 2012 behavioral risk factor surveillance system. *Cancer Epidemiology, Biomarkers & Prevention* 25 (10): 1402–1410.

11 Ross, M. and Wakefield, J. (2015). Bayesian hierarchical models for smoothing in two-phase studies, with application to small area estimation. *Journal of the Royal Statistical Society, Series A Statistics in Society* 178 (4): 1009–1023.

12 Ha, N.S., Lahiri, P., and Parsons, V. (2014). Methods and results for small area estimation using smoking data from the 2008 national health interview survey. *Statistics in Medicine* 33 (22): 3932–3945.

13 Nandram, B., Sedransk, J., and Pickle, L.W. (2000). Bayesian analysis and mapping of mortality rates for chronic obstructive pulmonary disease. *Journal of the American Statistical Association* 95 (452): 1110–1118.

14 Stroup, W.W. (2013). *Generalized Linear Mixed Models: Modern Concepts, Methods and Applications*. Boca Raton, FL: CRC Press.
15 McCulloch, C.E., Searle, S.R., and Neuhaus, J.M. (2008). *Generalized, Linear, and Mixed Models*, 2e. Hoboken, NJ: Wiley.
16 Fay, R.E. and Herriot, R.A. (1979). Estimates of income for small places–application of James-Stein procedures to census-data. *Journal of the American Statistical Association* 74 (366): 269–277.
17 Liu, B.M., Lahiri, P., and Kalton, G. (2014). Hierarchical bayes modeling of survey-weighted small area proportions. *Survey Methodology* 40 (1): 1–13.
18 Guadarrama, M., Molina, I., and Rao, J.N.K. (2016). A comparison of small area estimation methods for poverty mapping. *Statistics in Transition New Series* 17 (1): 41–66.
19 Battese, G.E., Harter, R.M., and Fuller, W.A. (1988). An error-components model for prediction of county crop areas using survey and satellite data. *Journal of the American Statistical Association* 83 (401): 28–36.
20 You, Y. and Rao, J.N.K. (2002). A pseudo-empirical best linear unbiased prediction approach to small area estimation using survey weights. *The Canadian Journal of Statistics–La Revue Canadienne De Statistique* 30 (3): 431–439.
21 Spiegelhalter, D.J., Best, N.G., Carlin, B.R., and van der Linde, A. (2002). Bayesian measures of model complexity and fit. *Journal of the Royal Statistical Society Series B-Statistical Methodology* 64: 583–616.
22 Gelman, A. and Little, T.C. (1997). Poststratification into many categories using hierarchical logistic regression. *Survey Methodology* 23 (2): 127–135.
23 Little, R.J. (1993). Post-stratification–a modelers perspective. *Journal of the American Statistical Association* 88 (423): 1001–1012.
24 Cox, D.R. and Donnelly, C.A. (2011). *Principles of Applied Statistics*. Cambridge: Cambridge University Press.
25 Mowery, P.D., Dube, S.R., Thorne, S.L. et al. (2015). Disparities in smoking-related mortality among American Indians/Alaska Natives. *American Journal of Preventive Medicine* 49 (5): 738–744.
26 Besag, J., York, J., and Mollie, A. (1991). Bayesian image-restoration, with 2 applications in spatial statistics. *Annals of the Institute of Statistical Mathematics* 43 (1): 1–20.
27 Ryan, L. (2009). Spatial epidemiology some pitfalls and opportunities. *Epidemiology* 20 (2): 242–244.
28 Congdon, P. (2007b). Mixtures of spatial and unstructured effects for spatially discontinuous health outcomes. *Computational Statistics & Data Analysis* 51 (6): 3197–3212.
29 Waller, L.A., Carlin, B.P., Xia, X., and Gelfand, A.E. (1997). Hierarchical spatio-temporal mapping of disease rates. *Journal of the American Statistical Association* 92 (438): 607–617.
30 Knorr-Held, L. (2000). Bayesian modelling of inseparable space-time variation in disease risk. *Statistics in Medicine* 19 (17–18): 2555–2567.
31 Congdon, P. (2007a). Bayesian modelling strategies for spatially varying regression coefficients: a multivariate perspective for multiple outcomes. *Computational Statistics & Data Analysis* 51 (5): 2586–2601.
32 Gelfand, A.E., Kim, H.J., Sirmans, C.F., and Banerjee, S. (2003). Spatial modeling with spatially varying coefficient processes. *Journal of the American Statistical Association* 98 (462): 387–396.

33 Congdon, P. (2004). A multivariate model for spatio-temporal health outcomes with an application to suicide mortality. *Geographical Analysis* 36 (3): 234–258.
34 Arab, A., Hooten, M.B., and Wikle, C.K. (2007). Hierarchical spatial models. In: *Encyclopedia of Geographical Information Science* (eds. S. Shekhar and H. Xiong). New York: Springer.
35 Holt, J.B., Zhang, X.Y., Presley-Cantrell, L., and Croft, J.B. (2011). Geographic disparities in chronic obstructive pulmonary disease (COPD) hospitalization among medicare beneficiaries in the United States. *International Journal of Chronic Obstructive Pulmonary Disease* 6: 321–328.
36 Zhu, L., Gorman, D.M., and Horel, S. (2006). Hierarchical Bayesian spatial models for alcohol availability, drug "hot spots" and violent crime. *International Journal of Health Geographics* 5: 54.
37 Hao, Y., Balluz, L., Strosnider, H. et al. (2015). Ozone, fine particulate matter, and chronic lower respiratory disease mortality in the United States. *American Journal of Respiratory and Critical Care Medicine* 192 (3): 337–341.
38 Dwyer-Lindgren, L., Bertozzi-Villa, A., Stubbs, R.W. et al. (2016). US county-level trends in mortality rates for major causes of death, 1980–2014. *Journal of the American Medical Association* 316 (22): 2385–2401.
39 Namazi-Rad, M.R. and Steel, D. (2015). What level of statistical model should we use in small area estimation? *Australian & New Zealand Journal of Statistics* 57 (2): 275–298.
40 Little, R.J. (2012). Calibrated Bayes, an alternative inferential paradigm for official statistics rejoinder. *Journal of Official Statistics* 28 (3): 367–372.

Further Reading

Banerjee, S., Carlin, B.P., and Gelfand, A.E. (2014). *Hierarchical Modeling and Analysis for Spatial Data*. Boca Raton, FL: Chapman and Hall/CRC.
Lawson, A.B. (2018). *Bayesian Disease Mapping: Hierarchical Modeling in Spatial Epidemiology*. Boca Raton, FL: Chapman and Hall/CRC.
Rao, J.N.K. and Molina, I. (2015). *Small Area Estimation*. Hoboken, NJ: Wiley.
Waller, L.A. and Gotway, C.A. (2004). *Applied Spatial Statistics for Public Health Data*. Hoboken, NJ.: Wiley.

14

Applications of Big Data Science and Analytic Techniques for Health Disparities Research

Irene Dankwa-Mullan[1], Xinzhi Zhang[2], Phuong-Tu Le[3], and William T. Riley[4]

[1] *IBM Watson Health, International Business Machines (IBM) Corporation, Bethesda, MD, USA*
[2] *Division of Clinical Innovation, National Center for Advancing Translational Sciences, National Institutes of Health, Bethesda, MD, USA*
[3] *Division of Scientific Programs, National Institute on Minority Health and Health Disparities, National Institutes of Health, Bethesda, MD, USA*
[4] *Office of Behavioral and Social Sciences, National Institutes of Health, Bethesda, MD, USA*

14.1 Introduction

In recent years we have seen a dramatic increase in health-related data resulting in enormous data sets and sources, typically described as big data. Big data refers to the complexity, challenges, and new opportunities presented by the combined analysis of diverse, multimodal, and unstructured data, usually with data volumes greater than one terabyte (10^{12}). Big data science is the generation and storage of large amounts of data from biospecimens, health records, medical imaging, devices and sensors from which disease-specific factors, patterns, and associations can be computationally identified and used to generate insights for clinical care, decision making, interventions and treatment [1]. In the broad sense, it includes electronic health records (EHRs), biologic data (genomics, proteomics, metabolomics), environmental and lifestyle data, and data from wearable technology and mobile phone applications. This chapter will discuss the foundations of big data, big data analytics, and opportunities for the conduct of health disparities research.

Big data analytics is an evolving interdisciplinary scientific field that has adopted aspects of many scientific disciplines, such as statistics, machine learning, computational techniques, engineering operations, and pattern recognition to drive decision making, management and interventions. The increasingly heterogeneous nature and rapidly evolving volume of big data renders traditional statistical analysis and evaluations difficult. Using big data and analytics to generate insights for health disparities research, which requires understanding of interrelated diverse and complex data, shows promise in advancing the field through a team science approach.

The Science of Health Disparities Research, First Edition. Edited by Irene Dankwa-Mullan, Eliseo J. Pérez-Stable, Kevin L. Gardner, Xinzhi Zhang, and Adelaida M. Rosario.

Embracing big data in health disparities research challenges our traditional assumptions. Surveys may not be the primary source of understanding about what our study participants think or believe or how they behave. How they interact with digital technologies each and every day, our "digital breadcrumbs," [2] including internet searches, social media posts, and cell phone texts provide a rich data set that can be analyzed to detect the thoughts and beliefs of study participants. Sensors in smartphones, wearable sensors, and sensors in the home and car may be able to characterize behaviors relevant to understanding and addressing health disparities. Pairing geographic information system (GIS) capabilities with location data from phones allows us to understand not only how health is affected by the neighborhood in which one lives but also the neighborhoods where one works, runs errands, and visits friends or relatives. Digital interactions with social networks as well as technologies that can sample social context in real time provide data on social factors relevant to health disparities [3].

14.2 Characteristics of Big Data

The increasing volume of data generated by larger sample sizes and more data sources per person or data volume, – is one of the five Vs of big data [4]. Data variety from different sources and in different formats, provides researchers with a richer and more complex set of potential determinants of health disparities. Data velocity allows researchers to obtain relevant data more rapidly and more intensively, making it possible to be more responsive to research stakeholders. Data value and veracity increase the likelihood that these data will provide accurate answers to meaningful questions (Figure 14.1).

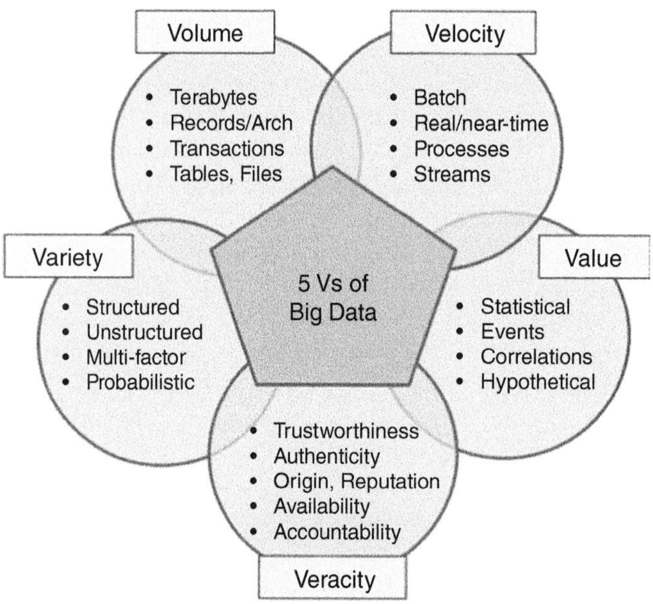

Figure 14.1 Big data characteristics.

The advent of big data produces not only new sources of data, but also new analytic approaches. Current traditional methods and analytic approaches are born from a data-poor environment in which much of our research resources were expended collecting the data. The science of these methods and analytic approaches must advance in this data-rich environment to accommodate the more complex and temporally dense data that is readily available [5].

The EHR has evolved from a system used primarily to document services for reimbursement into a source of information about healthcare encounters and health status on a variety of indices, including diagnoses, vital signs, laboratories, imaging, and patient-reported outcomes. Accessing EHR data has been hampered by proprietary conflicts between EHR systems, but efforts such as Fast Healthcare Interoperability Resources (FHIRs) [6] are producing interoperability standards and application protocol interfaces (APIs) that allow the data from these EHRs to be exchanged, not only between providers to improve continuity of care, but also with researchers who have patient consent to obtain data.

Among federally qualified health systems that disproportionately serve those experiencing health inequities, efforts such as the OCHIN [7] center in Portland, Oregon link EHR data with social determinant data, providing a rich data set for researchers. Machine learning approaches challenge traditional hypothesis-testing approaches and may generate findings that are potentially informative. Computational-modeling approaches extend our statistical modeling approaches, providing much greater specificity for our models of health disparities (e.g., not only that a change in x is associated with a change in y, but the magnitude and timing of that change), and allowing a greater understanding of the complex and dynamic systems that influence health disparities.

14.3 Importance of Big Data for Minority Health and Health Disparities Research

Like clinical research, health disparities science is rapidly entering into a data-rich environment in which core data will be available in an integrated network [5]. EHRs, mobile health technologies, and patient-generated data have greatly augmented the quantity and quality of data for disparity research. Social and environmental data are becoming available and openly accessible. Big data technology provides a juncture to easily collect and link contextual factors such as neighborhood segregation, food insecurity, neighborhood safety, social network, and community disempowerment to personal health information. These factors can be linked to the EHR to further explore the system level influences on differences in health outcomes.

Not surprisingly, efforts to incorporate social determinant information into EHRs have intensified [8, 9]. Accounting for social risk factors may have a significant impact on future reimbursement policies [10].

For many years, minority populations such as American Indians, Asians, and Pacific Islanders have not received needed public attention because of insufficient scientific evidence on diseases and conditions disproportionally affecting them. Gathering this evidence for major clinical trials or national surveys has been complicated by methodological issues, such as small sample sizes [11]. Lack of evidence can lead to a lack of appropriate resources dedicated to reduce disparities among these populations.

To assure that health disparities populations, especially those of lower socioeconomic status, are not neglected in the big data technology development and resulting research efforts and health benefits, recommendations have been made by the NIH to take proactive steps during the planning and implementing stages of new big data resource to address disparities reduction [12, 13]. The EHR, which is becoming widely adopted across healthcare systems, is a major source of big healthcare data. Assuring not only healthcare access, but also standardization of data among health disparities populations, becomes increasingly critical, not only for their health but also to ensure that their social and biological data is included in these EHR big data analyses [14].

The demand to improve the usability and interoperability of biomedical research data is growing exponentially. The FAIR principle of data storage (Findable, Accessible, Interoperable, Reusable) is widely endorsed by industry and research communities [15]. However, under-resourced minority-serving institutions and institutions most likely serving students from underrepresented backgrounds are less equipped and prepared to take full advantage of this great opportunity in data science. FAIR principles may be difficult to achieve due to poor data accessibility and interoperability, the lack of appropriate tools, and insufficient training.

Big data and health information technology currently faces a debate on a challenging issue: whether its advancement is an opportunity for health disparities populations. Are big-data's associated benefits reaching everyone at the same time, or are they delayed for certain group populations? How do we address these challenges to achieve health-data equity, and not exacerbate health inequities?

14.4 Goals of Big Data Analytics and Opportunities for Health Disparities Research

A major goal of big data analytics in health disparities research is to uncover hidden patterns, trends, and correlations of useful information that can help researchers make more informed decisions to guide health disparities interventions. Big data technology provides the opportunity and resources to link and merge data from different sources and systems through harmonization and standardization. Combining data from several sources or multiple years makes it possible to generate sufficient sample sizes in the study of otherwise small racial/ethnic groups [16]. Moderator and mediator effects that impact causal inference of health disparities among these small sample populations can then be studied (see Figure 14.2).

Using big data technology, evidence can be collected in real time and visualized in ways previously unimaginable, while interventions and policy changes to reduce disparities can be designed and continuously monitored to simulate population-level changes. For example, big data simulation modeling considers the multilevel and multidimensional influences on health disparities, can be used to study their causes, and can guide the development and implementation of multilevel interventions. Choi, Seligman, and Basu applied data from National Health and Nutrition Examination Survey and Supplemental Nutrition

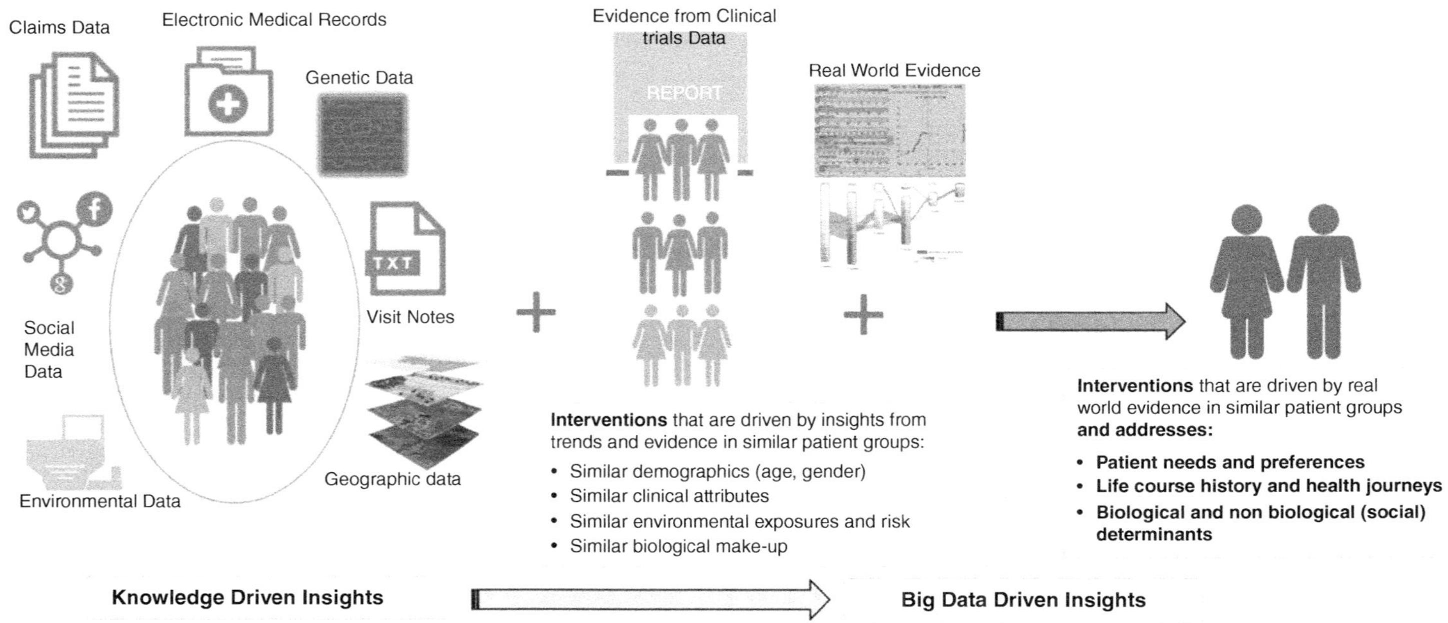

Figure 14.2 Goals and opportunities for big data in health disparities: evidence to inform interventions.

Assistance Program (SNAP) (formerly the Food Stamp Program) through a stochastic microsimulation model and concluded that nationwide SNAP fruits and vegetables subsidies would reduce chronic disease morbidity, mortality, and costs, particularly for non-Hispanic Blacks [17].

Some examples of the potential research opportunities presented by big data analytics for health disparities are to: (i) develop understanding of disease pathophysiology; (ii) inform epidemiology of disease and conditions; (iii) analyze the role of genomics or environmental determinants in disease onset and development; (iv) understand patient behaviors and the influence of social determinants; (v) identify real-world treatment practices that have been successful in various patient cohorts; (vi) facilitate new clinical trials designs, such as pragmatic trials, and align study protocol to real world practices; (vii) inform clinical decision making for improved quality; and (viii) guide tailored interventions and practices in health disparities populations.

14.5 Research Methods in Big Data

14.5.1 Types of Big Data Research Methods and Approaches

Big data and big data analytics attempt to uncover patterns and identify critical relations in observed phenomena or real-world experiences using the associations between data elements discovered through observation. Health disparities research can integrate the emerging applications and technologies for analyzing big data to understand relevant variables, or correlations that influence disparities, and adopt appropriate interventions or treatment. This section provides a review of common big data methods used to generate insights from a collection of individual (person-level) patient encounters and experiences to inform collective (population-level) knowledge about risk, causality, patterns of incidence and prevalence, and interventions that shape health disparities and outcomes.

Figure 14.3 shows the spectrum of analytic applications and technologies for understanding and generating insights from big data. Methods range from less complex analytics, such as descriptive, that do not involve artificial intelligence, to the advanced and more complex analytic methods that involve artificial deep machine learning methods. Descriptive, prescriptive, predictive, and similarity analytics are commonly used approaches and may or may not involve machine learning. In this section we will describe some common advanced big data statistical techniques, including analytics that combine artificial intelligence and machine learning to inform the research.

Artificial intelligence tasks used to analyze big data may include machine learning or other computational models. Machine learning is a subfield of computer science typically used to define artificial intelligence learning, which occurs without explicit programming. There are two forms of machine learning: conventional, "shallow" learning (neural networks with a single hidden layer or support vector machines [SVM]), and deep learning (neural networks with many hierarchical layers of nonlinear information processing).

Machine learning (ML) is an evolving scientific discipline that draws from advances in artificial intelligence, computational complexity theory, probability and statistics,

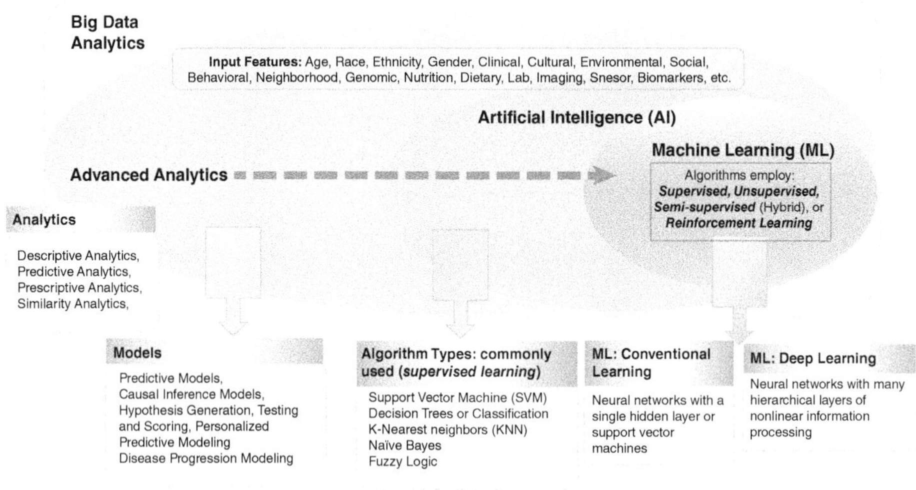

Figure 14.3 Spectrum of applications and techniques to analyzing big data. (See color representation of this figure in color inserts section.)

neurobiology, and concepts in other related disciplines [18]. ML methods used to generate insights in big data include decision trees (DT), artificial neural networks (ANN), genetic algorithms (GA), and SVM. While any of these methods can be used to generate evidence and information from big data, advanced analytics that include deep machine learning are of increased interest for their potential to generate actionable processes for complex solutions. Deep learning techniques show particular promise in extracting high-level abstractions from the raw data of very large, heterogeneous, high-dimensional data sets, all of which are likened to the diverse and complex nature of health disparities data.

The potential of big data analytics in disparities research is substantial but limited by the availability of technologies, tools, and skills available for performing various types of advanced analytics and modeling. For big data to generate data-driven knowledge and insights to inform evidence-based decision making, researchers need efficient methods to process large volumes of structured or unstructured data into meaningful information [19].

14.5.2 Commonly Used Big Data Science Analytics and Models

Various big data processing approaches used to uncover or generate insights that address disparities are illustrated in Figure 14.4. These analytic approaches require large datasets from either retrospective or prospective cases to achieve higher accuracy and precision.

14.5.2.1 Descriptive Analytics

Descriptive analytics is a preliminary stage of data processing that provides an illustrative summary of historical or retrospective data. The analytics examine the data and relevant information to understand patterns, trends, similarities, or connections [20]. It involves the description of derived patterns using simple statistical methods, such as mean, median, mode, standard deviation, variance, and frequency measurement of specific events in various big data streams. Large volumes of historical data are often used in descriptive

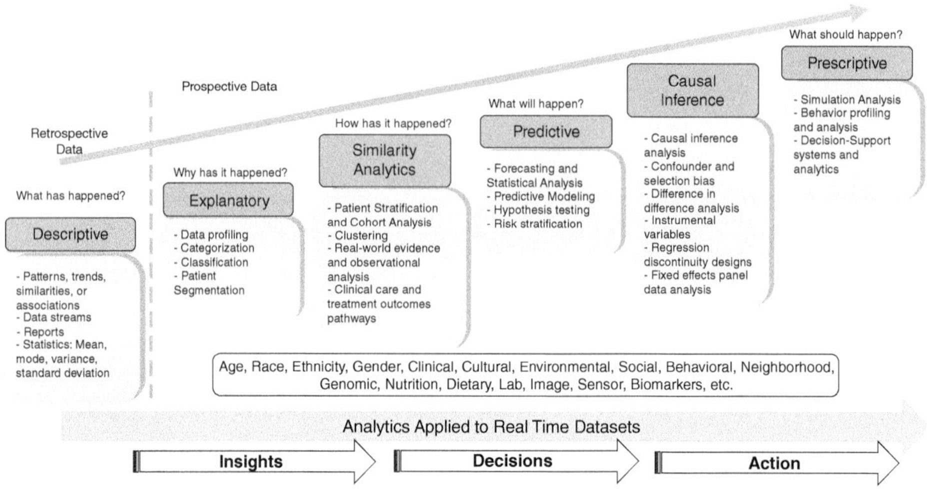

Figure 14.4 Schematic framework of advanced analytics for health disparity research. (See color representation of this figure in color inserts section.)

analytics to identify patterns and create descriptions. Most big data analytics are descriptive or exploratory in nature. The use of descriptive statistical methods (data mining tools) allows researchers to discover useful patterns or unidentified correlations that could be used to design effective interventions.

14.5.2.2 Predictive Analytics

Predictive analytics is concerned with forecasting and developing statistical models to determine future possibilities [20–22]. Predictive analytics are principally based on statistical methods and seek to uncover patterns and capture relationships in the data.

Some statistical approaches attempt to identify historical patterns in the outcome variable(s) and extrapolate from them to the future. Other statistical approaches, such as linear regression, seek to capture the interdependencies between outcome variable(s) and explanatory variables, and use them to make predictions.

14.5.2.3 Prescriptive Analytics

Prescriptive analytics is performed to model potential cause–effect relationships among analytic results and interventions. It identifies the necessary processes and actions needed to achieve an outcome. Organizations can optimize their research implementation or interventions based on the feedback provided by prescriptive analytic models.

14.5.2.4 Similarity Analytics

Patient similarity analytics can identify individual patients or patient groups who display similar health, clinical, or social characteristics to a reference patient of interest. The idea is that patients with similar clinical, demographic, genetics, environmental, and/or molecular and histopathological profiles or characteristics would have similar health outcomes or treatment intervention experiences. The technique examines distances between a variety of components of the patients' data and develops clusters and linkages based on short distances between some of the clinical, demographic, or other characteristics of interest [23]. Similarity analytics can also be used to track the clinical prognosis of a disease or condition, risk stratification, treatment management, or for care coordination pathways for patients with multiple chronic conditions.

The goal of these analytic techniques is to derive clinically meaningful insights from the data records of similar patients to provide personalized information for clinical decision making or health interventions. Patient similarity analytics can be used to relate to any kind of intervention or prevention, clinical treatment information from a reference, or "training patients." Ideally the reference patients are those with the outcomes of interest (e.g., treatment outcomes) or effective health outcomes based on the intervention treatments extrapolated to an identified population or health disparity group (i.e., with no known effective treatment information or for whom information from clinical studies may not be relevant due to under representation).

The challenge in patient similarity-based analytics is the lack of a predefined list of relevant predictors or assessment tools for research on health disparities populations. The choice of predictors and extraction of that data will affect the performance of the model. Identifying the most relevant and important patient characteristics for similarity assessment – and understanding the significance of the predictors to the intervention – is important (Table 14.1).

14.5.3 Applied Big Data Analytics and Artificial Intelligence

In recent years, several machine learning methods have been proposed to evaluate big data and provide advanced analytics. Machine learning algorithms can capture the best metrics from observed data and evidence provided by experts or extracted through advanced analytics. The algorithm can create multiple networks among subgroups of patients. Data mining extracts clinically relevant information hidden in clinical notes and embedded in other areas of the EHR coupled with International Classification of Disease codes. The result is a systematic individualized analysis of a subset of patients that can improve outcome prediction and help guide interventions for a particular patient or for various health disparity groups. Machine learning and deep-learning methods are increasingly being applied in several biological science fields including as genomic medicine, biomarker development, and proteomics as well as computational biology and drug discovery (see Table 14.2). Next generation sequencing (NGS) technology has resulted in massive quantities of genetic data that cannot be analyzed with traditional techniques. Some of these techniques include the structural annotation of genomes including coding regions, regulatory sequences, and protein-binding sites.

New technologies are still necessary to improve, scale, and expedite the handling and processing of large data that are increasing in size and complexity. Big data science and analytic approaches are still evolving as research disciplines and not fully established for addressing health disparities. The science of addressing health disparities using big data and analytics has the potential to generate new knowledge to inform innovative and actionable insights. The key is to have access to registries and databases that include large cohorts of patients with predictors and clinical features or conditions relevant for health disparities populations.

Table 14.1 Examples of questions that have helped drive big data research in the following areas of advanced analytics.

Type of analytics	Research applications and methods also involves
Descriptive	
• *Who has the highest readmission risk rates and what are the health outcomes?* • *Who is likely to incur the poorest health outcomes? And what interventions have been implemented?*	• Patient stratification and cohort construction and analysis • Retrospective cohort analysis • GIS mapping and hot spotting • Identification of healthcare utilization patterns
Predictive	
• *What are the triggering events for behavioral changes in disparity populations?* • *What is the optimal treatment pathway?* • *What will be the utilization rate or uptake of treatment for different patient cohorts?*	• Cohort construction and analysis • Feature engineering, a core computational task to infer phenotypic patterns and salient features from real-world evidence population-scale clinical data • Cross validation • Feature selection algorithms • Categorization and model selection • Risk stratification • Comparative effectiveness research
Prescriptive	
• *How do we optimize care for different patient segments?* • *How do we apply real-world and practice-based evidence for similar health disparities populations?* • *How do we detect real-world deviation in standards of care that result in improved outcomes?*	• Care management or treatment pathway analytics • Machine learning methods to address challenges with disease modeling • Visual analytic tools that extract clinical event sequences from EMR and other relevant databases to construct temporal networks or a visual flow of cooccurring events to model the disease

14.6 Selected Examples Illustrating Best Practices

14.6.1 Example 1. Electronic Health Record and Its Applications in the Reduction of Health Outcome Disparities

The application of EHR data has the potential to exert a powerful influence on health disparities. In 2004, Kaiser Permanente (KP), which serves 11.8 million members, established a comprehensive EHR system connecting inpatient and outpatient records, clinical-decision support, real-time lab connections, pharmacy, radiology, and electronic interprovider care messaging. Between 2004 and 2007, improved efficiency reduced annual office visits by 26%, primary care visits by 25%, and specialty care visits by 21% [55]. In Northern California, use of outpatient EHR resulted in an annual 2.9% reduction in emergency department visits by patients with diabetes [56].

Table 14.2 Selected examples: applications of big data analytics and artificial intelligence techniques.

Types	Applications	References
Descriptive analytics		
	Description of defect free care over time across race-ethnicity across 75 sites in the Florida-Puerto Rico Stroke Registry (66 Florida and 9 Puerto Rico) after research team recorded 58 864 ischemic stroke cases over 4 years.	Sacco, R. et al 2017 [24]
	Short- and long-term rehospitalization and mortality for heart failure in four racial/ethnic populations	Vivo et al. 2014 [25]
Predictive analytics		
Disease and risk predictions	Predicting primary myocardial infarction from EHR	Weiss et al. 2012 [26]
	Prediction of congestive heart failure and chronic obstructive pulmonary disease from longitudinal EHRs	Cheng et al. 2015 [27]
	Predictive modeling of hospital readmission and mortality from congestive heart failure using EHR and machine learning	Shameer et al. 2016 [28]
	Prediction of cardiac arrest in critically ill patients presenting to the emergency department using a machine learning score incorporating heart rate variability compared with the modified early warning score	Ong et al. 2012 [29]
	Machine learning methods to predict child post-traumatic stress: a proof of concept study	Saxe et al. 2017 [30]
Early detection and diagnosis	Machine-learning techniques for building a diagnostic model for very mild dementia	Jiao, et al. 2010 [31]
	Predicting asthma exacerbations before they occur	Finkelstein et al. 2017 [32]
Clinical decision support systems	Medical decision support for early detection of sepsis in emergency department	Horng et al 2017 [33]
	Computer advisor with a neural network model to assist multidisciplinary team in clinical decision making between CABG and PCI in order to achieve a higher five-year survival rate in patients with angina	Buzaev et al. 2016 [34]
	Computational prediction of a multidisciplinary team decision making for adjuvant breast cancer therapies	Lin et la. 2016 [35]
Artificial intelligence and machine learning techniques		
Cardiovascular medicine	To explore novel genotypes and phenotypes in existing diseases, improve the quality of patient care, enable cost-effectiveness, and reduce readmission and mortality rates	Krittanawong et al. 2017 [36] Duca et al. 2017 [37]

(*Continued*)

Table 14.2 (Continued)

Types	Applications	References
Diabetes	Diabetic retinopathy detection, decision support for gestational diabetes, applications of machine learning, data mining techniques and tools in the field of diabetes research with respect to (i) prediction and diagnosis, (ii) diabetic complications, and (iii) care management	Rigla et al. 2017 [38] Gulshan et al. 2016 [39] Takahashi et al. 2017 [40]
Hematology	Artificial intelligence platform on mobile devices in measuring and increasing medication adherence in stroke patients on anticoagulation therapy	Labovitz et al. 2017 [41]
	An anemia decision support system to help in predicting anemia outcomes of hemodialysis patients, minimizing erythropoietin-stimulating agent use, decrease in the number of transfusions with the potential to reduce the cost of treatment	Barbieri et al. 2016 [42]
Orthopedics	Use of standard deep learning networks can be trained to identify fractures in orthopedic radiographs. Also examined whether deep machine learning could be used to determine additional features such as body part, exam view, and laterality	Olczak et al. 2017 [43]
Dermatology	Skin cancer detection	Esteva et al 2017 [44]
	Diagnosis of melanoma skin lesions	Premaladha et al. 2016 [45]
Stroke	Determining therapeutic methods and prognosis prediction in stroke patients, artificial intelligence in stroke imaging	Lee et al. 2017 [46]
Sepsis	Prediction of patient mortality, patient outcomes for patients at risk for sepsis	Taylor et al. 2016 [47]
	Machine learning basis for a clinical decision support system EHR based decision support system to identify patients at high risk for hyperlactatemia based upon routinely measured vital signs and laboratory studies	Gultepe et al. 2014 [48]
Deep learning techniques to genetic data		
Biomarker development	Biomarkers are used to measure the progress of disease or the physiological effects of therapeutic intervention in the treatment of disease. Examples of computational biology application in research on biomarkers discovery	Yousef et al. 2014 [49]
Drug development and repurposing	Drug-induced liver injury prediction	Xu et al. 2015 [50]
	Prediction of cancer cell sensitivity to drugs based on genomic and chemical properties	Menden et al. 2013 [51]
Cancer biology	Cancer prediction, detection, and classification from a variety of gene expressions data, prediction of cancer recurrence, prediction of cancer survivability	Kourou et al. 2015 [52] Fakoor et al. 2013 [53] Cruz et al. 2006 [54]

The table illustrates selected examples that have used big data analytics including advanced artificial intelligence and machine learning technologies. References to these examples are provided to guide and inform research opportunities for health disparities science and are not included in this chapter's notes.

Analysis of disparities using Geographically Enriched Socio-demographics (GEMS) at KP has provided individual health information with imputed race/ethnicity and population data. After implementation of the Equitable Care Health Outcomes program from 2011 to 2014, the White–Black gap of hypertension control rate narrowed from 6.3% to 2.8% [57]. A healthcare team approach was adopted to strengthen programs for hypertension management, and culturally tailored communication tools were utilized among African American patients. The KP online healthcare system has advanced the idea of patient-centered healthcare and demonstrated growing consumer interest in further transformation of the healthcare system [58].

14.6.2 Example 2. National and Global Surveillance for Health Inequities

Mobile devices have been the source of a huge amount of information, including personal health data, which can be used to monitor and address geographic disparities. From 2011 to 2016, smartphone ownership in the United States increased from 35% per capita to 77%, according to the Pew Research Center. Althoff and coauthors used smartphone data with built-in accelerometry to measure global physical inactivity, and found different activity-related associations with obesity between countries [59]. For example, the United States has greater physical inequality than Mexico, even though Americans take a similar average number of daily steps. The authors obtained 68 million days of minute-by-minute step recordings between July, 2013 and December, 2014 from 717 527 anonymized Apple iPhone users of the Azumio Argus app, a free application for tracking physical activity and other health behaviors. The data were verified with prior surveillance data and population-based studies. Sample correction, stratification, outlier, and balance testing were used to correct for possible biases attributable to missing data. Activity inequality was measured using the Gini coefficient of the population activity distribution. These data can be used to show activity inequality between countries, assess their association with obesity, explore aspects of the built environment such as walkability, and implement simulation models for policy implications. For example, their findings suggest that activity inequality is lowest in San Francisco, the city with the highest walkability.

The availability of big data from smartphones provides measurable and dynamic information with a large geographic range and offers a powerful surveillance tool to monitor geographic disparities and trends among millions of people. Findings from Althoff and others highlight the importance of activity-friendly built environments in encouraging physical activity and improving health. Similar conclusions were found from the National Longitudinal Study of Adolescent Health, suggesting that lower socioeconomic status (SES) and high-minority census-block groups had reduced access to physical activity facilities, which in turn was associated with decreased physical activity and increased numbers of overweight individuals [60].

14.6.3 Example 3. Using the Internet and Social Media to Understand and Address Health Disparities

Despite variations in access, more than 70% of homes in the United States has broadband internet access [61], which makes the internet a valuable source for big data research. Internet query-based data from Google, for example, have been used to analyze and predict flu epidemics [62]. Most recently, Chae and colleagues used an internet query-based measure to understand the relationship between racism and birth outcomes among African

Americans in the United States [63]. Compared to traditional self-reported racism measures, internet query-based data are a promising method to capture information about population or structural-level characteristics that are not socially sanctioned, such as discrimination and prejudice [64].

In this study, Chae used US birth records to gather pregnancy outcomes among non-Hispanic Black women and linked the data to Google Trends. Internet queries at the media market or designated market area level were measured using Federal Information Processing Standard (FIPS) code. Area racism was measured as the proportion of internet search queries containing the N-word. Of the total of 3142 counties in the United States 3061 were successfully linked with 2 350 114 single live births to Black women between 2005 and 2008. The regression models were adjusted for maternal age, census region, county-level urbanicity, and percent of the African American population, which was further broken down into the percent of African Americans with less than a high school education and African Americans living below the federal poverty line.

The findings from this study suggest that each standard deviation increase in area racism corresponded to a 5% increase in both pre-term birth (prevalence ratio [PR] = 1.05, 95%; confidence interval [CI] = 1.02, 1.09) and low birthweight (PR = 1.05, 95% CI = 1.02, 1.07) after adjusting for all factors.

Internet query-based data provide a novel tool for population science and research in structural racism. Such data can be combined with other social media tools like Facebook or Twitter to better understand the causes of disparities and intervene to reduce them [65, 66].

14.6.4 Example 4: Chronic Disease Management

The Cardiovascular Health in Ambulatory Care Research Team (CANHEART) is a population-based observational research initiative aimed at measuring, tracking, and improving cardiovascular health in ambulatory care settings in Ontario, Canada [67]. Tu and colleagues created a study cohort of 9.8 million Ontario adults over 20 years of age by linking databases such as electronic surveys, health administration records, clinical laboratory records, community health surveys, and other administrative databases of vital statistics. It is currently being used to address primary and secondary prevention research initiatives and important disparity issues, such as (i) identifying the contributing factors to large regional and ethnic variations in the incidence rates of major cardiovascular events, (ii) examining important patient, physician, and community characteristics that influence the development of cardiovascular events, and (iii) understanding how the Ontario health system performs on various quality indicators related to cardiovascular prevention practices.

14.7 Challenges and Opportunities on Big Data Approaches in Health Disparities Research

14.7.1 Training in Big Data Approaches

Employing big data science approaches in health disparities presents a number of challenges. First and foremost, many health disparities researchers are not familiar with them. While these researchers should not be required to be experts in machine learning, social network analysis, or computational modeling, it is important to be sufficiently knowledgeable about these approaches to understand what is possible from big data and how different approaches can be used to answer health disparities questions or to pose new questions. Researchers need to become more comfortable with data that they did not collect, that are messy, and that may be too large to handle without additional computing capabilities.

Training opportunities in big data analytic approaches are increasingly being offered across academic institutions. The Big Data to Knowledge (BD2K) initiative at NIH has a significant training effort. Many statistical and data science programs are increasing their coverage of big data approaches. For health disparities researchers interested in big data training, it is important to look beyond one's own discipline for both training and collaboration. Computer science and engineering programs offer training in the use of big data. These programs provide not only the opportunity for health disparities researchers to learn new data science and analytic techniques, but also to find potential collaborators for whom applying big data approaches to health disparities research is also a challenge.

Access to such training, however, may be difficult for minority students and researchers, and less available in under-resourced academic institutions. Since health disparities research is conducted disproportionately by under-represented minority researchers, and disproportionately in under-resourced institutions, ensuring that minority researchers have access to such training, and that this training is conducted in under-resourced institutions is critical to advancing big data approaches in health disparities research.[1]

14.7.2 Big Data Access

Training alone, however, is insufficient. Researchers with these skills need to have access to big data sets that can be used to answer health disparities questions. The variety of data that needs to be accessed to answer important health disparities questions is extensive. Traditional data sources, such as large surveys, cohort studies, patient repositories, biobanks, EHRs, and claims data have served as the core data for health disparities research. Increasingly, big data from nontraditional data sets, including administrative data sets, social media interactions, and other digital footprints left as individuals interact with digital technologies, will be sources for understanding many behavioral and social determinants of health disparities. The Kavli Human Project,[2] as one such example, extracts not only health data but also education, housing, financial, and social data to characterize the lives of their participants and to understand how this broad range of social determinants affect health and well-being.

Being a big-data health disparities researcher will require digital investigative skills – learning how to seek out and identify nontraditional data sets that are relevant to the research question; negotiate with those who manage these data sets to obtain access without compromising the privacy, security, and proprietary concerns of those who hold these data; and determining a method to extract the relevant data and link it to other data sets.

1 https://datascience.nih.gov/bd2k/funded-programs/enhancing-training
2 http://www.thehumanproject.org

Issues of privacy and security are particularly important to consider in health disparities research given the history of violated trust many minority communities have experienced in prior research efforts [68].

Many of the big data sets of interest to health disparities researchers were not designed for research purposes, and permission to use them for such purposes varies. Most of these data sets are archival in nature and exempt from Institutional Review Board (IRB) review, or are subject to limited IRB review to ensure adequate anonymization. Even when used in this archival manner, however, it is prudent for the researcher to convene a panel of the health disparate groups under study and obtain input on the sensitivity of the data being extracted, even in anonymized form. For example, tribal nations will require input and oversight in research involving data from their members, even if the data is archival and available without individual permissions [69].

14.7.3 Representation of the Under-served in Big Data Sets

Although the promise of big data for health disparities research is considerable, it is important to recognize that some of the most vulnerable and disparate among us may not be represented in these big data sets. This issue is best highlighted by the digital divide of internet access, which is more precisely considered as a digital readiness gap [70]. While some technologies are great equalizers (e.g., feature or non-smart cell phones and some forms of social media), other technologies worsen the divide. Although smartphone penetration rates are growing rapidly, approximately one-quarter of US adults does not own a smartphone.

Wearable sensors (e.g., Fitbit) are predominately used by higher socioeconomic status population groups. As noted previously, limited healthcare access among medically under-served populations also means limited EHR data from these populations. Even common administrative data sets systematically exclude individuals who do not regularly receive services from these systems, as the debate about requiring voters to have a government ID attests.

14.7.4 Emerging Opportunities

Despite these challenges, there are growing opportunities for health disparities researchers to access and analyze big data sets. NIH alone has generated numerous epidemiologic data sets. Extending from its seminal work with Framingham, the National Heart, Lung, and Blood Institute (NHLBI) has expanded its epidemiologic research with a specific focus on under-served populations, including African Americans (Jackson Heart Study),[3] Hispanics[4] (Hispanic Community Health Study/Study of Latinos), Multi-Ethnic Study of Atherosclerosis (MESA),[5] and Native Americans (Genetic of Coronary Artery Disease in Alaska Natives – GOCADAN) [71]. National Institute on Aging (NIA), NICHD, and other NIH institutes also have generated epidemiologic data sets that span age ranges (e.g.,

3 http://www.jacksonheartstudy.org
4 https://sites.cscc.unc.edu/hchs
5 http://www.mesa-nhlbi.org

Health and Retirement Survey,[6] Midlife in the United States – MIDUS,[7] National Health and Aging Trends Study – NHATS),[8] and the Add Health Study.[9]

Three large-scale cohort efforts are emerging that will greatly expand the databases relevant to health disparities research:

- Environmental Influences of Child Health Outcomes (ECHO)[10]
- Adolescent Brain and Cognitive Development (ABCD)[11]
- All of Us (Precision Medicine Initiative)[12]

These three cohorts span the age ranges; are committed to diverse samples; are obtaining important biomedical, behavioral, and social determinants of health disparities; and will provide exceptionally large longitudinal data sets.

In the EHR arena, efforts to link or merge data from various vendor systems continues. FIHR has developed a system that makes EHRs from different vendors more easily sharable [6]. Under a current pilot project in cooperation with All of Us, FIHR is also developing an API that will allow participants to donate their EHR data for research purposes with the click of a button. OCHIN, working with the Federally Qualified Health Centers (FQHCs), has linked the EHRs of these centers to GIS social determinant data [72]. These data provide practitioners not only with data linking healthcare utilization in these community health centers to GIS determinants, but also with data they need to address these determinants.

In the technology realm, social media and search engine data have been used to detect population-level trends and patterns [73]. As this research area matures, the tools to access and analyze these types of data become more available and easier to use. Data from other digital sources, such as wearable and home-based sensors, are extending our ability to unobtrusively and directly observe behavior and context.

Clearly, the opportunities to use big data to answer health disparities questions are increasing. Health disparities researchers need to be ready to access, manage, and use these data to answer important health disparities questions, to move the field beyond descriptive phenomenology to a more mediational or mechanistic perspective, and to find testing strategies to ameliorate these disparities.

14.8 Summary and Conclusion

The combination of technologies generating voluminous and varied data rapidly (or at high velocity) along with analytic techniques that can structure, curate, and analyze these data, has ushered in the big data revolution. Although there will continue to be important scientific findings from small data, these big data sets provide opportunities to examine the

6 http://hrsonline.isr.umich.edu
7 http://midus.wisc.edu
8 www.nhats.org
9 http://www.cpc.unc.edu/projects/addhealth
10 http://www.nih.gov/echo
11 https://abcdstudy.org
12 https://allofus.nih.gov

complex determinants of health disparities and the mechanisms and interventions to ameliorate disparities in ways previously unimaginable. Health disparities research has long been limited by small sample sizes of the relevant disparate groups. Big data sources and analytics provide the potential to generate samples adequate to model complex relations among determinants, even among small populations that experience health inequities, and better identify moderators and mediators of these inequities.

As big data tools and procedures are refined, their use for predictive and prescriptive modeling and other analyses will increase. Recent examples of big data approaches applied to health disparities research questions illustrate how these approaches can be used to answer novel questions. It is incumbent on health disparities researchers to become more familiar with big data sources and analytics, understand their capabilities, access these data, and grasp their use in order to address critical questions of health inequities. A cadre of health disparities researchers with such expertise is needed not only to answer important research questions, but also to address the challenges of using big data for health disparities research. These challenges include the under-representation of health disparate groups in some of the available big data sources. Despite such challenges, big data approaches are and will continue to be a significant source of future findings that advance health disparities research.

14.9 Key Points

- Big data science has the potential to address many of determinants of health disparities.
- Applying big data to generate insights for health disparities research requires a team science approach.
- Interventions and policy changes to reduce disparities can be continuously monitored using big data.
- Ensuring that minority researchers have access to big data science training is critical.

Disclaimer

The views and opinions expressed in this chapter are those of the authors only and do not necessarily represent the views, official policy, or position of the U.S. Department of Health and Human Services or any of its affiliated institutions or agencies.

References

1 Luo, J., Wu, M., Gopukumar, D., and Zhao, Y. (2016). Big data application in biomedical research and health care: a literature review. *Biomedical Informatics Insights* 8: 1–10.

2 Pentland, A. (2013). The data-driven society. *Scientific American* 309 (4): 78–83.

3 Harari, G.M., Lane, N.D., Wang, R. et al. (2016). Using smartphones to collect behavioral data in psychological science: opportunities, practical considerations, and challenges. *Perspectives on Psychological Science: A Journal of the Association for Psychological Science* 11 (6): 838–854.

4 Baro, E., Degoul, S., Beuscart, R., and Chazard, E. (2015). Toward a literature-driven definition of big data in healthcare. *BioMedical Research International* 2015: 1–9. https://doi.org/10.1155/2015/639021.

5 Riley, W.T. (2016). A new era of clinical research methods in a data-rich environment. In: Oncology Informatics, vol. 18 (eds. B.W. Hesse, D.K. Ahern and E. Beckjord), 343–355. Boston, MA: Academic Press.

6 Mandel, J.C., Kreda, D.A., Mandl, K.D. et al. (2016). SMART on FHIR: a standards-based, interoperable apps platform for electronic health records. *Journal of the American Medical Informatics Association* 23 (5): 899–908.

7 Wallace, L.S., Angier, H., Huguet, N. et al. (2016). Patterns of electronic portal use among vulnerable patients in a nationwide practice-based research network: from the OCHIN practice-based research network (PBRN). *Journal of the American Board of Family Medicine* 29 (5): 592–603.

8 Committee on the Recommended Social and Behavioral Domains and Measures for Electronic Health Records (2015). Capturing Social and Behavioral Domains and Measures in Electronic Health Records: Phase 2. Washington DC: National Academies Press.

9 Board on Population Health and Public Health Practice; Institute of Medicine (2014). Capturing Social and Behavioral Domains in Electronic Health Records: Phase 1. Washington DC: National Academies Press.

10 Steinwachs, D.M., Stratton, K., and Kwan, L.Y. (2017). Accounting for Social Risk Factors in Medicare Payment. Washington, DC: National Academies Press.

11 Islam, N.S., Khan, S., Kwon, S. et al. (2010). Methodological issues in the collection, analysis, and reporting of granular data in Asian American populations: historical challenges and potential solutions. *Journal of Health Care for the Poor and Underserved* 21 (4): 1354–1381.

12 Zhang, X., Perez-Stable, E.J., Bourne, P.E. et al. (2017). Big data science: opportunities and challenges to address minority health and health disparities in the 21st century. *Ethnicity & Disease* 27 (2): 95–106.

13 Onukwugha, E., Duru, O.K., and Peprah, E. (2017). Foreword: big data and its application in health disparities research. *Ethnicity & Disease* 27 (2): 69–72.

14 Gottlieb, L.M., Tirozzi, K.J., Manchanda, R. et al. (2015). Moving electronic medical records upstream: incorporating social determinants of health. *American Journal of Preventive Medicine* 48 (2): 215–218.

15 Wilkinson, M.D., Dumontier, M., Aalbersberg, I.J. et al. (2016). The FAIR guiding principles for scientific data management and stewardship. *Scientific Data* 3: 160018.

16 Elliott, M.N., McCaffrey, D.F., Finch, B.K. et al. (2009). Improving disparity estimates for rare racial/ethnic groups with trend estimation and Kalman filtering: an application to the National Health Interview Survey. *Health Services Research* 44 (5, Pt 1): 1622–1639.

17 Choi, S.E., Seligman, H., and Basu, S. (2017). Cost effectiveness of subsidizing fruit and vegetable purchases through the supplemental nutrition assistance program. *American Journal of Preventive Medicine* 52 (5): e147–e155.

18 Mamoshina, P., Vieira, A., Putin, E., and Zhavoronkov, A. (2016). Applications of deep learning in biomedicine. *Molecular Pharmaceutics* 13 (5): 1445–1454.

19 Gandomi, A. and Haider, M. (2015). Beyond the hype: big data concepts, methods, and analytics. *International Journal of Information Management* 35 (2): 137–144.

20 Joseph, R.C. and Johnson, N.A. (2013). Big data and transformational government. *IT Professional* 15 (6): 43–48.

21 Waller, A.M. and Fawcett, S.E. (2013). Data science, predictive analytics, and big data: a revolution that will transform supply chain design and management. *Journal of Business Logistics* 34 (2): 77–84.

22 Dinov, I.D. (2016). Methodological challenges and analytic opportunities for modeling and interpreting big healthcare data. *GigaScience* 5: 12.

23 Chae, D.H., Clouston, S., Martz, C.D. et al. (2017). Area racism and birth outcomes among blacks in the United States. *Social Science & Medicine* 1982.

24 Sacco, R.L., Gardener, H., Wang, K. et al. (2017). Racial-ethnic disparities in acute stroke care in the Florida-Puerto Rico collaboration to reduce stroke disparities study. *Journal of the American Heart Association* 6 (2): e004073.

25 Vivo, R.P., Krim, S.R., Liang, L. et al. (2014). Short- and long-term rehospitalization and mortality for heart failure in 4 racial/ethnic populations. *Journal of the American Heart Association* 3 (5): e001134.

26 Weiss, J.C., Natarajan, S., Peissig, P.L. et al. (2012). Machine Learning for Personalized Medicine: Predicting Primary Myocardial Infarction from Electronic Health Records. *AI Magazine* 33 (4): 33.

27 Cheng, Y., Wang, F., Zhang, P. et al. (2015). Deep Learning Approach, Risk Prediction with Electronic Health Records (2016). In: Proceedings of the 23rd ACM SIGKDD International Conference on Knowledge Discovery and Data Mining August 2017, vol. 22(5), 993–1000.

28 Shameer, K., Johnson, K.W., Yahi, A. et al. (2017). Predictive modeling of hospital readmission rates using electronic medical record–wide machine learning: A case-study using Mount Sinai heart failure cohort. *Pacific Symposium on Biocomputing* 22: 276–287.

29 Ong, M.E.H., Lee Ng, C.H., Goh, K. et al. (2012). Prediction of cardiac arrest in critically ill patients presenting to the emergency department using a machine learning score incorporating heart rate variability compared with the modified early warning score. *Critical Care* 16: R108.

30 Saxe, G.N., Ma, S., Ren, J., and Aliferis, C. (2017). Machine learning methods to predict child posttraumatic stress: a proof of concept study. *BMC Psychiatry* 17 (1): 223.

31 Jiao, Y., Chen, R., Ke, X. et al. (2010). Predictive models of autism spectrum disorder based on brain regional cortical thickness. *Neuroimage* 50 (2): 589–599.

32 Finkelstein, J. and Jeong, I.C. (2017). Machine learning approaches to personalize early prediction of asthma exacerbations. *Annals of the New York Academy of Science* 1387 (1): 153–165.

33 Horng, S., Sontag, D.A., Halpern, Y. et al. (2017). Creating an automated trigger for sepsis clinical decision support at emergency department triage using machine learning. *PLoS ONE* 12 (4): e0174708.

34 Buzaev, I., Plechev, V., Nikolaeva, I. et al. (2016). Artificial Intelligence: Neural network model as the multidisciplinary team member in clinical decision support to avoid medical mistakes. *Chronic Disease and Translational Medicine* 2 (3): 166–172.

35 Lin, F.P.Y., Pokorny, A., Teng, C. et al. (2016). Computational prediction of multidisciplinary team decision-making for adjuvant breast cancer drug therapies: a machine learning approach. *BMC Cancer* 16 (1): 929.

36 Krittanawong, C., Zhang, H., Wang, Z. et al. (2017). Artificial intelligence in precision cardiovascular medicine. *Journal of the American College of Cardiology* 69 (21): 2657–2664.

37 Duca, F., Zotter-Tufaro, C., Kammerlander, A.A. et al. (2017). Cardiac extracellular matrix is associated with adverse outcome in patients with chronic heart failure. *European Journal of Heart Failure* 19 (4): 502–511.

38 Rigla, M., García-Sáez, G., Pons, B., and Hernando, M.E. (2018). Artificial intelligence methodologies and their application to diabetes. *Journal of Diabetes Science and Technology* 12 (2): 303–310.

39 Gulshan, V., Peng, L., Coram, M. et al. (2016). Development and validation of a deep learning algorithm for detection of diabetic retinopathy in retinal fundus photographs. *Journal of the American Medical Association* 316 (22): 2402–2410.

40 Takahashi, H., Tampo, H., Arai, Y. et al. (2017). Applying artificial intelligence to disease staging: Deep learning for improved staging of diabetic retinopathy. *PLoS ONE* 12 (6): e0179790.

41 Labovitz, D.L., Shafner, L., Gil, M.R. et al. (2017). Using artificial intelligence to reduce the risk of nonadherence in patients on anticoagulation therapy. *Stroke* 48 (5): 1416–1419.

42 Barbieri, C., Molina, M., Ponce, P. et al. (2016). An international observational study suggests that artificial intelligence for clinical decision support optimizes anemia management in hemodialysis patients. *Kidney International* 90 (2): 422–429.

43 Olczak, J., Fahlberg, N., Maki, A. et al. (2017). Artificial intelligence for analyzing orthopedic trauma radiographs. *Acta Orthopaedica* 88 (6): 581–586.

44 Esteva, A., Kuprel, B., Novoa, R. et al. (2017). Dermatologist-level classification of skin cancer with deep neural networks. *Nature* 542: 115–118.

45 Premaladha, J. and Ravichandran, K.S. (2016). Novel approaches for diagnosing melanoma skin lesions through supervised and deep learning algorithms. *Journal of Medical Systems* 40 (4): 96.

46 Lee, E.-J., Kim, Y.-H., Kim, N. et al. (2017). Deep into the brain: Artificial intelligence in stroke imaging. *Journal of Stroke* 19 (3): 277–285.

47 Taylor, R.A., Pare, J.R., Venkatesh, A.K. et al. (2016). Prediction of in-hospital mortality in emergency department patients with sepsis: a local big data-driven, machine learning approach. *Academic Emergency Medicine* 23 (3): 269–278.

48 Gultepe, E., Green, J.P., Nguyen, H. et al. (2014). From vital signs to clinical outcomes for patients with sepsis: a machine learning basis for a clinical decision support system. *Journal of the American Medical Informatics Association* 21 (2): 315–325.

49 Yousef, M., Najami, N., Abedallah, L., and Khalifa, W. (2014). Computational approaches for bio-marker discovery. *Journal of Intelligent Learning Systems and Applications* 6: 153–161.

50 Xu, Y., Dai, Z., Chen, F. et al. (2015). Deep learning for drug-induced liver injury journal of chemical information and modeling. *Journal of Chemical Information and Modeling* 55 (10): 2085–2093.

51 Menden, M.P., Iorio, F., Garnett, M. et al. (2013). Machine learning prediction of cancer cell sensitivity to drugs based on genomic and chemical properties. *PLoS One* 8 (4): e61318.

52 Kourou, K., Exarchos, T.P., Exarchos, K.P. et al. (2015). Machine learning applications in cancer prognosis and prediction. *Computational and Structural Biotechnology Journal* 13: 8–17.

53 Fakoor, R., Huber, M. (2013). Using deep learning to enhance cancer diagnosis and classification. In: Proceedings of the 30th International Conference on Machine Learning, Atlanta, GA. https://www.researchgate.net/publication/281857285_Using_deep_learning_to_enhance_cancer_diagnosis_and_classification (accessed June 29, 2020)

54 Cruz, J.A. and Wishart, D.S. (2007). Applications of machine learning in cancer prediction and prognosis. *Cancer Informatics* 2: 59–77.

55 Chen, C., Garrido, T., Chock, D. et al. (2009). The Kaiser Permanente electronic health record: transforming and streamlining modalities of care. *Health Affairs* 28 (2): 323–333.

56 Reed, M., Huang, J., Brand, R. et al. (2013). Implementation of an outpatient electronic health record and emergency department visits, hospitalizations, and office visits among patients with diabetes. *JAMA* 310 (10): 1060–1065.

57 Bartolome, R.E., Chen, A., Handler, J. et al. (2016). Population care management and team-based approach to reduce racial disparities among African Americans/blacks with hypertension. *The Permanente Journal* 20 (1): 53–59.

58 Silvestre, A.-L., Sue, V.M., and Allen, J.Y. (2009). If you build it, will they come? The Kaiser Permanente model of online health care. *Health Affairs* 28 (2): 334–344.

59 Althoff, T., Sosic, R., Hicks, J.L. et al. (2017). Large-scale physical activity data reveal worldwide activity inequality. *Nature* 547 (7663): 336–339.

60 Gordon-Larsen, P., Nelson, M.C., Page, P., and Popkin, B.M. (2006). Inequality in the built environment underlies key health disparities in physical activity and obesity. *Pediatrics* 117 (2): 417–424.

61 File, T. and Ryan, C. (2014). Computer and Internet Use in the United States: 2013. American Community Survey Reports, ACS-28. Washington, DC: US Census Bureau.

62 Lazer, D., Kennedy, R., King, G., and Vespignani, A. (2014). The parable of Google flu: traps in big data analysis. *Science* 343 (6176): 1203–1205.

63 Chae, D.H., Clouston, S., Martz, C.D. et al. (2018). Area racism and birth outcomes among Blacks in the United States. *Social Science & Medicine* 2 (199): 49–55.

64 Stephens-Davidowitz, S. (2014). The cost of racial animus on a black candidate: evidence using Google search data. *Journal of Public Economics* 118: 26–40.

65 Hudnut-Beumler, J., Po'e, E., and Barkin, S. (2016). The use of social media for health promotion in Hispanic populations: a scoping systematic review. *JMIR Public Health and Surveillance* 2 (2): e32.

66 Burke-Garcia, A. and Scally, G. (2014). Trending now: future directions in digital media for the public health sector. *Journal of Public Health* 36 (4): 527–534.

67 Tu, J.V., Chu, A., Donovan, L.R. et al. (2015). The Cardiovascular Health in Ambulatory Care Research Team (CANHEART): using big data to measure and improve cardiovascular health and healthcare services. *Circulation Cardiovascular Quality and Outcomes* 8 (2): 204–212.

68 Scharff, D.P., Mathews, K.J., Jackson, P. et al. (2010). More than Tuskegee: understanding mistrust about research participation. *Journal of Health Care for the Poor and Underserved* 21 (3): 879–897.

69 Harding, A., Harper, B., Stone, D. et al. (2012). Conducting research with tribal communities: sovereignty, ethics, and data-sharing issues. *Environmental Health Perspectives* 120 (1): 6–10.

70 Horrigan, J.B. (2016). Digital Readiness Gaps. Washington, DC: The Pew Research Center.

71 Howard, B.V., Devereux, R.B., Cole, S.A. et al. (2005). A genetic and epidemiologic study of cardiovascular disease in Alaska natives (GOCADAN): design and methods. *International Journal of Circumpolar Health* 64 (3): 206–221.

72 Devoe, J.E. and Sears, A. (2013). The OCHIN community information network: bringing together community health centers, information technology, and data to support a patient-centered medical village. *Journal of the American Board of Family Medicine* 26 (3): 271–278.

73 Ji, X., Chun, S., Cappellari, P., and Geller, J. (2017). Linking and using social media data for enhancing public health analytics. *Journal of Information Science* 43 (2): 221–245.

15

Complex Systems Science

Matt Kasman[1], *Nancy Breen*[2], *and Ross A. Hammond*[1, 3]

[1] *Center on Social Dynamics and Policy, Brookings Institution, Washington, DC, USA*
[2] *National Institute on Minority Health and Health Disparities, National Institutes of Health, Bethesda, MD, USA*
[3] *Brown School, Washington University in St. Louis, St. Louis, MO, USA*

15.1 Introduction

A complex adaptive system is one in which elements interact and change over time, generating system-level patterns that may not be uniform or intuitive. This often creates challenges for traditional quantitative analytic techniques, which rely on strict assumptions about independence [1]. Many social phenomena are best characterized as complex adaptive systems. This perspective allows for a deeper understanding of how social phenomena operate, and why policies may have effects that are unexpected or differ across contexts.

Health disparities involve a tangled web of influences, defined in terms of populations of individuals who evolve and coevolve over time in varied and dynamic contexts [2]. This makes analytical tools from complex systems science well suited to improving our understanding of both determinants of, and solutions to, these problems. In this chapter, we describe how such tools can be used to address critical questions in minority health and health disparities research.

15.2 Unique Properties of Minority Health and Health Disparities Research

Health disparities and minority health research is distinct from most biomedical research conducted at the National Institutes of Health. While the majority of biomedical research is designed to isolate a single cause, the complex forces that shape minority health and health disparities require research that may benefit from a different analytic approach. Determinants in biological, behavioral, sociocultural, environmental, and healthcare domains may all simultaneously contribute to shaping minority health and health disparities. Moreover, multiple pathways, operating on multiple levels and timescales, may interact over time and in ways that shift across population groups. How health disparities

The Science of Health Disparities Research, First Edition. Edited by Irene Dankwa-Mullan, Eliseo J. Pérez-Stable, Kevin L. Gardner, Xinzhi Zhang, and Adelaida M. Rosario.

accumulate over the life course and are distributed over geographic space are important factors in identifying, understanding, and addressing the causes of disparities.

15.3 Data and Methodological Challenges in Health Disparities Research

Understanding causes of health disparities has been limited not only by analytic methods but also by a lack of data. Developments in data and methods in the twenty-first century offer the opportunity to address these challenges. A key advantage of complex systems modeling is its ability to incorporate data from a variety of sources into a common framework.

Detailed information on geography and household composition in the United States is collected in the US Census Bureau's American Community Survey and in the Current Population Survey (CPS). While these large surveys represent a well-defined majority of the US population, little or no health-related data are collected. The largest survey of health knowledge and behaviors is the much smaller National Health Interview Survey. Linking mortality data with the CPS [3] and with cancer registry data [4] was a critical breakthrough in population science. Advances in geographic information (GI) technologies, which include geographic information systems (GISs), remote sensing, and global position systems (GPSs), as well as concurrent advances in spatial analysis and visualization, have created new possibilities for generating and gathering geo-referenced data that can support health disparities research. Information and communication technology (ICT), which uses sensors to capture and report information by location, can collect data about activities, behaviors, and attitudes of individuals, specific to time and place.

Early efforts to confront the methodological challenge of combining data for meaningful analysis of health disparities involved adopting new tools. Multilevel analysis was a first step toward analyzing how residential characteristics are associated with health outcomes [5, 6]. For example, Diez Roux and colleagues concluded that interventions to reduce health disparities may need to combine approaches that address both individual needs and residential environments [7]. Debate over the utility of linking socioeconomic status (SES) averages as a substitute for lack of SES data in individuals [8–10] eventually concluded that SES data collection, especially conducted on more granular geographic scales, can help capture key aspects of disparities and health trends [11]. This finding led to widespread linkages of individual healthcare data with census averages, especially socioeconomic averages (e.g., income or education) at the county, tract, or block level, in studies [12] and public use data sets [13]. Incorporating spatially situated socioeconomic averages into population health studies allowed the field of health disparities to focus research on where disparities occur while allowing for more accurate analyses than were possible using stratification of health outcomes by race alone.

Complex systems modeling offers another step forward in consideration of spatial context. Activities, behaviors, and attitudes of individuals are mediated by the *built environment*, a social science term that refers to the design, construction, management, and use of places, spaces, and systems in which human activity takes place. Examples of built environments include buildings, parks, transportation, and environmental systems (including water, soil, and air) that are critical to health and well-being. In recent years, public health research has expanded measurement of the built environment to include metrics related to access to healthy food, community gardens, safe neighborhoods, multimodal transportation systems, and other features that promote a healthy lifestyle. Research has shown that how neighborhoods are built affects both the physical activity and mental health of residents [14]. Studies have shown that environments expressly designed to improve physical activity are linked to higher rates of physical activity [15] and that community gardens can be linked to increased fruit and vegetable consumption [16]. Another study combined smartphone use with built-in accelerometry to measure physical activity in 111 countries. The study highlights the role of the built environment in reducing inequality in activity, especially for women and girls [17].

Spatial context can also refer to social space. Social network analysis (SNA) with longitudinal data identifies the ways in which important relationships are created, modified, or dissolved over time. An important demonstration of the use of SNA to understand health disparities is a study of the effect of racism on the all-cause Black mortality rate [18].

A key advance for improving our understanding of health disparities in context will be combining spatial information into a single analytic framework. This is where complex systems modeling is particularly promising. Such models have already been used to study links between elements of the built or social environment and obesity and tobacco use. Similar approaches offer promise for understanding how spatial environments may lead to health disparities.

15.4 Strong Alignment Between Hypothesized Causes of Minority and Health Disparities and Complex Systems Science Approaches

Complex adaptive system models allow researchers to characterize and study dynamic and interdependent relationships between components within a system, which may occur at or between multiple levels of scale. These models can help explain system-level nonlinear behaviors; incorporate consideration of spatial effects, social network structures, and process timing; consider heterogeneity in effect pathways; and act as “policy laboratories,” allowing researchers to test interventions that cannot be feasibly explored in the real world due to time, cost, or ethical constraints [19].

In sum, complex adaptive system models may be able to elucidate the pathways of exposures that increase disease risk and lead to minority health status or health disparities.

15.5 Types of Questions that Complex Adaptive Systems Models Can Help Answer

Complex systems science provides a way to move beyond simply identifying associations between determinants and observed levels of minority health and health disparities outcomes. Researchers can use complex systems science to explore mechanisms and answer important questions concerning why observed levels of health disparities occur, whether there are key effect pathways or leverage points, why past or existing policies and interventions have observed effects in a given context, and how proposed policies or interventions might affect communities (including heterogeneous effects and unintended negative consequences). Although there are some exceptions (e.g., purely pedagogical models), complex adaptive systems models relevant to public health topics generally fall into one or more of the following categories: etiological, retrospective, or prospective. These model types are shown in Table 15.1 and are described in greater detail below.

Etiological models generally do not contain explicit representations of policies or interventions. Rather, they explore how pathways operate within a given system, answering fundamental questions about how the system works and how specific determinants work through defined pathways to drive outcomes. Findings from etiological models can enhance our understanding of the causes of existing patterns of behavior and outcomes. Although etiological models are not intended to test policies or interventions, identifying mechanisms, tipping points, path dependence (i.e., the impact of prior system states), and process timing can help clarify the leverage points needed for developing policies and interventions and for building future models that can be used to guide policies and intervention decisions.

One example of an etiological model appears in Hammond and Ornstein [20]. This model is based upon the hypothesis that individuals will attempt to gain or lose weight based on their observations of others. Starting with a simple characterization of this "follow the average" behavior, the paper incrementally adds more sophisticated and empirically grounded elements to a model that describes how patterns of population-level body mass index (BMI) emerge from individual behaviors. The model eventually includes individuals whose observations of others are dictated by their position in a social network, are willing to accept BMI that is "close enough" to the observed average, and are limited in their efforts to gain or lose weight at a given point in time by realistic physiological processes. The output of this model is then compared to longitudinal data to test the explanatory power of the core social influence mechanism.

Another example is a model of college enrollment [2, 21]. This model provides insight into the overlapping processes that "sort" higher-SES candidates into elite schools and present an imposing barrier to higher education and its benefits, including lifelong financial stability and health that go with it, for lower-SES individuals.

A third example is a model of physical activity among African American adults [22]. This model can be used to help explain differences in physical activity levels between men and women. The relatively low levels of physical activity experienced by women may be a key driver of their higher incidence of chronic illnesses.

Retrospective and prospective models both answer questions about the effects of policies and interventions. Retrospective models explore why specific policies, interventions, and/or

Table 15.1 Description and examples of three types of complex systems models.

Model type	Definition	Example	Example description
Etiological	Models that explore the influence of specific mechanisms on the operation of a system	Follow the average	Model of social influence on obesity status; agents change their BMI values based on the mean BMI observed among their social network neighbors
Retrospective	Models that examine why past policy changes or interventions had observed effects	COMPACT	Agents in a community update their knowledge about and engagement with a childhood obesity prevention intervention based on interactions with their social network neighbors
Prospective	Models that explore the potential effects of policies or interventions	Tobacco Town	Agents are smokers whose purchase behaviors are influenced by retailer environment

natural experiments had observed effects, and can be used to explain how these effects might have differed across populations or contexts. There are many notable examples of retrospective models outside the field of public health [23]. Researchers have recently begun using retrospective modeling to explain why large-scale public health initiatives have met with success, failure, or mixed results within and across communities [24]. The Childhood Obesity Modeling for Prevention and Community Transformation (COMPACT) study has produced a model that is being used to help explain the success of several whole-of-community child obesity prevention interventions. The model explores how overall increases in knowledge about and engagement with child obesity prevention efforts are driven by interactions between community members [24]. Retrospective models have the potential to build on the substantial resource investments of large-scale initiatives to guide future initiatives.

Prospective models explore policies and interventions that are under consideration but not yet implemented, and can provide important insight into the effects that they may have, including the potential for different impacts within communities. One example of a prospective model is described in Luke et al. [25]. This "Tobacco Town" model can provide insight into the potential effects of an array of tobacco retail control policies (and combinations of policies) on tobacco purchasing behavior in different communities. Figure 15.1 shows how policies that affect tobacco retailer density are projected to result in different total per-pack costs (i.e., the actual pack cost as well as indirect costs, such as those related to time and travel) in different communities. Another prospective model is a college enrollment model that builds on the etiological model described above [2, 21] to explore the effects of affirmative action policies on college composition and underrepresented minority applicants and enrollees [2]. Similarly, retrospective models can be employed as the basis for prospective models of similar policies and interventions. As we discuss below, this has been done as part of the COMPACT study.

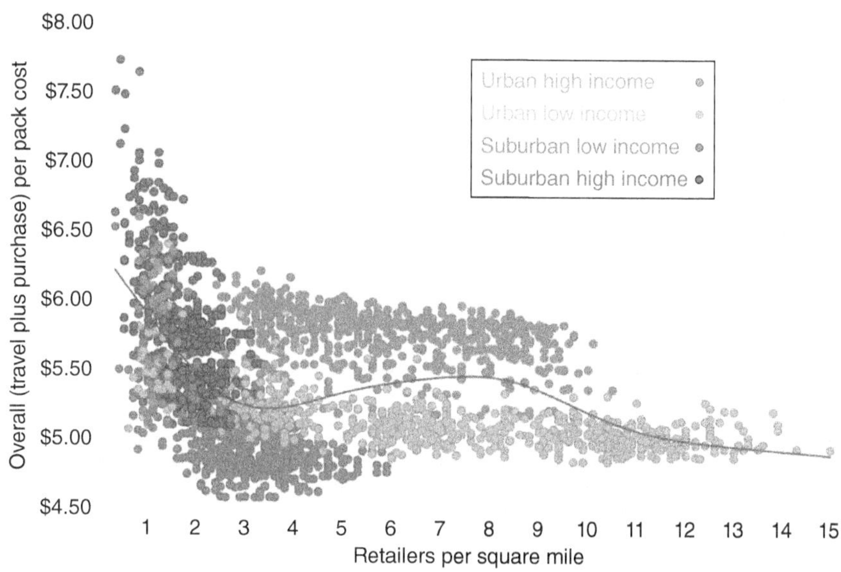

Figure 15.1 Relationship between tobacco retailer density and average, total per-pack cigarette cost (i.e., direct and indirect costs) across community contexts. *Source:* Luke et al. [25]. © 2017 American Public Health Association. (See color representation of this figure in color inserts section.)

15.6 Elements of a Successful Complex Adaptive Systems Model

Although a detailed description of each element is beyond the scope of this chapter, we discuss three elements that we believe are at the core of a successful modeling effort: engagement with theory, selection of a modeling approach, and use of available data, as summarized in Table 15.2 and described below. For each of these elements, we use as an illustrative example an iteration of the Tobacco Town model that incorporates demographic attributes of agents and can be used to explore health disparities that arise from differences in smoking prevalence and behavior across groups.

15.6.1 Engagement with Theory

Complex systems science research is strongly intertwined with theory. Theory generates hypotheses about causal effects that remain untested; dynamic modeling provides a particularly well-suited approach for testing some of these and can generate new hypotheses as well. In turn, complex systems approaches are grounded in theory; these models are built upon and explicate existing theory.

Researchers rely on existing theory to form initial hypotheses that are the basis of their research. For example, the "follow the average" mechanism that appears in the Hammond and Ornstein model described above tests hypotheses about social influences on obesity rates. Similarly, the Schelling model described below tests hypotheses about the relationship between individuals' preferences for neighbors of the same race and the overall level of residential segregation in the population.

When designing models, theory guides the selection of model assumptions that are used to select and characterize model elements (e.g., simulated individuals), as well as how these elements dynamically interact with one another during simulations. For example, in the college enrollment models described above, applications to colleges are based on rational choice theory (i.e., that applicants maximize expected utility).

For health disparities research, the National Institute on Minority Health and Health Disparities (NIMHD) has developed a framework of elements that are theorized to underlie health disparities. The domains, levels of influence, and dynamic aspects suggested can be used to develop specific, testable hypotheses about minority health and health disparities. For example, complex systems modeling is currently being applied in this way to examine inequality in healthcare utilization and lower levels of physical activity among

Table 15.2 Description of elements of a successful complex adaptive systems model.

Element	Description
Engagement with theory	Use of existing theory to form hypotheses that models will test and to characterize model elements and dynamics
Model approach	Selection of a specific modeling method that is well-suited to accomplishing research goals
Use of available data	How data are used as inputs into a model and to compare against model outputs

Example 15.1 Model-Building Example: Tobacco Town

The Tobacco Town model uses rational choice theory to inform its central assumptions about how smokers make tobacco purchase decisions. Policies that decrease retailer density may affect purchase decisions by increasing the cost of cigarettes either directly, by removing the cheapest options, or (more importantly in this model) indirectly, by forcing consumers to travel further. An iteration of the model that could be used to explore health disparities would disaggregate agents by demographic characteristics and incorporate between-group differences in smoking rates, product preferences, or geographic locations, but would continue to use rational choice theory to characterize purchase decisions.

African American women [19, 22]. It is anticipated that complex systems modeling will continue to engage with theory in order to advance health disparities research.

15.6.2 Selection of a Modeling Approach

The types of complex adaptive systems models include system dynamics, agent-based modeling, discrete event simulation, and hybrid models [19]. No single approach is superior in all cases, so researchers must select the modeling approach that is best suited for their research goals. Five considerations that can be used are outlined in Table 15.3 and described in further detail below.

15.6.2.1 The Perspective from Which the Research Question Is Analyzing the System: Top-down, Bottom-up, or Process-oriented

A research question with a top-down perspective focuses on how behavior results from the way system components are structured and connected to one another through casual pathways and feedback loops. For example, a research question about addiction might be addressed with a model that depicts the connections between substance use and physiological and psychological elements.

A research question with a bottom-up perspective is concerned with how individual entities within the system interact with one another and their environment to shape system-wide outcomes. The Hammond and Ornstein model takes a bottom-up perspective, with individuals making decisions about weight change that are potentially influenced by their peers' decisions.

Table 15.3 Comparisons of three complex systems approaches.

Axis of difference	Agent-based models	Discrete event simulations	System dynamics models
Perspective	Bottom-up	Process-oriented	Top-down
Heterogeneous entities	Yes	Yes	No
Entity agency	Yes	No	No
Geographic and social space	Yes	No	No
Handling of time	Discrete	Discrete	Continuous

A research question with a process-oriented perspective is concerned with how allocation of a finite resource changes system behavior and output. For example, an investigator may examine how changing the allocation of hospital beds between different wings in the hospital affects the flow of patients and waitlist times.

15.6.2.2 The Importance of Heterogeneity Among System Entities

Heterogeneity is important to consider if a researcher is interested in how the dynamics of a system change across different populations or how a policy intervention might have different effects in a variety of discrete locations. For example, if researchers want to create a model that examines the effects of a soda tax on soft drink consumption, then heterogeneity across factors like race, income, and urbanicity may be important, particularly if interventions are being considered across a spectrum of distinct geographic regions. In other cases, entities may be characterized by average patterns of behavior, without distinguishing between particular types of entities. This may be sufficient for showing how general trends emerge from system dynamics, particularly in cases in which entities are homogeneous across key dimensions or when more granular data are unavailable.

15.6.2.3 The Importance of Active Agency

Researchers must consider whether entities in the system are "active," in the sense that their perceptions and decision making are central to model dynamics, or "passive" and operated on by the processes within the system. In the former case, entities have "agency," meaning they behave in unique ways according to a set of properties (which may be heterogeneous across entities). Moreover, behavior may depend on that of other entities, as well as on changes in the environment, and agents can adapt and learn over time.

15.6.2.4 The Importance of Geographic and Social Space

Researchers must consider whether their model should include explicit representation of geographic or social space. For example, research into physical activity levels might be conducted with a model that depicts individuals traversing built environments. The Hammond and Ornstein model incorporates social network ties between individuals. Inclusion of spatial elements can be important (as described above), but comes with increased data and computational demands.

15.6.2.5 The Handling of Time

Investigators must consider whether updates to the system occur instantaneously (continuous) or at specific intervals over time (discrete). These intervals may reflect the passage of a fixed amount of time, such as entities updating their preferences every simulated day. Alternatively, intervals may reflect specific transitions between one state and other, such as patients entering and exiting a hospital ward.

Example 15.2 Model-Building Example: Tobacco Town

Tobacco Town is an agent-based model, which has a bottom-up perspective. Heterogeneity is important, more so when agents have demographic characteristics. Individual agency (i.e., decision making) and geographic space are central to the model. Time is handled as continuous, with agents traversing their environment and making purchase decisions during each simulated day.

15.6.3 Use of Available Data

Some complex adaptive systems models are "data free." Data-free models are useful for pedagogical (e.g., demonstrating emergence of macro-level patterns from micro-level behaviors) or theoretical (e.g., testing the feasibility or operational implications of theory) purposes. However, most models utilize data to some extent. There are three broad categories of data for use in complex systems modeling: direct, indirect, and expert guidance. Direct (or "primary") data refers to quantitative, qualitative, spatial, or social network data that are relevant for the specific research subject or context. Indirect (or "secondary") data refers to empirical evidence obtained from existing literature. Expert guidance refers to observations and estimates obtained from content experts.

Data can be used either as input for a model or for comparison with model output. Data used as input can help determine initial values of model elements (e.g., populations or a spatial environment) used in simulations, or can be used to characterize model dynamics (e.g., changes in model elements over time or relationships between model elements).

For example, the COMPACT model uses both direct data and expert guidance to characterize community conditions at the start of simulations of an intervention in a given community, and all three forms of data to represent the operation of the intervention over time. Comparing data with model output is critical for the testing, calibration, and validation of a model. Although there is no consensus among complex system scientists concerning what each of these activities specifically entails, there is broad agreement that testing involves observing model behavior under one or more conditions, calibration involves manipulation of elements to model specific behavior, and validation involves comparison of model behavior to expectations. For example, the COMPACT model was tested using expected output trends derived from a combination of retrospective surveys of participants and estimates provided by key intervention personnel.

Example 15.3 Model-Building Example: Tobacco Town

The Tobacco Town model is constructed using census data in conjunction with state and national surveys of smokers and tobacco retailers. These "input" data are used to characterize the environment (e.g., distributions of retailers, cigarette prices, home, and work locations) as well as individuals (e.g., transportation behavior, smoking rate, product preferences) within the model. Different data (e.g., survey and sales data) are used to test and calibrate the model. These "outcome" data describe patterns that we expect the model to reproduce, such as the proportion of cigarettes that are purchased in carton form by individuals from a given group in a specific community. Any extension of the Tobacco Town model would similarly need to draw on sources of "input" and "outcome" data.

15.7 Systems Science as Iterative Research

Complex systems science is inherently iterative. Models frequently complement other methods and are refined through repeated use, while the process of generating complex systems models identifies gaps in knowledge and understanding. Importantly, gaps can lead to development of theory and identify the need for additional data.

Complex adaptive system models complement other methods. When models use data as inputs or to compare to model output, those data are frequently generated with other tools and methods, including econometric analysis, GIS, SNA, surveys, interviews, and group model building [26]. Similarly, results from complex adaptive systems models can be used in conjunction with other tools to provide important guidance for researchers, policymakers, and stakeholders. For example, Banerjee et al. [27] use a model to simulate the diffusion of knowledge about microfinance loans between socially connected households and their participation in a microfinance intervention. They subject their findings to rigorous statistical analysis and propose a statistical measure that can be used to select the households in a community network who maximize the reach of the intervention.

Although a model can become sophisticated over the course of a long-term project, a best practice is to start with a relatively simple model. Researchers can benefit from starting with a model with one or two dynamic elements in order to understand their operation and effects through repeated testing before adding more. Additional mechanisms should be added one at a time so that their operation can be tested and understood [28]. Simple, etiological models can not only build confidence prior to proceeding to prospective policy modeling, but can also guide decisions about which policies to study [2]. Similarly, prospective policy models can build on retrospective models and can be gradually expanded during a project [28]. The COMPACT model of childhood obesity prevention interventions described above is a good example of this: it was applied retrospectively to study completed interventions in two sites (one in the United States and one in Australia). It was then used to prospectively provide guidance for a new intervention (also in the United States), with plans to use a more sophisticated iteration of the model to do the same for additional interventions.

Models can be used across contexts. The general model of tobacco retail policies described above [25] has been applied to several real-world settings. When applying the model to new settings, not only were pertinent data carefully selected and processed for the model, but the model itself was adapted to handle unique attributes of the setting (e.g., adding public transit as a transportation option for simulated residents of a large urban community).

The process of building and deploying complex adaptive system models can contribute to future advances in theory and data. Schelling proposed an early example of an agent-based model that demonstrated how relatively minor levels of preference for same-race neighbors can result in high levels of residential segregation [29]. This model provided a major contribution to theories of segregation by inspiring the collection of data about distributions of same-race preferences and spurring empirical examination of the role of racial preferences in residential decision making [30]. These new data have, in turn, been used in subsequent complex adaptive systems models [31, 32].

15.8 Limitations and Challenges

For findings to provide meaningful insight, complex adaptive systems models must use theory and data appropriately. This means starting with a clear research goal, characterizing model elements and dynamics with that goal in mind, using the highest quality data appropriately, and subjecting uncertainty in model characterization to rigorous sensitivity analyses. A complex adaptive systems model needs to be built with a team that has sufficient expertise, including in modeling (which may include programming acumen, depending on the type of model), content (i.e., familiarity with relevant current theories and evidence), and data (i.e., ability to appropriately handle quantitative or qualitative data that will be used and produced by the model).

We suggest two approaches. The first is frequent engagement between modeling, content, and data experts. This entails taking advantage of opportunities for interaction in person (e.g, conferences and meetings) as well as in burgeoning online settings [33]. The second is to take advantage of the opportunity presented by this challenge. Those with modeling expertise can become more proficient working in specific topic settings relevant to minority health and health disparities, while content and data experts can become familiar with complex systems approaches. Although true expertise may not be an easy or even feasible goal, such cross training allows for more effective collaborations.

A major challenge in complex systems modeling is dissemination. Model findings are frequently misunderstood in one of two ways: they are either disregarded as lacking sufficient rigor due to unfamiliarity with the method, or they are relied upon too heavily by policymakers or practitioners who view them as providing a sort of clairvoyance. To guard against these misconceptions, it is necessary for systems scientists to carefully describe their model's relationships to theory and data, the testing processes they engaged in, and the limitations and caveats associated with their model. For example, when discussing the Tobacco Town model described above, the authors were careful to note that although the model could provide suggestive evidence about the relative efficacy of different potential tobacco retailer policy combinations in a given setting, it could not provide accurate forecasting of the number of retailers of different types who would operate in a specific municipality after implementing actual policies. Nor was it designed to provide "downstream" effects (e.g., smoking cessation rates, cancer incidence levels).

Complex systems science has the potential to allow researchers to build theory, guide data collection, explore mechanisms, answer important questions, and uncover new ones about why observed levels of health disparities occur—whether there are key effect pathways or leverage points, why past or existing policies and interventions have observed effects in a given context, and how proposed policies or interventions might affect communities. Realizing this potential will require developing research capacity and interdisciplinary teams of researchers who can build models that will further our knowledge of the pathways and mechanisms for understanding how health disparities occur and what leverage points need to be addressed to reduce them.

15.9 Key Points

- Complex adaptive system models can help explain system-level nonlinear behaviors; incorporate consideration of spatial effects, social network structures, and process timing; consider heterogeneity in effect pathways; and act as "policy laboratories," allowing researchers to test interventions that cannot be feasibly explored in the real world due to time, cost, or ethical constraints.
- Researchers can use complex systems science to explore mechanisms and answer important questions concerning why observed levels of health disparities occur; whether there are key effect pathways or leverage points; why past or existing policies and interventions have

observed effects in a given context; and how proposed policies or interventions might affect communities (including heterogeneous effects and unintended negative consequences).

- Complex systems science research is strongly intertwined with theory. Theory generates hypotheses about causal effects that remain untested; dynamic modeling provides a particularly well-suited approach for testing some of these, and can generate new hypotheses as well. In turn, complex systems approaches are grounded in theory; these models are built upon and explicate existing theory.
- Complex systems science is inherently iterative. Models frequently complement other methods and are refined through repeated use, while the process of generating complex systems models identifies gaps in knowledge and understanding. Importantly, gaps can lead to development of theory and identify the need for additional data.
- For findings to provide meaningful insight, complex adaptive systems models must use theory and data appropriately. This means starting with a clear research goal, characterizing model elements and dynamics with that goal in mind, using the highest quality data appropriately, and subjecting uncertainty in model characterization to rigorous sensitivity analyses. A complex adaptive systems model needs to be built with a team that has sufficient expertise, including in modeling, content, and data.

Disclaimer

The views and opinions expressed in this chapter are those of the authors only and do not necessarily represent the views, official policy, or position of the U.S. Department of Health and Human Services or any of its affiliated institutions or agencies.

References

1 Miller, J.H. and Page, S.E. (2009). *Complex Adaptive Systems: An Introduction to Computational Models of Social Life*. Princeton, NJ: Princeton University Press.

2 Kasman, M. and Klasik, D. (2017). Individual, social, and institutional practices determining school and college choice. In: *Growing Inequality: Bridging Complex Systems, Population Health and Health Disparities* (eds. G.A. Kaplan, A.V. Diez Roux, C.P. Simon and S. Galea), 151–172. Washington, DC: Westphalia Press.

3 Backlund, E., Rowe, G., Lynch, J. et al. (2007). Income inequality and mortality: a multilevel prospective study of 521 248 individuals in 50 US states. *International Journal of Epidemiology* 36 (3): 590–596.

4 National Cancer Institute Surveillance (n.d.). Epidemiology and end results program. https://seer.cancer.gov/statistics/types/mortality.html (accessed June 19, 2020).

5 Diez-Roux, A.V. (1998). Bringing context back into epidemiology: variables and fallacies in multilevel analysis. *American Journal of Public Health* 88 (2): 216–222.

6 Diez-Roux, A.V., Nieto, F.J., Muntaner, C. et al. (1997). Neighborhood environments and coronary heart disease: a multilevel analysis. *American Journal of Epidemiology* 146 (1): 48–63.

7 Diez Roux, A.V., Merkin, S.S., Arnett, D. et al. (2001). Neighborhood of residence and incidence of coronary heart disease. *The New England Journal of Medicine* 345 (2): 99–106.

8 Krieger, N. (1992). Overcoming the absence of socioeconomic data in medical records: validation and application of a census-based methodology. *American Journal of Public Health* 82: 703–710.

9 Geronimus, A.T., Bound, J., and Neidert, L.J. (1996). On the validity of using census geocode data to proxy individual socioeconomic characteristics. *Journal of the American Statistical Association* 91: 529–537.

10 Susser, M. (1994). The logic in ecological: I. the logic of analysis. *American Journal of Public Health* 84 (5): 825–829.

11 Krieger, N., Chen, J.T., Waterman, P.D. et al. (2002). Geocoding and monitoring of US socioeconomic inequalities in mortality and cancer incidence: does the choice of area-based measure and geographic level matter?: the public health disparities geocoding project. *American Journal of Epidemiology* 156 (5): 471–482.

12 Brodish, P.H. and Hakes, J.K. (2016). Quantifying the individual-level association between income and mortality risk in the United States using the national longitudinal mortality study. *Social Science & Medicine* 170: 180–187.

13 Yu, M., Tatalovich, Z., Gibson, J.T., and Cronin, K.A. (2014). Using a composite index of socioeconomic status to investigate health disparities while protecting the confidentiality of cancer registry data. *Cancer Causes & Control* 25 (1): 81–92.

14 Renalds, A., Smith, T.H., and Hale, P.J. (2010). A systematic review of built environment and health. *Family & Community Health* 33 (1): 68–78.

15 Carlson, C., Aytur, S., Gardner, K., and Rogers, S. (2012). Complexity in built environment, health, and destination walking: a neighborhood-scale analysis. *Journal of Urban Health* 89 (2): 270–284.

16 Litt, J.S., Soobader, M.J., Turbin, M.S. et al. (2011). The influence of social involvement, neighborhood aesthetics, and community garden participation on fruit and vegetable consumption. *American Journal of Public Health* 101 (8): 1466–7143.

17 Althoff, T., Sosic, R., Hicks, J.L. et al. (2017). Large-scale physical activity data reveal worldwide activity inequality. *Nature* 547 (7663): 336–339.

18 Chae, D.H., Clouston, S., Hatzenbuehler, M.L. et al. (2015). Association between an internet-based measure of area racism and black mortality. *PLoS One* 10 (4): e0122963.

19 Hammond, R.A., Osgood, N., and Wolfson, M. (2017). Using complex systems simulation modeling to understand health inequality. In: *Growing Inequality: Bridging Complex Systems, Population Health and Health Disparities* (eds. G.A. Kaplan, A.V.D. Roux, C.P. Simon and S. Galea). Washington, DC: Westphalia Press.

20 Hammond, R.A. and Ornstein, J.T. (2014). A model of social influence on body mass index. *Annals of the New York Academy of Sciences* 1331: 34–42.

21 Reardon, S., Kasman, M., Klasik, D., and Baker, R. (2016). Student resources and stratification among colleges: an agent-based simulation of five mechanisms of the college sorting process. *Journal of Artificial Societies and Social Simulation* 19 (1): 8.

22 Kumanyika, S., Kasman, M., Whitt-Glover, M. et al. (2017). A prototype for identifying policy-relevant reasons for gender differences in physical activity. In: *Growing Inequality: Bridging Complex Systems, Population Health and Health Disparities* (eds. G.A. Kaplan, A.V.D. Roux, C.P. Simon and S. Galea). Washington, DC: Westphalia Press.

23 Axtell, R.L., Epstein, J.M., Dean, J.S. et al. (2002). Population growth and collapse in a multiagent model of the Kayenta Anasazi in Long House Valley. *Proceedings of the National Academy of Sciences* 99 (Suppl 3): 7275–7279.

24 Economos, C.D. and Hammond, R.A. (2017). Designing effective and sustainable multifaceted interventions for obesity prevention and healthy communities. *Obesity* 25 (7): 1155–1156.

25 Luke, D.A., Hammond, R.A., Combs, T. et al. (2017). Tobacco town: computational modeling of policy options to reduce tobacco retailer density. *American Journal of Public Health* 107 (5): 740–746.

26 Hovmand, P.S. (2014). Group model building and community-based system dynamics process. In: *Community Based System Dynamics*, 17–30. New York: Springer.

27 Banerjee, A., Chandrasekhar, A.G., Duflo, E., and Jackson, M.O. (2013). The diffusion of microfinance. *Science* 341 (6144): 1236498.

28 Hammond, R.A. (2015). *Considerations and Best Practices in Agent-Based Modeling to Inform Policy*. Washington, DC: National Academies Press.

29 Schelling, T. (1978). *Micromotives and Macrobehavior*. New York: W.W. Norton.

30 Bruch, E.E. and Mare, R.D. (2002). Methodological issues in the analysis of residential preferences, residential mobility, and neighborhood change. *Sociological Methodology* 42 (1): 103–154.

31 Bruch, E.E. (2014). How population struscture shapes neighborhood segregation. *American Journal of Sociology* 119 (5): 1221–1278.

32 Xie, Y. and Zhou, X. (2012). Modeling individual-level heterogeneity in racial residential segregation. *Proceedings of the National Academy of Sciences* 109 (29): 11646–11651.

33 Systems Science in Health (Beta): Online Community for Researchers, (2020) Teachers and practitioners. http://healthmodeling.org (accessed June 23, 2020).

16

Improving Equity in Healthcare through Multilevel Interventions

Margarita Alegria[1,2], Jacqueline J. Lloyd[3], Naomi Ali[1], and Karissa DiMarzio[1]

[1] *Department of Medicine, Disparities of Research Unit, Massachusetts General Hospital, Boston, MA, USA*
[2] *Departments of Medicine and Psychiatry, Harvard Medical School, Boston, MA, USA*
[3] *National Institute on Drug Abuse, National Institutes of Health, Rockville, MD, USA*

16.1 What Are Multilevel Interventions?

Multilevel interventions simultaneously address more than one aspect of the socioecological framework of health (i.e., individual; social support/network; provider; healthcare setting, such as a clinic or department; healthcare system, such as a hospital; or community level, such as a community agency) [1, 2]. The socioecological frameworks of health that guide multilevel interventions are a way to chart key relationships between the social and environmental context and health, and design a multilevel intervention approach.

Most researchers concentrate on several focal areas. Gee and Payne-Sturges [3] focus on the multiple ways in which neighborhoods could have an influence on health through structural factors, such as zoning laws, infrastructure investment, and tax incentives; psychosocial conditions, such as economic deprivation, discrimination, and crime; and neighborhood resources encompassing social capital, collective efficacy, and political influence to solve problems.

There is no consensus on the definition of multilevel interventions or of the designs that best capture the multiplicity of factors that might influence health. This ambiguity gives rise to diverse models. A commonly used socioecological framework includes five distinct levels from the narrowest to the broadest spheres of influence: intrapersonal, interpersonal, organizational, community, and policy [1, 4]. The order of the levels may change depending on the health issue of interest [5]. In another model, five levels become four, with a "micro" level that incorporates both intrapersonal and interpersonal factors [2]. For certain health conditions such as cancer, some have separated intrapersonal factors into two levels: the individual level that pertains to psychological and sociodemographic factors, and biological factors that pertain to genetic markers and metabolic processes [6]. Others have expanded the framework to differentiate between providers and the greater organizational network, as well as to separate state and national policy levels [7].

The Science of Health Disparities Research, First Edition. Edited by Irene Dankwa-Mullan, Eliseo J. Pérez-Stable, Kevin L. Gardner, Xinzhi Zhang, and Adelaida M. Rosario.

Some researchers construct a hierarchical model of micro, meso, macro/exo, and global/macro system levels that represent networks that systematically grow in size and sphere of influence [7, 8]. Stevens et al. [8] suggest a micro-, meso-, exo-, and macro-system-level framework for understanding socioecological influences in obesity: micro, individual (biological and behavioral/psychosocial factors); meso, social environment (peers, family, healthcare providers); exo, physical environment (work, school, home); and macro, community (recreation centers, cardiology clinics).

Conceptual models used in multilevel intervention design expand the socioecological framework of health and may focus on quality improvement. Three such models are: the PRECEDE model (Predisposing, Reinforcing and Enabling Constructs in Educational Diagnosis and Evaluation); the Put Prevention Into Practice Implementation Model (PPIP); and the Agency for Healthcare Research and Quality (AHRQ) model to improve delivery of care [9].

The PRECEDE planning model for multilevel intervention design utilizes a series of assessments of factors related to the health outcome, which could include social, epidemiological, behavioral, environmental, educational, ecological, administrative, and policy assessments to determine the relationship of these factors with the desired health outcome [10]. Once these relationships are identified, a multilevel intervention can be developed that addresses the different relationships between these factors and the health outcome.

To improve the delivery of care in primary care settings [11], PPIP has been used for taking a stepwise approach to understanding the complex dynamics within the organizational level of an intervention. PPIP first assesses attitudes from providers and staff about what changes are needed in organizational procedures and clinical care. It subsequently assesses the organizational and staff capacity for the identified and desired changes in practice. Finally, it involves physicians and staff at all organizational levels in planning, implementation, and evaluation procedures. In some circumstances, investigators will develop a causal model that represents the associations between different behaviors and identified social determinants and the health outcome of interest in order to identify moderators and mediators and their unique relation to the selected health behavior and health outcome [11].

In all of the formulations of the socioecological framework, a multilevel approach is based on the premise that an individual's health and healthcare interact multidirectionally with their relationships, social networks, and greater societal structures [12]. At each level of intervention, there is a different target or objective for change [2]. There is a lack of consensus as to the minimum number of levels a multilevel intervention must target (e.g., at least two levels of influence [13, 14], or three [7, 15]). A multilevel intervention thus differs from a multicomponent intervention which targets various aspects of the same level in one intervention (i.e., for obesity control, providing nutrition education and exercise training sessions) [8]. A multilevel intervention targets a health issue in a multipronged approach aimed at multiple levels of influence.

Using the example of smoking cessation, interventions may target multiple levels by engaging health behavior change at the individual level (smoking cessation) and through policy change (increased taxes on cigarette sales), improvement in health services delivery (improved screening practices for tobacco use disorder and rapid referral), and environmental justice (reduced disproportionate cigarette advertising in low-income neighborhoods) [13].

Some interventions may take both multilevel and multicomponent approaches simultaneously, as in the case of Robinson et al. [16], a randomized controlled trial (RCT) to treat obese and overweight children. The intervention tests a multilevel approach of a community-based after-school sports program (to increase exercise in children), a home-based environmental and behavioral intervention, and a primary care behavioral counseling intervention. Some strategies included under these interventions address individual, family, and community factors related to obesity, including helping participants change personal preferences for certain foods and activities, family eating habits, the physical home environment, family income, provider capacity for counseling, and community center programs for after-school team sports [16].

There has been considerable emphasis on implementing multilevel approaches and frameworks of health that address various spheres of health, such as stigma and population health [12], health education and behavior [4], mental illness [17], human immunodeficiency virus (HIV) and acquired immune deficiency syndrome (AIDS) [5], family planning [2], child and adolescent health [18], obesity [8], and cancer [6]. In some settings, multilevel approaches to health promotion have been more well established and researched, such as identifying, preventing, and improving physical and mental health outcomes in schools, where multilevel approaches may be referred to as multitier [18]. In 2012, Golden and Earp conducted a comprehensive literature search on health promotion interventions, classifying each intervention on the level(s) targeted, and found that about two-thirds of the studies only addressed one or two levels, and that there was no significant increase in the number of multilevel health promotion intervention studies in the 2000s compared to the 1990s [4]. There is minimal literature from the past 5 years on multilevel interventions across disciplines of health and healthcare that directly address disparities such as health promotion, disease prevention, and the reduction of mortality among people with chronic illness [1, 2, 5, 17]. Among the studies that have been done, most are based on health behavior change and illness prevention.

The general assumption is that multilevel interventions are likely to result in more long-lasting or synergistic change, or at least additive changes, from the multiple levels of influence in the expected direction. Few study designs can easily separate or evaluate the unique contributions of the individual intervention components from the synergistic effects or influences from the different levels. It is important to note that single-level interventions that target a particular level of the socioecological framework, such as the interpersonal level, can affect other levels that influence health and/or healthcare in a reciprocal manner [12, 17]. This phenomenon is important to the development of successful multilevel interventions [1], because some intermediary outcomes (i.e., community screening) might affect other domains (i.e., providers doing motivational interviewing to engage the participants in behavior change).

The complementarity principle and risk compensation theory emphasize the importance of paying attention to the connections between levels rather than only distinguishing between the levels themselves. In particular, the complementarity principle posits that intervening at one level can feed back and affect another level [1]. For example, zoning laws that improve housing quality and enable greater access to space for recreation might augment social cohesion and improve social interactions that consequently affect children's health. The risk compensation theory purports that intervening and producing

Examples of process and outcome measures used in selected multilevel intervention studies are described below.

16.2.3 Methodology and Evaluation of Multilevel Interventions

A *Health Affairs* article states that "conducting multilevel interventions designed to address health disparities is often complex and messy" [13]. Changes in policy and the perception of policymakers on the importance of a health issue must occur concomitantly with multilevel interventions targeting a combination of individual, interpersonal, organizational, and community levels of influence in order for those interventions to succeed [13]. With its emphasis on engaging and developing strong partnerships between researchers and multiple, diverse stakeholders in the community [29], CBPR could improve implementation and sustainability of multilevel interventions. Combining multilevel interventions with CBPR principles that actively engage community partners and build community capacity may be especially effective in reducing health disparities [14, 15]. Notably, Andrews et al. describe how a CBPR framework allowed for a decade-long collaboration between community stakeholders in public housing neighborhoods and their research team. This led to implementation of a sustainable multilevel tobacco-cessation intervention. Extensive CBPR collaboration empowered local figures to significantly improve the feasibility and adaptation of the intervention. Combining individual, interpersonal, and neighborhood-level approaches, and utilizing neighborhood resources such as the formation of a neighborhood advisory board, increased smoking cessation rates in the community [30].

Even with the potential benefits of community-based multilevel interventions, there are significant considerations for program implementation and the dissemination of results. Five considerations to community-based, multilevel interventions include: building collaborative partnerships with community members; deciding the intensity and dose of the intervention; maintaining consistency across activities and targeted levels; synchronizing activities across levels; and creating program sustainability post study [31]. A cross-evaluation of three multilevel intervention studies found that these challenges can be addressed by engaging with the community before, during, and after the study; identifying community leaders who can act as ambassadors of the intervention; conducting frequent visits to sites to ensure compliance with the intervention protocol; and hosting trainings to increase community capacity [31]. CBPR might be central for adopting a sustainable intervention and then evaluating its impact.

Since multilevel interventions target more than one level of influence on health, they are believed to have at least an additive, if not a synergistic, effect on the health outcome of interest [32]. A synergistic effect could result from the nuanced, multidirectional relationship between the different levels of influence in the socioecological framework [32]. A mixed-method evaluation of an international, multisite suicide prevention intervention found that multilevel and single-level interventions may also stimulate catalytic effects, such as unplanned activity from stakeholders or external parties responding to the intervention's efforts to improve the targeted health outcome. Due to the complexity of intervention effects, researchers must decide whether to compare the combined effect of the multiple interventions together to no intervention, to separately analyze the impact of each single-level intervention, or to assess whether there are interactive effects [32]. Perhaps because of the expense and the complexity of designing an intervention to study potential interactive effects, it is more common for studies to examine the combined effect, or overall impact, of a multilevel intervention [32]. This leads to not being able to establish the impact of the different intervention components, which can be important to funders and implementers who are making implementation decisions.

Despite the methodological challenges of designing, conducting, and evaluating multilevel interventions, they may be more effective than single-level interventions, as they target multiple determinants of health simultaneously and, thus, have the potential to improve population health and the health of underserved populations from different targets [13]. More research is needed to determine whether multilevel interventions are more effective than single-level interventions [23] and whether they actually reduce disparities. Although some multilevel intervention studies include or select for racial/ethnic minorities, these studies predominantly neglect to examine whether the intervention successfully reduces disparity in the targeted health outcome. Since there are many methodological issues in developing and evaluating multilevel interventions [13], recent literature has focused on acknowledging and addressing issues related to study design, participant recruitment and retention, and intervention evaluation [8, 32, 33].

In developing appropriate measures to evaluate multilevel intervention effects, researchers should include measures assessing change in racial disparity among heterogeneous racial/ethnic samples, such as Project ReD CHiP, a Baltimore-based, multilevel initiative to reduce racial/ethnic disparities in hypertension [26]. Upon completion, Project ReD CHiP will quantify change in racial disparities through outcomes measures that characterize differences in the percent of participants with blood pressure control across racial/ethnic groups and, subsequently, the degree of disparity post intervention. This is the exception rather than being typical of most studies [26].

RCTs and natural experiments are two methods used to evaluate the impact of multilevel interventions [13]. While RCTs are considered the gold standard, they might not always be ideal for evaluating multilevel interventions. In contrast to individually targeted interventions, multilevel interventions may have a greater impact at the population level, and the magnitude of their impact may not be realized until after the study concludes. To assess population-level effects, large sample sizes and studies of longer duration would be required [13]. Unlike RCTs, natural experiments do not manipulate exposure to the intervention of interest and may be more useful for studying the impact of policy or environmental-level interventions that target rare health outcomes of interest, such as suicide prevention. Rare health outcomes would require large sample sizes to garner enough power for analysis in RCT designs [33]. Nonetheless, RCTs and quasi-experimental designs are better able to control for selection bias and confounding variables and, if possible, are the recommended design for intervention evaluation [32, 34].

The literature contains quantitative, qualitative, and mixed-methods approaches to analyze synergistic effects in data from multilevel intervention studies. Harris et al. developed the QuaSIC approach, a longitudinal process evaluation that analyzes context, mechanisms, and outcomes of a multilevel intervention in a mixed-methods approach of semi-structured interviews, focus groups, field notes, and questionnaires [35]. Each method of evaluation provides unique input that allows researchers to map synergistic interactions between levels targeted by the intervention, those being: interviews and focus groups, which provided input from health professionals and community-based professionals (i.e., police, social workers); questionnaires, which tracked implementation of the intervention; and field notes, which recorded intervention team meetings as a way to track processes as they unfolded [35].

16.3 Multilevel Interventions: 2012–2017

Table 16.1 outlines a selection of multilevel interventions conducted around the world with original research articles published between the years 2012 and 2017. The interventions targeted a range of outcomes, including childhood obesity, sexually transmitted infections, hypertension self-management and control, smoking cessation, mortality, and hepatitis B virus screening.

16.4 Future Directions

1) The absence of multilevel interventions that demonstrate a reduction in health disparities is a major gap in the research. Future research should test multilevel interventions for disparities reduction.
2) Because of the complexity of multilevel interventions, achieving acceptability, feasibility, and sustainability in diverse settings is challenging. Trying to disentangle what levels of the intervention are likely to work and which are indispensable is an important area of inquiry.
3) Brief measures that capture the most important dimensions of multilevel interventions are also necessary, given the staff and participant burden that many of these studies require. Use of computerized adaptive testing might help identify ways of shortening these assessments while maintaining precise and effective measurements of the myriad constructs involved in multilevel interventions.
4) Although CBPR is recommended to accelerate acceptability and sustainability, collaborating with communities to identify and quantify the types of measures and processes to evaluate the impact of multilevel interventions is an area for expansion.
5) Transportability of multilevel interventions from one context to another should be investigated so that it does not take 17 years to integrate successful multilevel interventions in environments where they are not developed. Marketing of these interventions to policymakers and consumers would be needed. Also, it might be necessary to identify the messengers to engage stakeholders in adopting the multilevel interventions that have worked in other settings.

16.5 Case Study to Illustrate Multilevel Interventions

We include here a case study to illustrate a multilevel intervention that addresses health disparities.

Table 16.1 Multilevel Interventions: 2012–2017.

Authors	Place/setting	Sample size	Study population	Targeted outcome	Study design and measures	Intervention levels targeted	Results
Crespo et al. [21]	13 elementary schools in San Diego, CA	808	Latino parent–child dyads; children were enrolled in kindergarten to 2nd grade	Prevention of childhood obesity	Randomized controlled trial Primary health outcome measures: child BMI z-score, BMI percentile, and weight status; parent BMI and weight status Secondary behavioral outcome measures: child daily fruit and vegetable consumption, daily snack consumption, daily sweetened beverage consumption, daily water intake, physical activity, sports participation, and TV viewing; total number of behavioral strategies parents used to reduce dietary fat Mediator measures: parent monitoring of diet and physical activity, parent control of diet and physical activity, parent reinforcement of diet and physical activity, and parental support for physical activity	Family, Community	No intervention effects on BMI z score. The family only intervention and combined family and community interventions significantly led to increased healthy eating and increased physical activity.

(Continued)

Table 16.1 (Continued)

Authors	Place/setting	Sample size	Study population	Targeted outcome	Study design and measures	Intervention levels targeted	Results
Lippman et al. [36]	Corumba, Brazil	420	Sex workers: 96.4% Brazilian, 3.6% Bolivian or Paraguayan	Condom use and sexually transmitted infections	Prospective cohort multilevel intervention Process measure: level of participation/exposure to the intervention (20-question survey assessing level of participation in different intervention components) Outcome measure: diagnosis of incident chlamydia and/or gonorrhea and reported condom use in the 3-month intervention period Social environmental correlates: perception of mutual aid, trust, support among sex workers (social cohesion, participation in social networks, and access to and management of social and material resources	Individual, interpersonal, community	Intervention participants had increased odds of consistent condom use with regular (OR: 1.9, 95%, CI: 1.1–3.3), new (OR: 1.6, 0.9–2.8) and nonpaying clients (OR: 1.5, 0.9–1.5). Odds of STI incidence were also reduced (OR: 0.46, 0.2–1.3).

Cooper et al. [26]	6 primary care practices in Baltimore, MD	42 845	Adult patients at six primary care clinics; n = 15 738, African American patients; n = 24 095, White patients	Disparities in hypertension control	Pragmatic trial design of three complementary multilevel interventions Process measures: percent with uncontrolled BP and uncontrolled BP with medication titration in the last 3 months, percent completion rate for those enrolled in care management, and change in provider self-reported use of assessment and partnership skills Outcome measures: percent of patients with controlled BP and racial disparity in controlled BP	Patient, provider/staff, clinic	Ongoing
Gorin et al. [37]	Providence, RI	201 pairs	Overweight and obese individuals	Weight loss	Randomized controlled trial Outcome measure: body weight Mediator/correlate measures: type and placement of food in the home, availability of exercise equipment, other items in home, social support, perceived helpfulness of home environment, dietary intake, and physical activity	Individual, interpersonal/ home environment	Baseline BMI was positively correlated for both treatment groups. In the group randomized to receive individual and home environment interventions, BMI changed positively. In the individual-treatment only group, change in BMI was negatively correlated with intervention exposure.

(*Continued*)

Table 16.1 (Continued)

Authors	Place/setting	Sample size	Study population	Targeted outcome	Study design and measures	Intervention levels targeted	Results
Ephraim et al. [10]	Primary care clinic in Baltimore, MD	336	Adult African American patients	Hypertension self-management and control	Randomized controlled trial Outcome measures: patients' blood pressure control at 12 months (two measures – the average of three measurements at 4, 8, and 12 months; the average of measurements at 8 and 12 months) Mediators/correlates: biomedical correlates (e.g., hyperlipidemia, glycemic control, medical comorbidities) psychological correlates (e.g., patients' self-efficacy, problem-solving skills, patient activation) environmental correlates (e.g., patients' food security, residential mobility, neighborhood cohesion) social and demographic correlates (e.g., self-reported age, gender, ethnicity/race)	Patient, family, community	Completed, results currently not published

Finch et al. [27]	3 childcare centers in New South Wales, Australia	459	Children between the ages of 3 and 5	Physical activity	Cluster randomized controlled trial Process measures: fundamental movement skill development activity sessions (frequency, length of session) staff delivery of structured physical activity (total minutes of structured physical activity) staff role modeling of active play and delivery verbal prompts (number of times staff participated in active play; number of times staff provided positive statements about physical activity) limiting small screen reaction and sedentary time (total minutes of television viewing) Process evaluation at the clinic level: monthly checks of intervention clinics for displayed educational materials and posters provider retention in educational intervention session Outcome measures: child step counts (measured at baseline and follow up, using pedometer)	Social, organizational, physical environment	No significant change in mean child step counts ($p = 0.07$)

(Continued)

Table 16.1 (Continued)

Authors	Place/setting	Sample size	Study population	Targeted outcome	Study design and measures	Intervention levels targeted	Results
Gittelsohn et al. [22]	30 low income zones surrounding recreation areas in Baltimore, MD	720	African American parent–child dyads; children were aged 10 to 14	Childhood obesity prevention	Group randomized study Process evaluation measures: policy – e.g., number of different sectors represented/meeting wholesaler – e.g., number of food items that meets the nutrition guidelines per phase recreation center – e.g., number of planned intervention sessions delivered by youth-leaders corner store/carryout – e.g., number of healthier options on menu family/SMS messaging, social media – e.g., percent of families that participate in intervention program's websites Impact evaluation measures: policy – percent of action items achieved/year wholesaler – e.g., percent of change in sales of promoted sales based on collected sales data from wholesalers	Individual, household, institutional, city/policy	Ongoing

					recreation center – e.g., changes in recreation center policies regarding food environment adult/household – e.g., household food purchasing children and youth leaders – e.g., healthiness of food preparation methods		
Lv et al. [38]	3 districts in central Hangzhou, China	2016	Adults, aged 18–64	Smoking, physical activity, and diet	Nonrandomized controlled before-after quasi-experimental trial Process measure: self-reported engagement in intervention activities Outcome measures: smoking status, vegetable and fruit consumption, and physical activity Correlates: knowledge and beliefs about health hazards of tobacco, the impact of diet on health, and the benefits of regular physical activity	Individual, interpersonal, physical environment, policy	Smoking prevalence significantly decreased by 25.8% ($p < 0.001$); fruit and veggie consumption significantly increased ($p = 0.036$); metabolic equivalent of physical activity significantly increased ($p = 0.023$).

(Continued)

Table 16.1 (Continued)

Authors	Place/setting	Sample size	Study population	Targeted outcome	Study design and measures	Intervention levels targeted	Results
Heart Healthy Lenoir Study (2010–2015)[39]	6 clinics in Lenoir County, NC	525	Adult African American and White patients	Reduce cardiovascular disease risk	Longitudinal cohort/randomized controlled trial patient level measures: e.g., blood pressure, weight, creatinine, cholesterol, and A1c levels, depression, smoking status, sociodemographic characteristics, quality of life, knowledge/ behavior/ beliefs, exercise, diet, social determinants of health Practice level process (providers/ clinical practice) measures: e.g., action taken if uncontrolled blood pressure, consequences of uncontrolled hypertension discussed, assess medication adherence, assess for understanding of instructions, and percent patients with hypertension and blood pressure in normal range	Individual, provider/staff	Black and White patients both experienced a reduction in blood pressure, but there was no change in the difference between Black and White blood pressure management (difference of 2.8 mmHg, $p = 0.26$).

Krok-Schoen et al. [28]	10 primary care clinics in Columbus, OH	527	Patients, ages 50+	Colon cancer screening	Group randomized trial Process evaluation measures: monthly checks of intervention clinics for displayed educational materials and posters, provider retention in educational intervention session, the number of patients who completed each step of the intervention and were exposed to the clinic-level intervention Patient-level primary outcome measure: whether patients completed any colorectal cancer screening test patient-level secondary outcome measures: reported barriers, intention to screen, knowledge, and beliefs related to colorectal cancer screening Provider-level outcome measure: CRC screening knowledge	Clinic, provider, patient	No significant effect on colorectal cancer screening (35.4% in intervention group compared to 35.1% in control)

(*Continued*)

Table 16.1 (Continued)

Authors	Place/setting	Sample size	Study population	Targeted outcome	Study design and measures	Intervention levels targeted	Results
Michels et al. [40]	Belgium, Cyprus, Estonia, Germany, Hungary, Italy, Spain, and Sweden	8543	Children, aged 2 to 9.9 years	Sleep duration	Pre-test post-test control group multilevel intervention Process evaluation measure: four items that assess school and community efforts and parent involvement in the intervention Primary outcome measure: parental report of child diet and sleep duration Secondary outcome measure: parental report of TV presence in child's room Correlates: socio-demographic variables and child body composition	Individual, school, family, community	No significant changes in sleep duration.
Andrews et al. [30]	14 public housing neighborhoods in Augusta, GA and Charleston, SC	409	Adult women smokers; 86% African American	Smoking cessation	Cluster randomized controlled trial Outcome measure: Biochemically verified 8-day point prevalence abstinence from smoking at 6 and 12 months Correlates: nicotine dependence, sociodemographic characteristics, smoking cessation self-efficacy, and social support	Individual, interpersonal, neighborhood	9% of participants remained abstinent from smoking at 12 month follow up compared to 4.3% of the control group ($p = 0.05$).

Coronado et al. [41]	4 health centers in Washington state	536	Latina patients, aged 42 to 74	Disparities in mammogram screening	Randomized controlled trial Primary outcome measure: completion of a mammogram within 1 year after randomization Secondary outcome measures: cost-effectiveness and neighborhood-level influences Moderators of program efficacy measures: e.g., sociodemographic characteristics, breast cancer screening behaviors, functioning and well-being, social norms, social support, perceived susceptibility of breast cancer, health care interactions, and perceived discrimination	Patient, clinic	19.6% of patients were screened in the patient intervention compared to 11.0% of patients in the control group ($p < 0.01$). No significant differences in screening between clinics randomized to clinic-level intervention and those who were not.
Hankonen et al. [42]	6 vocational schools in Helsinki, Finland	267	Students, aged 15 to 17	Physical activity and sedentary behavior	Pragmatic, cluster randomized parallel group trial Process evaluation measures: mediators (behavior change techniques use and teacher sitting reduction behaviors) as well as intervention fidelity, responses to intervention and effectiveness in subgroups Primary outcome measures: changes in moderate-to-vigorous intensity physical activity and sedentary behavior	Individual, interpersonal	Ongoing

(Continued)

Table 16.1 (Continued)

Authors	Place/setting	Sample size	Study population	Targeted outcome	Study design and measures	Intervention levels targeted	Results
					Secondary outcome measures: students – e.g., muscle and fat mass and physical and mental well-being teachers – self-reported sitting reduction activities and observed student behavior		
Kerrigan et al. [43]	Santo Domingo, Dominican Republic	250	Female adult sex workers with HIV	HIV treatment outcomes	Longitudinal cohort Outcome measures: numbers of and consistent condom use, nonpaying steady sexual partners in the last 30 days, sexually transmitted infections, and linkage and engagement in HIV case services (attended or missed any clinic visits in the last 6 months; attendance and adherence to antiretroviral therapy) Correlate measures: sociodemographic characteristics and drug and alcohol use	Patient provider clinic	There was a significant increase in treatment adherence (72% to 89%, $p < 0.001$) and protected sex (71% to 81%, $p < 0.002$).

Keyserling et al. [39]	Eastern North Carolina	291	Adults recruited from communities in Lenoir County, as part of the Heart Healthy Lenoir Project	Dietary and physical change; weight loss	Randomized controlled trial Outcome measures: overall diet quality, fruit and vegetable intake, dietary fat quality, changes in physical activity, quality of life, weight, and cholesterol levels	Patient provider	There were substantial improvements in diet measures [4.3 units (95% CI 3.7 to 5.0)], walking [64 min wk^{-1} (95% CI 19 to 109)], and systolic blood pressure [−6.4 mmHg (95% CI −8.7 to −4.1)] that were maintained through 24 months of follow up. At 24 months, weight loss was −2.1 kg (95% CI −4.3 to 0.0). Outcomes for African American and Whites were similar.
Paskett et al. [44]	24 clinics across 12 counties in Appalachiand, OH	337	Parent–daughter dyads; children were aged 9 to 17	HPV vaccination	Group randomized trial Provider level outcome measure: knowledge of HPV and HPV vaccine Patient level outcome measure: receipt of HPV vaccine Clinic level structural measures: patient volume, revenue sources, number of insured patients, and number of providers prior to patient and provider interventions	Parent, provider, clinic	7.7% of daughters in the intervention group received HPV shot within 3 months compared to 3.2% of daughters in comparison group.

(Continued)

Table 16.1 (Continued)

Authors	Place/setting	Sample size	Study population	Targeted outcome	Study design and measures	Intervention levels targeted	Results
Seguin et al. [45]	Eligible communities will be classified as rural–urban commuting area code 7 or higher and be designated as a medically underserved area in the United States		Female, 40 years old or older, with a BMI of 25 or greater, and currently sedentary	Cardiovascular disease prevention	Randomized controlled efficacy trial Process evaluation measures: leader and participant-level process evaluation (questions related to attendance, program delivery, and fidelity) and economic evaluation (information on salaries, cost of facilities, equipment) Outcome measures: behavioral measures – e.g., nutrition and physical activity behaviors, health behaviors (smoking) psychosocial measures – e.g., social support, depression, anxiety, and eating behaviors demographic variables – e.g., age, race/ethnicity anthropometric and physiologic measures – e.g., BMI, body fat, blood pressure physical activity, diet, and functional fitness measures – e.g., physical activity, diet	Individual, community, environmental	Ongoing

Gittelsohn et al. [46]	Six rural reservations in New Mexico and Wisconsin	240	American Indian adults	Prevention and reduction of obesity	Stratified, group randomized trial process evaluation measures: policy – e.g., number different sectors represented/meeting schools – e.g., percent teachers trained/school food Stores – e.g., percent promoted healthy foods stocked worksites – e.g., number of participating worksites/ community community media – e.g., number of radio announcements/week social media – e.g., number of Facebook, Instagram, Twitter posts/day/platform impact evaluation measures: policy – e.g., number of health-related issues put on tribal policymaker's agenda/year schools – e.g., number of policy changes made to improve the school food and physical activity environments	Individual, community, policy	Ongoing

(*Continued*)

Table 16.1 (Continued)

Authors	Place/setting	Sample size	Study population	Targeted outcome	Study design and measures	Intervention levels targeted	Results
					food stores – e.g., number of units of promoted foods sold worksites – e.g., number of policy changes made to improve workplace food and physical activity environments adult/household – e.g., household food purchasing		
Go et al. [19]	32 communes in Thai Nguyen Province	455	Males drug injectors with HIV	Mortality	Randomized controlled trial Outcome measures: uptake of antiretroviral therapy and mortality Correlate measures: sociodemographic characteristics, sexual risk behaviors, injecting behaviors, and HIV and drug use-related stigma	Individual, community	Survival at 24 months was the highest in the combined community and individual intervention group compared to the standard care, community only, and individual only groups (87% vs. 82%, 68%, and 73%, $p = 0.001$).

Komro et al. [47]	6 communities in northeastern Oklahoma	1623	High school students; 46% American Indian; 46% White only	Underage alcohol use prevention	Randomized controlled trial Implementation evaluation process measures: e.g., percent schools implemented intervention program Outcome measures: individual level measures – drinking behavior (quantity/frequency and riskiness of drinking) and alcohol-related risks/harms (academic, social) School and community measures: alcohol purchase attempts (the likelihood of stores to sell alcohol to underage youth)	School Community	Compared to the control condition, the community only, school only, and combined interventions all reduced the probability over time of 30-day alcohol use (25%, 22%, and 12% reduction) and heavy episodic drinking (24%, 19%, and 13% reduction).
Ma et al. [48]	18 community-based organizations in Pennsylvania, New Jersey, and New York City	2337	Vietnamese American adults	Disparities in hepatitis B virus screening	Cluster randomized trial Process evaluation measures: participant acceptance of and satisfaction with the intervention and use of patient navigation services as well as quality control and treatment fidelity Outcome measure: hepatitis B virus screening at 6-month follow up Correlate measures: sociodemographic characteristics (including indicators of acculturation)	Individual, community, health care system	Intervention participants had a higher screening rate than control participants (88.1% to 4.6%). The rate ratio of intervention participants who received hepatitis B virus screening by a 6 month follow up were significantly higher (RR = 19.12; 95% CI = 12.75, 24.77; $p < 0.001$).

16.5.1 Case Study: Heart-Healthy Lenoir High Blood Pressure Study

The Lenoir High Blood Pressure Study is one of three studies conducted under the Heart -Healthy Lenoir project in Lenoir County, North Carolina [39]. Lenoir County is a rural region of eastern North Carolina and is in the middle of the "Stroke Belt," an area in the southeastern United States that has a higher risk for heart attack and stroke compared to other parts of the United States. In addition, obesity rates in Lenoir County are higher compared with the rest of North Carolina.

For the High Blood Pressure study, a CBPR approach was used to conduct a cohort study of a multilevel intervention that included both a practice- and a patient-level intervention to improve patient blood pressure control and narrow health disparities gaps between Black and White patients' blood pressure control [34]. The project was a collaborative effort between community agencies, community members, and two universities (University of North Carolina at Chapel Hill and East Carolina University). Practices and study staff worked collaboratively to develop and test hypertension management strategies, and providers met regularly to discuss effective strategies. The multilevel intervention for patients with hypertension in primary care practices incorporated enhanced activities, including integration of a community coach, use of home blood pressure monitoring in clinical decision making, standardization of care delivery processes, and efforts to improve medication adherence.

The study included 525 African American and White primary care patients with uncontrolled hypertension. Both Black and White patients experienced a reduction in blood pressure, but there was no change in the difference between Black and White blood pressure management (difference of 2.8 mmHg, $p = 0.26$). The gap between Blacks and Whites in blood pressure management was not reduced. Most of the impact of blood pressure reduction occurred in the first year of the study, and the reduction was sustained in the second year.

Including African American and White patients allowed for examination of the multilevel intervention effects by race, and the researchers tested whether the intervention led to reductions in the health disparities gaps between African American and White patients in blood pressure control. The finding that the multilevel intervention did not narrow the gap in health disparities, and was equally effective for both African American and White patients, is important for informing future strategies and studies to address health disparities in blood pressure control. It is important for future studies to include sufficient sample composition and size to test subgroup outcomes and to test whether interventions reduce gaps in health disparities for targeted health outcomes.

16.6 Key Points

- When considering a multilevel approach, the important focus should be on hypothesizing what the theory-based spheres of influence are that might affect an individual's or population's health, considering that the levels of influence might vary depending on the health condition of interest and/or the target group for the intervention. At each level of intervention, there is a different target or objective for change, which calls for a multipronged approach.

- There are substantial challenges that need to be solved when addressing health disparities through multilevel interventions. These include: the time-intensive nature and cost of evaluating multilevel interventions; the lack of generalizability and replicability; the incompatibility of using rigorous scientific designs in many contexts; the effort and time needed to develop collaborative partnerships with community stakeholders; budgetary caps by grant funding agencies (e.g., the NIH); and difficulty in coordinating multiple actors necessary to implement multilevel interventions. There are also obstacles in the adoption and actual dissemination and implementation of multilevel interventions. These include building collaborative partnerships with community members, deciding the intensity and dose of the intervention, maintaining consistency across activities and targeted levels, coordinating multiple actors involved in on-the-ground implementation, synchronizing activities across levels, and creating program sustainability post-study.
- The evidence shows that many multilevel interventions have positive outcomes in one or multiple ethnic/racial minority groups, but only a few studies examine or measure whether the multilevel intervention reduces disparity. Most studies have an implicit target to reduce health disparities but do not evaluate if they do so.
- To reduce health disparities, integrating multilevel interventions with CBPR principles that coalesce with community partners and build community capacity may be a successful way of ensuring adoption and sustainability of the intervention.
- When evaluating the effects of multilevel interventions, researchers must decide whether to compare the combined effect of the intervention components that make up the multilevel intervention to no intervention, to separately analyze the impact of each level of intervention, or to assess whether there are interactive effects. Disentangling the unique contributions of the levels of interventions or synergistic effects of these interventions is not only expensive but requires a complex intervention design to study potential interactive effects.

Disclaimer

The views and opinions expressed in this chapter are those of the authors only and do not necessarily represent the views, official policy, or position of the U.S. Department of Health and Human Services or any of its affiliated institutions or agencies.

References

1 Schölmerich, V.L. and Kawachi, I. (2016). Translating the socio-ecological perspective into multilevel interventions gaps between theory and practice. *Health Education & Behavior* 43 (1): 17–20.

2 Schölmerich, V.L. and Kawachi, I. (2016). Translating the social-ecological perspective into multilevel interventions for family planning how far are we? *Health Education & Behavior* 43 (3): 246–255.

3 Gee, G.C. and Payne-Sturges, D.C. (2004). Environmental health disparities: a framework integrating psychosocial and environmental concepts. *Environmental Health Perspectives* 112: 1645–1653.

4 Golden, S.D. and Earp, J.A.L. (2012). Social ecological approaches to individuals and their contexts twenty years of health education and behavior health promotion interventions. *Health Education & Behavior* 39 (3): 364–372.

5 Kaufman, M.R., Cornish, F., Zimmerman, R.S., and Johnson, B.T. (2014). Health behavior change models for HIV prevention and AIDS care: practical recommendations for a multilevel approach. *Journal of Acquired Immune Deficiency Syndromes* 66: S250–S258.

6 Lynch, S.M. and Rebbeck, T.R. (2013). Bridging the gap between biologic, individual, and macroenvironmental factors in cancer: a multilevel approach. *Cancer Epidemiology, Biomarkers & Prevention* 22 (4): 485–495.

7 Taplin, S.H., Price, R.A., Edwards, H.M. et al. (2012). Introduction: understanding and influencing multilevel factors across the cancer care continuum. *Journal of the National Cancer Institute. Monographs* 44: 2–10.

8 Stevens, J., Pratt, C., Boyington, J. et al. (2017). Multilevel interventions targeting obesity: research recommendations for vulnerable populations. *American Journal of Preventive Medicine* 52 (1): 115–124.

9 Charns, M.P., Foster, M.K., Alligood, E.C. et al. (2012). Multilevel interventions: measurement and measures. *Journal of the National Cancer Institute Monographs* 44: 67–77.

10 Ephraim, P.L., Hill-Briggs, F., Roter, D.L. et al. (2014). Improving urban African Americans' blood pressure control through multilevel interventions in the achieving blood pressure control together (ACT) study: a randomized clinical trial. *Contemporary Clinical Trials* 38 (2): 370–382.

11 Ling, B.S., Schoen, R.E., Trauth, J.M. et al. (2009). Physicians encouraging colorectal screening: a randomized controlled trial of enhanced office and patient management on compliance with colorectal cancer screening. *Archives of Internal Medicine* 169 (1): 47–55.

12 Cook, J.E., Purdie-Vaughns, V., Meyer, I.H. et al. (2014). Intervening within and across levels: a multilevel approach to stigma and public health. *Social Science & Medicine* 103: 101–109.

13 Paskett, E., Thompson, B., Ammerman, A.S. et al. (2016). Multilevel interventions to address health disparities show promise in improving population health. *Health Affairs* 35 (8): 1429–1434.

14 Trickett, E.J. and Beehler, S. (2013). The ecology of multilevel interventions to reduce social inequalities in health. *American Behavioral Scientist* 57 (8): 1227–1246.

15 Gorin, S.S., Badr, H., Krebs, P., and Das, I.P. (2012). Multilevel interventions and racial/ethnic health disparities. *Monographs-Journal of the National Cancer Institute* 44: 100–111.

16 Robinson, T.N., Matheson, D., Desai, M. et al. (2013). Family, community and clinic collaboration to treat overweight and obese children: stanford goals—a randomized controlled trial of a three-year, multi-component, multilevel, multi-setting intervention. *Contemporary Clinical Trials* 36 (2): 421–435.

17 Liu, N.H., Daumit, G.L., Dua, T. et al. (2017). Excess mortality in persons with severe mental disorders: a multilevel intervention framework and priorities for clinical practice, policy and research agendas. *World Psychiatry* 16 (1): 30–40.

18 Kilgus, S.P., Reinke, W.M., and Jimerson, S.R. (2015). Understanding mental health intervention and assessment within a multi-tiered framework: contemporary science, practice, and policy. *School Psychology Quarterly* 30 (2): 159–165.

19 Go, V.F., Frangakis, C., Le Minh, N. et al. (2017). Increased survival among HIV-infected PWID receiving a multilevel HIV risk and stigma reduction intervention: results from a randomized controlled trial. *Journal of Acquired Immune Deficiency Syndromes* 74 (2): 166–174.

20 Ramey, S.L., Schafer, P., DeClerque, J.L. et al. (2015). The preconception stress and resiliency pathways model: a multilevel framework on maternal, paternal, and child health disparities derived by community-based participatory research. *Maternal and Child Health Journal* 19 (4): 707–719.

21 Crespo, N.C., Elder, J.P., Ayala, G.X. et al. (2012). Results of a multilevel intervention to prevent and control childhood obesity among Latino children: the Aventuras Para Niños study. *Annals of Behavioral Medicine* 43 (1): 84–100.

22 Gittelsohn, J., Steeves, E.A., Mui, Y. et al. (2014). B'More healthy communities for kids: design of a multilevel intervention for obesity prevention for low-income African American children. *BMC Public Health* 14 (1): 942.

23 Purnell, T.S., Calhoun, E.A., Golden, S.H. et al. (2016). Achieving health equity: closing the gaps in health care disparities, interventions, and research. *Health Affairs* 35 (8): 1410–1415.

24 Ndumbe-Eyoh, S. and Moffatt, H. (2013). Intersectoral action for health equity: a rapid systematic review. *BMC Public Health* 13 (1): 1056.

25 Klein, K.J. and Kozlowski, S.W. (eds.) (2000). Multilevel Theory, Research, and Methods in Organizations: Foundations, Extensions, and New Directions. Hoboken, NJ: Wiley.

26 Cooper, L.A., Marsteller, J.A., Noronha, G.J. et al. (2013). A multilevel system quality improvement intervention to reduce racial disparities in hypertension care and control: study protocol. *Implementation Science* 8 (1): 60.

27 Finch, M., Wolfenden, L., Morgan, P.J. et al. (2014). A cluster randomized trial of a multilevel intervention, delivered by service staff, to increase physical activity of children attending center-based childcare. *Preventive Medicine* 58: 9–16.

28 Krok-Schoen, J.L., Young, G.S., Pennell, M.L. et al. (2015). Testing interventions to motivate and educate (time): a multilevel intervention to improve colorectal cancer screening. *Preventive Medicine Reports* 2: 306–313.

29 Lindamer, L.A., Lebowitz, B., Hough, R.L. et al. (2009). Establishing an implementation network: lessons learned from community-based participatory research. *Implementation Science* 4 (1): 17.

30 Andrews, J.O., Mueller, M., Dooley, M. et al. (2016). Effect of a smoking cessation intervention for women in subsidized neighborhoods: a randomized controlled trial. *Preventive Medicine* 90: 170–176.

31 Mikkelsen, B.E., Novotny, R., and Gittelsohn, J. (2016). Multilevel, multi-component approaches to community based interventions for healthy living—a three case comparison. *International Journal of Environmental Research and Public Health* 13 (10): 1023.

32 Cleary, P.D., Gross, C.P., Zaslavsky, A.M., and Taplin, S.H. (2012). Multilevel interventions: study design and analysis issues. *Monographs-Journal of the National Cancer Institute* 44: 49.

33 Craig, P., Cooper, C., Gunnell, D. et al. (2012). Using natural experiments to evaluate population health interventions: new medical research council guidance. *Journal of Epidemiology and Community Health*: 2011–200375.

34 Halladay, J.R., Donahue, K.E., Hinderliter, A.L. et al. (2013). The heart-healthy Lenoir project – an intervention to reduce disparities in hypertension control: study protocol. *BMC Health Services Research* 13 (1): 441.

35 Harris, F.M., Maxwell, M., O'Connor, R. et al. (2016). Exploring synergistic interactions and catalysts in complex interventions: longitudinal, mixed methods case studies of an optimised multilevel suicide prevention intervention in four European countries (Ospi-Europe). *BMC Public Health* 16 (1): 268.

36 Lippman, S.A., Chinaglia, M., Donini, A.A. et al. (2012). Findings from Encontros: a multilevel STI/HIV intervention to increase condom use, reduce STI, and change the social environment among sex workers in Brazil. *Sexually Transmitted Diseases* 39 (3): 209–216.

37 Gorin, A.A., Raynor, H.A., Fava, J. et al. (2013). Randomized controlled trial of a comprehensive home environment-focused weight-loss program for adults. *Health Psychology* 32 (2): 128–137.

38 Lv, J., Liu, Q.M., Ren, Y.J. et al. (2013). A community-based multilevel intervention for smoking, physical activity and diet: short-term findings from the community interventions for health programme in Hangzhou, China. *Journal of Epidemiology and Community Health* 68 (4): 333–339.

39 Keyserling, T.C., Samuel-Hodge, C.D., Pitts, S.J. et al. (2016). A community-based lifestyle and weight loss intervention promoting a Mediterranean-style diet pattern evaluated in the stroke belt of North Carolina: the heart-healthy Lenoir project. *BMC Public Health* 16 (1): 732.

40 Michels, N., De Henauw, S., Eiben, G. et al. (2015). Effect of the IDEFICS multilevel obesity prevention on children's sleep duration. *Obesity Reviews* 16 (S2): 68–77.

41 Coronado, G.D., Beresford, S.A., McLerran, D. et al. (2016). Multilevel intervention raises latina participation in mammography screening: findings from ¡Fortaleza latina! *Cancer Epidemiology, Biomarkers & Prevention* 25 (4): 584–592.

42 Hankonen, N., Heino, M.T., Araujo-Soares, V. et al. (2016). 'Let's move it'–a school-based multilevel intervention to increase physical activity and reduce sedentary behavior among older adolescents in vocational secondary schools: a study protocol for a cluster-randomised trial. *BMC Public Health* 16 (1): 451.

43 Kerrigan, D., Barrington, C., Donastorg, Y. et al. (2016). Abriendo puertas: feasibility and effectiveness a multilevel intervention to improve HIV outcomes among female sex workers living with HIV in the Dominican Republic. *AIDS and Behavior* 20 (9): 1919–1927.

44 Paskett, E.D., Krok-Schoen, J.L., Pennell, M.L. et al. (2016). Results of a multilevel intervention trial to increase human papillomavirus (HPV) vaccine uptake among adolescent girls. *Cancer Epidemiology, Biomarkers & Prevention* 4: 593–602.

45 Seguin, R.A., Eldridge, G., Graham, M.L. et al. (2016). Strong hearts, healthy communities: a rural community-based cardiovascular disease prevention program. *BMC Public Health* 16 (1): 86.

46 Gittelsohn, J., Jock, B., Redmond, L. et al. (2017). Oprevent2: design of a multi-institutional intervention for obesity control and prevention for American Indian adults. *BMC Public Health* 17 (1): 105.

47 Komro, K.A., Livingston, M.D., Wagenaar, A.C. et al. (2017). Multilevel prevention trial of alcohol use among American Indian and White high school students in the Cherokee Nation. *American Journal of Public Health* 107 (3): 453–459.

48 Ma, G.X., Fang, C.Y., Seals, B. et al. (2017). A community-based randomized trial of hepatitis B screening among high-risk Vietnamese Americans. *American Journal of Public Health* 107 (3): 433–440.

17

Using Implementation Science to Move from Knowledge of Disparities to Achievement of Equity

Lisa A. Cooper[1], Tanjala S. Purnell[1], Michael Engelgau[2], Kristina Weeks[1], and Jill A. Marsteller[1]

[1] *Johns Hopkins Center for Health Equity, Johns Hopkins University, Baltimore, MD, USA*
[2] *National Heart, Lung, and Blood Institute, National Institutes of Health, Bethesda, MD, USA*

17.1 Introduction

17.1.1 Overview of Implementation Science

17.1.1.1 Definitions

Implementation can be defined as a set of activities designed to put an intervention into practice [1]. When we discuss implementation, we are describing intentional, planned, or coordinated activities that occur at levels higher than the individual. This could be at the team, group, system, organization, program, practice, unit, or other level [1], its purpose being to effect a change in process or practice, such as in professional behavior. In public health and healthcare, interventions are programs, processes, or practices that have been demonstrated by research to be effective in accomplishing a health or healthcare goal. We will refer to intervention and implementation together as the "change effort." Implementation science or implementation research is the study of the process by which research is translated from knowledge into practice [1]. It focuses on questions of whether, how, to what extent, and why or why not interventions are applied or used. Its investigations are not limited to public health and healthcare, although that is our focus here.

17.1.1.2 Principles

Among the fundamental principles underpinning implementation science is the point that implementation and its effects are distinct from an intervention and its effects. As Fixsen and colleagues caution, "When thinking about implementation the observer must be aware of two sets of activities (intervention-level activity and implementation-level activity) and two sets of outcomes (intervention outcomes and implementation outcomes)" (p. 5) [1]. Thus, failure to change patient-reported or clinical outcomes in any given change effort may be due to failure of the intervention or to failure of the implementation. As an

The Science of Health Disparities Research, First Edition. Edited by Irene Dankwa-Mullan, Eliseo J. Pérez-Stable, Kevin L. Gardner, Xinzhi Zhang, and Adelaida M. Rosario.

extension of that principle, an intervention must be implemented as designed to realize its expected effect. Adaptations may weaken effects, although, in some cases, if the adaptations are unexpectedly superior to the original intervention, they also could strengthen effects. Another principle of implementation science is that a range of factors influence implementation success, such as readiness to change, culture, and characteristics of both the intervention and the person being asked to change [2, 3]. These factors not only can, but must, be measured in order to understand success or failure of change efforts. Finally, implementation is complex and involves multilevel processes [4]. To implement effectively, the change agent must pay attention to processes, facilitators, and barriers within the regulatory environment, the surrounding community, the organization, the department, the unit or practice, the clinician or staff member, the family, and the patient.

17.1.1.3 Approaches and Methods

Typically, the evaluative approaches and methods used in implementation science are broadly similar to the methods used in health services research and improvement science. Implementation science often relies on mixed-method designs that include quantitative data collection and analysis, as well as qualitative approaches such as interviews, focus groups, and observation. Change efforts are guided by: (i) a conceptual framework or logic model for the production of patient outcomes and (ii) a framework guiding how to implement effectively (e.g., the Institute for Healthcare Improvement [IHI] Model for Improvement [5] or the 4Es – Engage, Educate, Execute, and Evaluate [6]). Specific implementation strategies are used to communicate the requirements of the intervention and encourage active participation. These include use of purposeful staff selection, audit and feedback, formal training, communication plans, engagement methods, performance evaluation, and staff coaching/practice facilitation [1]. Provision of information alone has been identified as a generally ineffective implementation strategy [1]. Outcomes studied in implementation science that are not necessarily considered in other research fields include: (i) fidelity (use of the intervention as designed); (ii) reach/penetration (proportion of eligible patients receiving the intervention); (iii) adoption/uptake (proportion of clinicians and staff using/complying with the intervention); (iv) acceptability (whether clinicians and staff perceive the intervention positively); and (v) feasibility (fit of the intervention with the setting and circumstances) [7]. In addition, implementation science concerns itself with the sustainment or maintenance of the intervention over time [8].

17.1.2 Why Health Disparities Research Needs Implementation Science

In the specific case of health disparities, we know that inequitable care is widespread and underaddressed in today's healthcare system. Several interventions and several state and national policy initiatives target improved population health and reduced health disparities. There is a growing evidence base regarding basic biomedical therapies and effective clinical and community-based approaches to reduce disparities [9, 10]. However, evidence that interventions proven in these studies will improve population health and reduce disparities is limited because few interventions have been successfully translated into real-world practices and communities. The gap is an implementation of evidence gap, and to bridge that gap, we need to couple transdisciplinary research and stakeholder engagement with dissemination

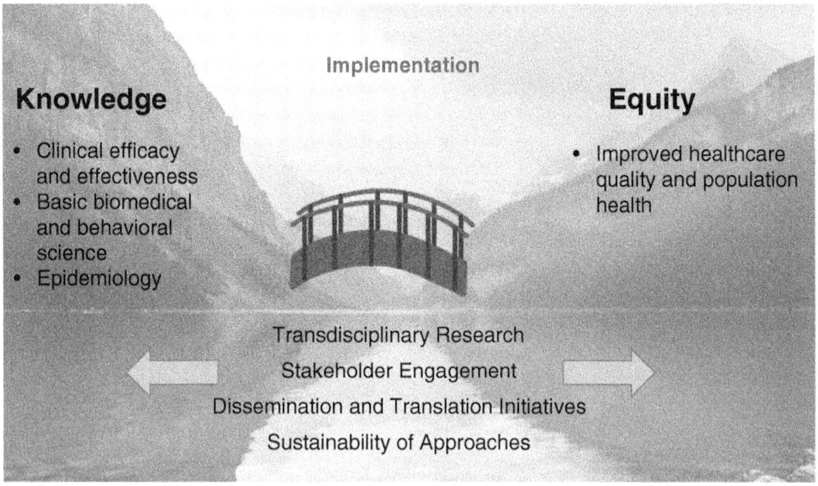

Figure 17.1 Population health and health disparities research: where we are and where we need to be. *Source:* Modified from The Colebrooke Centre for Evidence and Implementation: https://www.slideshare.net/TheSRAOrg/jane-lewis. (See color representation of this figure in color inserts section.)

and translation initiatives in order to enhance the sustainability of evidence-based approaches (see Figure 17.1). "Systematic implementation practices are essential to any national attempt to use the products of science—such as evidence-based programs—to improve the lives of its citizens" (p. vi) [1]. As with other change efforts in health and healthcare, a lack of attention to appropriate implementation (and to the diversity within and across targeted populations and settings) can doom an otherwise promising intervention. We cannot afford to continue to fail populations that already face overwhelming barriers to accessing and benefitting from the best evidence-based care. In the words of Fixsen and colleagues:

> "As a field, we have discovered that all the paper in file cabinets plus all the manuals on the shelves do not equal real world transformation of human service systems through innovative practice. While paperwork and manuals do represent what is known about effective interventions, these tools are not being used effectively to achieve behavioral health outcomes for children, families, and adults nationally" (p. vi) [1].

17.2 Selected Implementation Frameworks Applied to Health Disparities Research

The development of many different theories, models, and frameworks by implementation researchers has made it challenging for investigators to determine which is the most appropriate in a given circumstance. A recent report has attempted to provide a better understanding of the entire suite of theories, models, and frameworks, describing three overarching approaches [11]: (i) models that guide the process of taking effective research findings and translating them into routine practices for prevention or treatment; (ii) deterministic frameworks along with classic behavioral theories and implementation theories, all aiming to elucidate key drivers of desired implementation outcomes [12]; and (iii) frameworks that focus on evaluating the implementation process.

In this section, we highlight selected approaches that are useful for health disparities implementation research: Community-based Participatory Research (CBPR) [13, 14]; the RE-AIM framework (Reach, Effectiveness, Adoption, Implementation, and Maintenance) [15]; the Practical, Robust Implementation and Sustainability Model (PRISM) [8]; the Consolidated Framework for Implementation (CFIR) [2]; and, finally, the Public Health Critical Race (PHCR) praxis, which uses an iterative methodology [16].

17.2.1 Community-based Participatory Research

According to the W.K. Kellogg Community Health Scholars Program, CBPR is "a collaborative approach to research that equitably involves all partners in the research process and recognizes the unique strengths that each brings. . . ." [17]. Traditional research prioritizes health issues based on epidemiologic data; selects study designs based primarily on scientific rigor and feasibility; and uses approaches to recruitment, measurement, intervention design, and data analysis that are based mainly on literature, theory, and scientific considerations. Historically, researchers rather than community members have driven research agendas as well as specific research processes. In contrast, in CBPR, community members participate actively in identifying health issues of greatest importance to them. They are involved in study design; funding of proposals; selection of recruitment and retention strategies and study measures; development of intervention approaches; and interpretation, dissemination, and translation of findings [18]. This approach fosters a long-term, equitable partnership between researchers and the community and shared ownership of the processes and outcomes of implementation.

The CBPR approach may draw on a full range of research designs and can include case studies, etiological and longitudinal studies, nonexperimental designs, and quasi-experimental designs to randomized controlled trials. In essence, CBPR is not a research design or method but, rather, an approach to research that can enhance research finding uptake and sustainability—both key implementation research outcomes. CBPR is a unique research approach in that it creates a functional link between the researchers and the community so that shared decision making occurs within all aspects of the research: planning, prioritization, methods, report development, and the communication of findings.

Health disparities implementation research depends on a high level of understanding from the community and of the environmental context within which implementation will be studied. Gaining context insights and understanding priority community issues and implementation barriers are all critical elements for successful implementation research for health disparities.

17.2.2 The RE-AIM Framework (Reach, Effectiveness, Adoption, Implementation, and Maintenance)

The RE-AIM (Reach, Effectiveness, Adoption, Implementation, and Maintenance) framework studies the health impact of interventions based on five factors: reach, effectiveness, adoption, implementation, and maintenance. Compatible with both socioecological and health systems concepts, RE-AIM can be used for a broad range of interventions, including public health- and clinic-based settings. When each dimension is given a scale rating (between 0 and 100%), the multipliable product of these ratings provides an estimate of impact within the community.

Reach is defined by the proportion of the population that actually participates in the intervention—reflecting genuine uptake—at the individual level (e.g., the proportion that participates).

Effectiveness is represented as the net success rate of the intervention; that is, the number of positive outcomes and less negative outcomes, measured at the individual level.

Adoption represents the proportion of potential sites that can implement the intervention (e.g., clinics, communities) that actually adopts the process. It is assessed at the organizational level.

Implementation describes the extent to which the intervention is delivered as intended and represents fidelity of intervention delivery. It is also assessed at the organizational level.

Finally, *Maintenance* assesses the extent to which delivery continues over time and captures sustainability elements. It is measured at either the individual or organizational level.

17.2.3 Practical, Robust Implementation and Sustainability Model (PRISM)

The PRISM model integrates concepts relevant to taking proven effective research interventions and understanding how to translate them into routine use across various settings. It includes elements and concepts from quality improvement[1] (models for improvement) [19]; chronic care delivery;[2] diffusion of innovations [20]; and measures of population-based impact (RE-AIM) [15]. Key elements within PRISM include the program (intervention delivery) from both the organizational (e.g., coordination across departments and specialties, complexity of the tasks, adaptability, and ability to see results) and patient perspectives (e.g., provides patient choice, addresses barriers, complexity of task, and feedback on results); the external environment (e.g., regulatory environment, community resources, and competition); and the implementation and sustainability infrastructure (e.g., dedicated team, adaptable protocols and procedures, and sustainability plan). Organizations are considered at three levels: leaders, managers, and staff.

17.2.4 Consolidated Framework for Implementation (CFIR)

CFIR [2] provides a broad typology that promotes implementation theory development and verification about what works where, and why, across multiple contexts. CFIR includes five major domains: intervention characteristics, outer setting, inner setting, characteristics of the individuals involved, and the process of implementation. These domains interact in complex ways and influence implementation effectiveness. Each domain has several constructs.

For *implementation characteristics*, the key constructs include intervention source, evidence strength and quality, adaptability, trialability, complexity, and costs.

1 http://www.ihi.org/IHI/Topics/Improvement/ImprovementMethods/HowToImprove
2 http://www.improvingchroniccare.org/index.php?p=

Outer settings constructs include patient needs and resources, organizational networking with external organizations, competitors and competition, and external policies and incentives.

Inner settings include structural characteristics (e.g., social structure, age, maturity, and size of the organization), networks and communications, culture (norms, values, and basic assumptions), and implementation climate capturing the capacity of the organization and individuals to adopt and sustain change.

Characteristics of individuals include constructs for knowledge and belief in the intervention, self-efficacy (self-belief that they have the capability to achieve the implementation goal), readiness to change, and their self-identification with the organization delivering the intervention.

Finally, the elements of *process of implementation* include planning, engaging stakeholders, having the ability to execute the intervention, and evaluating what has been accomplished. Ultimately, CFIR can be used to organize and promote synthesis of research findings, studies, and settings through clear and consistent terminology that will result in stimulating theory development.

17.2.5 Public Health Critical Race Praxis (PHCR)

Critical Race Theory (CRT) [21] has dominated racial studies since the 1980s. However, one of its limitations is that it is rooted primarily in the law enforcement and judicial systems. Because of CRT's methodological complexity and the nonhealth-related focus, it has not been used widely for health applications. The PHCR praxis [16] praxis has been developed to improve the ease and fidelity of CRT's use by researchers conducting health equity (and disparities) research. The PHCR praxis maintains the scientific integrity of CRT and taps into the extensive research efforts done outside the health domain. The PHCR praxis (i.e., an iterative methodology) uses science, theory, and experiential knowledge to address race-based inequities and informs research on the causes of health disparities. This approach guides researchers through four focus areas:

1) *Contemporary patterns of racial relations*, which capture the dynamic nature of racism and how it changes over time. It is essential to include current patterns to interpret and understand research results.
2) *Knowledge production*, which considers how racialization can reinforce existing racial beliefs or go the other direction and help shape the project structure. This challenges the assumption that research methods are uniformly objective and bias will not influence the research.
3) *Conceptualization and measurement*, which addresses the study's race constructs and their relationship to social context.
4) *Action*, which applies findings to tackle health inequities. This approach can move researchers from documenting disparities and inequities to a deeper understanding of root causes and, possibly, to their active involvement in determining solutions.

17.3 Best Practices in Health Disparities Implementation Science: Selected Examples

How can we best use implementation science in research and intervention for health disparities reduction? One example of work that marries concepts from both fields is Project ReD CHiP (Reducing Disparities and Controlling Hypertension in Primary Care) [22].

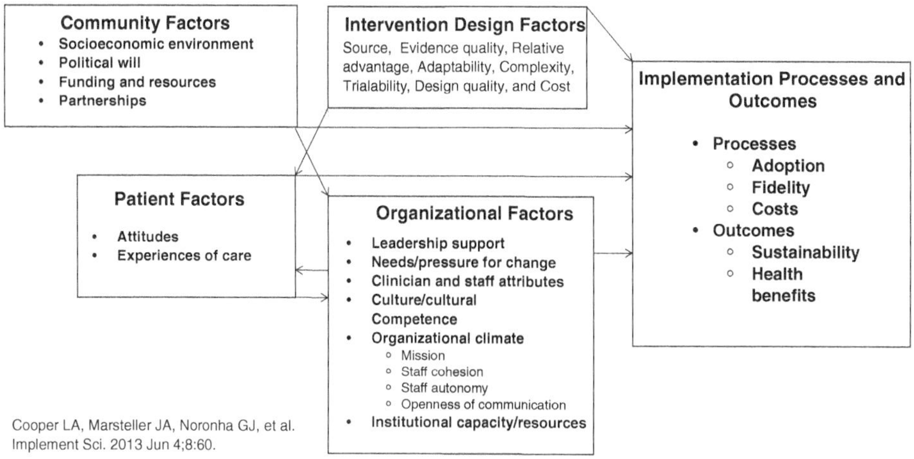

Figure 17.2 Project ReD CHiP's conceptual model. *Sources:* Cooper et al [22]; Shediac-Rizkallah and Bone [23]; Simpson [24]; Damschroder et al. [2]

Project ReD CHiP was a pragmatic trial funded by the National Heart, Lung, and Blood Institute from 2010 to 2015 that effectively used implementation science and CBPR principles to design and implement a multimethod health system quality improvement program targeting patients, providers, clinical staff, and organizational factors to improve guideline-concordant hypertension (HTN) care and reduce disparities [22]. Specifically, the project sought to improve blood pressure (BP) control among African Americans in six primary care practices in Baltimore, Maryland, using a registered dietician (RD)/pharmacist team to counsel patients with uncontrolled BP. Here, we use the project to offer supporting examples of best practices in implementation science as applied to health disparities reduction. The conceptual model used in Project ReD CHiP, as adapted from the work of Shediac-Ritzkallah and Bone [23], Simpson [24], and Damschroder et al. [2], demonstrates that implementation and outcomes are influenced by patient, provider, organizational, and community factors, as well as by intervention characteristics (Figure 17.2).

17.3.1 Asking Relevant Questions

The first step in conducting high-quality work in both implementation science and disparities research is to ask the right questions. Change agents must identify meaningful gaps in care, in both disparate care and evidence-based practice known to be beneficial but not fully implemented. Disparities reduction begins with the identification of an important deficit in a critical health outcome for a historically disadvantaged population. In Project ReD CHiP, the critical gap identified was between African Americans and Whites in BP control.

Research questions in implementation science that advance knowledge are becoming more difficult to identify but must reach beyond identification of barriers and facilitators for intervention adoption [25]. Thus, in implementation science, key questions focus on how researchers can effectively communicate about interventions with the implementers to encourage active participation. Implementation strategies in Project ReD CHiP included training in accurate BP measurement and in effective clinician-patient communication, as well as stakeholder engagement via a community advisory board (CAB) and continuous contact with the implementing practices [22].

An important area at the intersection of implementation science and disparities research may be the investigation of the efficacy of various implementation strategies where there is a need to "lift some boats" more than others. That is, in the presence of the need to target and tailor to specific populations, do different implementation strategies still serve, and which serve better?

17.3.2 Testing Multilevel Interventions

Both implementation science and disparities research emphasize the importance of addressing multiple levels of the healthcare system simultaneously [26]. The sources of health disparities occur at many levels, from individual biases to institutional policies that impact populations differentially. Similarly, implementation science also reasons that barriers and facilitators to uptake and fidelity, as well as other important implementation outcomes, emanate from multiple levels. Project ReD CHiP intervened with patients using expanded education and repeated consultation; with frontline staff and clinicians through training; at the system level by adding roles to the care team (RD and PharmD); and at the community level by engaging stakeholders from local community organizations [22].

17.3.3 Leveraging Technology, Existing Data Sources, and Natural Experiments

Implementation studies in health disparities should seek, where possible, to leverage technology to reduce care gaps and tap into existing data sources or natural experiments (such as local policy changes) to enhance feasibility. As an example, Project ReD CHiP took advantage of automated BP measurement devices to improve the accuracy of BP measurements, a critical step in persuading clinicians to act more assertively to control BP in their patients. In addition, the project relied on existing electronic medical record (EMR) data to evaluate changes in BP control rates rather than conducting *de novo* data collection that would have been burdensome to clinicians and staff and could have reduced program adoption.

17.3.4 Study Designs and Methods

Implementation science brings its mixed-methods approach to health disparities research and argues for interviews, focus groups, and, possibly, observation methods to complement quantitative measures and to better understand implementation progress and success. A good example of this is in the focus groups we conducted at the beginning of Project ReD CHiP to better understand how frontline clinicians and staff would react to the interventions planned. We gained important insight into clinicians' reasons for failure to escalate BP treatment. The focus groups also presaged workflow barriers to successful implementation of the BP measurement protocols, which largely focused on time constraints [22].

17.3.5 Measures of Implementation Success

Project ReD CHiP took advantage of several implementation outcome measures to help drive reduction in BP control disparities. In particular, we used data from the EMR to assess device uptake and fidelity to the BP measurement protocol by estimating terminal digit preference (TDP) [27]. The final digit in BP readings taken manually is often 0, suggesting that clinicians and staff round down or up in their measurement. An automated device does not round; thus, if the devices are in regular use, the occurrence of 0 as the terminal digit in EMR BP measurements should decline to close to 10% (the frequency with which any single digit from 0 to 9 should occur in a random sample of data). In our intervention sites, TDP decreased from approximately 33% to 11% [27].

17.3.6 Innovative Approaches to Data Analysis

Implementation science offers data analytic methods that may not have been used extensively to date in health disparities research. As one example, Project ReD CHiP used an approach that dates back to early quality improvement work: the statistical process control chart. This approach provides an over-time view of changing data points (in this case, BP control rates and systolic and diastolic measurements) with statistical cues to distinguish meaningful change from natural variation.

17.3.7 Dissemination of Findings

Health disparities research and translation may be aided by implementation science's emphasis on reach (proportion of eligible patients served) [8]. A critical but oftentimes overlooked aspect of research is that upon completion of a study, the evidence-based practice activity ceases. Many interventions are never disseminated or spread to all eligible recipients, leaving many untreated by the intervention. Implementation science insists that reach be considered and actively addressed. For example, the 4Es framework described earlier has been broadened in application to also include "Expand," spreading the intervention to additional populations and areas [3]. In the case of health disparities research and practice, we often must reach all vulnerable patients in order to effectively reduce gaps between minority and majority populations.

17.4 Challenges and Opportunities for Implementation Science in Health Disparities Research

17.4.1 Stakeholder Engagement

It is important to engage a broad group of stakeholders in formulating the research questions and informing the design of the studies to address health equity. Relationship-building is paramount in building partnerships between researchers and community stakeholders. Thus, implementation researchers who are testing interventions to achieve health equity in socially at-risk populations and underserved clinical and community-based settings often use CBPR principles [14]. The most successful partnerships highlight the importance of a "fit for purpose" approach and of combining collaboration with solid research methods [28]. Relationship-centered principles, such as clear communication; respect for diverse perspectives; knowing or familiarity with the individuals, organizations,

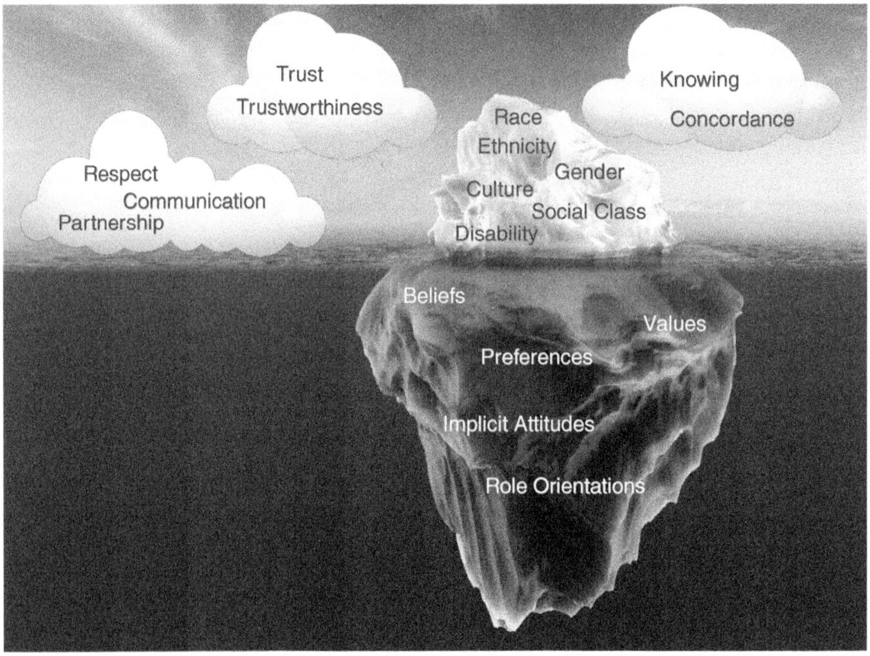

Figure 17.3 Iceberg concept of culture applied to relationship-centeredness in health disparities research. *Source:* Modified from Cooper et al. [29]. (See color representation of this figure in color inserts section.)

and communities involved in the research; authentic partnership; efforts to identify concordance in values and desired outcomes; and trust and trustworthiness, all enhance the likelihood of success in implementation of the research [29]. These dimensions are shown in Figure 17.3, and applications of some of these dimensions to address research challenges are described in more detail below.

17.4.1.1 Improving Researcher and Interventionist Communication Skills and Cultural and Structural Competence

Improving communication skills includes a focus on enhancing information exchange through more effective data-gathering, education and counseling, and rapport-building behaviors [30]. Effective data-gathering skills include reflective listening, open-ended questions to probe others' concerns, and posing of specific questions to better understand barriers to the achievement of goals. Effective education and counseling skills include the use of clear and concise language, avoidance of technical jargon, and the use of teach-back (asking others to explain in their own words what they understood from the interaction). Rapport-building skills include the use of empathic statements, expressions of concern or caring, self-disclosure, and offering of emotional support. Cultural and structural competence involves the use of behaviors that indicate respect for diverse perspectives, interest in knowing more about the experiences and situations of others (particularly social determinants of health (SDOH) and cultural and contextual factors), and the avoidance of bias and stereotyping in interpersonal interactions.

17.4.1.2 Partnering with and Aligning Priorities and Incentives of Diverse Stakeholders

Participatory approaches in decision making enhance the likelihood of success in implementation processes and outcomes. This means researchers must make efforts to engage stakeholders in problem solving and decision making by eliciting their opinions and preferences and negotiating mutually beneficial agreements. Aligning priorities and incentives of diverse stakeholders can be time-consuming and challenging, and implementation researchers often must be patient and flexible, communicating clearly the importance of certain decisions for the integrity of the research while at the same time respecting the concerns of stakeholders and working with them to arrive at mutually agreeable decisions.

17.4.1.3 Building Trustworthiness

Many individuals from socially at-risk groups (e.g., ethnic minorities, religious and cultural minorities, sexual minorities, persons with low income, inner-city residents, rural residents, persons with low health literacy, and persons with physical and mental disabilities) and persons working in community-based institutions serving these groups have a strong mistrust of healthcare professionals and researchers [31]. This mistrust stems from historical events, such as the Tuskegee syphilis study and the experience of Henrietta Lacks, the African American woman who was the unwitting source of the "HeLa" cancer cell line that has led to many research breakthroughs. It is reinforced by healthcare system issues and discriminatory events that continue to this day. Disparities in healthcare delivery are often manifested by a lack of cultural sensitivity, a lack of respect for persons and for their autonomy, bias and stereotyping behaviors of clinicians and researchers in their decision making, and poor communication with persons from at-risk groups and historically disadvantaged communities [29]. As a result, demonstrating trustworthiness to stakeholders may be even more important in implementation research to address health equity than it is in other types of implementation research.

Scholars define trustworthiness as the ability to be relied upon by others based on benevolence, integrity, and ability or competence by persons and/or institutions [32]. Trustworthiness is important because it leads to positive behavioral outcomes. One might suggest the following organizational and interpersonal strategies to enhance trustworthiness in implementation science: open information systems or communication, person- and community-centered framing of messages, participatory approaches, and authentic leadership and monitoring of the quality of activities.

17.4.1.4 Framing and Education about Disparities to Engage Diverse Audiences

Although SDOH have been studied and discussed extensively by academics, the Robert Wood Johnson (RWJ) Foundation Vulnerable Populations Portfolio[3] discovered that the concept did not work with policymakers, nor did it work with many of their grantees from community-based organizations. RWJ identified challenges in communicating the relationship between SDOH and social factors such as poverty, housing, education, and health-challenges made even greater by the fact that persons most affected by SDOH include groups who are marginalized by society. Thus, the Foundation worked with communications professionals to

3 http://www.rwjf.org/en/library/research/2015/05/vulnerable-populations-research-and-policy-briefs.html

develop a framework that describes SDOH plainly, without political overtones. The guide they developed is a resource for best practices in the language and framing of SDOH.[4] It provides different ways to talk about SDOH, a glossary of other terms to use in place of typical phrasing, and ways to communicate facts and data with this new frame around SDOH in an effective manner. The suggested key message is "Health starts where we live, learn, work, and play." Implementation researchers addressing health disparities may find this frame useful in their efforts to engage with diverse audiences, including community members, healthcare professionals, executives, and policymakers.

17.4.1.5 Case Study: The Johns Hopkins Center for Health Equity Community Advisory Board

The Johns Hopkins Center for Health Equity Community Advisory Board (JHCHE-CAB) is one example of an effective academic/community partnership to support implementation science targeting health equity [33]. The JHCHE-CAB is led jointly by the Center director and a well-known and highly regarded leader of a community-based organization. All partners share their expertise and assume responsibilities and coownership of outputs from the research. The CAB uses CBPR principles to guide the research through all phases, including planning, implementation, evaluation, translation, and dissemination. Stakeholders are involved in all phases of the Center's research, including development of grant proposals, piloting and refinement of research procedures and materials, recruitment of study participants, staff training, and manuscript authorship. Researchers keep clinical and community partners informed of all research activities by keeping the Center and study-specific websites up to date, disseminating a monthly electronic newsletter, reaching out to individual partners by telephone or email, and arranging project-related meetings or attending existing stakeholder meetings as needed to address specific questions and concerns related to the research. CAB meetings are held in the community on a quarterly basis each year. The partnership also holds strategic planning retreats where members devote time to reviewing and adapting the mission and vision of the partnership and identifying priorities for future research.

17.4.2 Contextual Considerations

17.4.2.1 Overcoming Limited Infrastructure and Resources

Implementation researchers conducting studies about health disparities or developing interventions to address them may encounter challenges in supporting stakeholder attempts to remain attentive to the project's demands while addressing the stakeholders' own competing priorities. Health system stakeholders who serve socially at-risk groups have administrative and reporting requirements that encroach on time dedicated to direct patient care. Given these pressures, leaders of these organizations feel responsible for protecting their providers' and frontline staff members' time. This lack of time for researchers to engage with frontline providers and staff may contribute to resistance to research initiatives.

Another key challenge may lie in limited financial resources provided by research grants to support salaries of staff participating in the research. Even when efforts are made to align the research with existing priorities and to use existing staff, many of these staff are already

4 http://www.rwjf.org/content/dam/farm/reports/reports/2010/rwjf63023

at capacity and cannot take on the additional responsibilities required by the research. Researchers must be attentive to such concerns and flexible in supporting different approaches to hiring and the allocation of research funds. Often, community-based organizations do not have previous experience in conducting research and completing appropriate regulatory paperwork and research training. Implementation researchers must help their partners to understand research procedures and facilitate completion of the necessary requirements through regular communication and support from research project managers.

17.4.2.2 Targeting and Tailoring Intervention Approaches to Context and Persons

It is widely accepted that health programs and materials that succeed in making information relevant to their intended audience are more effective than those that do not. Implementation research focused on socially at-risk populations can be informed by two behavioral intervention approaches: targeting and tailoring. While targeted interventions involve messages that are intended to reach population subgroups based on a specific set of shared characteristics, tailored interventions are intended to change an individual's or individuals' behaviors in a particular setting based on specific characteristics of the individual or setting as measured in a formal assessment process [34]. Experts suggest that targeting or adapting interventions to address cultural factors includes incorporation of surface and deep structures [35]. Surface structures include language and literacy level; popular food, music, and clothing choices; as well as educational materials and messages and stakeholder involvement in designing materials. Focusing on surface structures increases the attractiveness of the intervention materials and processes and deepens connections between researchers and study participants.

Deep structures are elements that tap into the cultural, historical, and political factors that have shaped the beliefs of the targeted population. Enhancing deep structures in research conveys relevance of the approach within a specific context and helps motivate behavior change in the targeted group. Examples of deep structures include recognition of spiritual values and the roles of family members in certain communities and acknowledgment of institutional values and mission in specific organizational settings. Cultural targeting strategies include: (i) gathering input from the targeted population to enhance acceptability and uptake (e.g., by using exploratory focus groups and in-depth interviews) and pretesting specific program materials; (ii) incorporating surface and deep structures; (iii) optimizing positive images (e.g., by featuring persons from the targeted group who have experienced the condition being studied, engaged in the proposed behaviors, and achieved good results); (iv) dispelling misconceptions; (v) discussing common myths and counteracting them with information; (vi) minimizing negative stereotypes; and (vii) reducing stigma. One approach to tailoring interventions to individuals in clinical settings was developed by cultural anthropologist Arthur Kleinman, who suggested using key questions to better explicate the individual patient's explanatory model of illness [36]. Similar strategies might be used to explore attitudes and readiness to make behavior changes among other types of stakeholders [37].

17.4.2.3 Addressing Social Determinants of Health

Socially at-risk populations are disproportionately impacted by SDOH. A review of the literature reveals several social risk domains that influence health disparities, including housing instability, food insecurity, transportation needs, utility needs, interpersonal safety, financial strain, education, employment, social isolation, childcare, and discrimination. Implementation research to address health disparities should include assessments of SDOH among research participants, as well as the ability of health and service providers to address these social determinants [38, 39]. Thus, implementation research to achieve health equity for these populations must address or intervene upon factors at multiple levels of the socioecological model [40].

17.4.2.4 Increasing Reach

Earlier in this chapter, we described the RE-AIM model, consisting of five elements, or dimensions, related to health behavior interventions. The first element, reach, is defined as the absolute number, proportion, and representativeness of individuals or organizations who are willing to participate in a given initiative, intervention, or program. Many implementation studies to address health disparities fail to achieve adequate reach. Barriers to reach include outreach approaches that are limited to healthcare settings, lack of culturally and linguistically appropriate communication strategies, insufficient attention to social determinants that may serve as barriers to engagement by the target population, and a lack of representation or meaningful engagement with community-based partners who might enhance the trustworthiness and relevance of programs being implemented. Many strategies enhance the reach of programs. Determining the best approach depends on the characteristics of the target population and organizations. Recommended strategies include: (i) building relationships with the targeted community and population by offering education, support, and service before asking them to engage in a research program; (ii) asking members of the target population to guide the research; (iii) involving members of the targeted population in actual recruitment efforts; (iv) identifying barriers to participation and eliciting feedback from stakeholders on ways to reduce these barriers; (v) determining the recruitment materials, methods, and locations that are best for your program; (vi) going to where the target population is (typically in community-based settings) instead of expecting them to come to you; and (vii) modifying the recruitment approach to fit the location, target group, and social and cultural context.[5]

17.4.3 Intervention Design and Delivery

17.4.3.1 Involving Implementers

Although we have argued that implementation must be considered distinct from intervention, benefits do accrue to designing interventions with implementation in mind. First, in selecting an intervention to reduce disparities, change agents or researchers must involve the implementers, or those who are being asked to change the way they deliver care. The needs, preferences, and facilitators and barriers faced by the frontline implementers should be understood and used to shape the intervention in order to improve its uptake/adoption and implementation success. Intervention characteristics that will reduce adoption include complexity, low relative advantage over other interventions, a lack of opportunity to try the intervention before committing to it, and a poor fit with the expectations and values of the implementers [41]. In Project ReD CHiP, we involved stakeholders early in providing feedback on the proposed interventions and in negotiating how implementation would be best carried out.

5 http://re-aim.org/about/what-is-re-aim/reach/improving-reach

17.4.3.2 Training, Technical Support, and Capacity-building

Implementation science makes it clear that for an intervention to achieve its expected effects, it must be implemented appropriately. Training, technical support or coaching, and building capacity among the implementers are all foundational aspects of appropriate implementation and should be built into every intervention at the design phase. To return to our example, each of these influential implementation strategies was used to advantage in Project ReD CHiP.

17.4.3.3 Evaluations, Adjustments, and Reiterations

Formative evaluation should be conducted during the implementation of the change effort (e.g., as described in the 4Es framework – Engage, Educate, Execute, and Evaluate). Otherwise, the opportunity to make adjustments and reiterations of previous implementation steps, such as retraining or returning to engagement efforts to address program weaknesses, could be missed, and essential course corrections might go unmade.

17.4.3.4 Measurement and Monitoring of Uptake and Fidelity

One example of such ongoing evaluation is the measurement and monitoring of uptake of and fidelity to the designed intervention. In Project ReD CHiP, careful monitoring of the automated BP measurement devices via site visits revealed that some devices were left in single-measurement mode instead of multiple-measurement mode as per protocol. This discovery permitted the research team to retrain and re-engage staff on the need to have the average of three readings to provide accurate measurements that clinicians could trust in their treatment decisions [29].

17.4.4 Planning for Sustainability

A persistent challenge for implementation science researchers has been the consideration of whether the benefits of successful programs are maintained after the research ends. One opportunity to overcome this challenge is through effective planning for the translation and sustainability of health disparities programs beyond the initial funding support. Sustainability, "the continued use of program components and activities for the continued achievement of desirable program and population outcomes," [42] promotes: long-term community health improvements; the development of new and more community-appropriate research questions; increased community support for future research; and increased cost-effectiveness.

17.4.4.1 Sustaining Program Benefits for Target Patient Populations

Sustainability planning is especially important in research targeting disparate care and outcomes for socially at-risk populations, as the initial funding for many programs may end prior to the achievement and demonstration of all potential program benefits. In addition, future health equity programs may receive less support from organizations and funders if the benefits of prior successful programs have not been sustained [25]. As such, it is important for implementation science and health equity researchers to determine core program elements to maintain for optimal health impact, as well as to determine factors that facilitate translation and long-term maintenance of programs and potential adaptations to interventions that encourage sustainability [42]. Planning for program sustainability can be guided by public health evaluation models and frameworks, such as PRISM and

RE-AIM. For example, the maintenance component of the RE-AIM framework focuses on evaluating the extent to which individual and organizational factors influence program continuation and dissemination over time [15]. PRISM evaluates core program elements (e.g., infrastructure) that facilitate long-term sustainability [8].

17.4.4.2 Sustaining Collaborative Partnerships with Key Stakeholders

Another important challenge has been that many health equity programs have not actively built sustainable collaborations among clinic-based health systems, patients, their social networks, and broader stakeholder groups (such as healthcare leaders and administrators, insurers, healthcare product manufacturers, community-based organizations, and faith communities) [40]. One opportunity for researchers to address this challenge is by early consideration and adoption of strategies to maintain ongoing communication and collaborative partnerships with representatives from key stakeholder groups that are needed to sustain health equity program benefits over time and that may facilitate the successful translation and adoption of findings into real-world clinical and public health practice.

17.4.4.3 Case Study: Project ReD CHiP

Project ReD CHiP, described earlier, was a pragmatic trial that implemented a multimethod health system quality improvement program to improve guideline-concordant HTN care and reduce disparities in BP control [22]. To enhance program adoption and sustainability, study investigators engaged CAB members throughout all phases of the research, and also conducted focus groups and directed interviews with key stakeholders (e.g., patients, insurers, organizational leaders, providers, and frontline staff) to elicit their perspectives about planned program components, which were refined based upon this feedback. Study investigators also held regular debriefings with medical directors at each of the clinical sites to identify and address real-time barriers to program implementation and to engage in collaborative problem-solving solutions. The project delivered a hypertension care management program to patients by adding pharmacists and dietitians to primary care teams, and both African American and White patients who received care management showed dramatic reductions in blood pressure [43]. Project ReD CHiP also implemented a standardized protocol to increase accuracy of blood pressure measurements taken by frontline clinical staff and introduced an audit and feedback process in which race-specific hypertension control data from the electronic health records were used to improve provider and clinic performance related to addressing disparities in hypertension control. These program components are now being sustained by the ReD CHiP practices and disseminated to other practices within the healthcare organization.

17.5 Summary and Implications for Future Research, Practice, Policy, and Social Change

Implementation science is the study of methods to promote the integration of research findings and evidence into healthcare policy and practice. It seeks to understand the behavior of healthcare professionals and other stakeholders as a key variable in the sustainable uptake, adoption, and implementation of evidence-based interventions, and aims to investigate and address major bottlenecks that impede effective implementation, test new approaches to improve health programming, and determine a causal relationship between the intervention and its impact.[6] Many implementation science models share principles and concepts with other behavioral science models and frameworks used in health disparities research. Despite extensive knowledge of health disparities and a growing evidence base for effective interventions to address these disparities, evidence that interventions proven in these studies will improve population health and reduce disparities is limited because few interventions have been successfully translated into real-world practices and communities. The gap is an implementation of evidence gap.

This chapter discussed best practices in implementation science applied to health disparities research, challenges, and opportunities to address these challenges through innovative approaches and case studies. We show that the use of implementation science principles, including transdisciplinary collaborations and broad stakeholder engagement, along with CBPR and relationship-centered principles, provides promise for facilitating dissemination and translation initiatives, enhances the sustainability of evidence-based approaches, and helps to make the vision of health equity a reality.

17.6 Key Points

- Despite extensive knowledge of health disparities and a growing evidence base for effective interventions to address these disparities, evidence that interventions proven in these studies will improve population health and reduce disparities is limited, because few interventions are translated into real-world practices and communities.
- Implementation science, the study of methods to integrate evidence into healthcare policy and practice, can be effectively applied to health disparities research, offering opportunities to address challenges in applying interventions to reduce disparities.
- Use of implementation science principles, including transdisciplinary collaborations and broad stakeholder engagement, along with community-based participatory research and relationship-centered principles, provides promise for facilitating dissemination and translation initiatives, enhances the sustainability of evidence-based approaches, and helps to make the vision of health equity a reality.

Disclaimer

The views and opinions expressed in this chapter are those of the authors only and do not necessarily represent the views, official policy, or position of the U.S. Department of Health and Human Services or any of its affiliated institutions or agencies.

6 http://www.nlm.nih.gov/hsrinfo/implementation_science.html

References

1 Fixsen, D.L., Naoom, S.F., Blase, K.A. et al. (2005). *Implementation Research: A Synthesis of the Literature*. Tampa, FL: University of South Florida FMHI Publication #231.

2 Damschroder, L.J., Aron, D.C., Keith, R.E. et al. (2009). Fostering implementation of health services research findings into practice: a consolidated framework for advancing implementation science. *Implementation Science* 4: 50.

3 Marsteller, J.A., Gurses, A.P., Ozok, A.A. et al. (2014). Clinician behavior change. In: *Health Behavior Change in Populations* (eds. S. Kahan, A. Gielen, P. Fagan and L. Green). Baltimore, MD: Johns Hopkins University Press.

4 Proctor, E.K., Landsverk, J., Aarons, G. et al. (2009). Implementation research in mental health services. *Administration and Policy in Mental Health* 36: 24–34.

5 Langley, G.J., Moen, R.D., Nolan, K.M. et al. (1996). *The Improvement Guide: A Practical Approach to Enhancing Organizational Performance*. San Francisco, CA: Jossey Bass.

6 Pronovost, P.J., Berenholtz, S.M., and Needham, D.M. (2008). Translating evidence into practice: a model for large scale knowledge translation. *British Medical Journal* 6: a1714.

7 Peters, D.H., Nhan, N.T., and Adam, T. (2013). *Implementation Research in Health: A Practical Guide. Alliance for Health Policy and Systems Research*. Geneva: World Health Organization. http://who.int/alliance-hpsr/alliancehpsr_irpguide.pdf (accessed June 18, 2020).

8 Feldstein, A.C. and Glasgow, R.E. (2008). A practical, robust implementation and sustainability model (prism) for integrating research findings into practice. *The Joint Commission Journal on Quality and Patient Safety* 34: 228–243.

9 Clarke, A.R., Goddu, A.P., Nocon, R.S. et al. (2013). Thirty years of disparities intervention research: what are we doing to close racial and ethnic gaps in health care? *Medical Care* 51: 1020–1026.

10 Mueller, M., Purnell, T.S., Mensah, G.A., and Cooper, L.A. (2015). Reducing racial and ethnic disparities in hypertension prevention and control: what will it take to translate research into practice and policy? *American Journal of Hypertension* 28: 699–716.

11 Nilsen, P. (2015). Making sense of implementation theories, models and frameworks. *Implementation Science* 10: 53.

12 Proctor, E., Silmere, H., Raghavan, R. et al. (2011). Outcomes for implementation research: conceptual distinctions, measurement challenges, and research agenda. *Administration and Policy in Mental Health* 38: 65–76.

13 Israel, B.A., Schulz, A.J., Parker, E.A., and Becker, A.B. (2001). Community-based participatory research: policy recommendations for promoting a partnership approach in health research. *Education for Health* 14: 182–197.

14 Israel, B.A., Schulz, A.J., Parker, E.A., and Becker, A.B. (1998). Review of community-based research: assessing partnership approaches to improve public health. *Annual Review of Public Health* 19: 173–202.

15 Glasgow, R.E., Vogt, T.M., and Boles, S.M. (1999). Evaluating the public health impact of health promotion interventions: the RE-AIM framework. *American Journal of Public Health* 89: 1322–1327.

16 Ford, C.L. and Airhihenbuwa, C.O. (2010). The public health critical race methodology: praxis for antiracism research. *Social Science and Medicine* 71: 1390–1398.

17 Griffith, D.M., Citrin, T., Jerome, N.W. et al. (2000). The origins and overview of the W. K. Kellogg Community Health Scholars Program. *Progress in Community Health Partnerships* 3 (4): 335–348.

18 Viswanathan, M., Ammerman, A., Eng, E. et al. (2004). Community-based participatory research: assessing the evidence. *Evidence Report/Technology Assessment* 99: 1–8.

19 Nolan, K., Schall, M.W., Erb, F., and Nolan, T. (2005). Using a framework for spread: the case of patient access in the Veterans Health Administration. *Joint Commission Journal on Quality and Patient Safety* 31: 339–347.

20 Rogers, E.M. (1995). Lessons for guidelines from the diffusion of innovations. *The Joint Commission Journal on Quality Improvement* 21: 324–328.

21 Delgado, R. and Stefancic, J. (2001). *Critical Race Theory: An Introduction*, 3e. New York, NY: New York University Press.

22 Cooper, L.A., Marsteller, J.A., Noronha, G.J. et al. (2013). A multi-level system quality improvement intervention to reduce racial disparities in hypertension care and control: study protocol. *Implementation Science* 8: 60.

23 Shediac-Rizkallah, M.C. and Bone, L.R. (1998). Planning for the sustainability of community-based health programs: conceptual frameworks and future directions for research, practice and policy. *Health Education Research* 13: 87–108.

24 Simpson, D.D. (2002). A conceptual framework for transferring research to practice. *Journal of Substance Abuse Treatment* 22: 171–182.

25 Proctor, E.K., Powell, B.J., Baumann, A.A. et al. (2012). Writing implementation research grant proposals: ten key ingredients. *Implementation Science* 7: 96.

26 Paskett, E., Thompson, B., Ammerman, A.S. et al. (2016). Multilevel interventions to address health disparities show promise in improving population health. *Health Affairs* 35: 1429–1434.

27 Boonyasai, R.T., Carson, K.A., Marsteller, J.A. et al. (2018). A bundled quality improvement program to standardize clinical blood pressure measurement in primary care. *Journal of Clinical Hypertension* 20: 324–333.

28 O'Mara-Eves, A., Brunton, G., Oliver, S. et al. (2015). The effectiveness of community engagement in public health interventions for disadvantaged groups: a meta-analysis. *BMC Public Health* 15: 129.

29 Cooper, L.A., Beach, M.C., Johnson, R.L., and Inui, T.S. (2006). Delving below the surface. Understanding how race and ethnicity influence relationships in health care. *Journal of General Internal Medicine* 21: S21–S27.

30 Roter, D. (2000). The medical visit context of treatment decision-making and the therapeutic relationship. *Health Expectations* 3: 17–25.

31 Schmotzer, G.L. (2012). Barriers and facilitators to participation of minorities in clinical trials. *Ethnicity & Disease* 22: 226–230.

32 Schnackenberg, A.K. and Tomlinson, E.C. (2016). Organizational transparency: a new perspective on managing trust in organization–stakeholder relationships. *Journal of Management* 42: 1784–1810.

33 Cooper, L.A., Purnell, T.S., Ibe, C.A. et al. (2016). Reaching for health equity and social justice in Baltimore: the evolution of an academic-community partnership and conceptual framework to address hypertension disparities. *Ethnicity and Disease* 21: 369–378.

34 Kreuter, M.W. and Wray, R. (2003). Tailored and targeted health communication: strategies for enhancing information relevance. *American Journal of Health Behavior* 27: S227–S232.
35 Resnicow, K., Baranowski, T., Ahluwalia, J.S., and Braithwaite, R.L. (1999). Cultural sensitivity in public health: defined and demystified. *Ethnicity & Disease* 9: 10–21.
36 Kleinman, A., Eisenberg, L., and Good, B. (1978). Culture, illness, and care: clinical lessons from anthropologic and cross-cultural research. *Annals of Internal Medicine* 88: 251–258.
37 Lin, M.K., Marsteller, J.A., Shortell, S.M. et al. (2005). Motivation to change chronic illness care: results from a national evaluation of quality improvement collaboratives. *Health Care Management Review* 30: 139–156.
38 Braveman, P., Egerter, S., and Williams, D.R. (2011). The social determinants of health: coming of age. *Annual Review of Public Health* 32: 381–398.
39 Woolf, S.H. and Braveman, P. (2011). Where health disparities begin: the role of social and economic determinants—and why current policies may make matters worse. *Health Affairs (Project Hope)* 30: 1852–1859.
40 Purnell, T.S., Calhoun, E.A., Golden, S.H. et al. (2016). Achieving health equity: closing the gaps in health care disparities, interventions, and research. *Health Affairs* 35: 1410–1415.
41 Rogers, E.M. (2003). *Diffusion of Innovations*, 5e. New York, NY: Free Press.
42 Scheirer, M.A. and Dearing, J.W. (2011). An agenda for research on the sustainability of public health programs. *American Journal of Public Health* 101: 2059–2067.
43 Hussain, T., Franz, W., Brown, E. et al. (2016). The role of care management as a population health intervention to address disparities and control hypertension: a quasi-experimental observational study. *Ethnicity and Disease* 21: 285–294.

18

Healthcare and Public Policy: Challenges and Opportunities for Research

Olveen Carrasquillo[1], Sonjia Kenya[1], Stuti Dang[2], Cynthia Lebron[3], and Tilda Farhat[4]

[1] Division of General Internal Medicine, Department of Medicine, Miller School of Medicine, University of Miami, Miami, FL, USA
[2] Division of Geriatrics, Department of Medicine, Miller School of Medicine, University of Miami, Miami, FL, USA
[3] School of Nursing and Health Studies, University of Miami, Coral Gables, FL, USA
[4] National Institute on Minority Health and Health Disparities, Office of Science Policy and Planning, National Institutes of Health, Bethesda, MD, USA

18.1 Summary

In this chapter, we introduce readers to research methods that can be used to examine the impact of public policy on disparities in health and healthcare. The chapter begins by introducing key concepts in public policy, health policy, and healthcare policy. This first section will also explain why this line of research is so important and provide concrete examples of how policies in each of these three domains can potentially impact health and healthcare disparities. In the next section, we use selected examples from our own work and that of others to describe various methods and approaches that can be used to examine the impact of such policies on health disparities. These include secondary data analysis, innovative sampling strategies, and some rare examples of experimental designs. The strengths and limitations of these approaches will be discussed. Finally, we discuss challenges to this line of work such as how findings are interpreted and disseminated, and potential opportunities to continue to advance the field.

18.2 Background/Context

The overarching goal of medical research is to prevent, detect, diagnose, and treat disease and disability, leading to better health for everyone. Similarly, the goal of health-oriented public policy research is to generate findings that can inform the development of public policies that will lead to better health for everyone. Such investigations are particularly important in health disparities research, which, as noted in previous chapters, most often results from social and structural factors [1], including the physical/built environment,

The Science of Health Disparities Research, First Edition. Edited by Irene Dankwa-Mullan, Eliseo J. Pérez-Stable, Kevin L. Gardner, Xinzhi Zhang, and Adelaida M. Rosario.

sociocultural environment, and health systems, as well as individual behavior and biology.[1] These social determinants are highly dependent (and often driven) by laws. Health disparities research has long attempted to inform the formulation of policies that improve the health of minorities in order to reduce health disparities.

Studies to inform public programs, such as Medicaid research, are examples of health-policy research. Research on problems associated with Medicaid turnover [2] and the development of innovative programs to facilitate the enrollment of children in health insurance programs are examples of such work [3]. With respect to healthcare delivery, research showing how language barriers affect patient satisfaction [4] and result in high rates of errors made by ad-hoc interpreters [5] is often used to help advocate for policies that support the provision of linguistically appropriate services for non-English speakers. Similarly, work that increases the evidence base on the effectiveness of community health workers in areas such as HIV [6] and diabetes [7] can be used to advocate for public policies that support inclusion of such staff in the formal healthcare system. While such work is critical in helping inform and formulate public policy, another equally important line of health-disparities public policy research is analysis of the impact of laws and public policies on minority and vulnerable populations. Such work becomes even more pressing during times of rapid political change, when large-scale initiatives are being legislated and implemented, which may have negative (or positive) impacts on minority health and health disparities.

18.3 Key Constructs

For the purposes of this chapter, we define public policy as an organized and systematic approach to decision making made by a governing body. Most commonly, this decision making can be at the federal, state, or local government level. However, public policy can sometimes also be set by other groups, like companies (e.g., internet service providers) or private entities (a neighborhood organization).

Thus, there are many research areas in which one can examine the impact of public policy on disparities. We group public policy into three domains: (i) healthcare policy (ii) health policy and (iii) nonhealth-related public policy.

Most familiar to health researchers is healthcare policy research, which examines healthcare policies that affect the formal medical system. Such policies include government decisions about how healthcare is organized, delivered, and financed. Sections 18.4.1–18.4.3 are examples of such work. In contrast, the broader idea of health policy research includes healthcare policy as well as other forms of research examining or seeking to inform policies and legislation to influence health but not necessarily through the formal healthcare system. This type of research is often included within the context of public health research and is highlighted in Sections 18.5.1 and 18.5.2.

While the bulk of public policy research on health outcomes is focused on health and healthcare policy, the largest impact of public policies on health disparities may be from those not specifically designed to address health. These includes laws, policies, and

1 https://www.nimhd.nih.gov/images/research-framework-slide.pdf

regulations on issues such as the minimum wage, housing, education, and transportation. As an example, take policies enacted in the mid-twentieth century regarding highways. Policymakers decided to build parts of Interstate 95 through inner-city neighborhoods in Miami (Overtown) and New York City (The Bronx). Many social scientists have attributed the social decay experienced in these communities to the massive disruption and displacement that had to be done to accommodate this highway [8, 9]. Sections 18.5.3 and 18.5.4 provide two examples of policy research seeking to assess the health impact of such decisions.

18.4 Selected Examples of Healthcare Policy Research

18.4.1 Immigration Reform

Initiatives to bar access to healthcare services for certain immigrant groups often have considerable bipartisan support. One recent example was the Affordable Care Act (ACA), which barred undocumented immigrants from receiving ACA health insurance benefits even if the immigrants paid for the full cost of insurance. Fourteen years earlier, President Clinton signed the Personal Responsibility and Work Opportunity Reconciliation Act (PRWORA). One component of that law barred states from using federal funds to provide Medicaid and the Children's Health Insurance Program (CHIP) for legal immigrants who had resided in the United States for less than 5 years. We sought to examine how many people were excluded from healthcare coverage by this policy [10].

Often, the most important challenge in answering such questions is simply finding the right dataset. An additional challenge was that insurance criteria for Medicaid and CHIP vary by state, with some states choosing to use state funds to provide coverage for such recent immigrants. We used the Current Population Survey (CPS), whose supplement collects detailed information on insurance coverage and also has large sample size, allowing for robust state-level estimates. The CPS also collects data on foreign birth status, length of time in the country, and citizenship.

However, as with most federal surveys, it does not distinguish among the many different categories of noncitizens, with the largest two being permanent residents and undocumented persons (who are not eligible for coverage). Thus, we had to impute undocumented status based on other state-specific estimates. We found that about 110 000 children and 250 000 adults were excluded from coverage because of the law. We also provided a sensitivity analysis based on varying assumptions on the proportion of the immigrant population that may be undocumented. In addition, as some states with large immigrant populations, like California and New York, used state-only funding to cover such immigrants, a major result of the legislation was not to ban such immigrants from coverage but to shift the costs to those states.

18.4.2 The Veterans Affairs Home-Based Primary Care Model

Home-bound elders are among the most vulnerable population groups. Often, such elders are frail, have high health needs, and have limited social and economic resources [11]. To meet their needs, the Department of Veterans Affairs (VA) developed a Home-Based Primary Care (HBPC) interdisciplinary team-based program. Internal evaluations showed the program resulted in cost savings [12]. However, one concern was whether the program shifted some costs from the VA health system to Medicare (elders received more care outside the VA system).

To address this question, a group of VA investigators examined total costs by not only including the VA costs but also obtaining Medicare Standard Analytical Files to aggregate Medicare Part A and B person-level costs. Using these cost files requires expertise. First, data use agreements between government agencies such as the VA and the Centers for Medicare and Medicaid Services (CMS) need to be established. Second, the complexity of the distinct data files must be understood. CMS cost data included obtaining both Medicare and Medicaid data, as well as nursing home assessment data and Medicare Home Health assessment expenditure data. In addition, the CMS data are only available after a time lag of a few years. The VA data sources are just as complex (see Figure 18.1).

Given the popularity of the program, a randomized controlled trial (RCT) assigning elders to a control group was not possible. Instead, VA investigators created an experimental design among the nearly 7000 elders dually enrolled in VA, HBPC, and Medicare. The analysis compared the actual health costs of elders in the program to theoretical models projecting what their costs may have been without the program [13]. To model projected costs, they used a method used by the CMS called the Hierarchical Condition Category (HCC) model. HCC scores demographic factors, diagnoses, and interaction of multiple comorbidities to yield projected costs, which were adapted to the VA and calibrated to the elderly population. Using this approach and combining it with both Medicare and VA cost data, they were able to show that enrollment in the program was associated with reductions of 16.7% in VA costs, 10.8% in Medicare costs, and 13.4% in total costs. Thus, using cost projection models, they were able to approximate the findings of an RCT.

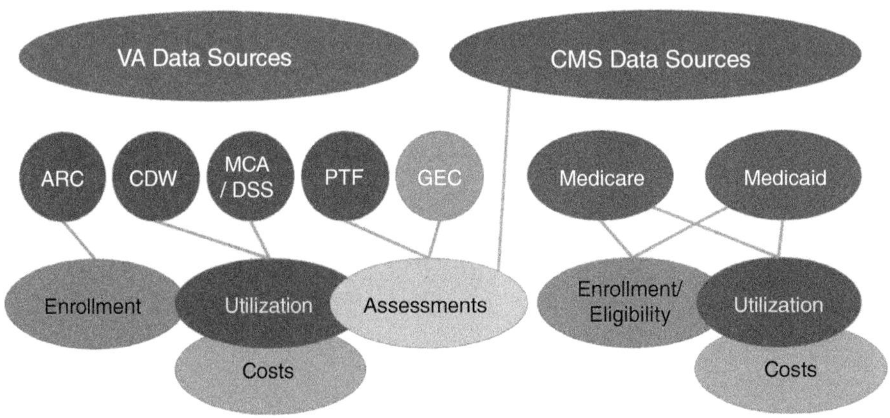

Figure 18.1 Distinct VA and CMS data sources needed to construct the VA HBPC model. ARC, Allocation Resource Center; CDW, Corporate Data Warehouse; CMS, Center for Medicare and Medicaid Services; GEC, VA Office of Geriatrics & Extended Care; MCA/DSS, Managerial Cost Accounting and Decision Support System; PTF, Patient Treatment File; VA, Veterans Affairs. *Source:* Courtesy of VA GEC IAGG Symposium Presentation, July 2017. (See color representation of this figure in color inserts section)

This study led CMS to implement its own Medicare Independence at Home (IAH) demonstration program authorized under the ACA. The program also showed costs savings, and modeling showed further cost savings if expanded to all Medicare beneficiaries [14]. These results have prompted legislation to convert IAH to a national Medicare program. This is an example where research in the VA setting informed a national health policy for non-VA elder beneficiaries.

18.4.3 Oregon Health Insurance Experiment

One of the main challenges in assessing policy effectiveness is that by their very nature, policies are rarely enacted in a way that allows assessment in an experimental design. A rare example is the Oregon Medicaid Experiment [15, 16]. In 2008, 3 years before the ACA, Oregon expanded its Medicaid program to low-income adults. But due to limited funding, it could only fund some of the potentially eligible 90 000 adults. The state allocated coverage by a lottery in which 30 000 names were drawn. This lottery presented a unique opportunity to study the impact of Medicaid with random assignment. The study showed that after 2 years, the low-income adults randomly assigned to Medicaid had improved self-reported health, increased healthcare utilization, improved mental health outcomes, and reported less medically related financial hardship than those whom were not assigned to Medicaid.

Yet, even in this randomized study, there were challenges. Persons were not automatically enrolled in Medicaid. Selected adults still had to apply and were only allowed to enroll if they met eligibility requirements. As expected, those who were selected and also enrolled in Medicaid were different with respect to baseline characteristics from those that did not enroll. This selection/participation bias is a common problem when evaluating public programs.

In response, the study team had to use instrument variable (IV) analysis. This analytic strategy, also called "two-stage models," is often used in economics and social science analyses. The first step is to find variables ("instruments") that are correlated with program participation, and then develop a regression model to predict the probability of program participation. In the second stage, the predicted probability is used in another model that examines the outcomes adjusted for the probability of program participation (from the first model). Using this two-stage approach, one can estimate outcomes that account for the selective participation in Medicaid. Thus, even in this experimental design, the differential characteristics of program participants choosing to enroll in Medicaid still had to be considered.

Despite the large sample size, the study was not able to detect changes in many of the examined health conditions, such as blood pressure, cholesterol, or diabetes control. The only exception was depression, which was prevalent in this group of low-income adults and for which positive changes were detected. The reason for positive findings with respect to overall health and service utilization and yet negative findings in most disease-specific outcomes may be due to the fact that the proportion of enrollees with these conditions was low. Thus, the study was underpowered to detect definite changes in these outcomes.

The study findings played a major role in deliberations and modeling of the ACA's Medicaid expansion. Advocates cited the positive findings, while opponents of expansion focused on null findings without mentioning that the study was not adequately powered to detect such effects.

18.5 Non-Healthcare Policy Research

18.5.1 Nutrition Program for Women, Infants, and Children

The Nutrition Program for Women, Infants, and Children (WIC) was established in the 1970s to address malnutrition among low-income pregnant mothers and their young children. The program now serves 7.7 million people per month at an annual cost of nearly $6 billion [17]. While there is consensus that the program is beneficial [18], selection bias has long been a challenge. Comparisons of outcomes of women in WIC to the general population are problematic, as women of higher socioeconomic status have better health status. Conversely, comparing low-income women who did not apply for WIC to those who did is also problematic, as women choosing to participate in WIC may have more health-seeking behaviors than eligible mothers who do not participate.

One approach to address this issue has been to compare WIC mothers who applied for WIC while pregnant to those who applied after having the baby, as both groups were still motivated to apply for WIC. In such analyses, investigators showed that WIC improved birth outcomes but to a lesser extent than reported in other studies [19]. However, selection bias remained a concern, as women who apply for WIC while pregnant may still differ on other unadjusted characteristics from those who apply after pregnancy.

One study addressed this issue using the instrumental variable approach [20] and data from the Pregnancy Risk Assessment Monitoring System (PRAMS) collected by individual states with support from the Centers for Disease Control and Prevention (CDC). Investigators first conducted logistic regression models to predict WIC participation. These models included several maternal and infant characteristics, such as mother's age, education, race and ethnicity, number of live births, smoker status, and maternal weight. In addition, as eligibility criteria vary by state and year, investigators also included dummy variables for state-year combinations to account for state-level characteristics. The analysis found that among women with Medicaid, those who were more disadvantaged were more likely to participate in WIC. In light of this finding, one would expect to find a negative effect if the probability of program participation were included in the models. Instead, they found that despite selective enrollment of the most at-risk mothers in WIC, program participation was associated with higher rates of initiation of early prenatal care and fewer low birthweights.

One limitation of using instrumental variables is that the reliability of the results is based on the variables in the first stage being highly predictive of program participation. In the study, even with use of changes in WIC regulations by state and year, the models still had limited ability to predict the probability of participating in WIC [21]. A subsequent study using more detailed socioeconomic and health behavior data (from the Early Childhood Longitudinal Study) was able to identify a better set of instrumental variables that was more likely to predict the probability of WIC participation. Using that approach, researchers also found that WIC participation significantly reduced the probability of having very premature births and very low birthweights. The effects were most pronounced among high-risk mothers. Thus, although selection bias is always a concern, techniques such as IV analysis and other statistical strategies can help address challenges in evaluating public policies.

18.5.2 Flint Water Crisis

Residents of more vulnerable communities are often exposed to environmental hazards at greater rates than those who live in wealthier communities [22]. Flint had been a financially successful city until automobile plants began to shut down in the 1980s. Since then, the city has faced significant financial challenges and population loss. At present, 41% of the 100 000 residents in the city live below the federal poverty level and 60% are Black.

In April 2014, Flint changed its water supply to the Flint River. This temporary move was projected to result in approximately $2 million in savings to the city. Shortly thereafter, residents began complaining about the water's appearance, taste, and odor, and about health problems. When *Escherichia coli* bacteria were identified in the drinking water, boil advisories were issued. However, increased water lead levels were also found, and boiling water does not affect these.

To determine whether increased water lead levels had led to increased blood lead levels (BLL) among children, researchers conducted a retrospective study that measured BLL in local children before (Jan–Sept 2013) and after (Jan–Sept 2015) the water supply was changed. The team used data from the laboratory that processed most lead levels in the area as well as advanced spatial analytic approaches to overcome limitations of simple geocoding by zip code (zip codes did not correlate to water supply sources) [23]. They found that prior to the change, 2.4% of children receiving Flint water had elevated BLL. After the change in water supply, 4.9% of children had such elevated levels. Among those whose homes were not receiving water from the Flint river, there were no changes. The changes in BLL were most pronounced in areas having the highest levels of lead in the water. In addition, they also examined the change in disparities. They found that even before the water supply change, children in Flint had increased levels versus those outside Flint (2.4% versus 0.7%). After the water source change, the disparity widened. Based on these findings, the county health department declared a public health emergency and, ultimately, the water supply was restored to Lake Huron. A subsequent study using more detailed spatio-temporal analysis of water-lead exposure made a stronger case for a causal relationship [24]. In addition, researchers also estimated that future lost income associated with decreased IQs among lead-water exposed children at $65 million. This conservative estimate did not include the many additional health and societal costs associated with this change in water source.

18.5.3 Minimum Wage Policies

One area of intense policy interest has been whether raising the minimum wage can reduce health disparities. Despite extensive debate, few well-done studies have empirically examined the impact of such policies on child health outcomes. One study reviewed 30 years of "natural experiments" involving minimum wage laws [25]. Investigators identified 206 changes in state minimum wages from 1980 to 2011. They then used the CDC's National Vital Statistics System (NVSS) to obtain rates of low birthweight, and postneonatal mortality by state for the same months for which they had minimum wage data. They showed a consistent pattern of improved child health outcomes when a state's minimum wages were higher than the federal minimum wage. Each dollar above the federal minimum wage was associated with a 1–2% decrease in low birthweight births. Another example of such research showed that from 1986 to 2004 a one dollar decrease in the real minimum wage was associated with a 0.06% increase in body mass index [26].

18.5.4 Residential Segregation

Residential segregation has long been considered a major social determinant of health. A classic example is that low-income people living in high-income neighborhoods have better health outcomes than low-income people living in low-income neighborhoods [27].

In 1976, the Supreme Court ordered the Chicago Housing Authority (CHA) to provide alternative housing for public housing residents in concentrated areas of poverty [28]. CHA used randomly assigned vouchers to help 7000 low-income African American families move to private apartments in either another low-income urban location or in a middle-income suburban location. Two studies examined the program's effect on mortality and drug-related crime. One study found that African American male youth who were relocated to areas with a higher proportion of residents with a college degree had lower all-cause and homicide-related mortality than those relocated to areas with a lower proportion of residents with a college degree [29]. Other neighborhood characteristics that differentiated both locations, including the percentage of White residents and poverty, were less strong predictors of these associations.

Another study found that males (but not females) relocated to the suburbs had fewer drug-related arrests than those placed in urban settings [30]. They had a 42% drop in the odds of being arrested and a 52% drop in the odds of being convicted of a drug offense. The federal Moving to Opportunity Demonstration Program (MTO) conducted in five large cities also found that those randomly assigned to move out of segregated areas experienced significantly better mental health outcomes and lower obesity rates than those in the control group [31]. Further subgroup analyses suggested that the positive health outcomes occurred mostly among girls.

18.6 Key Challenges

As previously mentioned, the implementation of public policies can be examined using rigorous research methods. The challenge is that policy effects on health disparities can be hard to evaluate depending on the study design (Table 18.1).

18.7 Future Directions and Opportunities

These studies highlight research addressing the impact of policies on different levels. Such research is not only possible but can also make a positive impact. Continuing to identify future opportunities for such research will be key. One clear opportunity is ongoing health reform. Our recent policy analysis of the ACA showed that while the ACA reduced the number of uninsured minorities [35], major gaps remain. For example, it is unclear

Table 18.1 Challenges and approaches in minority health and health disparities public policy research.

Potential challenges

1) Policy implementations often do not follow a randomized experimental design.
2) Policies (or components of policies) are often implemented over time and in distinct geographic areas, making it difficult to define what would constitute an exposure and to link a particular policy or component to health disparity outcomes. The effect may be gradual and confounded by evolving social, environmental, or behavioral factors.
3) The enactment of policies is beyond researchers' control. They cannot develop timely data collection instruments to gather information from populations affected by these policies, both prior to and after the policies' enactment.
4) Pathways linking policies to reductions in health disparities are poorly understood. This lack of data limits our ability to link policy changes to outcome changes, as well as to scale up and disseminate best practices.
5) Structural factors, such as income and educational inequalities, residential segregation, and stigma and discrimination, are important drivers of health disparities. Yet, there often remains limited recognition of the role that these nonhealth factors can have on disparities [32].

Potential approaches

1) Policy evaluations require innovative methodological and measurement strategies that can accommodate the challenges of linking policies to desired outcomes.
2) Researchers need to carefully consider the selection of a comparison group. Characteristics might include historical controls, pre-post comparisons, use of nearby unaffected populations, or other statistical approaches (e.g., instrumental variables).
3) Researchers need to be creative in finding existing and timely population-based sources of data that include valid and reliable measures for the exposures and outcomes of interest and allow for tracking over time to monitor changes.
4) Moving the field forward will require a non-siloed approach of interdisciplinary research teams having diverse expertise and research skills [33, 34].

whether such reductions led to improved access to care and health outcomes among minorities. Another research opportunity is the large-scale demonstration and policy intervention projects CMS has introduced over the past few years [36, 37]. It will be critical that investigators with a more nuanced understanding of minority health and health disparities populations help focus some of that work on the impact of such interventions on vulnerable populations.

Perhaps the most important area in need of increased research relates to the neoliberal polices recently adopted (increased free market capitalism, lower taxes on the wealthy, diminished role of government, and enacting of barriers to accessing government services, such as health insurance). It is widely acknowledged that these will have a major detrimental impact on structural factors such as income and educational inequalities and residential segregation, and that this will lead to a widening of health disparities [38]. Assessing how these policies affect minority health and health disparities will be an increasingly important role for researchers, who will need more sophisticated methods and approaches to conduct such rigorous research.

18.8 Key Points

- Public policies can have a major impact on minority health and health disparities.
- Various research methods and approaches can be used to examine how public policies impact health disparities. These include secondary data analysis, innovative sampling strategies, and, in rare instances, experimental designs.
- Moving forward, the field will require a cadre of well-trained researchers using increasingly more sophisticated methods and having a highly nuanced understanding of health disparities.

Disclaimer

The views and opinions expressed in this chapter are those of the authors only and do not necessarily represent the views, official policy, or position of the U.S. Department of Health and Human Services or any of its affiliated institutions or agencies.

References

1 World Health Organization (2008). Closing the gap in a generation: health equity through action on the social determinants of health. Final report of the Commission on Social Determinants of Health. WHO, Geneva.
2 Carrasquillo, O., Himmelstein, D.U., Woolhandler, S., and Bor, D. (1998). Can Medicaid managed care provide continuity of care to new Medicaid enrollees? An analysis of tenure on Medicaid. *American Journal of Public Health* 88: 464–466. PMID: 9518985.
3 Flores, G., Abreu, M., Chaisson, M. et al. (2005). A randomized, controlled trial of the effectiveness of community-based case management in insuring uninsured Latino children. *Pediatrics* 116 (6): 1433–1441.
4 Carrasquillo, O., Brennan, T.A., Orav, E.J., and Burstin, H.R. (1999). The impact of language barriers on patient satisfaction. *Journal of General Internal Medicine* 14: 82–87. PMID: 10051778.
5 Flores, G., Laws, M.B., Mayo, S.J. et al. (2006). Errors in medical interpretation and their potential clinical consequences in pediatric encounters. *Pediatrics* 111: 6–14.
6 Kenya, S., Jones, J., Arheart, K. et al. (2013). Using community health workers to improve clinical outcomes among people living with HIV: a randomized controlled trial. *AIDS and Behavior* 17 (9): 2927–2934.
7 Carrasquillo, O., Lebron, C., Alonzo, Y. et al. (2017). Effect of a community health worker intervention among Latinos with poorly controlled type 2 diabetes: the Miami healthy heart initiative randomized clinical trial. *JAMA Internal Medicine* 177 (7): 948–954.
8 Dunn, M. (1997). *Black Miami in the Twentieth Century*. Miami: University Press of Florida.
9 Jonnes, J. (2002). *South Bronx Rising: The Rise, Fall, and Resurrection of an American City*, 2e. New York, NY: Fordham University Press.

10 Carrasquillo, O., Ferry, D.H., Edwards, J., and Glied, S. (2003). Eligibility for government insurance if immigrant provisions of welfare reform are repealed. *American Journal of Public Health* 93 (10): 1680–1682.

11 Blumenthal, D., Chernof, B., Fulmer, T. et al. (2016). Caring for high-need, high-cost patients—an urgent priority. *New England Journal of Medicine* 375 (4): 1–3.

12 Beales, J.L. and Edes, T.E. (2009). Veteran's affairs home based primary care. *Clinical in Geriatric Medicine* 25: 149–154.

13 Edes, T., Kinosian, B., Vuckovic, N.H. et al. (2014). Better access, quality, and cost for clinically complex veterans with home-based primary care. *Journal of the American Geriatric Society* 110: 1–14.

14 Kinosian, B., Taler, G., Boling, P. et al. (2016). Projected savings and workforce transformation from converting independence at home to a medicare benefit. *Journal of the American Geriatric Society* 64 (8): 1531–1536.

15 Baicker, K., Taubman, S.L., Allen, H.L. et al. (2013). The Oregon experiment — effects of Medicaid on clinical outcomes. *New England Journal of Medicine* 368: 1713–1722.

16 Taubman, S.L., Allen, H.L., Wright, B.J. et al. (2014). Medicaid increases emergency-department use: evidence from Oregon's health insurance experiment. *Science* 343: 263–268.

17 Oliveira, V., (2017) The Food Assistance Landscape, US Department of Agriculture, Economic Research Service FY 2016 Annual Report. Washington, DC.

18 Oliveira, V., Racine, E., Olmsted, J. and Ghelfi, L.M. (2002) The WIC Program: Background, Trends, and Issues. Food Assistance and Nutrition Research Report No. 27, US Department of Agriculture Economic Research Service. Washington, DC.

19 Joyce, T., Racine, A., and Yunzal-Butler, C. (2008). Reassessing the WIC effect: evidence from the pregnancy nutrition surveillance system. *Journal of Policy Analysis and Management* 27 (2): 277.

20 Bitler, M.P. and Currie, J. (2005). Does WIC work? The effects of WIC on pregnancy and birth outcomes. *Journal of Policy Analysis and Management* 24 (1): 73–91.

21 Gai, Y. and Feng, L. (2012). Effects of federal nutrition program on birth outcomes. *Atlantic Economic Journal* 40 (1): 61–83.

22 Schulz, A. and Northridge, M.E. (2004). Social determinants of health: implications for environmental health promotion. *Health Education Behavior* 31 (4): 455–471.

23 Hanna-Attisha, M., LaChance, J., Sadler, R.C., and Schnepp, A.C. (2016). Elevated blood lead levels in children associated with the Flint drinking water crisis: a spatial analysis of risk and public health response. *American Journal of Public Health* 106 (2): 283–290.

24 Zahran, S., McElmurry, S.P., and Sadler, R. (2017). Four phases of the Flint water crisis: evidence from blood lead levels in children. *Environmental Research* 157: 160–172.

25 Komro, K.A., Livingston, M.D., Markowit, S., and Wagenaar, A.C. (2016). The effect of an increased minimum wage on infant mortality and birth weight. *American Journal of Public Health* 106 (8): 1514–1516.

26 Meltzer, D.O. and Chen, Z. (2011). The impact of minimum wage rates on body weight in the United States. In: *Economic Aspects of Obesity* (eds. M. Grossman and N.H. Mocan). Cambridge, MA: National Bureau of Economic Research.

27 Diez Roux, A.V., Merkin, S.S., Arnett, D. et al. (2001). Neighborhood of residence and incidence of coronary heart disease. *New England Journal of Medicine* 345 (2): 99–106.

28 US Supreme Court 1976. Hills v. Gautreaux (425 US 284).
29 Votruba, M.E. and Kling, J.R. (2009). Effects of neighborhood characteristics on the mortality of black male youth: evidence from Gautreaux, Chicago. *Social Science & Medicine* 68 (5): 814–823.
30 Keels, M. (2008). Second-generation effects of Chicago's Gautreaux residential mobility program on children's participation in crime. *Journal of Research on Adolescence* 18 (2): 305–351.
31 Kling, J.R., Liebman, J.B., and Katz, L.F. (2007). Experimental analysis of neighborhood effects. *Econometrica* 75 (1): 83–119.
32 Lieberman, L., Golden, S.D., and Earp, J.A.L. (2013). Structural approaches to health promotion: what do we need to know about policy and environmental change? *Health Education & Behavior* 40 (5): 520–525.
33 Aboelela, S.W., Larson, E., Bakken, S. et al. (2007). Defining interdisciplinary research: conclusions from a critical review of the literature. *Health Services Research* 42 (1 Pt 1): 329–346.
34 Little, M.M., St. Hill, C.A., Ware, K.B. et al. (2016). Team science as interprofessional collaborative research practice: a systematic review of the science of team science literature. *Journal of Investigative Medicine* https://doi.org/10.1136/jim-2016-000216.
35 Carrasquillo, O. and Muller, M. (2018). Refinement of the Affordable Care Act: a progressive perspective. *Annual Review of Medicine* 69: 81–88.
36 Howell, B.L., Conway, P.H., and Rajkumar, R. (2015). Guiding principles for Center for Medicare & Medicaid innovation model evaluations. *JAMA* 313 (23): 2317–2318.
37 US Government General Accountability Office (2017) Telehealth and Remote Patient Monitoring Use in Medicare and Selected Federal Programs. Report # GAO-17-365 Washington, DC. US Government General Accountability Office.
38 Nkansah-Amankra, S., Agbanu, S.K., and Miller, R.J. (2013). Disparities in health, poverty, incarceration, and social justice among racial groups in the United States: a critical review of evidence of close links with neoliberalism. *International Journal of Health Services* 43 (2): 217–240.

19

Addressing Disparities in Access to High-quality Care

Milda R. Saunders[1], Regina James[2], Kesi Williams[3], and Marshall Chin[1]

[1] Department of Medicine, The University of Chicago, Chicago, IL, USA
[2] 2M Research, Arlington, VA, USA
[3] National Institute on Minority Health and Health Disparities, National Institutes of Health, Bethesda, MD, USA

19.1 Racial Disparities as a Quality Problem

Many studies have shown that the US healthcare system faces challenges providing consistent high-quality care [1–3], with especially severe problems across the dimension of equity [1, 4]. In this chapter, racial and ethnic healthcare disparities will be used to elucidate the various issues involved in researching disparities in access to high-quality care.

Racial and ethnic minorities have worse outcomes and receive lower quality care even when taking insurance and income into account [4]. There are two ways that racial and ethnic minorities can receive lower quality care than their nonminority counterparts (Figure 19.1). First, racial and ethnic minorities utilize the same care providers as their nonminority counterparts yet receive lower quality care (Figure 19.1a, Disparities by race). Second, racial and ethnic minorities are over-represented at healthcare facilities or with clinicians that provide lower quality care (Figure 19.1b, Disparities by site of care).

Minorities are more likely to attend facilities of lower quality due to geography, insurance, and referral patterns [5]. Minorities may have less access to higher quality facilities because those facilities are not nearby, are not perceived as welcoming, or are not affordable. Facilities that predominately serve minority patients may be under-resourced due to inadequate reimbursement and payer mix. They may also be overburdened with patients with higher medical comorbidities and social challenges. In addition, minority-serving facilities may be less likely to attract and retain high-quality staff due to location, working conditions, or compensation.

However, even within the same healthcare facility, minorities may receive different care due to implicit biases of providers, such as perceived difference in need, benefit, or adherence to treatments, and structural problems such as inadequate interpreter services and systems of care not designed to meet the needs of minority patients [6].

The Science of Health Disparities Research, First Edition. Edited by Irene Dankwa-Mullan, Eliseo J. Pérez-Stable, Kevin L. Gardner, Xinzhi Zhang, and Adelaida M. Rosario.

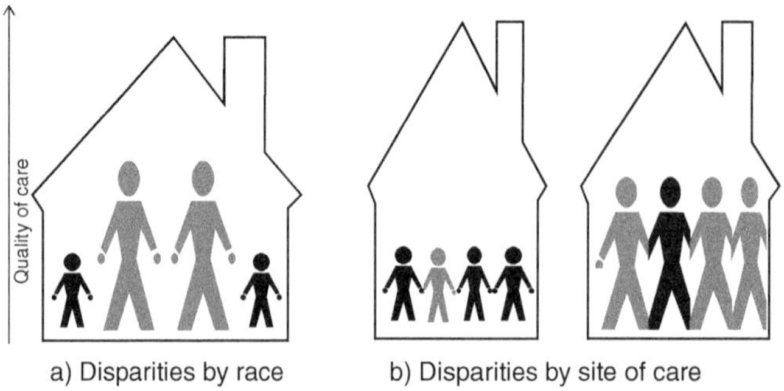

Figure 19.1 Healthcare disparities by race or site of care. (See color representation of this figure in color inserts section.)

The causes—and, thus, the solutions—likely vary by clinical condition, site of care, and racial/ethnic group. In this chapter, we discuss ways to reduce health disparities in access to high-quality care through research and practice, addressing key constructs and using examples from the literature.

19.2 Defining Quality and Access

Donabedian, the father of quality assurance, developed a model of structure, process, and outcome to evaluate the quality of healthcare [7]. Structure refers to the settings, providers, and administrative system in which care takes place; processes are the components of care delivered; and outcomes include health status, quality of life, maintenance of health, recovery or restoration of function, and patient experience and satisfaction [7, 8].

High-quality care is both effective and accessible. Effective care helps patients achieve the greatest improvements in health status based on current science and requires both technical and interpersonal skills [9]. Access is the first step to receiving high-quality care. Access has multiple dimensions [10, 11]. Aday and Anderson posit that access is "those dimensions which describe the potential and actual entry of a given population to the health delivery system." Donabedian (Figure 19.2) describes the multiple domains of accessibility to include geographic/physical access, affordability, and availability. These models apply to access to a usual regular source of care (e.g., primary care clinic), care provider (dentist), or specific technology or treatment (e.g., kidney transplant or cardiac defibrillator).

Affordability is essential, but not sufficient, to ensure access to care. The Affordable Care Act of 2010 (ACA) decreased the number of uninsured overall and among racial and ethnic minorities by expanding eligibility for Medicaid, creating health insurance marketplace exchanges, and subsidizing insurance premiums for low-income residents [12]. Between 2010 and 2016, the percentage of Blacks and Latinos with a regular place for care increased; however, these groups still are less likely than Whites to have a usual place for medical care [12]. Many newly insured individuals are still without a usual source of care or they may seek care at institutions that already disproportionately serve minority patients and provide lower quality of care, or at institutions that are ill-equipped to serve racial/ethnic minority patients.

In the Agency for Healthcare Research and Quality *2016 National Healthcare Quality and Disparities Report*, poor individuals had worse access than nonpoor people on nearly

	Care		
	Health care system (Structure)	*Patient-centred care* (Process)	*Consequences of care* (Outcome)
Quality			
Accessibilty	Geographic / physical access Affordability Availability	Affordability Availability	Health status User evaluation
Effectiveness		Effectiveness of clinical care Effectiveness of inter-personal care	Health status User evaluation

Figure 19.2 Donabedian model: structure process outcome. *Source:* Campbell et al [13].

all measures of access, including having health insurance, having a usual regular source of care, encountering difficulties when seeking care, and receiving care as soon as wanted. On three-fourths of these measures of access, Latinos had worse access to care than Whites. African Americans had worse access than Whites for one-half of the measures; other racial minority groups (Asians and American Indians/Alaskan Natives) had worse access to care than Whites on one-third of the measures [12].

Another goal of the ACA was to improve quality of care. Nearly 60% of broad measures of healthcare quality have improved [12]. However, despite overall improvements in access to care and quality of care, persistent disparities remain in access to high-quality care across race/ethnicity, socioeconomic status (SES), and other social risk factors [2, 4, 12].

19.3 Examples of Racial Disparities as a Quality Problem

19.3.1 Cardiovascular Care and Implantable Defibrillators

Multiple studies document disparities in the quality of care for cardiovascular (CV) disease. CV disease is common and costly [14]. Evidence-based clinical guidelines exist for much CV care, and quality of care can be assessed with clinical performance measures based on these guidelines [15]. New technologies and clinical trials may rapidly change the clinical landscape. In addition, CV care is also subject to clinical judgment and, thus, can be prone to clinicians' implicit biases and variations in local practice.

The use of implantable cardiac defibrillators (ICDs) provides an example of racial and ethnic disparities in cardiac care. Cardiac defibrillators are recommended for primary prevention in patients at high risk of sudden cardiac death, such as patients with ejection fraction <35% after optimization of medical therapy [16]. Although ICDs are expensive, several trials have shown them to be cost-effective when used appropriately. Since 2005, the Centers for Medicare and Medicaid Services (CMS) has reimbursed for ICDs for primary prevention in high-risk patients. Racial, ethnic, and gender disparities are seen in receipt of cardiac defibrillators [17]. Clinically appropriate women and racial and ethnic minorities are less likely to be counseled on cardiac defibrillators [17]. Among patients counseled on defibrillators, women were as likely to receive an ICD; however, African American and Latino patients were still less likely to receive an ICD. Additional work is needed to determine whether racial and ethnic minorities experience ICD-related complications, including shock-related anxiety, and how best to mitigate those complications [18]. Interventions to identify appropriate candidates and to increase the effectiveness of patient counseling and acceptance of ICDs are all needed [17]. In addition, less is known about whether these differences are due primarily to different providers, hospitals, or regions where racial and ethnic minorities receive care versus disparities at the same site of care [19].

19.3.2 End-Stage Renal Disease

End-stage renal disease (ESRD) is a high-cost, high-morbidity disease. Racial and ethnic minorities are disproportionately affected. Compared to non-Hispanic Whites, African Americans have three times the incidence, Native Americans and Asians/Pacific Islanders both have 1.2 times greater incidence, and Latinos have 1.3 times greater incidence of ESRD [2]. While some of the disparities in ESRD incidence are associated with biological and social factors, racial and ethnic minorities also experience disparities in the quality of pre-ESRD care, such as blood pressure control, nephrology referral, and treatment of anemia and bone mineral disease that could prevent ESRD and its complications [20].

Among patients with ESRD, disparities persist in quality of dialysis care. These disparities have been most consistently demonstrated for African Americans with ESRD [21]. One dialysis quality indicator is urea reduction, a measure of dialysis adequacy. Across multiple studies, African Americans have been less likely than Whites to receive an adequate dialysis dose [21, 22]. A 2003 quality improvement (QI) intervention disproportionately improved rates of dialysis adequacy for African Americans, thus reducing disparities [22]. In addition, after years of quality reporting and pay-for-performance (P4P) for dialysis facilities, quality has improved overall. However, African American patients and facilities in African American communities disproportionately experience lower quality dialysis facilities based on multiple measures, including dialysis quality adequacy, anemia management, and fistula use [21, 23]. Findings of lower quality do not extend to all minority-serving dialysis facilities [24].

Examining dialysis facility quality is instructive because it highlights both the ability of, and the challenges to, the government to affect quality through both funding and legislation. Medicare funds most patients with ESRD, so CMS has both the incentive and means to improve quality and reduce costs. In 2012, CMS introduced P4P for dialysis facilities and bundled care payments [25]. P4P was designed to reduce payment to facilities that did not meet benchmarks on anemia management and dialysis adequacy. Bundled payments aggregated all dialysis-related drugs and services into the capitated payment to facilities. However, dialysis facilities are also subject to segregated markets that reflect residential segregation as well as patient sorting, likely due to patient preference, physician network, and center acceptance [24, 26]. Prior work has found that African Americans bypass closer facilities that may be of higher quality in order to reach their own dialysis facilities [26]. Early examination of P4P showed that quality of care increased; however, minority-serving facilities were more likely to be penalized [27].

19.4 Addressing Disparities in Access to Quality Care

Prior work has highlighted challenges and successes in improving access to high-quality care for disparity populations [28]. Early-childhood immunizations provide an example of success. For years, racial/ethnic minority and low-income children had lower immunization rates.

In 1994, the Vaccines for Children (VFC) program was created to eliminate cost barriers to early-childhood immunizations. The program supports vaccine distribution infrastructure at the state and local levels and provides vaccines to publicly insured, uninsured, and underinsured children, as well as to American Indians and Alaskan Natives [29]. Local public health agencies worked with communities and clinics to increase awareness of the need for vaccines and to ensure that clinics are able to meet those needs. Through targeted federal funding and increased education and outreach at the local level, racial disparities in vaccine coverage for measles-mumps-rubella and poliovirus in the United States have been eliminated [30].

What makes an intervention successful may vary by type of disparity, population, and where the intervention is directed, so disparities must often be addressed at multiple levels. For example, QI methods that a single dialysis facility uses to improve outcomes and reduce disparities in dialysis adequacy differ from the policy and regulatory tools CMS would use nationally to improve dialysis adequacy across facilities. In addition, QI programs may improve overall quality but may not reduce disparities if the intervention is not tailored to the needs of the population with social risk factors. Overall quality indicators may improve, but particular subgroups may not improve at the same rate; thus, the disparity remains or increases. Institutional or local disparity reduction initiatives may not be able to address resource allocation or make systemic changes. The National Quality Forum (NQF) "Four I's for Health Equity" (Figure 19.3) are linked to the principles highlighted by Chin et al. and Fiscella et al. to identify the clinical outcome and disparity population, and the intervention level used to address the disparity [31–33].

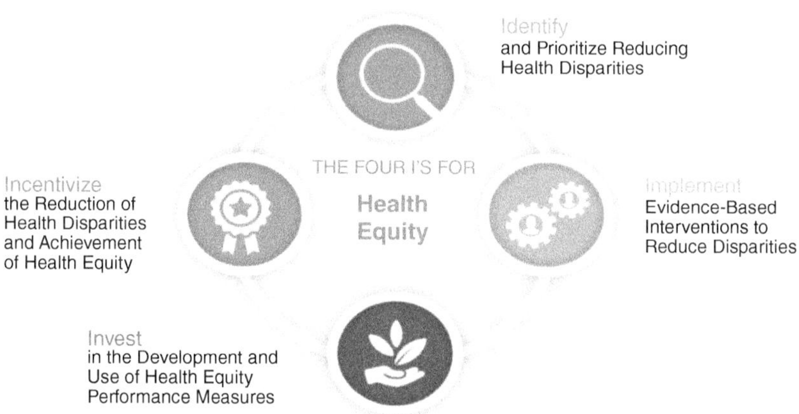

Figure 19.3 National Quality Forum Roadmap for Promoting Health Equity and Eliminating Disparities. *Source:* National Quality Forum [33].

19.5 Steps to Addressing Healthcare Disparities

19.5.1 Identify and Prioritize Reducing Healthcare Disparities

19.5.1.1 Relevant and Reliable Data are Needed

Efforts to reduce disparities require defining and measuring racial differences in high-quality care [32].

19.5.1.1.1 Measuring Race and Ethnicity Patient registries and medical records must have accurate information about race and ethnicity in order to compare processes of care and outcomes by racial and ethnic group. The ACA has codified best practices standards for collection of race and ethnicity data. It is important that research and health information systems use the standard categories [34]. It is equally important to implement administrative procedures that allow patient self-reporting rather than the "eyeball" test of the clinical or clerical staff. Race is largely a social construct, and it has profound implications on how patients are perceived by medical providers and how patients experience the world.

19.5.1.1.2 Other Measurement Challenges Collecting information on social risk factors, including sex, primary language, disability status, insurance status, and zip code is essential for the review of stratified healthcare quality data. Failure to consider these additional categories may mask disparities and potential ways to target quality improvement interventions.

19.5.1.2 To Improve Patient Outcomes, Basic Quality Improvement Structures and Processes Must be in Place

Institutions must include providing high-quality care to all patients in their missions. Quality metrics, goals, and teams should be in place or developed. Resources should enable staff to measure, report, and improve quality [31].

19.5.1.3 Disparities Must be Recognized as a Quality Problem

Understanding and eliminating inequities in healthcare must be mission-critical to clinics, institutions, and government. Healthcare disparities are too often relegated to a specific researcher, administrator, or office rather than seen as a larger quality issue [32].

19.6 Implement Evidence-based Strategies to Eliminate Disparities

19.6.1 After Prioritizing Health Equity and Identifying Existing Health Disparities, the Next Challenge is to Develop an Intervention

Healthcare disparities are the result of complex interactions of personal, social, historical, and institutional inequities. Ameliorating these inequities is daunting and ultimately requires collaboration among individuals, institutions, and sectors. However, health equity can be increased incrementally through appropriate interventions.

19.6.2 Determine the Root Cause(s) of the Disparity

In most cases, determining the root cause of the disparity requires careful study, including the use of chart reviews, provider interviews, surveys, administrative dataset analyses, and policy analyses. However, there is no substitute for talking with patients to understand their challenges and the issues that must be addressed to reduce health disparities.

19.6.3 Consider the Levels of Influence of Patients, Providers, the Microsystem (Immediate Care Team), Organizations, Communities, and Policy

Often, interventions must target multiple levels of influence to address the root causes of disparities. For example, minority patients may not be offered guideline-recommended cardiac interventions (e.g., percutaneous coronary intervention [PCI] with drug-eluting stent) at the same rate as White patients. If the root cause at a particular hospital is implicit bias by physicians, one solution would be cultural competence training to improve access to clinically appropriate treatment [31].

Another intervention at the organization level would be to create a chest pain protocol driven by objective clinical data. While physician implicit bias could still play a role in patient care, physicians would have to be more explicit in their thought processes and choices, which could reduce disparities in treatment. As part of quality improvement efforts, hospitals could analyze rates of PCI with drug-eluting stents for patients fitting guideline criteria, stratify rates by patient race/ethnicity, and implement interventions until disparities are eliminated. Financially rewarding physicians or hospitals that improve or achieve equity in the provision of guideline-recommended care could prove to be an effective policy-level intervention.

19.6.4 Build on Prior Work and Modify for Specific Context

There is a long history of interventions that have affected health disparities in quality of care, both to reduce them and, unintentionally, to increase them [22]. Many quality improvement interventions are not published. Discussing plans and concerns with peer institutions can help an organization learn from prior interventions. However, even the best program will need to be tailored to new situations, such as a disease, institutional and political context, and patient population.

19.6.5 Implement, Evaluate, and Adjust the Intervention

While aiming for the correct initial targets and intervention, it is important to implement the plan, evaluate the results, acknowledge missteps, and make necessary adjustments [31]. Healthcare institutions are subject to changes in staff, leadership, medical knowledge, government regulations, and patient flow. A change in any of these can make a plan more difficult to implement. The Consolidated Framework for Implementation Research (CFIR) provides a checklist of factors to consider when implementing an intervention [35]. Evaluating the intervention requires looking for intended outcomes and costs, but also unintended consequences in other disease areas, facilities, or patient groups. If intended outcomes are not as desired or the unintended consequences are undesirable, the intervention or implementation must be modified.

19.7 Invest in Health Equity Performance Measures

19.7.1 One Must Define What Is Meant by Equity in Quality

Agreeing on the goal is an important step in developing a quality improvement/disparity reduction intervention. Is the goal equity of opportunity or equity of outcome? That is, is the goal defined by equitable standard-of-care process measures, such as making the human papillomavirus (HPV) vaccine available to teens, referral to stress test for chest pain, or provision of kidney transplant information? Or is equity defined by the outcome, such as the actual delivery of HPV vaccine, documented stress test with associated chest pain diagnosis, or appearance on the kidney transplant waitlist? Equity of opportunity is an important first step, but, ultimately, equity of outcome improves health. While QI programs may use process measures as an early indicator, outcomes should be prioritized as performance measures.

19.7.2 Measure Capacity for Health Equity

Any quality improvement program, general or for specific diversity groups, requires investment in resources, including clinician and administrator time, data infrastructure, and clinical services. Institutions should be aware of the resources in place that support a culture of equity. Resources required may vary by type of institution and diversity population. One helpful resource to improving health equity is the employment of physicians, administrators, and staff from the group with health disparities. While having diverse staff is important both for face validity and to offer varied opinions and approaches, staff diversity alone is not sufficient to achieve health equity. Reducing disparities is everyone's responsibility. Interprofessional training on QI, communication, and cultural competency empowers the entire staff to contribute to institutional change.

19.8 Incentivize the Reduction of Health Disparities

19.8.1 Design Quality Improvement and Payment Interventions to Achieve Equity

As they proliferate, it is important to consider if and how quality improvement programs will impact disparities. The standard quality improvement approach may improve quality of care for all groups, but the effect may be variable across different racial/ethnic groups.

Say an outpatient practice wants to increase HPV vaccination rates among preteens, so they initiate a QI program. At baseline, HPV coverage for Whites is 30% but just 20% for Latinos. The program components include passive patient education through brochures in the waiting area and posters in the exam room and a computer alert for the physician. In clinic 1, the materials are only in English and the posters display White teen girls. Due to lack of cultural tailoring, lack of language concordance, and focus on insurance barriers, more White patients than Latino patients agree to HPV vaccine, and the disparity increases.

In clinic 2, passive patient education is also available through brochures in Spanish and English, and display a more diverse group of teens. In addition, the case manager has been trained in overcoming cost barriers for Medicaid patients. The uptake is the same for White and Latino patients, but since Whites had a higher baseline HPV vaccination rate, the disparity persists.

Clinic 3 has the same culturally tailored, language-concordant education materials, a trained social worker, and a health educator who talks to each of the Latino parents about HPV vaccine. He or she specifically addresses concerns that the physicians are offering the HPV vaccine because of presumed promiscuity or that agreeing to HPV vaccine at age 12 is encouraging the child to have intercourse. In addition, the clinic has evening sessions for parents who cannot take time off work for appointments.

Because clinic 3 has a multidimensional approach that addresses patient education, cultural tailoring, and specific structural barriers (insurance and hours), the Latino uptake is greater than for White patients and the disparity is reduced. Green et al. advocate for culturally competent QI interventions that target all groups but have additional program components for disparity groups to, at a minimum, equalize or, ideally, narrow disparities (Figure 19.4) [36]. Only by removing barriers that are more likely to affect the disparity group can disparities be reduced.

19.8.2 Consider the Impact of Social Risk Factors on Public Reporting and Pay-For-Performance

When examining process or outcomes by race, is it appropriate to adjust clinical performance measures by social risk factors such as income, insurance type, and neighborhood deprivation? The National Academies of Sciences, Engineering, and Medicine Committee on Accounting for SES in Medicare Payment Programs has grappled with the issue of adjustment for social risk factors [37]. Physicians and healthcare facilities who disproportionately see more

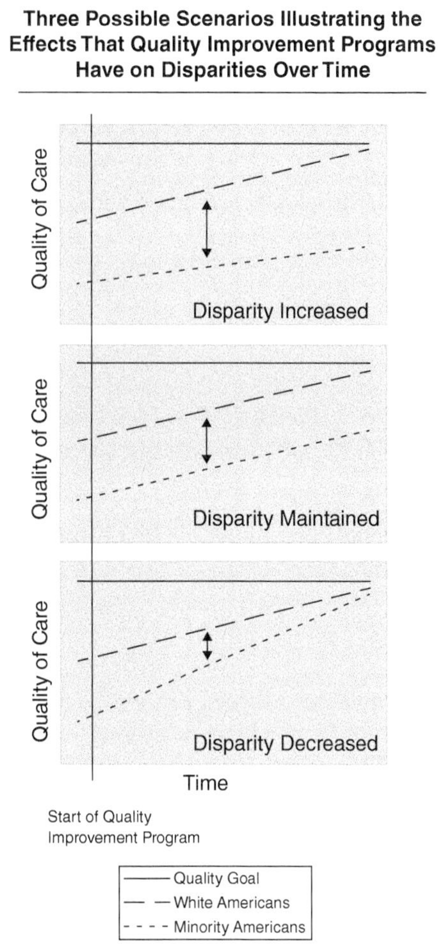

Figure 19.4 Possible impacts of QI on disparities. *Source:* Green et al. [36].

socially disadvantaged patients may be at risk for underpayment through P4P, since they care for populations for whom it is more difficult to attain good clinical performance scores. Reduced payment could lead to "white [coat] flight" out of disadvantaged communities or lower quality of care due to reduced resources [37]. The Committee acknowledges that not adjusting may cause physicians or insurers to avoid serving patients with social risk factors. On the other hand, the Committee notes that accounting for social risk factors could adjust away disparities resulting from true differences in the quality of care. Adjusting for social risk factors could also support the belief that patients with social disadvantages should expect lower quality care and reduce incentives to improve care for socially disadvantaged patients.

Potential solutions vary by method of quality improvement. For example, for public reporting programs, there could be stratification of performance scores by patient characteristics or type of institution (e.g., private versus safety net). P4P programs have additional possible solutions. Well-planned P4P programs have the ability to change behavior and reallocate resources. It is important that they do so in a way that improves quality and reduces inequality while not unfairly penalizing minority-serving providers. P4P programs' ability to reduce disparities has been mixed. When Massachusetts implemented a disparity reduction component within its Medicaid P4P program, one challenge was that two-thirds of minority patients were clustered in one-sixth of the hospitals, limiting statistical analyses [38]. Research shows that minority-serving providers are more likely to be penalized under P4P [27]. For P4P programs, incentives could include rewarding improvement rather than solely hitting a specific benchmark, or adjusting payment based on social risk factors. Additional resources could be given to minority-serving institutions that face a higher cost of providing quality care to patients with social challenges [39].

19.8.3 Develop a Business Case to Enable Organizations to Implement and Sustain Health Equity Interventions

The NQF, National Academy of Medicine, and Department of Health and Human Services Assistant Secretary for Planning and Evaluation (ASPE) have all recommended providing greater incentives and support for health equity efforts and rewarding better outcomes for patients with social risk factors [33, 37]. Clinical performance measures can be used for accountability through public reporting and payment. Two themes in the 2017 NQF's *A Roadmap for Promoting Health Equity and Eliminating Disparities: The Four I's for Health Equity* were prioritizing performance measures in the domains of equitable access and equitable high-quality care for accountability purposes, and redesigning payment models to support health equity [33]. For example, capitated payment schemes could direct funding to infrastructure required for health equity interventions implementing culturally tailored team-based care that addresses medical and social determinants of health. P4P could reward improvement in quality or achievement of absolute levels of quality for minority populations, and/or reduction in disparities among different racial/ethnic groups. The integration of quality improvement, clinical performance measurement, and payment reform represents an exciting, high-impact frontier for efforts to achieve health equity [40].

19.9 Advice for a Healthcare Disparities Researcher

19.9.1 Pick an Important Research Question

Topics of interest can be motivated by clinical observations or general curiosity, or guided by clinical expertise. Develop a list of questions whose answers could advance the field. Start by brainstorming, or with a question that presented itself in prior work.

Healthcare disparities work has a rich history of documenting disparities and causes. Many research efforts are focused on reducing or eliminating disparities through interventions that address root causes or moderate negative downstream effects of drivers of disparities. Interventions can act at multiple levels, including patient, provider, clinic/institution, community, and policy (see Figure 19.5). For chronic kidney disease (CKD) disparities, we have pursued research questions such as: How does CMS P4P policy for dialysis facilities affect disparities in quality of care? Does a culturally tailored inpatient education intervention for African American patients with advanced CKD improve knowledge of renal replacement options?

Good research questions are "SMART": Specific, Measurable, Achievable, Relevant, and Time-bound.

19.9.2 Review the Relevant Literature

After picking topics or a specific question, review the relevant literature for what work has already been done in the field, areas of uncertainty, and questions that have not been addressed. The background helps us understand how questions are framed and what conceptual models are applicable. The methods sections highlight the statistical methods and analytic approaches commonly used. As authors discuss their limitations and next steps, glean ideas about the next logical steps. Read widely about history, sociology, economics, policy, health beliefs, and the intersection of race/ethnicity with SES, gender, lesbian, gay, bisexual, transgender, and questioning (LGBTQ) status, and age.

Intervention Level	Definition	Examples
Patient	Change the knowledge and/or behaviors of patients to improve their health outcomes	Symptom monitoring Incentives Culturally targeted education and outreach Self-management and goal setting Patient narratives
Provider	Change the knowledge and/or behavior of providers to improve patient outcomes	Cultural competency training Disparity report cards
Microsystem	Add new members to or shift responsibilities among the immediate care team, such as the primary care provider, nurse, and staff	Integration of community health workers or other staff into the care team Nurse-led interventions
Organization	Change organization operations; may require coordination among management, providers, information technology, and/or human resources	Redesigning the system of operations Instituting new forms of technology Improving the clinical encounter Systematic literacy screening of patients
Community	Work with people and organizations outside traditional health settings such as churches, schools, and social service agencies	Engaging local stakeholders to integrate community and health care resources
Policy	Influence laws, regulations, and/or resource allocation on a regional or national basis	Medicare reimbursement regulations Accreditation standards for providers and health care organizations Breast cancer screening laws

Figure 19.5 Levels of influence of an intervention. *Source:* Chin et al. [31].

19.9.3 Choose, Adapt, or Develop a Conceptual Model to Guide Your Research

Conceptual models guide the description of research questions, the data collected through instruments or secondary datasets, variable selection in statistical analyses, and intervention components.

Cooper et al. provide a conceptual model for barriers to and mediators of equitable healthcare that can be used for formative work to understand a particular healthcare disparity or, later, to design an intervention [41]. Purnell et al. have built a model to elucidate the multilevel influences on health disparities and their intervention targets, as well as the key stakeholders and outcomes affected by the intervention [28]. A model popularized in the National Academy of Medicine's *Unequal Treatment* report[1] used a conceptual model to demonstrate the domains under healthcare disparities, such as discrimination and the operation of the health system, and to define healthcare disparities as differences in quality of care not due to clinical need or preferences.

19.9.4 Select the Appropriate Quantitative and/or Qualitative Approach for Your Study Question

The choice of approach is based on the type of question and the group studied. Qualitative work, such as through focus groups or in-depth interviews of patients, providers, or administrators, is important for hypothesis generation, deep exploration of complex issues, and determining acceptability or usability of an intervention. Quantitative methods include chart reviews, surveys, administrative dataset analysis, and policy analyses. After hypothesis generation, hypothesis testing can occur using observational studies, natural experiments, and randomized trials. Often, quantitative and qualitative approaches complement each other, and a mixed-methods approach is used [42]. For example, conducting quantitative analysis using an administrative dataset, one determines that African Americans are likely to attend lower quality dialysis facilities. With the available variables, the quantitative analysis cannot determine why. In this case, the researcher could conduct in-depth interviews of dialysis patients about why they chose their current facility, and with nephrologists and clinic social workers about their roles in the selection process. The work determines that social workers do not want to steer patients to a particular facility and that patients are hesitant to select an unfamiliar facility. Based upon these mixed-method findings, an intervention could be developed to help social workers present patients with information about facility quality and proximity. In addition, patients could tour higher quality facilities prior to dialysis so they are more familiar with them.

19.9.5 Use the Appropriate Analytical Methods

Within the broad categories of qualitative, quantitative, or mixed methods, a range of analytic techniques can be used. The choice of analytic method is guided by the research

---

1 https://www.nap.edu/catalog/10260/unequal-treatment-confronting-racial-and-ethnic-disparities-in-health-care

question and should lead to valid and reliable results. Cook et al. highlight three statistical ways to examine racial and ethnic disparities in healthcare and note that the choice of approach should be guided by the specific question one would like to answer.

The first method would be analyzing unadjusted rates (racial groups only). This method is best used in situations where confounding variables such as SES are less important. For example, a public health department or particular clinic might be interested in knowing absolute immunization rates stratified by race. The analysis can be further stratified by education, poverty status, or insurance, answering questions such as: How does the vaccination rate for privately insured Whites compare to that of privately insured Latinos?

The second statistical method looks at residual direct effects of race and ethnicity after adjusting for other SES effects (e.g., education, poverty status, and insurance) to examine the direct effect of race only. An example of this type of question is examining whether racial and ethnic minorities have equal access to cardiac defibrillators after adjusting for comorbidities, insurance, education, income, and site of care.

For the final method, race by health status (treatment indication and comorbidities), the condition varies by race/ethnicity but receipt of treatment may also be determined by other SES factors that may also vary with race/ethnicity (e.g., education, poverty status, insurance). These analyses use methods such as a propensity score, with rank and replace adjustment, to construct samples of minority and White populations that have similar distributions of health status (indication and comorbidities) but maintain their "real world" observed differences in SES variables [43]. Using the previous example, this method could answer whether racial and ethnic minorities have equal access to cardiac defibrillators when they have the same clinical need, while measuring the effect of the SES differences that vary by race [43].

19.9.5.1 Qualitative Research as Part of Root Cause Analysis to Identify Drivers of Disparities and Design Solutions.

Root cause analysis (RCA) is common in the patient safety/quality improvement sphere. RCAs are a way medical systems investigate errors and poor-quality care and design solutions to prevent future errors and suboptimal care by asking: What happened? Why did it happen? and What can be done to prevent it from happening again? [44]. Qualitative research can be used in a similar manner to examine the root cause of poor outcomes in diversity populations. Qualitative work starts with a broad set of questions for a particular group [45]. Initial questions include: Does a problem exist? What is it? What are contributors? What are potential solutions? The initial answers can help researchers frame the issue in ways they would not have done previously, guide language and themes for subsequent focus groups or surveys, and point to solutions. For example, qualitative research determined the barriers to shared decision making for African American patients with diabetes on the South Side of Chicago. Interviews and focus groups investigated how African Americans defined shared decision making and explored how racial issues such as mistrust may contribute to inadequate shared decision making [46]. This qualitative research informed development of a successful multicomponent intervention that included a culturally tailored patient education and communication skills training course for African Americans with diabetes, community partnerships and outreach, and more traditional health center quality improvement [47].

19.9.5.2 Multilevel Modeling Exploring Individual and Place Effects

In the United States, racial and ethnic minorities are overrepresented within certain regions, neighborhoods, hospitals, and providers [24, 38, 48]. Many of these units have lower overall quality outcomes. It is important to determine whether observed racial disparities are due to who people are (different outcomes within a healthcare setting) or where they receive care (different outcomes due to the healthcare setting) (Figure 19.1). Multilevel modeling (MLM) is helpful in these situations [49]. MLM adjusts the standard error for correlated values, and although the point estimate is similar to other models, the standard error is larger as confidence becomes lower. This is because individuals who are similar on many attributes provide less information than independent observations. Other methods can be used to account for correlated values, such as using a robust standard error to adjust for clustering. MLM is also helpful in assessing the relative impact of level (individual, institution, or region) after adjusting for relevant factors within each level.

One example is Barnato et al. and the examination of hospital-level racial disparities in acute myocardial infarction outcomes and treatment [48]. Using a type of hierarchical linear model (HLM), a general estimating equation (GEE) with a separate intercept for hospital-level factors, they found that the effect of hospital on racial disparities varied by outcome. For example, African American patients had lower rates of beta blocker use than Whites, but adjustment for hospital fixed effect narrowed that gap. This indicates that African Americans and Whites at the same hospital were treated similarly and differences were attributable to the site (Figure 19.1b, Disparities based on site of care). However, for cardiac catheterization, adjustment for hospital fixed effects widened the disparity, which indicates that at the same hospitals Africans Americans had lower rates of catheterization than their White counterparts (Figure 19.1a, Disparities based on race) [48]. Use of MLM helps to elucidate better the level where the disparity occurs, which helps guide interventions.

19.9.5.3 Mixed-Method Research that Covers Clinical Outcomes, the Process of Organizational Change, and Economic Outcomes

Decision makers and policy makers attempting to improve access to high-quality care need data on whether a given intervention improves clinical performance measures; the economic costs and benefits of the initiative from the perspectives of healthcare organizations, payers, and society; and implementation lessons for organizations attempting to reduce healthcare disparities. An example of mixed-method research that provided this critical information was an evaluation of the Health Resources and Services Administration Health Disparities Collaboratives, a nationwide effort using quality improvement collaboratives to improve care and reduce disparities among patients cared for in federally qualified health centers [50]. A series of studies including chart review, survey, qualitative interviews, business case cost analysis, and societal cost-effectiveness analysis found that the Health Disparities Collaboratives improved clinical processes and outcomes and were cost-effective from a societal perspective, but the initiative did not have consistent revenue streams for uninsured and underinsured patients. The research identified important health center priorities, including funding for direct patient care and quality improvement, and assistance with patient self-management, health information technology, and facilitation of provider behavioral change.

19.10 Conclusion

Healthcare disparities are a quality issue. However, the causes and solutions are not one size fits all. Healthcare disparities cannot be eliminated without improving the care that disparity populations receive and the places where disparity populations receive care. Research is important not only to document the disparities but also to determine the underlying causes and test interventions to eliminate these disparities. Evidence-based interventions can then be translated into larger scale policies and programs. By integrating and aligning QI with policy and payment reform, we can incentivize and support health equity.

19.11 Key Points

- Healthcare disparities, differences in quality of care or access between two groups not justified by differences in need or preferences, exist for groups with social disadvantages such as race, ethnicity, socioeconomic status, gender, sexual orientation, age, or national origin.
- Quality improvement efforts should be "disparity sensitive"; that is, the programs should be tailored to maximize effectiveness for disparity groups, and outcomes should be monitored to ensure interventions decrease disparities and improve care for all.
- Health disparity researchers can help eliminate disparities through research that documents the problem; elucidates mechanisms of the disparity; and develops, implements, and evaluates interventions.
- The research question and conceptual model will help determine appropriate methods and analytic approaches.

Disclaimer

The views and opinions expressed in this chapter are those of the authors only and do not necessarily represent the views, official policy, or position of the U.S. Department of Health and Human Services or any of its affiliated institutions or agencies.

References

1 Haider, A.H., Schneider, E.B., Sriram, N.N. et al. (2015). Unconscious race and social class bias among acute care surgical clinicians and clinical treatment decisions. *JAMA Surgery* 150 (5): 457–464.

2 Agency for Healthcare Research and Quality (2016). *National Healthcare Quality and Disparities Report*. Rockville, MD: Agency for Healthcare Research and Quality.

3 Beal, A.C. (2011). High-quality health care: the essential route to eliminating disparities and achieving health equity. *Health Affairs* 30 (10): 1868–1871.

4 US Department of Health and Human Services (2013). *National Healthcare Disparities Report, 2012*. Rockville, MD: AHRQ.

5 Gaskin, D.J., Spencer, C.S., Richard, P. et al. (2008). Do hospitals provide lower-quality care to minorities than to Whites? *Health Affairs* 27 (2): 518–527.

6 Zestcott, C.A., Blair, I.V., and Stone, J. (2016). Examining the presence, consequences, and reduction of implicit bias in health care: a narrative review. *Group Processes & Intergroup Relations* 19 (4): 528–542.

7 Ayanian, J.Z. and Markel, H. (2016). Donabedian's lasting framework for health care quality. *The New England Journal of Medicine* 375 (3): 205–207.

8 Donabedian, A. (2005). Evaluating the quality of medical care. *Milbank Quarterly* 83 (4): 691–729.

9 Donabedian, A. (1988). The quality of care: how can it be assessed? *JAMA* 260 (12): 1743–1748.

10 Aday, L.A. and Andersen, R.M. (1981). Equity of access to medical care: a conceptual and empirical overview. *Medical Care*: 4–27.

11 Aday, L.A., Andersen, R., and Fleming, G.V. (1980). *Health Care in the US: Equitable for Whom?* SAGE Publications.

12 Agency for Healthcare Research and Quality (2015). *National Healthcare Quality and Disparities Report and 5th Anniversary Update on the National Quality Strategy*. Rockville, MD: AHRQ.

13 Campbell, S.M., Roland, M.O., and Buetow, S.A. (2000). Defining quality of care. *Social Science & Medicine* 51 (11): 1611–25.

14 Turenne, M., Baker, R., Pearson, J. et al. (2017). Payment reform and health disparities: changes in dialysis modality under the new Medicare dialysis payment system. *Health Services Research* 53 (3): 1430–1457.

15 Fihn, S.D., Blankenship, J.C., Alexander, K.P. et al. (2014). ACC/AHA/AATS/PCNA/SCAI/STS focused update of the guideline for the diagnosis and management of patients with stable ischemic heart disease. A report of the American College of Cardiology/American Heart Association Task Force on Practice Guidelines, and the American Association for Thoracic Surgery, Preventive Cardiovascular Nurses Association, Society for Cardiovascular Angiography and Interventions, and Society of Thoracic Surgeons. *Journal of the American College of Cardiology* 130 (19): 1749–1767.

16 Russo, A.M., Stainback, R.F., Bailey, S.R. et al. (2013). ACCF/HRS/AHA/ASE/HFSA/SCAI/SCCT/ SCMR appropriate use criteria for implantable cardioverter-defibrillators and cardiac resynchronization therapy: a report of the American College of Cardiology Foundation Appropriate Use Criteria Task Force, Heart Rhythm Society, American Heart Association, American Society of Echocardiography, Heart Failure Society of America, Society for Cardiovascular Angiography and Interventions, Society of Cardiovascular Computed Tomography, and Society for Cardiovascular Magnetic Resonance. *Journal of the American College of Cardiology* 61 (12): 1318–1368.

17 Hess, P.L., Hernandez, A.F., Bhatt, D.L. et al. (2016). Sex and race/ethnicity differences in implantable cardioverter-defibrillator counseling and use among patients hospitalized with heart failure: findings from the Get With The Guidelines-Heart Failure Program. *Circulation* 134: 517–526.

18 Wilson, M.H., Engelke, M.K., Sears, S.F. et al. (2013). Disease-specific quality of life–patient acceptance: racial and gender differences in patients with implantable cardioverter defibrillators. *Journal of Cardiovascular Nursing* 28 (3): 285–293.

19 Groeneveld, P.W., Heidenreich, P.A., and Garber, A.M. (2005). Trends in implantable cardioverter-defibrillator racial disparity: the importance of geography. *Journal of the American College of Cardiology* 45 (1): 72–78.

20 Norton, J.M., Moxey-Mims, M.M., Eggers, P.W. et al. (2016). Social determinants of racial disparities in CKD. *Journal of the American Society of Nephrology* 27 (9): 2576–2595.

21 Saunders, M.R. and Chin, M.H. (2013). Variation in dialysis quality measures by facility, neighborhood and region. *Medical Care* 51 (5): 413.

22 Sehgal, A.R. (2003). Impact of quality improvement efforts on race and sex disparities in hemodialysis. *JAMA* 289 (8): 996–1000.

23 Zarkowsky, D.S., Arhuidese, I.J., Hicks, C.W. et al. (2015). Racial/ethnic disparities associated with initial hemodialysis access. *JAMA Surgery* 150 (6): 529–536.

24 Hall, Y.N., Xu, P., Chertow, G.M., and Himmelfarb, J. (2014). Characteristics and performance of minority–serving dialysis facilities. *Health Services Research* 49 (3): 971–991.

25 Swaminathan, S., Mor, V., Mehrotra, R., and Trivedi, A. (2012). Medicare's payment strategy for end-stage renal disease now embraces bundled payment and pay-for-performance to cut costs. *Health Affairs* 31 (9): 2051–2058.

26 Saunders, M.R., Lee, H., Maene, C. et al. (2014). Proximity does not equal access: racial disparities in access to high quality dialysis facilities. *Journal of Racial and Ethnic Health Disparities* 21 (4): 291–299.

27 Saunders, M.R., Lee, H., and Chin, M.H. (2017). Early winners and losers in dialysis center pay-for-performance. *BMC Health Services Research* 17 (1): 816.

28 Purnell, T.S., Calhoun, E.A., Golden, S.H. et al. (2016). Achieving health equity: closing the gaps in health care disparities, interventions, and research. *Health Affairs* 35 (8): 1410–1415.

29 Smith, P.J. and Singleton, J.A. (2011). County-level trends in vaccination coverage among children aged 19–35 months – United States, 1995–2008. *Morbidity and Mortality Weekly Report* 60 (4): 1–86.

30 Walker, A.T., Smith, P.J., and Kolasa, M. (2014). Reduction of racial/ethnic disparities in vaccination coverage, 1995–2011. *MMWR Supplements* 63 (1): 7–12.

31 Chin, M.H., Clarke, A.R., Nocon, R.S. et al. (2012). A roadmap and best practices for organizations to reduce racial and ethnic disparities in health care. *Journal of General Internal Medicine* 27 (8): 992–1000.

32 Fiscella, K., Franks, P., Gold, M.R., and Clancy, C.M. (2000). Inequality in quality: addressing socioeconomic, racial, and ethnic disparities in health care. *JAMA* 283 (19): 2579–2584.

33 National Quality Forum (2017). *A Roadmap for Promoting Health Equity and Eliminating Disparities: The Four I's for Health Equity*. Washington, DC: Department of Health and Human Services.

34 Goddard, M. and Smith, P. (2001). Equity of access to health care services: theory and evidence from the UK. *Social Science & Medicine* 53 (9): 1149–1162.

35 Damschroder, L.J., Aron, D.C., Keith, R.E. et al. (2009). Fostering implementation of health services research findings into practice: a consolidated framework for advancing implementation science. *Implementation Science* 4: 50.

36 Green, A.R., Tan-McGrory, A., Cervantes, M.C., and Betancourt, J.R. (2010). Leveraging quality improvement to achieve equity in health care. *The Joint Commission Journal on Quality and Patient Safety* 36 (10): 435–442.

37 National Academies of Sciences, Engineering and Medicine (2016). *Accounting for Social Risk Factors in Medicare Payment: Criteria, Factors, and Methods: Report in Brief.* Washington, DC: The National Academies Press.

38 Blustein, J., Weissman, J.S., Ryan, A.M. et al. (2011). Analysis raises questions on whether pay-for-performance in Medicaid can efficiently reduce racial and ethnic disparities. *Health Affairs* 30 (6): 1165–1175.

39 Alberti, P.M., Bonham, A.C., and Kirch, D.G. (2013). Making equity a value in value-based health care. *Academic Medicine* 88 (11): 1619–1623.

40 DeMeester, R.H., Xu, L.J., and Nocon, R.S. (2017). Solving disparities through payment and delivery system reform: a program to achieve health equity. *Health Affairs* 36 (6): 1133–1139.

41 Cooper, L.A., Hill, M.N., and Powe, N.R. (2002). Designing and evaluating interventions to eliminate racial and ethnic disparities in health care. *Journal of General Internal Medicine* 17 (6): 477–486.

42 Johnson, R.B., Onwuegbuzie, A.J., and Turner, L.A. (2007). Toward a definition of mixed methods research. *Journal of Mixed Methods Research* 1 (2): 112–133.

43 Cook, B.L., McGuire, T.G., and Zaslavsky, A.M. (2012). Measuring racial/ethnic disparities in health care: methods and practical issues. *Health Services Research* 47 (3pt2): 1232–1254.

44 Wu, A.W., Lipshutz, A.M., and Pronovost, P.J. (2008). Effectiveness and efficiency of root cause analysis in medicine. *JAMA* 299 (6): 685–687.

45 Pope, C., van Royen, P., and Baker, R. (2002). Qualitative methods in research on healthcare quality. *Quality and Safety in Health Care* 11 (2): 148–152.

46 Peek, M.E., Wilson, S.C., Gorawara-Bhat, R. et al. (2009). Barriers and facilitators to shared decision-making among African-Americans with diabetes. *Journal of General Internal Medicine* 24 (10): 1135.

47 Peek, M.E., Harmon, S.A., Scott, S.J. et al. (2012). Culturally tailoring patient education and communication skills training to empower African-Americans with diabetes. *Translational Behavioral Medicine* 2 (3): 296–308.

48 Barnato, A.E., Lucas, F.L., Staiger, D. et al. (2005). Hospital-level racial disparities in acute myocardial infarction treatment and outcomes. *Medical Care* 43 (4): 308–319.

49 Goldstein, H. and Leyland, A.H. (eds.) (2001). *Multilevel Modelling of Health Statistics.* West Sussex: Wiley.

50 Chin, M.H. (2010). Quality improvement implementation and disparities: the case of the health disparities collaboratives. *Medical Care* 48 (8): 668–675.

20

Health Communication as a Mediator of Health and Healthcare Disparities

Neda Ratanawongsa[1, 2, 3, 4], Benyam Hailu[5], and Dean Schillinger[1, 2, 4]

[1] *Division of General Internal Medicine, Department of Medicine, Zuckerberg San Francisco General Hospital, University of California, San Francisco, San Francisco, CA, USA*
[2] *UCSF Center for Vulnerable Populations, University of California, San Francisco, San Francisco, CA, USA*
[3] *San Francisco Department of Public Health, San Francisco, CA, USA*
[4] *Richard Fine People's Clinic, San Francisco, CA, USA*
[5] *Division of Scientific Programs, Clinical & Health Services Branch, National Institute on Minority Health and Health Disparities, National Institutes of Health, Bethesda, MD, USA*

20.1 Introduction: Scope and Conceptual Framework

20.1.1 Prevalence of the Problem

The quality of communication between patients and their care providers is an important factor influencing healthcare and healthcare outcomes [1]. Evidence demonstrates that individual, interpersonal, and contextual health communication risk factors—such as limited health literacy (LHL) among patients, the patient-centeredness of providers, and the lack of professional interpreters in the healthcare environment—play roles in disparities in care and health for vulnerable populations [2]. Communication risk factors are prevalent in US healthcare settings, particularly in safety-net healthcare settings such as Federally Qualified Health Centers. Close to half of the US adult population has LHL [3], with nearly 1 in 10 having limited English proficiency (LEP) [4]. Implicit bias among healthcare providers and healthcare teams is as common as it is in the general population [5]. In combination with other contextual factors at the interpersonal, health system, and community level, these communication risk factors influence health disparities.

In this chapter, we review the evidence for several modifiable health communication risk factors, as well as contextual factors that influence the quality of communication, healthcare, and health for vulnerable populations. We describe evidence for interventions to improve the quality of communication and care for populations with communication risk factors. Finally, we discuss important gaps in our current knowledge of how poor communication contributes to healthcare disparities; our understanding of clinician-, patient-, and system-level solutions; and our appreciation of how the rapidly changing clinical communication landscape is influencing healthcare disparities.

The Science of Health Disparities Research, First Edition. Edited by Irene Dankwa-Mullan, Eliseo J. Pérez-Stable, Kevin L. Gardner, Xinzhi Zhang, and Adelaida M. Rosario.

20.1.2 Narrative, Elicitation, Explanatory, and Relationship Aspects

The relationship between health communication risk factors and communication-sensitive health outcomes is complex. Just as any individual patient may have multiple biologic risk factors for their health, any individual patient may face multiple communication barriers related to his/her race/ethnicity, gender, language, health literacy (HL), and social status; those of his/her provider; and the context of their healthcare setting.

Cooper and Roter offer an insightful conceptual model for the effect of race/ethnicity on patient-provider communication [6]. In their chapter and model, they describe how many patient sociodemographic factors, physician sociodemographic factors, interpersonal dynamics, and institutional resources influence the quality of communication, the quality of diagnostic and treatment decisions arising from that communication, and the quality of health outcomes arising from those decisions.

Using the example of HL, Schillinger also describes the multiple mechanisms by which a single communication risk factor may affect healthcare quality and health outcomes for vulnerable populations [7]. These mechanisms include five hypotheses:

1) *Confounding*: Health communication risk factors may be markers for sociodemographic and behavioral factors or experiences that directly or indirectly lead to morbidity and mortality.
2) *Reverse causation*: Health communication risk factors may be consequences of high disease burden or poor disease control and, thus, associated with worse health trajectories (cyclical effect).
3) *Mediation*: Health communication risk factors may affect health through behavioral and exposure-related factors at the individual and community levels that directly or indirectly lead to morbidity and mortality.
4) *Effect modification at the individual level*: Health communication risk factors may provide buffers that help maintain patients' function in the face of the physical and emotional consequences of acute and chronic disease.
5) *Effect modification at the system level*: Health communication risk factors may affect healthcare and related health outcomes through a demand-capacity mismatch, with the healthcare system placing inappropriate demands on patients, or communication resources are poorly distributed for the population with the greatest needs. This hypothesis suggests that modifications at the health system level provide intervention targets for mitigating health disparities related to health communication risk factors.

20.1.3 Communication Domains Affecting Health and Healthcare Outcomes

Three key relational constructs that mediate the path from communication to healthcare processes and outcomes are respect, trust, and participation in decision making.

20.1.3.1 Respect

Demonstrations of respect communicate to vulnerable patients that clinicians are willing to enter a relationship based on equality and dignity. Beach et al. defined respect as "a recognition of the unconditional value of patients as persons" [8]. In other words, it is the healthcare provider's duty to both feel and demonstrate respect equally to all, independent of patients' personal characteristics, behaviors, and values. Using direct observation of chronic care ambulatory visits with high sampling of African American patients, researchers have found that higher provider ratings of respect for particular patients are associated with more positive affective tone by providers and patients, less verbal dominance by providers, and more information giving and patient-centered content. Perceptions of being treated with dignity and respect have been associated with higher patient satisfaction among all racial/ethnic groups, with significant associations with self-reported adherence to doctors' treatment recommendations and greater appointment adherence among racial/ethnic minorities [9].

Meanwhile, studies have also demonstrated significant associations between patients' reports of feeling disrespected and discriminated against by healthcare providers. Perceived disrespect and discrimination due to race or socioeconomic status (SES) is associated with lower satisfaction with the healthcare system and worse health outcomes among patients with chronic illnesses [10]. In a 2001 national survey, compared with 9% of Whites, 14–20% of African Americans, Latinos, and Asians reported being treated with disrespect or unfairly or receiving worse care because of their position in society, a feeling associated with reduced self-reported adherence to recommended advice and increased delay for chronic illness care [11]. However, within particular conditions that disproportionately affect minority populations, such as sickle cell disease, perceptions of disrespect in care exceed those of a comparable African American population, suggesting an interplay of multiple factors contributing to disparities in this vulnerable population's healthcare experience [12].

Conceptions of respect may also vary across populations. African American, Latino, and White focus groups voiced similarity in the importance of feeling treated like a person and as an equal [8]. However, the behaviors recognized as demonstrating respect differed, with African American participants particularly voicing their value of physicians hearing (not dismissing) their statements and knowledge of themselves [8].

20.1.3.2 Trust

A patient who trusts a healthcare provider believes that this provider will act, both interpersonally and technically, to support the patient's interests [13]. Trust has been a significant predictor of patient satisfaction with care, continuity with a physician, and self-reported adherence [14]. Meanwhile, trust can be a moderating influence to reduce medication nonadherence due to cost [15].

Trust in physicians has been shown to be lower among racial/ethnic minorities. In a 2000–2001 national survey, minority users of healthcare services had similar responses on direct measures of trust ("to put my medical concerns above all other concerns") compared with non-Hispanic White users of healthcare services, but had lower scores on indirect measures of trust in physicians, such as when referred to specialty care, when asked to take tests of questionable necessity, or when having to deal with insurance. Investigators concluded that measuring trust in providers may yield differences across race/ethnicity and language depending on expectations about specific physician behaviors [16].

Meanwhile, challenges exist in disentangling trust in the technical vs. the interpersonal caring of a particular provider, as well as distinguishing interpersonal trust in a particular provider from overall trust in providers or the healthcare system as a whole. Black respondents in the General Social Surveys were less likely than White respondents to trust the technical judgment and interpersonal competence of doctors, and Latinos were less likely than White respondents to trust the fiduciary ethics, technical judgment, and interpersonal competence of doctors. Meanwhile, Blacks and Latinos with two parents born in the United States reported similar levels of trust in physicians, but Latinos with at least one non-native parent reported lower levels of trust [16].

20.1.3.3 Participation in Decision Making

Shared decision making—a collaborative process between patients and providers for generating treatment plans that meet patients' needs and preferences—includes mutual information sharing, discussion of the benefits and risks of different treatment options, and reaching of a decision endorsed by both the patient and provider [17]. As with earning trust, providers must demonstrate a capacity for technical skills (knowing and effectively communicating evidence-based information) and interpersonal skills (eliciting and demonstrating respect for a patient's values and concerns) [18].

Research using patient reporting has demonstrated that minority patients are less likely to report participation in decision making compared with White patients. African American patients report less participatory visits and are more likely to report that their healthcare providers made the final decision for certain chronic disease treatments compared with White patients [19]. Similarly, Hispanics and Asian American populations have been shown to be less likely to report participation and control over treatment decision making [20]. Research with direct observation suggests that both patients' and providers' perceptions of the quality of shared decision making may not match observed communication behaviors. In one study using simulated recall interviews after video-recorded visits, some low-income English- and Spanish-speaking patients rated as noncollaborative encounters in which collaborative decision making occurred, whereas some patients rated as collaborative encounters with no observed collaborative decision making [21].

Studies examining potential differences in preference for participation by race/ethnicity have yielded mixed findings. A 2002 study found that African Americans and Hispanics were more likely to prefer that physicians make final treatment decisions in general [22].

20.1.4 An Operational and Contextual Framework for Organizing Communication in the Clinical Encounter

Building on prior research [23], we have developed a framework for describing communication in the clinic setting, with a focus on chronic disease management (see Figure 20.1). This framework defines the cocreation of "shared meaning" as the most proximal visit outcome. This outcome reflects attempts to achieve patient-provider agreement in shared meaning in two domains: (i) *elicitation domains*, in which we assess disease state and symptom burden and uncover barriers to adherence, including social vulnerabilities and

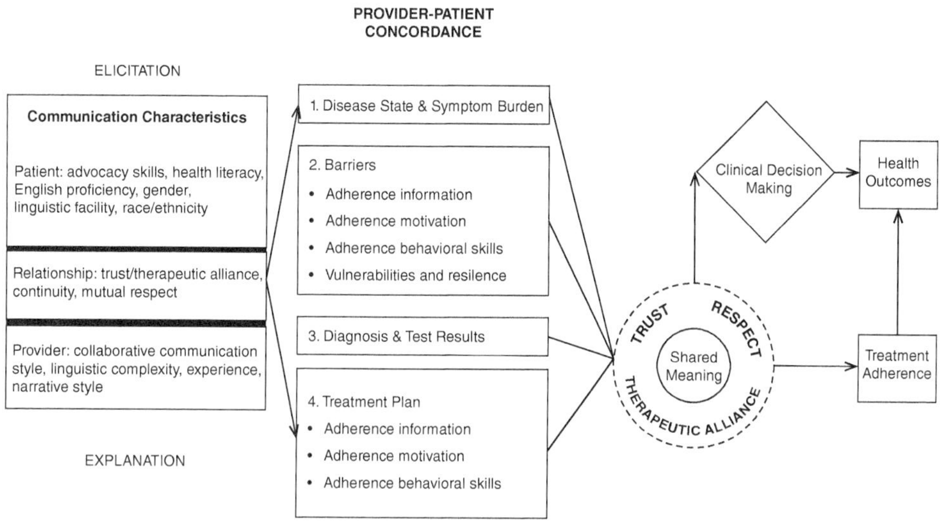

Figure 20.1 Framework for communication with vulnerable patients in the outpatient clinical encounter. *Source:* Schillinger et al. [24].

Table 20.1 The seven essential communication skills in the care of vulnerable and underserved patients.

1) Show authentic interest	Show authentic interest in patients as whole people—that you are curious about them beyond their symptoms or illnesses
2) Employ narrative methods	Coach the patient into telling his or her story
3) Utilize reflective listening	Listen for and/or explicitly elicit information about social vulnerabilities and resilience factors
4) Show empathy	Express support and compassion around the health concerns and the social vulnerabilities that may be interfering with his or her health
5) Build on strengths	Call out the resilience factors that provide hope and confidence for patients and for you, as their clinician
6) Integrate vulnerabilities and resilience	Acknowledge the social vulnerabilities and generate a shared plan to address them; integrate resilience factors into the treatment plan
7) Ensure recall and comprehension	Avoid clinical jargon; use the teach-back method

Source: Schillinger et al. [25].

resilience factors, as well as treatment-related preferences and value; and (ii) *explanatory domains*, in which we convey diagnoses and results and discuss treatment plans. Achieving shared meaning in these domains requires a combination of communication skills and a commitment to the relational aspects of communication that are mutually reinforcing (see Table 20.1). Showing authentic interest in the patient as a whole person (that you are curious about him or her beyond his or her symptoms or illnesses) and coaching the patient into telling his or her story promotes disclosure of barriers and narrows social distance while also fostering trust and a more therapeutic alliance with the patient. In what is considered a benevolent cycle, the therapeutic alliance can engender greater degrees of shared meaning, which can enhance the therapeutic alliance. The interplay between the use of narrative approaches, the cocreation of shared meaning, and the deepening of the therapeutic alliance can improve clinical decision making and resource acquisition, promote adherence, and enhance overall health. A collateral benefit is that such relationship-centered care appears to enhance clinician well-being and serve as both a preventive measure against, and a tonic to, clinician burnout [25].

The framework (see Figure 20.1) shows how communication and relational barriers such as limited HL, LEP, social distance, racial/ethnic discord, and implicit bias can impair the development of shared meaning along the path to achieving optimal health and well-being: trust and therapeutic alliance → appropriate clinical decision making → optimal adherence → health and well-being.

20.2 Risk Factors for Health Communication Disparities and Impact on Communication and Health Outcomes

20.2.1 Limited Health Literacy

It is estimated that close to half of the US adult population has LHL [3], defined by the National Academy of Medicine (NAM) as a limited capacity to obtain, process, and understand the basic health information and services needed to make informed health decisions [26]. LHL affects individuals across the spectrum of sociodemographic characteristics. However, it disproportionally affects vulnerable populations, including the elderly, the disabled, people of lower SES, ethnic minorities, those with LEP, and people with limited education [27]. In some cases, LHL has been shown to be a factor in observed healthcare disparities [28].

LHL should not be considered a diagnosis but a final common pathway and marker for or manifestation of a number of life circumstances, including limited access to education, access to poor-quality education, LEP, learning differences and disabilities, and cognitive impairment. Patients with LHL are more likely to have poor health, higher rates of chronic disease, and a nearly twofold higher mortality rate as compared to health-literate patients [29]. They are also more likely to experience disparities in health and healthcare access and have lower rates of screening and preventive services. Compared to patients with adequate HL, patients with limited LHL exhibit patterns of utilization of care reflecting a greater degree of unmet needs, such as excess emergency room visits and hospitalizations, even when comorbid conditions and health insurance status are held constant. Patients with LHL are more likely to have poorer knowledge of their disease processes, medication regimens, and medication adherence [29], and skills and methods for managing their disease [28]. LHL also has a negative effect on doctor-patient communication. Patients with LHL more often use a passive communication style with their physicians, are less likely to engage in shared decision making, and report that interactions with their physicians are not helpful or empowering. It has been estimated that LHL leads to excess health expenditures of greater than \$100 billion annually [30].

20.2.2 Limited English Proficiency and Patient-Provider Language Discordance

According to the latest US Census, 8.7% of the US population have LEP, the highest proportion being in low-income, less educated, and underserved populations [4].

The NAM-endorsed [31] screening for LEP begins by asking the US Census-LEP item: "How well do you speak English?" with response options of "not at all, not well, well, or very well." Among participants who report speaking English less than "very well," language preference is assessed by asking: "In general, in what language do you prefer to receive your medical care?" This approach was validated against patients' self-reported ability to communicate effectively in English with their physicians in a 2008 study of Spanish-speaking ambulatory and hospitalized patients [32].

Research has validated the relationship of LEP with communication-sensitive health outcomes. LEP patients with language-discordant physicians were more likely to have poor glycemic control. National Health and Nutrition Examination Survey participants with LEP were more likely to have poorly controlled hypertension than those with adequate English proficiency. LEP patients have been shown to be more likely to have an unplanned emergency department revisit within 72 hours [33].

The interactions between LHL and LEP are complex and, because of their frequent co-occurrence, are worthy of special mention. One study assessed the effects of LHL among different language categories, including English speakers, Spanish speakers with Spanish-speaking physicians (Spanish-concordant), and Spanish speakers without Spanish-speaking physicians (Spanish-discordant) [34]. Overall, LHL was associated with poor receptive and proactive patient communication. Spanish concordance and discordance, overall, were associated with poor interactive (bidirectional) communication. However, among English speakers, LHL was associated with poor receptive and proactive, but not interactive, communication. Among Spanish-concordant participants, LHL was associated with poor proactive and interactive, but not receptive, communication. Spanish-discordant participants reported the worst communication for all types, independent of HL status.

These findings suggest that LHL impedes patient-physician communication, but its effects vary with language concordance and for communication subtypes. For language discordant dyads, language barriers may supersede LHL in impeding interactive communication. Patient-clinician communication interventions for diverse populations need to consider HL, language concordance, and communication type.

20.2.3 Discordance in Race/Ethnicity

Concordance is defined as "a similarity or shared identity between physician and patient based on a demographic attribute." Researchers theorize that the association between concordance and patient outcomes is mediated through "perceptions of relational similarity" (i.e., that patient assumptions of shared beliefs and values improve patient trust in the provider) [35]. In addition to language concordance, patient–provider concordance on other characteristics, including trust in providers, continuity with a care provider, and satisfaction with the quality of communication and care, has emerged as an important mediator of the patient outcomes.

Decades of research have examined the role of concordance in race/ethnicity, particularly in African American patients. On postvisit surveys, African American patients rated their physicians lower in participatory decision making style, and those in racially concordant relationships had more participatory visits than those in discordant relationships. In studies analyzing actual content of ambulatory visits, race-concordant visits were longer and characterized by more positive affective tone of voice, as rated by external observers. Patients in race-concordant visits reported greater participation and higher satisfaction with their providers, independent of the externally rated quality of communication demonstrated during the visits [36].

Research has examined the impact of race concordance on process outcomes. Studies have found that Black respondents with racially concordant physicians were more likely to rate their physicians as excellent and to report receiving both preventive care and all needed medical care. They were also more likely to report using needed health services, less likely to postpone or delay seeking care, and more likely to report the use of a higher volume of health services [36].

20.2.4 Implicit Bias

Implicit or unconscious bias refers to positive or negative beliefs and feelings about group attributes, such as race/ethnicity or gender, that exist outside conscious awareness and that may influence a person's judgments and behaviors [5]. Conclusions about groups or ethnicities are influenced by an individual's past experiences and perceptions of others rather than through reasoning or active thought. Implicit biases are normal and common, as evidenced by an implicit preference for White people by 68% of respondents in a nonrandom US online study. Implicit bias is rooted in stereotyping, a cognitive process in which individuals use a social category to acquire, process, and recall information about people. This processing pattern functions unconsciously to help individuals organize complex information, particularly in stressful or time-pressured situations [37].

Implicit bias among clinicians is increasingly recognized as another important factor contributing to racial disparities in healthcare. A key study highlighting this risk was the 1999 study by Schulman et al., which revealed that clinicians were more likely to refer for cardiac catheterization men and White patients than women and Blacks, regardless of their clinical characteristics. While this study used video vignettes to assess potential bias, the majority of recent, high-quality published studies have measured implicit bias using the Implicit Association Test (IAT), a computerized categorization task that measures the relative strength of associations between pairs of concepts by asking participants to quickly sort images/words on the screen into one of two categories [38].

A 2015 review included 15 US studies with convenience sampling of healthcare providers and trainees from multiple disciplines and fields. Fourteen studies showed low to moderate levels of implicit racial/ethnic bias, with similar scores to the general population. Evidence was stronger for associations with patient–provider communication than for clinical decision making or outcomes [37].

A 2017 systematic review of implicit bias among physicians and nurses included 42 eligible international studies, with implicit bias found in 35. In these: (i) rates often were similar to rates of implicit bias in the general population; (ii) significant interaction effects were related to other sociodemographic (e.g., gender or SES) or behavioral (e.g., challenging) characteristics; (iii) rates of implicit bias were related to the sociodemographic characteristics of the physicians/nurses, including male and international medical graduate status; and (iv) stronger real-world evidence exists for the relationship between implicit bias and negative communication behaviors, while evidence for direct impact on clinical decision making is mixed [5].

Systematic reviews noted future research directions to improve the scientific literature on implicit bias in healthcare [5]. The majority of research to date is cross-sectional and includes small sizes drawn using convenience sampling. In addition, IAT is only one measure of implicit bias, with good internal consistency but only moderate test-retest reliability, raising concerns about the uncaptured contextual factors that may influence implicit bias [37]. Also, the use of vignettes to assess clinical decision making may have limited validity as a proxy for real-world decision making, particularly if participants are made aware of the patient characteristic under study [5]. Most research focuses on bias against Black Americans, requiring further study involving other racial/ethnic populations and across other dimensions. Future research should include rigorous documentation of the research methodology (e.g., response rates, power calculations); longitudinal studies; more representative sampling strategies; use of validated outcomes measures, including patient ratings of the quality of communication and care, objective measurement of communication behaviors, objective and validated process measures, and patient-important clinical outcomes; quantifying the potential interactions between different characteristics, such as race/ethnicity, gender, and socioeconomic status; and, ultimately, rigorous evaluations of the impact of implicit bias interventions on clinical decision making and patient-important outcomes [5].

Research is less clear on the effective methods for reducing implicit bias in healthcare settings. Improving mindfulness and cross-cultural communication skills may benefit patients in discordant relationships by enhancing trust. A variety of undergraduate, graduate, and continuing medical education courses aim to enhance the capacity of healthcare providers to engage in empathic and effective therapeutic alliances with patients, bridge relational discordance, increase cultural humility, and build provider resilience in underresourced clinical settings. High-quality evidence is needed to determine whether such interventions can affect patient-important outcomes.

20.3 An Example of Systems-based Approaches: Creating Health-Literate Healthcare Systems

20.3.1 Defining Health-Literate Organizations

HL is a dynamic state that represents the balance (or imbalance) between an individual's capacity to comprehend and apply health-related knowledge to health-related decisions and to acquire health-related skills and the HL-related demands and attributes of the healthcare system. Healthcare organizations need to develop and implement HL strategies to enable patients and families to optimally access and benefit from healthcare services and to successfully interact with healthcare entities.

An NAM paper identified and described a set of activities and attributes that diverse healthcare organizations can engage in and aspire to in order to mitigate the negative consequences of LHL and improve access, quality, safety, and value of healthcare services [39]. Those organizations that have committed to improving and re-engineering to better accommodate the communication needs of populations with LHL are known as "health-literate healthcare organizations," reinforcing the notion that the healthcare sector shares significant responsibility in promoting HL. The most proximate goals of these organizational investments are to maximize the extent of patients' and families' capacities to:

a) Comprehend and engage in recommended preventive health behaviors and receive preventive healthcare services if desired
b) Recognize changes in health states that require attention and access healthcare services accordingly
c) Develop meaningful, ongoing relationships with healthcare providers based on open communication and trust
d) Obtain timely and accurate diagnoses for both acute and chronic health conditions
e) Comprehend the meaning of their illness, their options for treatment, and anticipated health outcomes
f) Build and refine the skills needed to safely and effectively manage their conditions at home and communicate with the healthcare team when their illness trajectory changes
g) Report their communication needs or comprehension gaps
h) Make informed healthcare decisions that reflect their values and wishes
i) Effectively navigate transitions in care.

In addition, these investments can enable people to make more appropriate healthcare coverage choices based on their personal or family health needs, better comprehend the range of benefits and services available to them and how to access them, and be more aware of the financial implications of their healthcare choices so as to improve decision making. Herein, we present intervention approaches that have shown benefit at the clinician-patient, system-patient, and community-patient levels (see Table 20.2).

20.3.2 Clinician-Patient Level Interventions

Employing effective communication techniques may be one of the most important interventions to reduce health disparities related to LHL. The basic principles of effective clinician-patient communication have been developed within the adult education fields and adopted by the HL scientific community. Recommendations include the following [40]:

- Patient-centered communication: This allows clinicians to tailor their messages to individual patients.
- Clear health communication: Clinicians should slow down their speech and use plain, or what is often called "living-room," language and avoid jargon.

Table 20.2 Interventions for patients with limited health literacy.

Clinician-patient level
Patient-centered communication
Assess what patients already know ("What do you already know about...?")
Identify "What do I need to do?" rather than "What do I need to know?"
Clear health communication
Slow down. Use plain language, avoid jargon (e.g., "not cancer" instead of "benign"). Attempt to match patient's vocabulary. Keep number of points to less than three.
Confirmation of understanding
"What questions do you have?" Teach back, teach-to-goal (does not increase visit duration).
Ask patients to say back or demonstrate what was just taught ("I've just said a lot of things. To make sure I did a good job and explained things clearly, can you describe to me...?").
Reinforcement
Use multiple modalities (pictures, graphs, drawings) and written information
Keep points to a minimum
Write legibly in a large font
Use the active voice (e.g., "Take your medicine at noon")
Use short sentences, less than eight words at a 5th-grade level
Use patients' support network/family
Numeracy and risk
Use consistent denominator
Keep time span ~10 years and consistent
Provide absolute risks, not relative risks
Present risk frequencies, not percentages
Avoid solely positive or negative framing
Medication reconciliation
Fit dosing schedule around patient's daily routine
Simplify regimens as much as possible
Confirm regimen dosage concordance (e.g., "Tell/show me *exactly* how you take your...")
Use pillboxes, Medi-sets, and other medication organizers
Use visual medication schedules with computer programs or affix pills to schedule
Make follow-up appointments to assess compliance and side effects
Include pharmacists if possible
Geriatrics
Accommodate for visual impairment. Updated eye examination. Use 14-point non-serif fonts (Arial, Helvetica).
Use nonglossy materials with bright contrasting color. Use light that cuts down on glare. Use magnifying lenses.
Accommodate for hearing impairment. Updated audiologist appointments.
Face the patient, put patient's back to wall
Use assistive devices (Pocket Talker or hearing aids)

(*Continued*)

Table 20.2 (Continued)

Accommodate for cognitive impairment
Teach back (beneficial for mild cognitive impairment)
For dementia, elicit help from family and multidisciplinary team
System–patient Level
Health education materials
Large font. Text at 5th-grade level. Pictures. Clear headings and layout. Include target population in design of tools.
Use of cocreated and user-tested video material
Use of narratives to teach
Leverage accessible health information technology
Tailor mHealth for literacy, language, and culture for target populations
Conduct mHealth usability testing with both patients and healthcare team members
Integrating mHealth platforms with existing healthcare system infrastructure
Position exam room computers and optimizing electronic health record content, interfaces, and workflows to promote patient-centered communication
Medication drug labels
Use concrete instructions ("Take at 8 a.m., 4 p.m., and midnight")
Label should focus on the drug and the dosage.
Use universal pictures that match the text
Drug label, pill bottle, and health education should be designed together and match
Employ the Universal Medication Schedule (UMS) for prescribing and labeling instructions
Employ language-concordant prescribing and labeling instructions
Disease self-management support
Proactive, used to prevent poor outcomes before they happen
Disease-specific (e.g., diabetes or congestive heart failure)
Tailored to literacy and patient-perceived barriers
Multidisciplinary disease management education followed by telephone call back
Automated telephone calls
Use of peer educators and navigators
Creating an empowering environment
Make signs clear
Make forms easy to read
Make the telephone triage system easy to navigate
Teach patients to ask: (i) What is my main problem? (ii) What do I need to do about the problem? (iii) Why is it important for me to do this?
Clinician training
Clear health education while clinicians are in training
Incentivize with continuing medical education credits
Require for board examinations
Community-patient level
Partner with community adult literacy organizations
Use mass media to disseminate health information

Source: Sudore and Schillinger [40].

- Confirmation of understanding: The clinician should ask patients to restate or demonstrate the knowledge or technique just taught.
- Reinforcement: In addition to clear verbal instruction, clinicians may use additional reinforcement techniques, such as drawing pictures or graphs.
  See Table 20.2 for more details.

20.3.3 System-Patient-Level Interventions

At the system level, helpful interventions for patients with LHL include designing and offering easy-to-understand health education materials, improving medication drug labeling, designing and offering chronic disease management programs, creating an empowering environment, and offering communication training to clinicians.

20.3.3.1 Employing Appropriate Health Education Materials

While knowledge gained from easy-to-read materials increases for patients with both limited and adequate HL, the size of the knowledge gap between them often remains the same [41]. No studies have demonstrated that better written materials improve long-term outcomes.

20.3.3.2 Leveraging Accessible Health Information Technology

With the appropriate design and implementation, health information technology (HIT) facilitates patient communication by improving patient confidence, self-management, quality of life, and disease control (see Chapter 22).

20.3.3.3 Standardizing Medication Drug Labels and Drug Information

As a result of difficult language and icons that do not match the text or do not have universal meaning, patients with LHL are most likely to misinterpret medication labels [42]. Research supports a need for implementing language-concordant prescription drug instructions in pharmacy practices to promote safe medication use among LEP patients.

20.3.3.4 Employing Disease Management Programs

Literacy-appropriate disease management programs, such as for heart failure or diabetes case management, can improve the health of patients with LHL by lowering hospitalization rates and all-cause mortality.

20.3.3.5 Creating an Empowering Environment

Given the shame experienced by many patients with LHL and the implicit power imbalance in patient-clinician relationships, creating an empowering environment within the clinic or hospital setting is recommended [43].

20.3.3.6 Clinician Communication Training

Many of the skills inherent to clear health communication can be offered as part of continuing medical education. Courses are often offered at national medical meetings. HL communication trainings are being incorporated in the Accreditation Counsel for

Graduate Medical Education curricula and board examinations on clinician-patient communication. Clinician educators have begun to include HL in medical school curricula, with promising results.

20.3.4 Community-level Interventions

Referrals to adult education and literacy programs, use of lay health educators and health navigators, and harnessing of the media may be helpful in reducing HL-related disparities. While literacy training for adults tends to increase literacy level by only one or two grade levels per year, such training may provide peer support and increase patients' confidence to ask questions within the healthcare context, including in English as a Second Language (ESL) settings. Lay health educators, or navigators, are often trusted community members who can communicate across culture, language, and literacy levels. This approach has been successful in improving knowledge, behavior change, and health outcomes and has been shown to deliver cost savings. Social media campaigns that bring together art and public health, when created by the target populations, have been shown to increase public HL in the context of diabetes prevention [43]. There is also good evidence to suggest that health information coverage by the mass media can reach communities of all HL levels and may be powerful educational tools for public health campaigns.

20.4 Interventions to Mitigate Other Communication Risk Factors

As described above for HL, systems-level factors influence the impact of communications risk factors. Below, we examine the evidence for potentially modifiable systems-level factors that may provide intervention opportunities for other communications risk factors.

20.4.1 Cultural Competence of Healthcare Providers and Organizations

With the growing recognition that disparities in healthcare quality may result from differences in the backgrounds, experiences, and values between patients and healthcare providers, the movement to improve the cultural competence of healthcare providers has infused health profession education. However, little research has examined how the cultural competence of providers is associated with the quality of care [44].

The cultural competence of a health system can be assessed by its workforce diversity; availability and quality of language assistance for LEP patients; continuous professional development in the delivery of culturally and linguistically appropriate services (CLAS); partnership with communities and inclusion of community health workers; and availability of quality data stratified by race/ethnicity, language, and other important population characteristics [44]. The 15 National Standards for CLAS in Health and Health Care, developed by the Office of Minority Health at the US Department of Health and Human Services,[1]

1 http://www.thinkculturalhealth.hhs.gov/clas/standards

have been adopted to degrees across healthcare systems, with only 13% of hospitals reporting adherence to all four of the language-related CLAS standards in a 2010 survey [45].

20.4.2 Diversity of Health Professional Workforce

Expanding workforce diversity among racial/ethnic minorities in the healthcare professions could be a tool to mitigate disparities in patient outcomes (see Chapter 26).

20.4.3 Building the Healthcare Team: Professional Medical Interpreters and Lay Health Educators and Patient Navigators

The environment in which care is delivered is strongly influenced by the healthcare team supporting and extending the care between patients and their healthcare providers. Professional medical interpreters and lay health educators who are healthcare team members with linguistic and cultural concordance with patients serve important roles in facilitating communication and empowering patients to be active participants in healthcare decision making.

20.5 Future Directions for Communications Research: Measurement and Implementation

20.5.1 Health Literacy

Research on HL, and on HL interventions, is largely in its infancy, and there are a number of opportunities for expanding our knowledge base regarding what works and how effective and scaled interventions can be implemented.

20.5.1.1 Measurement

How best to measure patient HL, and whether or not HL measures are detecting true differences in capacities and skills, is controversial. A recent review of all HL research measures found that at least 51 unique measures have been created and employed, including a number in Spanish, with virtually all requiring paper-and-pencil responses requiring up to 1 hour each to administer. Of the 51 measures, 26 concern general HL, 15 measure disease- or content-specific HL, and 10 measure specific subpopulations [46].

Health disparities are produced and perpetuated by multilevel forces operating at the individual, family, health system, community, and public policy levels that mutually reinforce each other to produce injustice and perpetuate disparities. Since conventional literacy assessments are bound by cultural and linguistic assumptions derived from the dominant, majority population, more research is needed to assess patient HL in a comprehensive, holistic, and unbiased manner, and to expand the assessment of reliability and validity across subgroups of interest so as to avoid misattributing health disparities solely to limited HL. For this reason, we recommend that investigators do *not* use HL measures that require proper pronunciation of medical terms as a mean to assess HL. In selecting which HL measures to employ, investigators need to carefully balance the focus of the research objective (general- vs. disease-specific HL) with the number of HL domains it purports to measure, the characteristics of the target population, and feasibility concerns [46]. While greater attention is being paid to the communication attributes of clinicians and healthcare organizations as they relate to patient HL, there has been little work to operationalize a measure of clinician or systems responsiveness to the needs of populations with LHL. This has hindered progress in reducing literacy and racial and ethnic disparities in healthcare [47].

20.5.1.2 Verbal, Auditory, and Other Cognitive Skills

There has been little research on how difficulties with speaking and listening can both constitute HL and result in LHL. Understanding more deeply how LHL impedes verbal communication in the clinical encounter is of paramount importance. Studying the impacts of learning differences and disabilities, and mild cognitive impairment, on the acquisition of health-related skills will significantly advance our understanding of LHL and enable more tailored intervention designs [48].

20.5.1.3 Study Designs

Many studies of interventional modalities are limited by a lack of stratification by HL level, randomized or blinded design, or the use of multiple interventions. Without this information, it is difficult to identify the component that was most helpful [49] and evaluate any subgroup effects. Future research should examine more robustly whether HL-sensitive interventions can reduce racial and ethnic disparities in healthcare outcomes.

20.5.1.4 Mobile Health Technology

There is much to be learned regarding the development and implementation of mobile health (mHealth) technologies that could be useful for patients with LHL and that could reduce HL-related disparities. Significant investments need to be made to learn whether cocreation, user-centered design, and training interventions can enhance the adoption and utility of such technologies. In addition, while mHealth-facilitated, informal (peer) support services appear feasible, their impact on patients' self-management or health outcomes has not been demonstrated. While access to smartphones with reliable data plans is becoming more common in low-income communities, no large studies of smartphone applications have shown that they can improve outcomes in vulnerable populations.

20.5.1.5 Health Communication Messengers and Vehicles

Social media provide an important opportunity to empower the public to participate more fully in health promotion. Whether cocreation of health-related content can be shown to be more appealing and effective than content created by health education specialists with HL design principles is a critical research question.

20.5.2 Sociodemographic Concordance

Various methodological challenges arise in research on patient-provider concordance. Thornton et al. generated an additive composite measure of social concordance in relation to "status homophily" based on four dimensions (gender, ethnicity, age, and educational attainment) [50]. Compared with patients with high social concordance with their providers, patients with low social concordance demonstrated less positive affect in interactions and offered lower ratings of the quality of care. Future studies should continue to examine interactions across types of concordance and social distance and their impact on communication and outcomes.

20.5.3 Implicit Bias

(See Section 20.2.4)

20.5.4 Direct Observation of Health Communication Behaviors

Self-reported communication behavior and postvisit satisfaction may differ from observable communication behaviors and obscure important differences that may exist, particularly in interventional research. Future research should attempt to understand why and how perceived experiences of communication often differ from directly observed differences.

20.5.5 Patient Experience and Patient-Reported Outcomes Measurement

Patient experience measures and patient-reported communication outcome measures need to be valid and reliable for diverse populations. Future research should continue to develop patient-reported outcomes measures with validity and reliability across different LHL and LEP populations.

20.6 Key Points

- Communication and relational barriers, such as LHL, language discordance, racial/ethnic discordance, and implicit bias, can serve as important mediators of healthcare disparities.
- A growing body of practice-based research provides insights into how clinicians can change their behaviors and healthcare systems can be re-engineered to mitigate or eliminate communication-related disparities.
- Further research can build on our growing understanding of the most effective individual-, system-, and community-level solutions to improve therapeutic alliances with vulnerable populations, and thereby advance health equity.

Disclaimer

The views and opinions expressed in this chapter are those of the authors only and do not necessarily represent the views, official policy, or position of the U.S. Department of Health and Human Services or any of its affiliated institutions or agencies.

References

1 Stewart, M.A. (1995). Effective physician-patient communication and health outcomes: a review. *Canadian Medical Association Journal* 152 (9): 1423–1433.

2 Schillinger, D., Barton, L.R., Karter, A.J. et al. (2006). Does literacy mediate the relationship between education and health outcomes? A study of a low-income population with diabetes. *Public Health Reports* 121 (3): 245–254.

3 Paasche-Orlow, M.K., Parker, R.M., Gazmararian, J.A. et al. (2005). The prevalence of limited health literacy. *Journal of General Internal Medicine* 20 (2): 175–184.

4 US Census Bureau (2013). *Language Use in the United States: 2011* (ed. C. Ryan) American Community Survey Reports, 1–16. Washington, DC: US Dept. of Commerce.

5 FitzGerald, C. and Hurst, S. (2017). Implicit bias in healthcare professionals: a systematic review. *BMC Medical Ethics* 18 (1): 19.

6 Cooper, L.A. and Roter, D.L. (2003). Patient-provider communication: the effect of race and ethnicity on process and outcomes. In: *Unequal Treatment: Confronting Racial and Ethnic Disparities in Healthcare*, 552–593. Washington, DC: National Academies Press.

7 Schillinger, D. (2007). Literacy and health communication: reversing the 'inverse care law'. *American Journal of Bioethics* 7 (11): 2.

8 Beach, M.C., Branyon, E., and Saha, S. (2017). Diverse patient perspectives on respect in healthcare: a qualitative study. *Patient Education and Counseling* 100 (11): 2076–2080.

9 Flickinger, T.E., Saha, S., Roter, D. et al. (2016). Respecting patients is associated with more patient-centered communication behaviors in clinical encounters. *Patient Education and Counseling* 99 (2): 250–255.

10 Lyles, C.R., Karter, A.J., Young, B.A. et al. (2011). Patient-reported racial/ethnic healthcare provider discrimination and medication intensification in the Diabetes Study of Northern California (DISTANCE). *Journal of General Internal Medicine* 26 (10): 1138–1144.

11 Blanchard, J. and Lurie, N. (2004). R-E-S-P-E-C-T: patient reports of disrespect in the health care setting and its impact on care. *Journal of Family Practice* 53 (9): 721–730.

12 Haywood, C., Bediako, S., Lanzkron, S. et al. (2014). An unequal burden: poor patient-provider communication and sickle cell disease. *Patient Education and Counseling* 96 (2): 159–164.

13 Thom, D.H. and Campbell, B. (1997). Patient-physician trust: an exploratory study. *Journal of Family Practice* 44 (2): 169–176.

14 Thom, D.H., Ribisl, K.M., Stewart, A.L., and Luke, D.A. (1999). Further validation and reliability testing of the Trust in Physician Scale. The Stanford Trust Study Physicians. *Medical Care* 37 (5): 510–517.

15 Piette, J.D., Heisler, M., Krein, S., and Kerr, E.A. (2005). The role of patient-physician trust in moderating medication nonadherence due to cost pressures. *Archives of Internal Medicine* 165 (15): 1749–1755.

16 Stepanikova, I., Mollborn, S., Cook, K.S. et al. (2006). Patients' race, ethnicity, language, and trust in a physician. *Journal of Health and Social Behavior* 47 (4): 390–405.

17 Peek, M., Lopez, F., Williams, H. et al. (2016). Development of a conceptual framework for understanding shared decision making among African-American LGBT patients and their clinicians. *Journal of General Internal Medicine* 31 (6): 677–687.

18 Hargraves, I., LeBlanc, A., Shah, N.D., and Montori, V.M. (2016). Shared decision making: the need for patient-clinician conversation, not just information. *Health Affairs* 35 (4): 627–629.

19 Ratanawongsa, N., Zikmund-Fisher, B.J., Couper, M.P. et al. (2010). Race, ethnicity, and shared decision making for hyperlipidemia and hypertension treatment: the DECISIONS survey. *Medical Decision Making* 30 (5 Suppl): 76S.

20 Wallace, L.S., DeVoe, J.E., Rogers, E.S. et al. (2007). The medical dialogue: disentangling differences between Hispanic and non-Hispanic whites. *Journal of General Internal Medicine* 22 (11): 1538–1543.

21 Saba, G.W., Wong, S.T., Schillinger, D. et al. (2006). Shared decision making and the experience of partnership in primary care. *Annals of Family Medicine* 4 (1): 54–62.

22 Levinson, W., Kao, A., Kuby, A., and Thisted, R.A. (2005). Not all patients want to participate in decision making. A national study of public preferences. *Journal of General Internal Medicine* 20 (6): 531–535.

23 Sarkar, U., Schillinger, D., Bibbins-Domingo, K. et al. (2010). Patient-physicians' information exchange in outpatient cardiac care: time for a heart to heart? *Patient Education and Counseling* 85 (2): 173–179.

24 Schillinger, D., McNamara, D., Crossley, S., and Karter, A.J. (2017). The next frontier in communication and the ECLIPPSE study: bridging the linguistic divide in secure messaging. *Journal of Diabetes Research* 1: 1–9.

25 Schillinger, D., Villela, T., Ratanawongsa, N., and Saba, G. (2016). Creating a context for effective intervention in the clinical care of vulnerable patients. In: *Medical Management of Vulnerable and Underserved Patients: Principles, Practice, & Populations* (eds. T.E. King Jr., M. Wheeler, A. Bindman, et al.), 104–114. New York, NY: McGraw-Hill Com.

26 Institute of Medicine (2004). *Health Literacy: A Prescription to End Confusion*. Washington, DC: National Academies Press.

27 Kutner, M., Greenberg, E., Jin, Y., and Paulsen, C. (2006). *The Health Literacy of America's Adults: Results from the 2003 National Assessment of Adult Literacy*. Washington, DC.: US Department of Education.

28 Cavanaugh, K., Huizinga, M.M., Wallston, K.A. et al. (2008). Association of numeracy and diabetes control. *Annals of Internal Medicine* 148 (10): 737–746.

29 DeWalt, D.A., Berkman, N.D., Sheridan, S. et al. (2004). Literacy and health outcomes: a systematic review of the literature. *Journal of General Internal Medicine* 19 (12): 1228–1239.

30 Vernon, J., Trujillo, A., Rosenbaum, S., and DeBuono, B. (2007). *Low Health Literacy: Implications for National Health Policy*. Washington, DC: Health Policy and Management Faculty Publications.

31 US Dept. of Health & Human Services (2009). *Race, Ethnicity, and Language Data: Standardization for Health Care Quality Improvement*. Washington, DC.: National Academies Press.

32 Karliner, L.S., Napoles-Springer, A.M., Schillinger, D. et al. (2008). Identification of limited English proficient patients in clinical care. *Journal of General Internal Medicine* 23 (10): 1555–1560.

33 Ngai, K.M., Grudzen, C.R., Lee, R. et al. (2016). The association between limited English proficiency and unplanned emergency department revisit within 72 hours. *Annals of Emergency Medicine* 68 (2): 213–221.

34 Sudore, R.L., Landefeld, C.S., Pérez-Stable, E.J. et al. (2009). Unraveling the relationship between literacy, language proficiency, and patient-physician communication. *Patient Education and Counseling* 75 (3): 398–402.

35 Street, R.L., O'Malley, K.J., Cooper, L.A., and Haidet, P. (2008). Understanding concordance in patient-physician relationships: personal and ethnic dimensions of shared identity. *Annals of Family Medicine* 6 (3): 198–205.

36. Cooper, L.A., Roter, D.L., Johnson, R.L. et al. (2003). Patient-centered communication, ratings of care, and concordance of patient and physician race. *Annals of Internal Medicine* 139 (11): 907–915.

37 Hall, W.J., Chapman, M.V., Lee, K.M. et al. (2015). Implicit racial/ethnic bias among health care professionals and its influence on health care outcomes: a systematic review. *American Journal of Public Health* 105 (12): 60.

38 Schulman, K.A., Berlin, J.A., Harless, W. et al. The effect of race and sex on physicians' recommendations for cardiac catheterization. *New England Journal of Medicine* 340 (8): 618–626.

39 Brach, C., Dreyer, B.P., and Schillinger, D. (2014). Physicians' roles in creating health literate organizations: a call to action. *Journal of General Internal Medicine* 29 (2): 273–275.

40 Sudore, R.L. and Schillinger, D. (2009). Interventions to improve care for patients with limited health literacy. *Journal of Clinical Outcomes Management* 16 (1): 20–29.

41 Davis, T.C., Berkel, H.J., Arnold, C.L. et al. (1998). Intervention to increase mammography utilization in a public hospital. *Journal of General Internal Medicine* 13 (4): 230–233.

42 Wolf, M.S. and Bailey, S.C. (2007). Improving prescription drug labeling. *North Carolina Medical Journal* 68 (5): 340–342.

43 Schillinger, D., Ling, P.M., Fine, S. et al. (2017). Reducing cancer and cancer disparities: lessons from a youth-generated diabetes prevention campaign. *American Journal of Preventive Medicine* 53 (3): S103–S113.

44 Saha, S., Beach, M.C., and Cooper, L.A. (2008). Patient centeredness, cultural competence and healthcare quality. *Journal of the National Medical Association* 100 (11): 1275–1285.

45 Diamond, L.C., Wilson-Stronks, A., and Jacobs, E.A. (2010). Do hospitals measure up to the national culturally and linguistically appropriate services standards? *Medical Care* 48 (12): 1080–1087.

46 Haun, J.N., Valerio, M.A., McCormack, L.A. et al. (2014). Health literacy measurement: an inventory and descriptive summary of 51 instruments. *Journal of Health Communication* 19 (Suppl 2): 302–333.

47 Schillinger, D. and Sarkar, U. (2009). Numbers don't lie, but do they tell the whole story? *Diabetes Care* 32 (9): 1746–1747.

48 Rossignol, L.N. and Paasche-Orlow, M.K. (2013). Empowering patients who have specific learning disabilities. *Journal of the American Medical Association* 310 (14): 1445–1446.

49 Pignone, M., DeWalt, D.A., Sheridan, S. et al. (2005). Interventions to improve health outcomes for patients with low literacy. A systematic review. *Journal of General Internal Medicine* 20 (2): 185–192.

50 Thornton, R.L., Powe, N.R., Roter, D., and Cooper, L.A. (2011). Patient-physician social concordance, medical visit communication and patients' perceptions of health care quality. *Patient Education and Counseling* 85 (3): 201.

21

Comparative Effectiveness Research in Health Disparity Populations

William V. Padula[1, 2], Rick A. Berzon[3], Priscah Mujuru[4], and David O. Meltzer[5]

[1] *Department of Pharmaceutical and Health Economics, School of Pharmacy, University of Southern California, Los Angeles, CA, USA*
[2] *Leonard D. Schaeffer Center for Health Policy and Economics, Los Angeles, CA, USA*
[3] *Division of Extramural Scientific Programs, National Institute on Minority Health and Health Disparities, National Institutes of Health, Bethesda, MD, USA*
[4] *National Institute on Minority Health and Health Disparities, National Institutes of Health, Bethesda, MD, USA*
[5] *Department of Medicine, The University of Chicago, Chicago, IL, USA*

21.1 Introduction

In this chapter, we introduce readers to essential and emerging methods and applications of comparative effectiveness research (CER) in populations that experience health disparities, such as racial and ethnic minorities and populations that experience social or economic disadvantage. CER is designed to inform healthcare decisions by providing evidence on the effectiveness of treatment options and healthcare delivery services, including drugs, medical devices, tests, and surgeries.

First, we provide a brief background on the importance of CER in health disparities research areas that have been identified as priorities. We highlight some of the most promising opportunities to apply CER to achieve more equitable healthcare in health disparity populations.

Second, we review healthcare interventions that CER has shown to be of value to minority subgroups and through which we have advanced our understanding of the root causes of health disparities. Examples of such root causes include conditions common among all patient groups (e.g., cancer, cardiovascular disease) for which analysis of subgroups affected by health disparities may be preferable, and other conditions (e.g., sickle cell) that are commonly experienced by minority subgroups and which necessitate their own CER studies.

Third, we define the social determinants of health and their associations to the multisectoral impact of targeted improvements of public health for health disparity populations.

Finally, we analyze the value of CER in the context of healthcare delivery to health disparities populations and examine whether CER may direct resources toward higher value care for minority populations and reduce the use of lower value care for these same groups.

The Science of Health Disparities Research, First Edition. Edited by Irene Dankwa-Mullan, Eliseo J. Pérez-Stable, Kevin L. Gardner, Xinzhi Zhang, and Adelaida M. Rosario.

21.2 Background

CER is used to evaluate the efficacy and effectiveness of multiple drugs, medical devices, diagnostic tests, medical and surgical treatments, or healthcare delivery methods without considering the cost [1]. CER has emerged as an important strategy in US healthcare to explain effective healthcare interventions that improve quality and reduce overall health costs. The United States has seen an increasing fraction of its economy spent on healthcare over the past two decades—currently at over 18% of Gross Domestic Product (GDP)—often on unnecessary procedures [2]. CER can reduce wasteful spending by limiting the use of interventions that are less effective and more costly with respect to both minority and general populations.

CER methods can be used to ensure that effective care is accessible for minority health and health disparity (MHHD) populations even when it may not benefit the health of the general population. Likewise, CER can help to identify effective care for general populations that may not benefit MHHD populations, since these applications of misuse in minority groups represent some of the causes of wasteful spending and poor outcomes. CER is unique to the United States, since many other nations use health cost-effectiveness analyses to restrict the availability of health interventions to only those strategies that are dominant. CER can benefit patients in the United States, especially MHHD populations, by determining ways to improve the accessibility of high-quality, effective treatments that hold greater value to all patients, as well as determine when older technologies should be phased out because they result in lower value care.

The application of CER in the study of MHHD subgroups is complex given that many advances present a heterogeneous profile of efficacy, safety, and effectiveness for different patient subgroups, including African Americans; Latinos; American Indians and Alaska Natives (AI/AN); Native Hawaiians and Pacific Islanders; and Lesbian, Gay, Bisexual, Transgender, and Questioning (LGBTQ) persons. Understanding the differences among these MHHD subgroups is important for policymakers as they determine the future of investment in healthcare advances. CER ensures the delivery of care to patients that holds good value for MHHD subgroups, information that may not be intuitively valuable to the general public.

21.3 CER Study Designs

CER should be conducted using a standard approach that takes into consideration a combination of experimental and observational methods, depending on the availability of data (Table 21.1). If a study design can reflect real-world outcomes in a controlled patient setting, randomized controlled trials (RCTs) are used to evaluate the efficacy, and sometimes the effectiveness, of healthcare advances. They can also be more difficult and costly to conduct in MHHD populations, depending on the sample size. Generalizability is always a concern in RCTs and may be a special concern in MHHD populations, to the extent to which the population recruited for an RCT may be less than fully representative of targeted subgroups due to issues such as mistrust of medical research or socioeconomic barriers to care. Prospective and retrospective observational studies, including cohort and

Table 21.1 Matrix of comparative effectiveness research study designs and types of interventions to consider.

CER study design	Classification of outcome measured	Example intervention types	Example limitations	Example study
Randomized controlled trial	• Safety • Efficacy • Limited effectiveness	• Pharmaceuticals and medical devices • Surgical intervention • Medical devices • Diagnostic tests	• May lack generalizability to real-world outcomes	The comprehensive care physician (CCP) model [3]
Cohort study (observational)	• Effectiveness	• Pharmaceuticals and medical devices • Health outcomes related to medical and surgical interventions	• Provider and patient bias to intervention is known • Selection bias potential without randomization	Outcome of sickle cell anemia: a four-decade observational study of 1056 patients [4]
Case-control (observational)	• Effectiveness	• Pharmaceuticals and medical devices • Health outcomes related to medical and surgical interventions	• Provider and patient bias to intervention is known • Selection bias potential without randomization	Characteristics associated with differences in survival among Black and White women with breast cancer [5]
Quasi-/natural experiment	Natural: design includes randomization Quasi: longitudinal or pre-post controls included • Safety • Effectiveness	• Health insurance • Quality improvement interventions • Public health interventions	• Study centers may self-select to the intervention that best fits their population	Reducing disparities and controlling hypertension in primary care (ReD CHiP) [6]
Big data	• Large-scale observational effectiveness • N-of-1 trial of efficacy	• Pharmacogenomics • Data mining to identify treatment patterns for unique patient phenotypes • Predictive analytics of patient outcomes	• Generalizability limited to study sample • Big data agnostic to causal inference of treatment effects	Chicago-area patient-centered outcomes research network (CAPriCORN) [7]

case-control designs, are not immune from these concerns but may be less affected by them to some degree, as it may be easier to recruit large populations in real-world contexts. In selected circumstances, "quasi-" or "natural" experimental designs may result in strong conclusions about causal effects from observational studies, but observational studies more commonly are at risk for biases that investigators cannot know with precision.

Quasi-/natural experiments reflect the use of longitudinal or pre-post study designs (i.e., quasi) sometimes in combination with randomization (i.e., natural) between cohorts that naturally identify with certain treatment patterns. These have the potential to estimate the causal impact of an intervention through an experiment designed without random assignment [8]. These approaches may be especially useful for CER of health policies in MHHD populations, since variations in health policies across states and over time is common and administrative claims and other routinely collected data can often be used to assess the effects of those policies. Obtaining unbiased estimates of causal effects of interventions based on such natural experiments depends on the identification of an appropriate comparison group, which is not always possible. In rare cases, such as the Oregon Health Insurance Experiment, in which lotteries were used to allocate access to health insurance, quasi-/natural experiments can provide control groups that are as robust as those in randomized trials [9]. When a decision maker lacks sufficient resources to offer an intervention to all persons who seem appropriate for one, lotteries may provide a valuable opportunity to produce rigorous evidence of effectiveness that can be used to leverage additional resources.

Finally, a promising new area of investigation is the use of big data in MHHD CER. Big data allows researchers to aggregate samples from a wide variety of clinical trial or observational research. Big data appears to have two clear purposes. First, it can be used to match existing patients with the treatment patterns of previous individuals who were successfully treated, known as "N-of-1" trials. This is valuable to minority groups for which medical research has not fully grasped the individualized impact that treatment regimens may have on patients with less common phenotypes. Second, big data can be used to pool observations in order to increase the power to detect patient outcomes. This may be helpful for cohorts of certain rare-disease patient groups for which little is currently known, such as the orphan disease of patients with sickle cell disease, which is common in African Americans. The Patient-Centered Outcomes Research Institute (PCORI) recently funded the development of a clinical database research network in order to build databases of patients with sickle cell disease between multiple health systems so that the combined number of samples can inform clinicians of effective treatment patterns and positive outcomes with greater statistical power [7]. This data aggregation in effect can be used to draw strong inferences in support of CER in a patient population for which there remains limited evidence or financial bandwidth to fund new clinical trials.

Regardless of the study design, a pivotal point in conducting CER on MHHD populations is whether to design the study primarily around target cohorts (e.g., patients with cancer), and then conduct subgroup analyses around MHHD populations within the general population, or to develop a study that specifically recruits minority patients. Sometimes, this depends on their health, since minorities are increasingly predisposed to specific conditions (e.g., African Americans with sickle cell disease). Sculpher describes the need to conduct CER of small populations based on age, race, ethnicity, etc., using subgroup analysis in RCTs since these studies are typically powered to test one or two primary endpoints of efficacy and safety in overall populations [10]. Subgroup analysis is necessary to determine whether there are patient group interactions relative to the treatment effects. However, studies do not always achieve statistical power for these interactions if the diversity of the trial is limited [10]. In order to achieve well-powered results, consideration should be given to sampling patients in a trial so that enough observations in a subgroup would be available to observe an important treatment effect at a statistically significant level (i.e., $p > 0.05$). For multisite trials, patients, especially minority subgroups, may be clustered in a health system intervention in order to control for: (i) unobserved environmental factors associated with that system, and (ii) clinical contexts of healthcare delivery between systems. Diggle and colleagues note that clustering reduces power, so multisite trials need additional samples of minorities to achieve the same power as single-site trials [11].

The external validity of trial designs needs to be considered as well when performing CER. For instance, Bach and colleagues have found that African Americans seek healthcare in facilities that are considered minority health institutions, such as Federally Qualified Health Centers (FQHCs); unfortunately, patients who use these resources face quality of care issues as a result of variable clinical qualifications in providers and patient access to clinical resources [12]. CER should consider the use of multicenter trial designs that allow for minority populations to access their usual facilities, as a quasi-experimental design would entail. However, there are some limitations to this concept. Research by Wennberg and Gittelsohn notes the importance of small area variations in care (i.e., the fact that different facilities offer different sets of services at varying degrees of quality, all clustered by local geographic regions) [13].

For example, consider the differences between Dartmouth-Hitchcock Medical Center in New Hampshire and Howard University Medical Center in the District of Columbia or Intermountain Health in Utah relative to University of Chicago Medicine on Chicago's South Side. These four major academic medical centers are all known for providing high-quality care, but each has vastly different local contexts. Dartmouth-Hitchcock and Intermountain provide care in rural regions that do not witness the spread of infectious disease or emergency department presentations of orthopedic trauma (e.g., gunshot wounds and other violent crimes) at nearly the same rate that Howard University or The University of Chicago experience in their densely populated, urban service areas. As a result, to conduct multicenter trials that designate subgroups by their local facility could drastically affect outcomes, given the variability of experience between facilities. Because of small area variation, we note another potential limitation of conducting quasi-experimental research in multiple centers: minorities tending to use a specific location for their healthcare may receive varying quality of care and, by extension, variable outcomes with respect to the broader population in a CER trial. This needs to be controlled for in study designs, such as adjusting by trial site case mixes.

21.4 CER Measures

Every CER study regardless of its design is focused on main outcome measures to track patient-level improvements [14]. When considering main outcomes measures with respect to CER, they should be truly reflective of treatment effectiveness. That is, the intervention should sustain improved health status and health function, and not simply yield improvement of a surrogate measure that is not linked to patient improvements in health-related quality of life (HRQOL).

HRQOL and main outcome measures of CER can be divided into two categories—disease-specific and global—and may not necessarily be unique to MHHD subgroups. Disease-specific measures pertain to the disease affecting the cohort in the study. Examples of disease-specific measures are widespread for each clinical area but should be carefully selected when conducting CER on MHHD subgroups. Some measures are intuitive to interpret (e.g., common cancer outcomes, including survival [e.g., 5-year survival] and recurrence and remission, which mainly depend on the timing and duration of observation). However, other disease-specific measures can be difficult to translate to minority subgroups because of potential language barriers or limited health literacy. Disease-specific measures important in MHHD subgroups that may present difficulty in interpretation include issues related to obesity, such as body mass index, as well as the clinical measures of downstream consequences, including:Hemoglobin A1c (HbA1c) in diabetes, blood pressure in cardiovascular disease, dyspnea severity in chronic obstructive pulmonary disease (COPD), or the Hamilton Depression Rating Scale (HAM-D). In many of these types of consequences or stand-alone chronic conditions, the disease-specific measures are vast and clinically complex and, therefore, difficult to interpret. Some infectious diseases are also of concern in MHHD populations because of higher prevalence rates and could also be classified by their severity. However, measures for issues such as human immunodeficiency virus (HIV), including CD4 levels or advancement to acquired immune deficiency syndrome (AIDS), need additional interpretation in order to fully capture the impact of the disease on MHHD subgroups.

Global measures of effectiveness are measures that are not necessarily pertinent to an individual disease cohort. Most of these measures, such as mortality, life expectancy, overall survival, and suicidality appear to be generally relatable, even to MHHD populations. Many measures of acute illness are also relevant, including hospitalization and readmissions, since these are markers of HRQOL as well as indicators of costly utilization. Quality improvement or public health interventions that focus on the uptake of new healthcare advances, such as a pharmacologic intervention, are often concerned with adherence, compliance, or counts of missed opportunities.

Patient-reported outcomes (PROs) are used to collect information from patients' perspective about their healthcare and are especially relevant in the conduct of CER for many health interventions. A large body of literature refers to instruments that quantify HRQOL based on general expressions of physical and mental functioning, such as the SF-36. This and some of its instrumental derivatives have been used to calculate health utilities, which are useful in economic evaluation and could be applied to CER as well. Such instrumental derivatives include quality-adjusted life years (QALYs, derived from the SF-6D, EQ-5D, HUI-3, etc.) as well as disability-adjusted life years (DALYs) [15]. While QALYs and DALYs are convenient for comparing detrimental HRQOL outcomes and improvements across cohorts for potentially any condition, there is ongoing debate about whether these scales synthesized in labs by economists fairly compare the effectiveness of treatments that impact patient lives [16]. This has led to the preservation of PROs as disease-specific measures in many contexts rather than QALYs and DALYs. Regardless of whether global or disease-specific measures are used, MHHD researchers need to be aware that outcome measures may be affected by language, culture, and other differences among populations. This may justify using measures tailored to specific populations.

21.5 Approaches to CER

Given the diversity of measures and study designs that enable CER, the approaches to measuring health outcomes and health intervention effectiveness can take on multiple constructs, be they quantitative, qualitative, or a combination of the two. Quantitative CER is designed to make direct comparisons between alternative interventions using statistical inference of the treatment effect. Statistical inference is applied to quantify change in the main outcome measures pre- versus postintervention, or between experimental and control groups of a cohort. Respective of MHHD research, the treatment effect in a minority subgroup could be isolated and compared to the general population or comparable subgroups between experimental and control study arms.

Qualitative research is also common in CER. Qualitative interviews and surveys are used to capture the impressions and preferences of stakeholders in the health system experiencing or monitoring the change in effectiveness taking place. Qualitative research can also be combined with quantitative research using a "mixed methods" construct. Typically, qualitative research is performed first to create a framework of impressions and preferences on the effectiveness of different health interventions. As experiments are conducted on patients, the sizes of treatment effects between exposures are measured using quantitative approaches.

21.6 Applications of CER in Health Disparity Populations

Health disparities affect many groups and may reflect hereditary differences or environmental factors, including variation in access to health services. It may also be that populations behave and respond differently to treatments. This variability makes subgroup analysis among a general population in CER studies critical in order to discern whether minorities require similar or different types of health interventions to achieve health improvements. CER also is important to inform decisions about policies that government and commercial payers can implement to pay for high-value services and ensure accessibility to minority populations even when other patients may not find benefit.

Minority populations in CER can be defined in a number of ways. First, there are racial minorities, including the following groups that commonly exhibit some variability of health outcomes in response to hereditary and environmental factors: African Americans; Asians; American Indians/Alaskan Natives; and Native Hawaiians and Pacific Islanders. Second are ethnic minorities such as Latinos, the largest minority in the United States. Third, many other socioeconomically disadvantaged races and ethnicities are at considerable risk for poor health outcomes, since a lack of affordability affects patients' ability to access high-quality, often expensive, care. In addition, some risk factors that impact long-term health and access to care are also tied to affordability, such as attending higher education, living in safe housing, or purchasing healthy foods. Fourth, there are under-served rural populations in the United States. People who live in less-dense areas of the country are often unable to access high-quality healthcare locally (or any healthcare for that matter). According to the Dartmouth Atlas,[1] a patient living in Jackson, Wyoming may have to drive over 250 miles to receive a complete

1 www.dartmouthatlas.org

array of secondary and tertiary health services in Salt Lake City, the nearest identified hospital-referral region. Fifth, other disparities are linked to gender identity and sexual minority status. Members of the LGBTQ community face different health services needs that are often met with limited flexibility in the political and geographic landscape [17].

Finally, there are well-studied interactions between these categories of minority populations. Racial minorities who have different gender identities, such as African American transgender people, may have significantly different health outcomes, such as the endemic rate of HIV in this population. In addition, many racial, sexual, and ethnic minorities are predisposed to socioeconomic disadvantages that can affect their employment, housing, education, and, ultimately, their access to healthcare, since health services are often covered through employer-based insurance plans, and employment is largely a function of education.

Studying multiple outcomes in rare disease cohorts may in itself be challenged by a lack of sampling in most trial recruitment strategies. Big data plays an increasingly important role in aggregating data across health systems in order to power sampling to subgroups with isolated concerns that require CER to indicate healthcare or policy interventions that are of high value. As subgroups are identified and organized for CER studies, consideration needs to be given to the real-world disadvantages that disparities reveal when accessing health services. For instance, an imaging scan for orthopedic trauma or a treatment for cancer that can cost a patient thousands of dollars in out-of-pocket copays may be comparatively more effective for a minority cohort. However, if the cohort faces extreme socioeconomic disadvantages on average, then presenting CER results on the effectiveness without consideration of the technology's value and affordability may not be a well-thought-out study.

21.6.1 Examples of CER Questions in Minority Health

In 2008, the Institute of Medicine (now the National Academy of Medicine, or NAM) identified 100 CER priority areas specific to minorities. These areas include critical CER questions that are continually relevant for minority populations and have resulted in a number of landmark studies worth highlighting (Table 21.2)

21.7 Social Determinants of Health

Social determinants of health are the conditions in which people are born, grow, live, work, and age [23]. Other factors include socioeconomic status, education, physical environment, employment, social support networks, and access to healthcare. According to the World Health Organization (WHO), these dynamics are interconnected [24]. The interconnectivity of social determinants may not be consistent across all domains at a point in time or across all individuals, but are meant to illustrate that social determinants can positively influence health. However, the interconnectivity of social determinants may not be consistent across all domains at a point in time or across all individuals. Health gains for MHHD populations can be limited because some minority individuals may face challenges posed by socially interconnected determinants. These challenges to improved health can be amplified if they are found in surrounding communities or across generations. Improvements in health happen when minority subgroups can overcome the destructive

Table 21.2 Comparative effectiveness research priority areas and examples of research studies conducted in this space.

Priority area of CER	Study	Minorities included	Design and intervention	Main outcome measure(s)	Key finding	Citation
Prevention of obesity, hypertension, diabetes, and heart disease	Hip-Hop to Health Jr. Obesity prevention effectiveness trial	African Americans	Quantitative: cluster-randomized controlled trial; school-based weight control and physical activity	Body-mass Index (BMI)	Positive influences found on physical activity and screen time	Fitzgibbon *et al.* [18]
Reduced incidence of infant mortality, pre-term births, and low birth rates	NSAID treatment for patent ductus arteriosus (PDA) in reducing preterm infant mortality	African American and Latino subgroups	Quantitative: observational cohort; proportion of NSAID-medicated infants born at each institution	Mortality; bronchopulmonary dysplasia (BPD) severity	NSAIDs do not curb preterm infant outlook for improved mortality if presenting with moderate or severe BPD	Slaughter *et al.* [19]
Quality improvement strategies in disease prevention, acute care, chronic care, and rehabilitation	The comprehensive care physician model	Oversampling of African American subgroup; Socioeconomic disadvantaged dual-eligible Medicare/ Medicaid patients	Quantitative: randomized controlled trial; hospitalist outpatient clinic follow-up in hospitalized Medicare patients at-risk for readmission	30/60/90-day readmission; mortality; patient satisfaction	Hospitalist assignment to outpatient follow-up reduces readmission and increases patient satisfaction	Meltzer *et al.* [3]
Smoking cessation strategies in smokers from understudied populations such as minorities	Consortium of Hospitals Advancing Research on Tobacco (CHART): smoking cessation interventions for urban hospital patients	Oversampling of African American and Latino subgroups; Asian; Native Hawaiians and Pacific Islanders; American Indians and Alaskan Natives	Quantitative: quasi-experiment with randomization; multisession telephone counseling, and state Quitline referral	6-month self-reported abstinence	Telephone counseling resulted in greater odds of abstinence than Quitline referral	Sherman *et al.* [20]
Delivery models for controlling hypertension	Improving patient-centered care and hypertension control	African Americans; socioeconomic disadvantage	Mixed-method: randomized controlled trial; counselor training to community providers	Physician communication behavior; patient satisfaction; medication adherence; blood pressure	Improved physician community positively affects patient-centered care and better controls hypertension	Cooper *et al.* [21]
Promoting breastfeeding among low-income African American women	The Maternal Mental Health Program at New York State Psychiatric Institute	Oversampling of African American and Latino subgroups	Quantitative: prospective case-control; interpersonal psychotherapy for antepartum depression (IPT-P; case), parent education program (PEP, control)	Breastfeeding rates at 4 weeks postpartum	IPT-P increased the rate of breastfeeding at 4 weeks postpartum greater than PEP alone	Spinelli *et al.* [22]

consequences of socioeconomic disadvantages and categorical discrimination. CER that introduces interventions that relieve such disadvantages, in addition to improving patient health, can sustain greater health gains in MHHD populations.

CER priority areas address some of the social determinants of health and attempt to stratify populations by these factors. For example, Cooper and colleagues address the issues of health literacy, low employment, and poor access to healthcare in dense urban Baltimore through a Reducing Disparities and Controlling Hypertension in Primary Care (ReD CHiP) study by training healthcare providers to improve how they communicate the importance of managing hypertension to prevent downstream consequences of cardiovascular disease [6]. CER studies require complex study designs and evaluation methods in order to capture the treatment effects of complicated interventions that attempt to control for multiple social determinants of health simultaneously, such as factors related to education and local environment. The study by Chung and colleagues on the Depression Collaborative Care Program (DCCP) provides an example of the multidimensionality of social determinants of health in MHHD populations, given the concern over the difficulty of treating depression symptoms in racial minorities in a dense urban area with limited access to consistent primary care or consistent employment [25]. As a result, DCCP focused on a community intervention to assist patients with depression, including help from faith-based organizations and social services.

21.8 Groups with Particular Needs for CER

In addition to the CER priority areas and conditions previously identified by the NAM, there are many other acute and chronic conditions faced by MHHD populations. Some conditions are not unique to MHHD populations (e.g., obesity, diabetes, cancer), but given difficulties that MHHD populations face in accessing high-quality healthcare, even common acute and chronic conditions can exacerbate HRQOL due to MHHD patients' inability to find needed healthcare. Targeted interventions that address the needs of subgroup minority populations' access to healthcare are critical. CER can play an important role in bridging knowledge gaps in our understanding of healthcare needs by engaging all stakeholders (e.g., patients, providers, payers, government) to participate in research that improves health outcomes of these conditions stratified by MHHD populations.

21.8.1 African Americans

Certain conditions in African Americans are exacerbated by hereditary and environmental factors. A hereditary condition that is highly prevalent among African Americans is sickle cell anemia traits or disease. HIV is also more prevalent in African Americans. In addition, African Americans face higher rates of diabetes and cardiovascular disease due to a combination of hereditary factors, as well as concomitant social determinants of health that include socioeconomic disadvantages resulting in low access to healthcare and healthy foods. Furthermore, African Americans face higher prevalence of cancer and higher rates of maternal and child health complications, including infant mortality, as indicated in the NAM CER priority areas. Targeted interventions need to be tested using CER frameworks that include stakeholder input to improve relevance to the health of this minority subgroup.

21.8.2 Hispanics/Latinos

While obesity and type 2 diabetes mellitus effect all US subpopulations, they are uniquely significant to Latinos. The rates of these conditions are proximal to the environmental factors that place Latino communities at risk—factors that include language barriers in immigrant communities; lower employment prospects; lack of access to information on social determinants of health, such as physical activity; and food nutritional literacy. The CER of interventions for health risks and optimal health in these minority groups could reduce the prevalence of these conditions.

21.8.3 Sexual and Gender Minorities

Members of the LGBTQ community are significantly predisposed to a number of infectious diseases and mental health issues, in part as a result of workplace discrimination, unemployment, biases, and uncertain access to healthcare and social services. Padula and colleagues have noted in previous CER studies that depression, addiction, and substance abuse are gateways to concerning conditions that may be fatal, including drug overdose, suicidality, and HIV contraction [26].

21.8.4 American Indians and Alaska Natives

AI/AN populations also face concerns over obesity and diabetes. Social determinants of health that include environmental and hereditary risk factors, literacy, and low income due to unemployment predispose AI/AN populations to risk factors that include addiction and substance abuse, depression, suicidality, and other mental health issues.

21.8.5 Underserved Rural Populations

Rural populations are likely to experience inadequate access to health and social services due to low incomes, distance, and adequate healthy food choices. Lacking access to these services can be detrimental to the consistent management of chronic conditions or acute threats that can quickly and significantly deplete a patient's health status. As employment opportunities recede from the rural United States, there has been a concerning increase in opioid addiction as well, making the CER of opioid cessation and aversion interventions a priority.

21.8.6 Considerations for the CER Needs of Health Disparity Populations

While the conditions outlined above represent substantial threats to minority communities, conditions including obesity, diabetes, and depression are known to be widespread, affecting all demographics. Still, the field of CER lacks studies designed to focus on aspects of subgroup populations with similar or common conditions. A convenient approach to conducting CER in minority groups is to persist with research on large numbers of observations, but with oversampling of subgroups to ensure that CER is powered to identify healthcare advances that improve MHHD patients' HRQOL.

21.9 Major Public Health/Policy Interventions as a Result of CER

Multitude government agencies and nongovernmental institutions are active in preventing increases in health disparities suffered by minorities. These agencies also support funding for CER studies so that the United States may improve its understanding of health disparities and innovate new technologies and policies to further reverse the impact of disparities on minority health. Some of these institutions and agencies are outlined below.

A number of policies and agencies fund new CER. The US Health Resources and Services Administration (HRSA) is the primary federal agency for improving healthcare for people who are geographically isolated or economically or medically vulnerable. Programs are meant to help people access high-quality primary care, especially those with HIV and pregnant women and mothers. It also oversees organ donation. HRSA services are protected under multiple laws that guarantee healthcare to the disenfranchised. One example is the Ryan White Comprehensive AIDS Emergency Resources (CARE) Act, which was designed as a safety net of coverage for high-cost care for low-income, uninsured, and underinsured victims of AIDS and their families. The Act provided funding for a three-way cost sharing program between patients, the federal government (e.g., Medicaid), and private insurers. Prior to its enactment, AIDS patients could expect to pay entirely out of pocket for treatment. This disenfranchised many people who were victims of workplace discrimination and unemployment as a result of HIV, especially members of the LGBTQ community and African Americans. Other federal agencies that play important roles in funding research for MHHD include: (i) the Agency for Healthcare Research and Quality, which is focused on quality improvement and health system interventions with a particular focus on CER; (ii) the US Centers for Disease Control and Prevention (CDC), which monitors the spread of infectious disease and sponsors studies to compare the effectiveness of technologies that reduce infection (e.g., vaccines); (iii) the Substance Abuse and Mental Health Services Administration, which supports treatments that help manage addiction and calm the effects of depression, suicidality, etc.; (iv) PCORI, which focuses on CER that directly engages patients, family caregivers, and providers; (v) the NAM, which convenes panels of academic researchers and policy experts to set priorities for CER; and (vi) the National Institutes of Health (NIH), 27 Institutes and Centers, including the National Institute on Minority Health and Health Disparities, which provide funding across an array of clinical and therapeutic areas that meet the conditions of each Institute. The ReD CHiP study was supported by another NIH Institute, the National Heart, Lung, and Blood Institute, to study hypertension prevention in an urban African American population.

In addition to federal agencies, state and local governmental bodies focus on improving public health research. State public health departments often support CER of treatments for minority populations through community-driven interventions funded by state and local grants, which are important in controlling for the small area variations that can exist in multicenter studies conducted across states and large geographic areas. International organizations such as WHO strongly consider the impact of minority health disparities on poor health outcomes internationally, including the effect that policies, socioeconomic status, and race and ethnicity have on the health of patients in the United States.

21.10 Multisectoral Impact of CER on Disparities

The studies cited throughout this chapter highlight that CER has focused on improved health outcomes across diverse subpopulations of racial and ethnic minorities, as well as those disenfranchised as a result of social determinants of health factors, including geography and gender identity. Studies that aim to conduct CER should continue to focus not only on sampling and power to detect a treatment effect in the overall population, but also on subgroups of minorities and other patients facing health disparities that would benefit from the results of CER with high-value care. The impact inventory presented by the US Panel on Cost-effectiveness in Health and Medicine indicates that healthcare interventions that lead to improvements in these multisectoral factors (e.g., employment, housing) yield high-value outcomes as well since, from a societal perspective, economics places value on patient quality of life and increased productivity in the workforce resulting from healthcare that leads to improved patient outcomes [27].

A common misconception of intervening in health disparities is that when research focuses on improving patients' health, especially those disenfranchised from healthcare systems, other social determinants can be positively impacted. A patient with depression who receives effective mental health care is more likely to avoid incarceration and maintain an improved outlook for employment opportunities. Another patient who can effectively return to employment with successful treatment for lower back pain can obtain health insurance coverage for an entire family.

As a result of the multisectoral impact of healthcare, it is important to consider the impact of an intervention that significantly improves minority health on each stakeholder or entity. That is, CER may not only be able to identify effects of healthcare on disease-specific and global outcomes, but also on other factors. For instance, CER can show that patients who have access to more effective alternatives of care have better employment opportunities, have increased access to housing, and achieve higher levels of education. Improving all of these outcomes across other sectors of the US government highlights the true value of healthcare, given its inherent link to many other social determinants of health. Furthermore, CER can have a stronger influence over policy and decision makers who must decide which health interventions to invest in to improve health equity.

21.11 Future Directions of CER

CER in MHHD populations is an important approach that introduces the beneficial value of new health interventions to minority subgroups and identifies when the use of interventions can be of limited value in these subgroups because of a lack of effectiveness. As a result, CER saves the United States significant costs that might otherwise be wasted on low-value, ineffective, and outdated health interventions. Nonetheless, the impact of CER begs the question of why cost is not explicitly considered an outcome measure, especially since many of the interconnected social determinants of health can exacerbate societal costs. While the US Panel on Cost-effectiveness in Health and Medicine recommends the use of cost-effectiveness analysis to weigh the impact of health interventions, the Patient Protection and Affordable Care Act restricts its use [27, 28].

Perhaps this course is appropriate, especially for MHHD populations, since health interventions that may not be cost-effective for the general population—and would normally be phased out—could have significant health benefits to minority subgroups and need to be preserved even if at high direct medical cost. On the other hand, there is a place for economic evaluation to make the argument that cost-effective services should be delivered to minority subgroups when there is not intuitive value. For example, the CDC Advisory Committee on Immunization Practices has used cost-effectiveness analysis to advocate for services that the US health system does not provide yet are needed in order to improve health outcomes. Such services include the availability of vaccines and other preventive treatments to reduce the spread of infectious diseases, which are often endemic in MHHD populations [29]. In another example, many insurance companies have begun covering the gender-specific needs of LGBTQ subgroups. These were previously denied because of the use of cost-effectiveness analysis based on discrepancies in gender identity listed in patient medical records [26]. These instances illustrate the positive impact that cost-effectiveness and low budget impact can have on the extension of services to minorities, even though these instances were already comparatively effective but lacked financial support.

Ultimately, CER is an important approach to studying health interventions, especially with respect to MHHD populations, so that the most effective care can be directed at patient subgroups that stand to gain. In so doing, CER not only promotes the improvement of quality and access to effective healthcare, but likely gains value through reduction in costs related to inefficient and wasteful care. The value of CER is not exceptional to healthcare. CER also promotes multisectoral impacts on the social determinants of health, thereby leading to overall improvements in the quality of life.

21.12 Key Points

- CER informs healthcare decisions by providing evidence regarding treatment option effectiveness and healthcare services; it explains differential benefits and harms of alternate methods to prevent, diagnose, treat, and monitor clinical conditions.
- CER can identify if a new healthcare intervention for a health disparity population is of limited value due to a lack of effectiveness or if it is a high-quality, effective treatment.
- CER can ensure that effective care is accessible for health disparity populations even when it may not benefit the health of the general population.

Disclaimer

The views and opinions expressed in this chapter are those of the authors only and do not necessarily represent the views, official policy, or position of the U.S. Department of Health and Human Services or any of its affiliated institutions or agencies.

References

1 Chandra, A., Jena, A.B., and Skinner, J.S. (2011). The pragmatist's guide to comparative effectiveness research. *Journal of Economic Perspectives* 25 (2): 27–46.

2 Berwick, D.M. and Hackbarth, A.D. (2012). Eliminating waste in US health care. *Journal of the American Medical Association* 307 (14): 1513–1516.

3 Meltzer, D.O. and Ruhnke, G.W. (2014). Redesigning care for patients at increased hospitalization risk: the comprehensive care physician model. *Health Affairs (Project Hope)* 33 (5): 770–777.

4 Powars, D.R., Chan, L.S., Hiti, A. et al. (2005). Outcome of sickle cell anemia: a 4-decade observational study of 1056 patients. *Medicine* 84 (6): 363–376.

5 Silber, J.H., Rosenbaum, P.R., Clark, A.S. et al. (2013). Characteristics associated with differences in survival among black and white women with breast cancer. *Journal of the American Medical Association* 310 (4): 389–397.

6 Cooper, L.A., Marsteller, J.A., Noronha, G.J. et al. (2013). A multi-level system quality improvement intervention to reduce racial disparities in hypertension care and control: study protocol. *Implementation Science* 8: 60.

7 Kho, A.N., Hynes, D.M., Goel, S. et al. (2014). CAPriCORN: Chicago area patient-centered outcomes research network. *Journal of the American Medical Informatics Association* 21 (4): 607–611.

8 Campbell, D.T. and Stanley, J.C. (1963). *Experimental and Quasi-Experimental Designs for Research*. Boston, MA: Houghton Mifflin Company.

9 Finkelstein, A., Taubman, S., Wright, B. et al. (2012). The Oregon health insurance experiment: evidence from the first year. *Quarterly Journal of Economics* 127 (3): 1057–1106.

10 Sculpher, M. (2010). Reflecting heterogeneity in patient benefits: the role of subgroup analysis with comparative effectiveness. *Value in Health* 13 (Suppl 1): S18–S21.

11 Diggle, P., Heagerty, P., Liang, K.Y., and Zeger, S. (2002). *Analysis of Longitudinal Data*. Oxford: Oxford University Press.

12 Bach, P.B., Pham, H.H., Schrag, D. et al. (2004). Primary care physicians who treat blacks and whites. *New England Journal of Medicine* 351 (6): 575–584.

13 Wennberg, J. and Gittelsohn, A. (1973). Small area variations in health care delivery. *Science* 182 (4117): 1102–1108.

14 Ahmed, S., Berzon, R.A., Revicki, D.A. et al. (2012). The use of patient-reported outcomes (PRO) within comparative effectiveness research: implications for clinical practice and health care policy. *Medical Care* 50 (12): 1060–1070.

15 Drummond, M.F., Sculpher, M.J., Torrance, G.W. et al. (2005) Cost-utility analysis, EQ-5D). *Methods for Economic Evaluation of Health Care Programmes*, 156–157. Oxford: Oxford University Press.

16 Williams, A. (1992). Cost-effectiveness analysis: is it ethical? *Journal of Medical Ethics* 18: 7–11.

17 Padula, W.V. and Baker, K. (2017). Coverage for gender-affirming care: making health insurance work for transgender Americans. *LGBT Health* 4 (4): 244–247.

18 Fitzgibbon, M.L., Stolley, M.R., Schiffer, L.A. et al. (2011). Hip-Hop to Health Jr. obesity prevention effectiveness trial: postintervention results. *Obesity* 19 (5): 994–1003.

19 Slaughter, J.L., Reagan, P.B., Newman, T.B., and Klebanoff, M.A. (2017). Comparative effectiveness of nonsteroidal anti-inflammatory drug treatment vs no treatment for patent ductus arteriosus in preterm infants. *JAMA Pediatrics* 171 (3): e164354.

20 Sherman, S.E., Link, A.R., Rogers, E.S. et al. (2016). Smoking-cessation interventions for urban hospital patients: a randomized comparative effectiveness trial. *American Journal of Preventive Medicine* 51 (4): 566–577.

21 Cooper, L.A., Roter, D.L., Carson, K.A. et al. (2011). A randomized trial to improve patient-centered care and hypertension control in underserved primary care patients. *Journal of General Internal Medicine* 26 (11): 1297–1304.

22 Spinelli, M.G., Endicott, J., and Goetz, R.R. (2013). Increased breastfeeding rates in black women after a treatment intervention. *Breastfeeding Medicine* 8 (6): 479–484.

23 Marmot, M., Friel, S., Bell, R. et al. (2008). Closing the gap in a generation: health equity through action on the social determinants of health. *The Lancet* 372 (9650): 1661–1669.

24 Solar, O. and Irwin, A. (2010). *A Conceptual Framework for Action on the Social Determinants of Health. Social Determinants of Health Discussion Paper 2 (Policy and Practice).* Geneva: World Health Organization.

25 Chung, B., Ong, M., Ettner, S.L. et al. (2014). 12-month outcomes of community engagement versus technical assistance to implement depression collaborative care: a partnered, cluster, randomized, comparative effectiveness trial. *Annals of Internal Medicine* 161 (10 Suppl): S23–S34.

26 Padula, W.V., Heru, S., and Campbell, J.D. (2016). Societal implications of health insurance coverage for medically necessary services in the US transgender population: a cost-effectiveness analysis. *Journal of General Internal Medicine* 31 (4): 394–401.

27 Sanders, G.D., Neumann, P.J., Basu, A. et al. (2016). Recommendations for conduct, methodological practices, and reporting of cost-effectiveness analyses: second panel on cost-effectiveness in health and medicine. *Journal of the American Medical Association* 316 (10): 1093–1103.

28 Methodology Committee of the Patient-Centered Outcomes Research Institute (2012). Methodological standards and patient-centeredness in comparative effectiveness research: the PCORI perspective. *Journal of the American Medical Association* 307 (15): 1636–1640.

29 Kim, J.J. (2011). The role of cost-effectiveness in US vaccination policy. *New England Journal of Medicine* 365 (19): 1760–1761.

22

The Role of Electronic Health Records and Health Information Technology in Addressing Health Disparities

Irene Dankwa-Mullan[1], Sherine El-Toukhy[2], Jessamine Winer-Jones[3], Nora Haney[4], Morgan Foreman[5], Kesi Williams[6, 7], and Anil Jain[8]

[1] IBM Watson Health, International Business Machines (IBM) Corporation, Bethesda, MD, USA
[2] Division of Intramural Research, National Institute on Minority Health and Health Disparities, Bethesda, MD, USA
[3] IBM Watson Health, International Business Machines (IBM) Corporation, Herndon, VA, USA
[4] The James Buchanan Brady Urological Institute, Johns Hopkins Hospital, Baltimore, MD, USA
[5] Center for Computational Health & Integrated Care Research, IBM Research, International Business Machines (IBM) Corporation, Cambridge, MA, USA
[6] Center for Research to Advance Community Health (REACH), UT Health San Antonio, San Antonio, TX, USA, and National Institutes of Health, National Institute on Minority Health and Health Disparities, Bethesda, MD, USA
[7] National Institutes of Health, National Institute on Minority Health and Health Disparities, Bethesda, MD, USA
[8] IBM Watson Health, International Business Machines (IBM) Corporation, Cleveland, OH, USA

22.1 Introduction

Demographic changes are transforming the US healthcare landscape, with implications for how we facilitate care and use health information technologies (HIT) to improve the healthcare system. The US population is becoming increasingly diverse, with racial and ethnic minorities projected to become a majority population by 2060 [1]. And as a result of increasing healthcare costs, consumers are seeking affordable care options, and healthcare organizations are reorienting themselves to be more transparent and to provide quality and value-based care. These demographic shifts, along with the societal challenges of health disparities exacerbated by the COVID-19 pandemic, an increased burden of chronic disease, and national technology priorities such as bolstering cybersecurity place demands on health information systems and other emerging technologies to support and safeguard the accumulation of data. These shifts also place responsibility on the healthcare system to enable culturally appropriate HIT-based care and decision-support systems that are driven by patient needs and preferences.

It is indisputable that there are disparities in health and in the availability of quality health care in the United States. The optimal strategy for reducing disparities is a topic of debate [2–4]. With the passage in 2009 of the Health Information Technology for Economic and Clinical Health (HITECH) Act (part of the American Recovery and Reinvestment Act), the US government took the position that the advancement and increased diffusion of HIT was critical to the national effort to improve healthcare quality and reduce health

The Science of Health Disparities Research, First Edition. Edited by Irene Dankwa-Mullan, Eliseo J. Pérez-Stable, Kevin L. Gardner, Xinzhi Zhang, and Adelaida M. Rosario.

disparities. The new law presupposed that HIT and the greater digital health ecosystem could facilitate an overhaul of the healthcare system in six areas: safety, effectiveness, patient-centeredness, timeliness, efficiency, and equity [5].

HIT is a subset of digital health, which broadly speaking is the application of technology to health, wellness, and healthcare. HIT interacts with aggregated, electronic patient data in four ways: collecting, storing, sharing, and analyzing [6]. Built to help achieve the goals of reduced costs, improved health, increased patient engagement, and enhanced provider experience, many digital health tools can also be applied to reduce health disparities. Digital health can facilitate research to examine the causes of health disparities and identify evidence-based solutions [6]. The evaluation and integration of new types of data, such as social and behavioral determinants of health and personal digital health tools, which are not routinely collected, may generate insights to inform clinical practice. This chapter will discuss the breadth of the digital health ecosystem, with a focus on HIT and its role in identifying and addressing disparities in health and healthcare.

22.2 Healthcare Data and Electronic Records

The foundation of HIT is digital health data from individual patients. Examples include diagnostic, clinical, and imaging data, as well as demographics, laboratory results, family history, personal wellness records, and medical claims. The value of stand-alone electronic medical record (EMR) systems, which are first-generation digital versions of the traditional paper chart, is in providing redundancy in record retention, ease of data access, and reduction of legibility issues. However, EMR systems are often highly customized and cannot easily share data between institutions, which is critical for delivering advanced value, such as population health insights from digital health data.

The electronic health record (EHR) is an evolving concept defined as a systematic collection of electronic health information about individual patients or populations. EHR systems are the standardized, interconnected alternative to EMR systems. EHR systems, combined with a compatible health information exchange (HIE), can ensure continuity of care over time and between institutions. When aggregated, the data can inform improvements in healthcare and enable research into the drivers of disparate health outcomes and economics. Figure 22.1 illustrates how individual patients generate different types of data during a clinical visit, and how that data can then be utilized, such as use of an individual's entire health record during a point-of-care visit, blood draw records for site statistics, or demographic data and medication records for research into factors that predict medication adherence.

The key feature required for achieving advances based on EHR systems and HIT is interoperability: the ability of two or more systems to exchange and use information [7]. Interoperability in this regard has two aspects: technical and semantic. Technical interoperability is required to move data from one system to another. A lack of technical interoperability is what prevents a word processing program from opening a music file. For medical records, technical interoperability relies on the use of data transfer standards such as Health Level Seven (HL7) and Fast Healthcare Interoperability Resources (FHIR) for hospital and physician records, Digital Imaging and Communications in Medicine (DICOM)

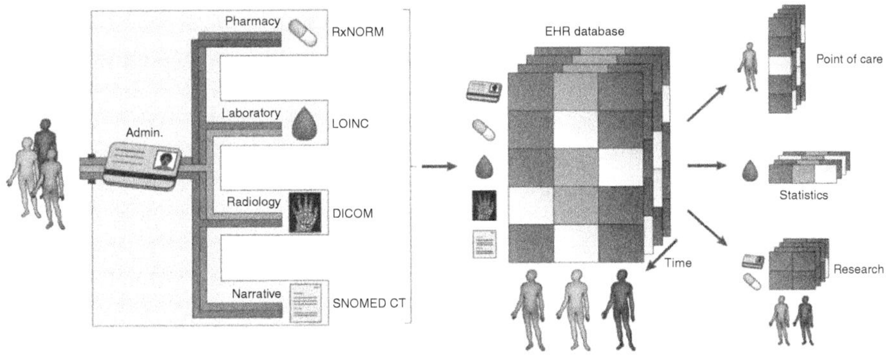

Figure 22.1 Electronic health records. *Source:* Adapted from Jensen et al. 2012. (See color representation of this figure in color inserts section.)

for medical images, and Accredited Standards Committee (ASC) X12 837 for billing data. The primary challenge to technical interoperability among HIT systems is that vendors have supported these standards but implementation has varied. Moreover, healthcare providers often desire customization that is not supported by existing transfer standards, while new data are added faster than the transfer standards can be updated.

Moving a file between providers is only half the challenge. The recipient must be able to correctly interpret the record of a patient they have never seen. This is achieved with semantic interoperability, which relies on the use of medical coding systems, such as the International Classification of Diseases for diagnoses reporting, Current Procedural Terminology (CPT) for clinical services reporting, Logical Observation Identifiers Names and Codes (LOINC) for medical laboratory observations, as well as patient questionnaires (e.g., PHQ-9) and Systematized Nomenclature of Medicine Clinical Terms (SNOMED CT) for comprehensive clinical record keeping (see Figure 22.2). Effective semantic interoperability requires that individuals who maintain the EHRs be conversant in both SNOMED CT, RxNorm, and LOINC, each of which requires substantial training and therefore can be a high barrier to achieving value.

According to the Office of the National Coordinator (ONC) for Health IT at the Department of Health and Human Services, adoption of EHR systems by US hospitals has increased from below 10% in 2008 to over 95% in 2016. However, as of 2014, only 23% of hospitals reported being able to find, send, receive, and use information through an HIE [8]. Adoption and utilization of EHR systems is lower for rural hospitals, small hospitals, and those with for-profit or public ownership [9]. Even acknowledging this limitation, digital health data is a boon to health disparities research because it enables the aggregation of standardized data across populations of interest.

22.3 Overview of Health Information Technologies

EMR and EHR systems add more than digitization of an existing paper system. The potential of HIT lies in leveraging collected data to advance the state of healthcare at the individual and population levels. To encourage the adoption of HIT, the HITECH Act for

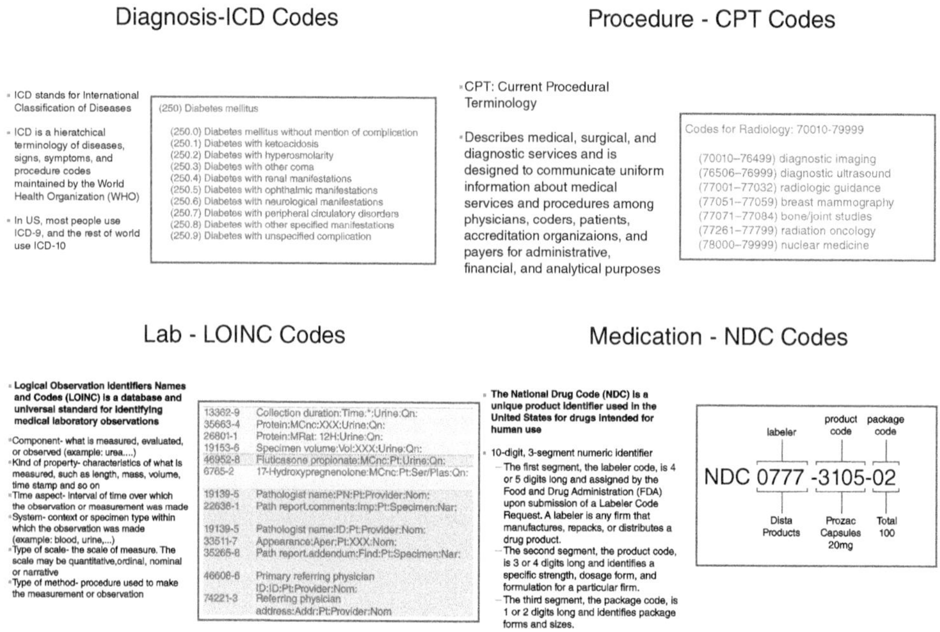

Figure 22.2 Medical coding systems in EHR.

implementing a certified EHR system provides financial incentives to providers. In order to be eligible for financial incentives from the Centers for Medicare and Medicaid Services (CMS), providers must demonstrate meaningful use. On HealthIT.gov, meaningful use is defined as using certified EHR technology to:

- improve quality, safety, efficiency, and reduce health disparities;
- engage patients and family;
- improve care coordination and population and public health; and
- maintain privacy and security of patient health information.

ONC promotes a national HIT infrastructure and oversees its development, including the Certified EHR Technology program. ONC has identified three stages for the meaningful use of HIT. Stage 1 (2011–2013) focused on the standardization and capture of patient data. Stage 2 (2014–2015) accelerated the exchange of health information among providers and between providers and their patients to facilitate information sharing and care coordination. Stage 3 (2016–2017) emphasized standards required to foster interoperable EHRs to benefit a growing ecosystem of certified HIT.

Despite the significant investment in the "meaningful use" program, meaningful actionable exchange of health information has been slow, resulting in further legislation in December 2016 to streamline sharing and penalize "information blocking." The 21st Century Cures Act created a 25-member HIT committee to ensure that the ONC is being advised by a multistakeholder group to accelerate interoperability and usability. The role of HIT in addressing health disparities is a focus area for several members of the committee.

The requirement to demonstrate meaningful use has helped drive the development of an expanded portfolio of HIT solutions. This includes tools to increase patient engagement, facilitate care coordination, enable care delivery and improve healthcare quality, and improve safety and efficiency. These technologies are primarily either provider/health system focused, such as clinical decisions support systems (CDSS) and electronic disease registries; or consumer/patient-focused, such as personal health records (PHRs) and telemedicine solutions. Benefits cross these category lines. Provider/system focused HIT has benefits for consumers/patients, and likewise consumer/patient focused HIT has benefits for provider/system.

22.3.1 Personal Health Record

PHR systems empower individuals to manage their own health by giving them access to their EHRs, where they can confidentially upload health-related information, such as diet and exercise, and data from monitoring devices [10]. PHR systems can be standalone applications, or they can be tethered to one or more provider EHR systems. As standalone applications, they can provide patients with tools to monitor their health and wellness while enabling access to curated health information. Their value is amplified when connected to EHR systems, enabling additional contact points between patients and providers. Through PHR-EHR connected systems, patients can view personal health information, such as test results, and perform administrative tasks, such as filling a prescription or requesting medical guidance (e.g., when to come in for a follow-up) [11]. A 2016 model based on Health Information National Trends Survey (HINTS) forecasted that PHR use in the United States would exceed 75% by 2020. Nevertheless, a 2016 survey-based study has shown PHR adoption varies among low-income and under-represented minorities [12].

22.3.2 Consumer Health IT Applications

Beyond PHR systems, there are many examples of consumer-focused HIT: from web-based risk calculators to wearable activity trackers [13]. Consumer HIT must generally meet three criteria. The technology needs to (i) collect data from the individual either automatically or by manual input, (ii) process that data, usually in combination with data from a validated source of health information, and (iii) provide the patient with informative, patient-specific content. Examples include diabetes self-management tools, electronic headache diaries, and sleep tracking devices [13, 14].

22.3.3 Clinical Decisions Support Systems

CDSSs are used by providers to ensure delivery of minimum standard of care for all patients at the point of care [15]. Integrated with the provider EHR system, CDSSs act as a second set of eyes to review medical records and bring relevant content to the attention of the provider. These may be rule-based alerts for things like drug-drug interactions, medication allergies, or abnormal test results. Or, they may be clinical care suggestions based on approved guidelines or dashboards for reviewing the patient record.

The meaningful use requirement for EHR system reimbursement has driven adoption of CDSS by providers in the United States. The majority of these platforms are embedded within the EHR system and provide suggestions based on a Boolean logic-based application of medical guidelines. Some advanced CDSSs apply predictive analytics to develop clinical suggestions based on patient features compared to a population, or use natural language processing to extract additional content from clinical notes to support decision making.

Implementation of CDSS remains a challenge when balancing the utility of supporting information needed by the clinicians in their decision-making process and the intrusion that alerts and other forms of CDSS may impose on the workflow. “Alert Fatigue” is a widely studied phenomenon that has led to clinicians ignoring important information because of perceived noise from CDSS systems. In an anonymous survey of 19 primary care programs, 37% reported at least one symptom of burnout with 75% of them attributing the burnout to the EHR [16]. EHR and other HIT vendors are looking at methods to reduce the adverse impact of EHRs in this manner, including investigating the role of scribes (in the exam room or virtual), technologies such as speech-to-text, and artificially intelligent systems to assist clinicians.

22.3.4 Electronic Disease Registries

Electronic disease registries are simplified EHRs for clinical decision support and outcome tracking of a defined patient population, such as those with a chronic condition like diabetes or hypertension [17]. Disease registries can exist as stand-alone applications or within a larger EHR system. Moreover, because they can be tailored to focus on chronic care for common conditions, disease registries have broad appeal to specialty clinics and general practitioners. They also provide a lower bar to entry than more comprehensive EMR systems, from both financial and operational standpoints, while enabling better care of high-risk populations. Many disease registries can be linked to larger state, country, or international registries facilitating broader healthcare initiatives, such as population health research and post-market surveillance. However, due to reimbursement penalties for not using certified HIT (e.g., EHRs), electronic disease registries are not an alternative in most care settings. EHR vendors and other vendors are enhancing their systems to create registry capabilities in their existing systems, including the ability to exchange data with other registries.

22.3.5 Telehealth, Telemedicine, and Mobile Health

While not a category of HIT, telehealth is a key component of the digital health ecosystem. Telehealth is the use of telecommunications technology to support health service delivery. Before exploring examples of telehealth solutions, it is important to distinguish between three terms with overlapping meaning: telehealth, telemedicine, and mobile health (mHealth).

Although occasionally used as a synonym for telehealth, telemedicine refers to the delivery of *clinical* care via a telecommunications system. mHealth refers to the use of *mobile* technology to improve clinical care or public health delivery. Many consumer HIT applications are also examples of mHealth solutions. Both telemedicine and mHealth fit under the broader umbrella of telehealth (see Figure 22.3).

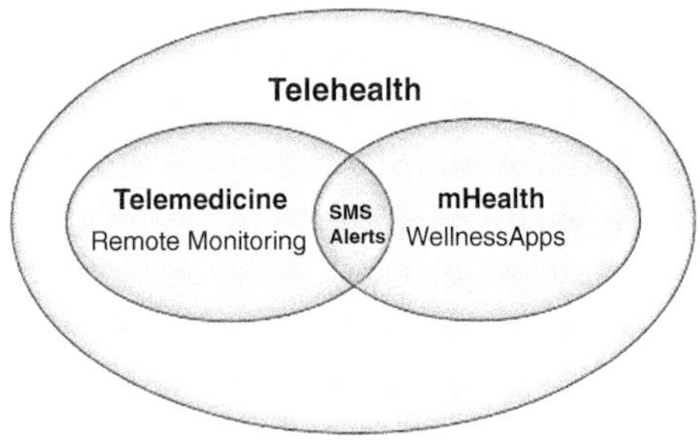

Figure 22.3 Types and examples of telehealth.

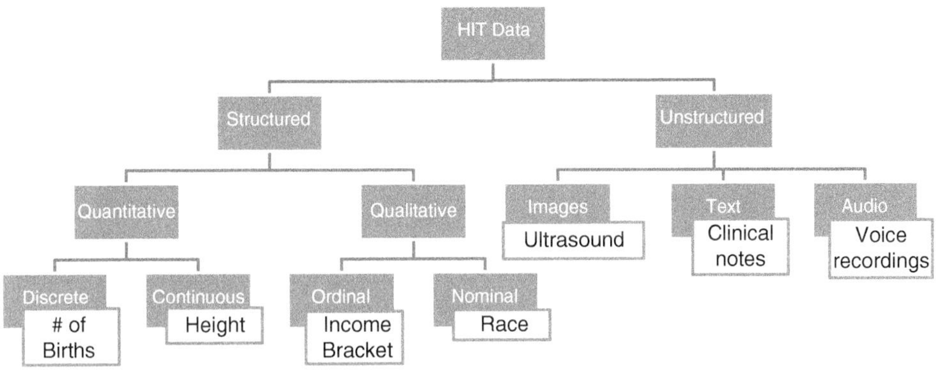

Figure 22.4 HIT data types and examples. *Source:* Murdoch and Detsky [20].

Examples of telehealth are now ubiquitous, from training webinars for practicing clinicians to wilderness first aid phone applications to suicide prevention hotlines [18]. And, while these examples do not require HIT connectivity, other types, such as teleradiology, depend on it. Teleradiology is the transmission of a radiological image to a remote location usually for diagnosis or consultation. It is only one of many examples of remote delivery of specialty care. Other examples include telepsychiatry, telenursing, and care coordination. Telehealth is viewed by many as the key to democratizing healthcare delivery, because it enables quality care at a distance.

mHealth applications enable the user to remotely capture data, eliminating the need to be physically linked to a network or restricted to a specific geographic location. This allows for healthcare practitioners to bring additional resources to the point of care or change the location of the point of care. Mobile devices also enable the provision of real-time decision support to patients. A patient-centered mHealth application, however, will make sure that the patient is a part of the decision-making process and not solely receiving instructions they have to follow. Patient-centered mHealth applications have targeted asthma, diabetes, hypertension, and cancer [17, 19].

22.3.6 Health Information Technologies Data

All of the above solutions rely on data collection and analysis to deliver value. HIT data come in four types: structured, unstructured, quantitative, or qualitative. Figure 22.4 shows the relationship between these types of data and examples of each. If data from multiple sources are going to be used in a HIT solution, such as a PHR, the data must be curated, normalized, and integrated before they can be interrogated. For unstructured data that are not machine-readable, machine learning methods, such as feature extraction for images and natural language processing for text, can independently derive meaning from unstructured data, making them available for analysis [20].

22.4 Application of Digital Health and Health Information Technologies to Addressing Disparities

In the following section, we present examples that showcase the impact of integrating a wide range of HIT and digital health devices on health disparities.

22.4.1 Population-level Health Outcomes

Many HIT tools are targeted to population health management and improving patient-centered health outcomes. Health disparities can be attributed, in part, to unequal access to health care and the lower quality of services rendered to minority populations. HIT affects health disparities by improving access to and quality of healthcare through increased adherence to protocols and guidelines, enhanced disease surveillance and monitoring activities, and reduction in medical errors [21]. Although it is often the case that HIT implementation improves healthcare quality on multiple levels, for purposes of illustration we will highlight cases of HIT impact on single quality indicators among underserved populations.

22.4.1.1 Access and Healthcare Delivery

Digital health applications, such as telemedicine and mHealth, can eliminate physical barriers to healthcare access/delivery. For example, telepsychiatry has the potential to enhance mental health care in underserved areas, such as rural locations and jails. Project ECHO GEMH (Extension for Community Healthcare Outcomes Geriatric Mental Health) relies on virtual clinics that enable clinicians in rural settings to connect with their counterparts in urban areas to provide evidence-based mental health care to elderly rural populations. Results of this project include a 20% reduction in emergency department visits and a 24% reduction in cost since 2014.[1] Further, evidence shows that virtual environments alleviate stigma and encourage free communication (compared to in-person communication) among disparity populations, such as those in lesbian, gay, bisexual, transgender, and queer (LGBTQ) communities and people suffering from psychiatric diseases, such as anxiety [22]. Geographic information systems (GISs) have been used to locate mothers with immigrant status in Brooklyn, NY and determine how easy it was to access prenatal healthcare in that area [23]. This could aid in city and hospital planning of mobile and community-based clinics, extending access to those with low clinic density in their area and limited transportation means.

---

1 https://www.ahrq.gov/news/newsroom/case-studies/201703.html

22.4.1.2 Enhanced Surveillance and Monitoring Activities

HIT allows automated surveillance and linkages to language and interpreter services often needed by racial/ethnic minorities, as well as standardization of quality monitoring among health disparity populations. For example, EHRs have been used for population-level surveillance of predictive factors (e.g., health insurance, racial/ethnic background) of health service utilization, such as cervical cancer screenings [24]. The Enabling Services Accountability Project (ESAP) provides a surveillance and monitoring system that allows automated collection of data on patients' needs, efforts to address these needs, and outcomes. ESAP serves low-income, uninsured Asian Americans, Native Hawaiians, and other Pacific Islanders [25].

22.4.1.3 Adherence to Protocols and Quality Care Metrics

HIT can potentially circumvent several provider- and patient-related causes of health disparities, such as providers' stereotyping biases and patients' misunderstanding of healthcare provider instructions. For example, Columbia Basin Health Association (CBHA) embedded a "check protocol" function in their EHR that alerts clinicians to services required by standard protocols. CBHA targets rural populations of the Columbia Basin in Washington, primarily Spanish-speaking, low-income individuals. The Cherokee Indian Hospital Authority (CIHA) implements a system where clinicians undergo quality reviews every month that incorporate clinical indicators, productivity, and patient feedback [25]. This system facilitates accountability and opportunities to improve performance through panel reports.

Although health disparity is inherently a population-level phenomenon, there is a shortage of studies examining the impact of HIT on closing the disparities gap between Whites and disadvantaged populations [25]. One example is the UPQUAL (Using Precision Performance Measurement to Conduct Focused Quality Improvement) initiative implemented in a Chicago ambulatory care practice [15]. To evaluate the impact of HIT on clinical care, the researchers tracked 17 quality metrics over 2 years. They found that while overall quality of care improved for both Whites and Blacks across a majority of metrics, there was minimal impact on the racial disparities in quality of care. Of the seven metrics with racial disparities at baseline, only two showed a narrowing of disparity after 2 years, while four stayed the same and one increased. In addition, two metrics which did not display racial disparity at baseline showed a widening gap at 2 years.

22.4.2 Consumer or Individual-level Health Outcomes

Health disparity populations often lack access to preventive care and treatment and have a higher prevalence of chronic conditions [3]. This chapter offers examples of HIT that improve patient-centered health outcomes through prevention, treatment, and management of chronic conditions in various settings. This classification of individual-level outcomes (i.e., prevention, treatment, and management) is only for illustrative purposes. HIT can have positive effects on one or more health outcomes, given that patient medical education and reminders can positively affect prevention, treatment, and management.

22.4.2.1 Prevention

HIT can improve preventive care by providing medical education, improving healthcare processes (e.g., appointment and vaccine reminders), and promoting behavioral change. For example, text messaging programs increase cancer screenings (e.g., colorectal and cervical cancers) among health disparity populations (e.g., Alaska Natives and Korean Americans) [26].

Social media are highly influential HIT vehicles for patient education and support systems. Studies have tallied 527 Facebook or Twitter groups for diabetes, 216 groups for breast cancer, and 171 for colorectal cancer [27]. These platforms provide an avenue to target specific communities in a familiar venue. GirlTrek, for example, is an online community that uses Facebook, Twitter, and the GirlTrek blog to promote walking for African American women and girls to improve their physical fitness [28].

22.4.2.2 Treatment

HIT can improve treatment by increasing medical adherence and improving the accuracy of data collection (via mobile sensors and wearables) and treatment interventions. For example, studies show that mHealth technologies positively affect physical activity among obese African American women [29]. TuDiabetes.org, and its Spanish-language counterpart EsTuDiabetes.org, is a combination crowdsourced and nonprofit social network created in part by the Centers for Disease Control and Prevention (CDC) for diabetic patients [14]. While studies have shown high patient engagement with TuDiabetes.org, there are currently no studies demonstrating reduction of health disparities in patient-centered outcomes in diabetes management using HIT [30].

22.4.2.3 Management

As with treatment outcomes, HIT can positively affect patients' ability to manage chronic conditions through increased medical adherence and improved data collection and behavioral interventions. A meta-analysis of 13 studies showed that HIT significantly improved glycemic control among underserved diabetic populations for an extended time (up to 12 months) through self-management education interventions [31]. Similarly, use of EHR to generate personalized goal reports was associated with greater reduction in hemoglobin A1c (HbA1c) levels among African American patients who received these reports versus those who did not. Other studies have shown favorable effects of technology-based interventions on adherence to hypertension medication and asthma controller medication among African Americans [19].

22.5 Conclusions: Challenges and Opportunities in Health Information Technologies Implementation

Issues implicated in the integration of HIT in healthcare include a myriad of social, political, and economic complexities [2, 32] related to four essential components: technological (e.g. see interoperability, Section 25.2), human, organizational, and managerial [21]. Main challenges and suggested opportunities for HIT to improve health disparities are highlighted in Table 22.1.

Table 22.1 Challenges and opportunities for HIT and health disparities research.

HIT issues	Challenges	Opportunities
Technical components: interoperability	Many EHR systems still lack interoperability, limiting their utility and viability for addressing health disparities.	Emphasize standardization and interoperability across HIT systems to capture all patient-generated data.
Human components: access, consumer/patient acceptance, and engagement	Health disparity populations lack access, knowledge, and skills necessary to benefit from HIT. Populations with limited access are likely to be poor, racial/ethnic minorities, and with lower health literacy. Acceptance of and engagement with HIT remain low among health disparity populations.	Collaborate with technology partners to increase rural broadband connectivity. Prioritize tailored or customized technologies for consumer use that summarize data and present it in forms that meet the literacy, numeracy, cultural, and educational needs of diverse patients and caregivers.
Organization and managerial components	High cost and operational barriers limit HIT adoption in under-served areas.	Incentivize and subsidize adoption of HIT among providers and health organizations in underserved areas.
Empirical research	Little empirical research has been performed specifically for HIT's effect on reducing health disparities.	Development of partnerships between for-profits, nonprofits, academia, and government presents opportunities for evidence-based research. Prioritize funding opportunities for HIT impact on health disparities.

Given its high cost, tools and resources from public and nonprofit entities for implementing and evaluating the impact of HIT integration in public health are emerging. Resources for HIT evaluation include the Agency for Healthcare Research and Quality (AHRQ) toolkit, which includes a step-by-step guide to developing an evaluation plan along with suggested evaluation measures in domains such as clinical outcomes and processes; provider and patient attitudes; and HIT adoption, workflow, and cost. Other resources and toolkits include telehealth policies and laws, standardized/centralized implementation and evaluation tools, and best practices to avoid/reduce common pitfalls of HIT integration. Common across these resources is the goal to increase the effectiveness of HIT integration in healthcare.

Although HIT may be created to reduce health disparities, they may contribute to the widening of disparities if they are not created with diverse individuals in mind. This is particularly true for technologies that depend on patient engagement and consumer adoption. Disparities exist by socioeconomic gradients, age, rural/urban communities, and health literacy in ownership of and access to new technologies [33].

For example, several studies documented disparities in use of patient portals by age, gender, and race/ethnicity [34]. Eliminating this digital divide remains the most important prerequisite to the ability of health disparity populations to reap the benefits of HIT. Another prerequisite for HIT to positively affect health disparity populations is digital literacy, which encompasses technological skills, linguistic and informational modality competencies, and learning predispositions and cognitive abilities. Low health literacy is associated with low cell-phone ownership and internet access; less adoption of HIT, such as mobile health applications and patient portals; and perceived difficulty seeking health information online [33]. Lower literacy, less web access, and inadequate leisure time, which are negative predictors of HIT adoption, are more prevalent in underserved communities [35].

Other recommendations for implementation and evaluation of HIT include the expansion of health disparity populations beyond racial and ethnic minorities. Data on sexual orientation and gender identity should be captured in EHRs to enable the provision of preventive and screening services specific to these groups [36]. While most information technology is not created with diverse populations in mind, there has been a recent push for more culturally appropriate or tailored HIT. A promising take on tailoring HIT to combat health inequalities with the use of PHR is Indivo X, which allows third-party developers to contribute to the PHR. This allows the PHR to be customizable based on patient needs and integrates online applications familiar to the patient, such as Facebook and Amazon. This concept is also termed Personally Controlled Health Record (PCHR). Additionally, implementation guidelines need to address issues such as the privacy and security of personal electronically transmitted health information through "The Nationwide Privacy and Security Framework for Electronic Exchange of Individually Identifiable Health Information."

From an organizational and managerial perspective, evidence is mixed in regards to adoption of HIT among providers caring for minority and health disparity populations. Studies show varying rates of adoption of EHRs [37]. Efforts should be dedicated to eliminating barriers to HIT adoption, such as implementation cost and time, staffing shortages, and limited broadband capabilities. Additionally, health management via data extraction needs constant tailoring and validation with external data sources for accuracy, especially with widespread misinformation on crowdsourcing and social networking platforms in healthcare.

Finally, there is limited empirical research into HIT's effect on population health and disparities. The majority of studies on mHealth are feasibility pilot studies [38]. Review articles consistently propose how these technologies could be applied to combat health disparities, but minimal evidence has been presented. More research is needed to document and standardize innovative ways to employ HIT to reduce health disparities. Partnerships between for-profits, nonprofits, academia, and government are required for such projects to become sustainable. There is wide variation in the types of HIT, as well as the types of populations and disparities, that have the potential to be affected. Reliable outcome measures on HIT adoption and success are crucial to investigating HIT effects on health disparities. Further, it is important to understand and address security and privacy concerns and distrust that are specific to health disparity populations, design culturally tailored interfaces, encourage engagement with HIT, and improve usability and increase acceptance of HIT among these populations [35].

The twenty-first century marked a shift in healthcare to an electronic environment where information technologies are utilized to improve the collection, assessment, and distribution of healthcare information [2]. More importantly, HIT technologies allow the integration of biomedical, sociobehavioral, individual, and environmental level data for better clinical and behavioral health outcomes [39]. Integration of HIT in healthcare systems has well-documented benefits from effectiveness, efficiency, and cost [21]. Of relevance to this chapter is the positive impact of HIT on health disparities, particularly by eliminating or reducing inequalities in the provision and quality of healthcare. The most recent (2015) Health, United States report highlighted persistent racial and ethnic disparities in morbidities and mortalities among minorities compared to non-Hispanic Whites [40]. These disparities can be attributed in part to unequal access to healthcare and the quality of services rendered to minority populations [41]. The integration of HIT into healthcare systems has the potential to address several underlying causes of health disparities rooted in those systems (e.g., workflow, economics), provider (e.g., patient–clinician communication, cultural competency), as well as patient factors (e.g., health literacy, healthcare coverage), which can help meet the national goal of reducing health disparities outlined in Healthy People 2010 and 2020 [42]. Meeting such goals, however, requires a concerted effort by policymakers, communities, health organizations, providers, and patients to realize the impact of HIT on individual and population health outcomes.

22.6 Key Points

- Health information technologies are a hallmark of the twenty-first century healthcare environment driven, in part, by policies that encourage their adoption to improve the collection, assessment, and distribution of biomedical, sociobehavioral, individual and environmental level data for better clinical and behavioral health outcomes.
- Health information technologies encompass system- and individual-level applications that afford opportunities to reduce or eliminate health disparities by addressing their root causes related to providers, patients, and the healthcare system generally.
- Health information technologies show promise in improving population- and individual-level health outcomes and reducing health disparities.
- Concerted efforts are needed to adapt the technological, human, organizational, and managerial aspects of health information technologies to the unique characteristics and needs of minorities and health disparity populations to fully achieve their intended purpose of reducing health disparities.

Acknowledgments

The authors wish to acknowledge Courtney Ferrell-Aklin, PhD, for assistance with the initial literature review and guidance on the topic outline, Amar Das, MD, PhD and Jose Morey MD for review and expert input into earlier versions of the chapter.

Disclaimer

The views and opinions expressed in this chapter are those of the authors only and do not necessarily represent the views, official policy, or position of the U.S. Department of Health and Human Services or any of its affiliated institutions or agencies.

References

1 Colby, S.L. and Ortman, J.M. (2015). *Projections of the Size and Composition of the US Population: 2014 to 2060*, 9e. Washington, DC: US Census Bureau.
2 National Research Council (2009). *Computational Technology for Effective Health Care: Immediate Steps and Strategic Directions*. Washington, DC: National Academies Press.
3 Lasser, K.E., Himmelstein, D.U., and Woolhandler, S. (2006). Access to care, health status, and health disparities in the United States and Canada: results of a cross-national population-based survey. *American Journal of Public Health* 96 (7): 1300–1307.
4 Phelan, J.C., Link, B.G., and Tehranifar, P. (2010). Social conditions as fundamental causes of health inequalities: theory, evidence, and policy implications. *Journal of Health and Social Behavior* 51 (1_suppl): S28–S40.
5 Institute of Medicine (2001). Committee on Quality of Health Care in America. To err is human: building a safer health system. In: *Crossing the Quality Chasm: A New Health System for the 21st Century* (eds. B.D. Smedley, A.Y. Stith and A.R. Nelson). Washington, DC: National Academies Press.
6 Jones, S.S., Rudin, R.S., Perry, T., and Shekelle, P.G. (2014). Health information technology: an updated systematic review with a focus on meaningful use. *Annals of Internal Medicine* 160 (1): 48–54.
7 Geraci, A., Katki, F., McMonegal, L. et al. (eds.) (1991). *IEEE Standard Computer Dictionary: Compilation of IEEE Standard Computer Glossaries*. New York: IEEE Press.
8 Adler-Milstein, J., Holmgren, A.J., Kralovec, P. et al. (2017). Electronic health record adoption in US hospitals: the emergence of a digital "advanced use" divide. *Journal of the American Medical Informatics Association* 24 (6): 1142–1148.
9 Adler-Milstein, J., DesRoches, C.M., Kralovec, P. et al. (2015). Electronic health record adoption in US hospitals: progress continues but challenges persist. *Health Affairs* 34 (12): 2174–2180.
10 Tang, P.C., Ash, J.S., Bates, D.W. et al. (2006). Personal health records: definitions, benefits, and strategies for overcoming barriers to adoption. *Journal of the American Medical Informatics Association* 13 (2): 121–126.
11 Tom, J.O., Chen, C., and Zhou, Y.Y. (2014). Personal health record use and association with immunizations and well-child care visits recommendations. *Journal of Pediatrics* 164 (1): 112–117.
12 Ancker, J.S., Hafeez, B., and Kaushal, R. (2016). Socioeconomic disparities in adoption of personal health records over time. *American Journal of Managed Care* 22 (8): 539–540.
13 Evenson, K.R., Goto, M.M., and Furberg, R.D. (2015). Systematic review of the validity and reliability of consumer-wearable activity trackers. *International Journal of Behavioral Nutrition and Physical Activity* 12 (1): 159.
14 Mandl, K.D., McNabb, M., Marks, N. et al. (2014). Participatory surveillance of diabetes device safety: a social media-based complement to traditional FDA reporting. *Journal of the American Medical Informatics Association* 21 (4): 687–961.
15 Jean-Jacques, M., Persell, S.D., Thompson, J.A. et al. (2012). Changes in disparities following the implementation of a health information technology-supported quality improvement initiative. *Journal of General Internal Medicine* 27 (1): 71–77.

16 Robertson, S.L., Robinson, M.D., and Reid, A. (2017). Electronic health record effects on work-life balance and burnout within the I(3) population collaborative. *Journal of Graduate Medical Education* 9 (4): 479–484.

17 Bodenheimer, T., Wagner, E.H., and Grumbach, K. (2002). Improving primary care for patients with chronic illness. *Journal of the American Medical Association* 288 (14): 1775–1779.

18 Weinstein, R.S., Lopez, A.M., Joseph, B.A. et al. (2014). Telemedicine, telehealth, and mobile health applications that work: opportunities and barriers. *The American Journal of Medicine* 127 (3): 183–187.

19 Buis, L., Hirzel, L., Dawood, R.M. et al. (2017). Text messaging to improve hypertension medication adherence in African Americans from primary care and emergency department settings: results from two randomized feasibility studies. *JMIR mHealth and uHealth* 5 (2): e9.

20 Murdoch, T.B. and Detsky, A.S. (2013). The inevitable application of big data to health care. *Journal of the American Medical Association* 309 (13): 1351–1352.

21 Chaudhry, B., Wang, J., Wu, S. et al. (2006). Systematic review: impact of health information technology on quality, efficiency, and costs of medical care. *Annals of Internal Medicine* 144 (10): 742–572.

22 Wangberg, S.C., Gammon, D., and Spitznogle, K. (2007). In the eyes of the beholder: exploring psychologists' attitudes towards and use of e-therapy in Norway. *Cyberpsychology & Behavior* 10 (3): 418–423.

23 McLafferty, S. and Grady, S. (2005). Immigration and geographic access to prenatal clinics in Brooklyn, NY: a geographic information systems analysis. *American Journal of Public Health* 95 (4): 638–640.

24 Cowburn, S., Carlson, M.J., Lapidus, J.A., and DeVoe, J.E. (2013). The association between insurance status and cervical cancer screening in community health centers: exploring the potential of electronic health records for population-level surveillance, 2008–2010. *Preventing Chronic Disease* 10: E173.

25 US Department of Health and Human Services (2010). *Understanding the Impact of Health IT in Underserved Communities and those with Health Disparities*. Chicago: National Opinion Research Center, University of Chicago.

26 Lee, H.Y., Koopmeiners, J.S., Rhee, T.G. et al. (2014). Mobile phone text messaging intervention for cervical cancer screening: changes in knowledge and behavior pre-post intervention. *Journal of Medical Internet Research* 16 (8): e196.

27 De la Torre-Diez, I., Diaz-Pernas, F.J., and Anton-Rodriguez, M. (2012). A content analysis of chronic diseases social groups on Facebook and Twitter. *Telemedicine Journal and E-Health* 18 (6): 404–408.

28 Gibbons, M.C., Fleisher, L., Slamon, R.E. et al. (2011). Exploring the potential of web 2.0 to address health disparities. *Journal of Health Communication* 16 (supl): 77–89.

29 Thomas, S., Yingling, L., Adu-Brimpong, J. et al. (2017). Mobile health technology can objectively capture physical activity (PA) targets among African-American women within resource-limited communities—the Washington, DC cardiovascular health and needs assessment. *Journal of Racial and Ethnic Health Disparities* 4 (876): 1–8.

30 Eggleston, E.M. and Klompas, M. (2014). Rational use of electronic health records for diabetes population management. *Current Diabetes Reports* 14 (4): 479.

31 Heitkemper, E.M., Mamykina, L., Travers, J., and Smaldone, A. (2017). Do health information technology self-management interventions improve glycemic control in medically underserved adults with diabetes? a systematic review and meta-analysis. *Journal of the American Medical Informatics Association* 24 (5): 1024–1035.

32 Blumenthal, D. (2011). Implementation of the federal health information technology initiative. *New England Journal of Medicine* 365 (25): 2426–2431.

33 Bailey, S.C., O'Conor, R., Bojarski, E.A. et al. (2015). Literacy disparities in patient access and health-related use of internet and mobile technologies. *Health Expectations* 18 (6): 3079–3087.

34 Goel, M.S., Brown, T.L., Williams, A. et al. (2011). Disparities in enrollment and use of an electronic patient portal. *Journal of General Internal Medicine* 26 (10): 1112–1116.

35 Valdez, R.S., Gibbons, M.C., Siegel, E.R. et al. (2012). Designing consumer health IT to enhance usability among different racial and ethnic groups within the United States. *Health and Technology* 2 (4): 225–233.

36 Graham, R., Berkowitz, B., Blum, R. et al. (eds.) (2011). *The Health of Lesbian, Gay, Bisexual, and Transgender People: Building a Foundation for Better Understanding*. Washington, DC: National Academies Press.

37 Hing, E. and Burt, C.W. (2009). Are there patient disparities when electronic health records are adopted? *Journal of Health Care for the Poor and Underserved* 20 (2): 473–488.

38 Gorski, I., Bram, J.T., Sutermaster, S. et al. (2016). Value propositions of mhealth projects. *Journal of Medical Engineering & Technology* 40 (7–8): 400–421.

39 Singer, B.H. and Ryff, C.D. (eds.) (2001). *New Horizons in Health: An Integrative Approach*. Washington, DC: National Academies Press.

40 National Center for Health Statistics (2016). *Health, United States, 2015: With Special Feature on Racial and Ethnic Health Disparities*. Hyattsville, MD: US Department of Health and Human Services.

41 Smedley, B.D., Stith, A.Y., and Nelson, A.R. (eds.) (2002). *Unequal Treatment: Confronting Racial and Ethnic Disparities in Health Care*. Washington, DC.: National Academies Press.

42 Gibbons, M.C. (ed.) (2007). *eHealth Solutions for Healthcare Disparities*. New York: Springer.

23

Precision Medicine and Health Disparities

Consuelo H. Wilkins[1, 2, 3], Nancy J. Cox[1], Sarah C. Stallings[1, 3], Maria F. Lima[4], Roy E. Weiss[5], and Nishadi Rajapakse[6]

[1] *Vanderbilt University Medical Center, Nashville, TN, USA*
[2] *Meharry Medical College, Nashville, TN, USA*
[3] *Meharry-Vanderbilt Alliance, Nashville, TN, USA*
[4] *CUNY School of Medicine, The City College of New York, New York, NY, USA*
[5] *L. M. Miller School of Medicine, University of Miami, Miami, FL, USA*
[6] *Division of Scientific Programs, National Institute on Minority Health and Health Disparities, National Institutes of Health, Bethesda, MD, USA*

23.1 The Promise of Precision Medicine

23.1.1 Precision Medicine Defined

The term precision medicine (PM), coined by a National Academies of Science committee [1], refers to approaches that target an individual's unique characteristics (e.g., genetic, cultural, or lifestyle) to maximize the benefits and lower the likelihood of adverse effects from medical interventions.

This concept originated from the observation that people vary in their response to external exposures, including food, medications, physical activity, and environmental conditions. Such variations are based on each individual's mix of genetic and environmental influences and manifest themselves across the population as a range of symptoms and pathogenicity. The underlying principle of precision medicine is that if the causes of variation in outcomes can be discovered, interventions, including prevention and medication, could be tailored to individual needs and administered based on the likelihood for best outcomes. Consider, for example, the genetic association with clinical symptom subclassifications of type 2 diabetes (T2D) patients [2]. As what underlies how people vary becomes better understood, physicians become able to identify and treat not just the population with "diabetes," but also the T2D subpopulations with one of three subtypes: subtype 1, which is characterized by diabetic nephropathy and diabetic retinopathy complications; subtype 2, where cancer and cardiovascular diseases (CVDs) predominate; and subtype 3, most strongly associated with CVDs, neurological diseases, allergies, and human immunodeficiency virus (HIV)

The Science of Health Disparities Research, First Edition. Edited by Irene Dankwa-Mullan, Eliseo J. Pérez-Stable, Kevin L. Gardner, Xinzhi Zhang, and Adelaida M. Rosario.

Table 23.1 Examples of PM that have been incorporated into clinical care.

Personalized diets	In Phenylketonuria (PKU), genetic polymorphisms in the gene for the phenylalanine hydroxylase enzyme result in lower levels of this enzyme causing accumulation of the amino acid phenylalanine and subsequent damage to the brain. Infant screening identifies those for whom phenylalanine is withheld from their diets, allowing normal or near normal brain development. This tailoring of diet to the genetic predisposition reduces the risk of adverse clinical consequences, and the approach is easy to implement in a typical clinic setting, can easily be explained to parents and is economically viable since the test is reimbursable by insurance or government payers.
Finding the proper dose of the proper drug	Pharmacogenomics is prescribing based on genetic [4] variations that affect drug response or adverse effects, such as genotype-guided antiplatelet therapy prescribing [5]. (see more on pharmacogenomics, in section 23.1.4)
Individualized cancer treatment	Cancer diagnostics [4] Cancer treatment targeted to molecular markers [6–8] and immunotherapy targeting patient's immune system [9]
Bone marrow transplant matching	HLA typing for improving the accuracy of the match for improved transplantation outcomes [10, 11]

infections. Specialized preventive or therapeutic interventions targeted to specific subdisease types can maximize benefits and minimize adverse effects [3].

While there are many new precision medicine examples, the most successful and low-cost precision medicine approach in clinical use to date is that of blood typing and blood type-specific transfusion. Table 23.1 includes examples of PM that have been incorporated into clinical care.

23.1.2 Natural History of Precision Medicine

Progress in PM proceeds with the availability of rapid, valid, and inexpensive methods for identifying the sources of variation. The breakthroughs in technology for PM emerged slowly until the late 1980s, when the Human Genome Project was launched.[1] As chronicled in historical accounts of PM's development [1, 11], a need for biomarkers and data analysis lagged behind collection [11]. Efforts to pursue modern precision medicine remained slow until 2015, when the Precision Medicine Initiative was launched by President Obama [12, 13]. Since then, we have witnessed rapid successes in PM efforts.

The concept of tailoring approaches that depend on the presence of specific characteristics in individuals or population subsets has expanded beyond genomics and biomarkers, and even beyond the individual. Precision medicine has given rise to the concept of precision public health, since large numbers of people in a unique population subgroup could be targeted for improved health outcomes using group-tailored interventions [14]. Precision

1 www.genome.gov/12011238/An-Overview-of-the-Human-Genome-Project

public health is anticipated to progress hand-in-hand with PM, especially in prevention and risk-reduction and in the use of genetic variability to address health disparities [15, 16].

23.1.3 Precision Medicine and Health Disparities

Specialized preventive or therapeutic interventions targeted to specific subdisease types can maximize benefits and minimize adverse effects. In addition to biological differences, there are health disparities (HD), or differences in healthcare outcomes across populations, that vary according to nonclinical contributors to health, including race/ethnicity, gender, societal stresses, and insurance status.[2] These social, behavioral, and cultural influences on the health status of populations across their lifespans have been called social determinants of health (SDH). Data-related SDH have not been regularly collected by health providers or health systems. In this context, if we apply the PM definition, the collection and use of nonclinical determinants can inform the causes for the disparities and suggest remedies that could lead to greater health equity.

An expanding body of literature regarding the impact of SDH on clinical outcomes and on quality metrics has led to calls to standardize the collection of SDH in electronic health records (EHR) [17–19]. The Institute of Medicine (IOM) and Centers for Medicare and Medicaid Services (CMS) have both proposed tools to collect social factors known to influence outcomes and contribute to health disparities. Table 23.2 provides examples of social determinants associated with health disparities and measures of those determinants, including those available in the PhenX Toolkit (for prospective data collection) and EHR data availability (for retrospective data analysis).

23.1.4 Pharmacogenomics-Biologic Predictors of Disparities in Treatment Response

While the growing body of pharmacogenomics research has significant potential for guiding treatment decisions, the persistent heterogeneity of observed treatment responses in many clinical situations suggests that additional genetic and other biologic factors may contribute to the success or failure of a given treatment approach in individuals of different racial and ethnic backgrounds. Such differences may significantly contribute to ongoing health disparities, particularly as they relate to potential decreased therapeutic efficacy or increased risk of side effects. Notable examples of variation in safety and efficacy in African Americans as compared with other demographic groups include: decreased virologic failure during efavirenz therapy in HIV-infected participants with a slow metabolizer genotype [20]; increased risk of treatment failure in African Americans treated for asthma [21]; and altered risk profile in African Americans receiving spironolactone for severe heart failure [22]. The implications of these examples are striking: each of these conditions represents a known area of health disparity among African Americans, and systematic variations in treatment response could significantly contribute to widening these disparities.

2 www.cdc.gov/minorityhealth/chdireport.html

Table 23.2 Social determinants of health measures that can be collected prospectively using standardized approaches and retrospectively using Electronic Health Records (EHR).

Key areas	Examples/elements	Measures	In EHR[b]
Race, ethnicity, culture	Demographics	Ethnicity[a]	Y
		Race[a]	Y
		Multigroup ethnic identity[a]	N
		Current age[a]	Y
		Sex	Y
		Gender[a]	N
		Current marital status[a]	Y
		Sexual identity / orientation[a]	P
		Years living in the United States[a]	N
Socioeconomic status	Income neighborhood	Annual family income[a]	N
Economic stability – (availability of resources to meet basic needs: food, housing, medical care, and heat) *also called: (financial resource strain)*		Current address[a]	Y
		Neighborhood concentrated	N
		Disadvantage[a]	
		Homelessness[a]	P
		Census tract median income	P
Education		Current educational attainment[a]	Y
		School risk and protective factors[a]	N
		Child-reported parental education attainment[a]	N
Social and community context *also called: (social connection/isolation)*	Social support	Social support[a]	N
		Social isolation[a]	N
	Social cohesion	School social environment[a]	N
		Neighborhood collective efficacy –community cohesion and informal social control[a]	N
	Home, work, recreational environmental exposures	Characteristics of current residence[a]	N
		Air contaminants in the home environment[a]	N
		Home and workplace exposures to floor and wall materials[a]	N
		Exposures from hobbies[a]	N
		Exposure at work and in daily life[a]	N
		Lifetime farm exposure[a]	N
Built environment	Physical activity	Total physical activity – comprehensive[a]	N
		Sitting/sedentary behavior[a]	N

Table 23.2 (Continued)

Key areas	Examples/elements	Measures	In EHR[b]
	Tobacco use	Self-reported smoking status	Y
		Point of sale environment for alcohol and tobacco[a]	N
		Tobacco retailer density/proximity[a]	N
	Alcohol use	Self-reported alcohol intake	Y
		Unmet need for alcohol or drug treatment[a]	N
		Family history of substance use problems[a]	Y
		Peer/partner substance use and tolerance of substance use[a]	N
	Exposure to crime and violence	Exposures to violence[a]	N
		Neighborhood safety[a]	N
		Adverse childhood events	P
	Stress	Perceived stress[a]	P
		Chronic stress[a]	N
		Job strain[a]	N
Mental health/quality of life	Depression	Depression[a]	Y
		Insomnia[a]	Y
	Discrimination/ racism	Discrimination	N
	Overall quality of life	Physical, social, and mental health functioning (SF-36v2®)[a]	N
		Current quality of life[a]	N
		Disability index[a]	N
Healthcare	Access to healthcare	Health insurance coverage[a]	Y
		Professional dental care[a]	N
		Health and wellness before, during, and after pregnancy[a]	N
	Healthcare services quality	Service utilization and cost[a]	Y
		Quality of care[a]	N
		Treatment availability and level of care[a]	N, N
		Substance abuse treatment quality[a]	N
		Health quality index for economic evaluations[a]	Y
	Health literacy	Brief health literacy screen	N

[a] available in PhenX toolkit.
[b] Y = Yes, N = No, P = Proposed.
Sources: Compiled from IOM, ONC, and PhenX sources, and Social Determinants of Health in the EHR project conducted by Vanderbilt-Miami-Meharry Precision Medicine and Health Disparities Collaborative.

The published literature, as a body of existing data, represents an important opportunity for employing analytic approaches to understanding safety and efficacy patterns in the literature that may further inform clinical practice and future research related to the care of individuals in different racial and ethnic backgrounds. To begin to explore potential biologic factors that may affect treatment response, including potential differences in safety and efficacy, we undertook a quasi-systematic literature review approach. The intent of this analysis was to inspire hypothesis formation to develop a deeper understanding of the biologic explanations underpinning some of these results, as well as to inform selection of therapeutic courses when feasible and appropriate.

We used data aggregated by the IQVIA Institute for Human Data Science[3] (formerly known as IMS Healthcare Informatics) to identify the most prescribed brand name and generic drugs in the United States [23], prioritizing for the literature review those with U.S. Food and Drug Administration (FDA) approved indications for treatment of conditions with known health disparities among African Americans.

Selected examples from our application of this method to the literature describing efficacy and safety of the antipsychotic class are outlined in Table 23.3. Antipsychotics are prescribed for a range of mental health issues, including schizophrenia, and reflect a number of conditions with known racial disparities in health and healthcare [30, 31]. In the studies restricted to analysis of outcomes by race and ethnicity, we observed notable signals in the literature comparing safety and efficacy outcomes among African Americans and other groups. In one example, African Americans were more likely to experience increased efficacy of antipsychotic treatment as compared with Hispanic patients. Conversely, African Americans exposed to an antipsychotic agent were more likely to experience moderate neutropenia [32] and weight gain [25] as compared with other groups.

23.1.5 Existing Gaps in Precision Medicine

Precision medicine and precision public health rely on the premise that determinants of disease incidence and differences in response to interventions can be identified and their biology well enough understood that applications to reduce risk of disease and improve treatment can be developed. Even when FDA guidelines for genotyping to improve therapeutic outcomes are followed, established factors explain only a small fraction of the inter-individual variability in both efficacy and adverse events for most drugs.

We have only just begun to integrate sequence variation with structural genomic changes (deletions, duplications, inversions) and to examine both rare and common genetic variants. We are also only beginning to understand how ribonucleic acids (RNAs)—including mitochondrial RNA (miRNA), long non-coding RNA (lncRNA), enhancer RNA (eRNA), and small interfering RNA (siRNA)—help to regulate gene expression and translation of messenger ribonucleic acid (mRNA) into protein and how environmental attributes (e.g., alcohol use, pollution, diet) affect response to exposures.

Even as we gain this understanding, we will still have to learn whether genomic (and other biomarkers) will add economic value to standard approaches, where healthcare

3 https://www.iqvia.com/institute

Table 23.3 Examples of disparate efficacy and safety signals related to antipsychotic therapy in African Americans.

Author (year)	Study design; data type(s)	Age group	Population description	Agents	<u>Efficacy</u> in African Americans	<u>Safety</u> in African Americans
Subgroup analyses of clinical data only						
Maher et al. 2013 [24]	Retrospective case series	Children/adolescents	African American (28); Caucasian (48); Asian (4) Native American (1)	Clozapine	Not analyzed	 African Americans more likely to develop moderate versus mild neutropenia compared with other groups
Krakowski et al. 2009 [25]	Secondary analysis of RCT	Adults	Schizophrenia or schizoaffective disorder African American (56); Caucasian (17); Hispanic (19)	Clozapine olanzapine	Not analyzed	 African Americans gained weight (mean 2.64 kg) while a combined Caucasian and Hispanic group lost weight (mean 0.68 kg)
Patel et al. 2006 [26]	Retrospective case series	Children/adolescents	Schizophrenia spectrum African American (38); Caucasian (30); Hispanic (37)	Risperidone	 Increased efficacy among African Americans at 90 days compared to Hispanic patients	Not analyzed

(*Continued*)

Table 23.3 (Continued)

Author (year)	Study design; data type(s)	Age group	Population description	Agents	Efficacy in African Americans	Safety in African Americans
Analyses combining clinical and genetics data						
Tiwari et al. 2010 [27]	RCT	Adults	African American (57); Caucasian (147); Other (13)	Clozapine, olanzapine	Not analyzed	↓ Allele significantly associated with increased risk of weight gain among African Americans was also significantly associated with decreased weight gain among Caucasians
Fijal et al. 2009 [28]	Secondary analysis of RCT	Adults	African American (78); Caucasian (65)	Risperidone	↑ Significant association between efficacy and 3 SNPs only in African Americans; 2 SNPs only in Caucasians; 1 significant SNP in both	Not analyzed
Muller et al. 2005 [29]	Pooled analysis of two cohorts	Adults	African American (57); Caucasian (77); Other (11)	Clozapine, haloperidol, olanzapine, risperidone	∅ Genotype associated with increased efficacy in Caucasians; no effect among African Americans	Not analyzed

Key: ↑ = Increased; ↓ = decreased; ∅ = no effect.
Sources: Original data from systematic literature review described above; Wilkins et al. manuscript in preparation.

providers choose interventions based on commonly understood patient characteristics and clinical intuition. The economic and clinical value of biomarker-based dosing has been challenged, for instance, in the selection of loading or maintenance doses for warfarin and related anticoagulants, such as clopidogrel [33]. The value proposition for using genetic variation in PM depends significantly on whether genetic information is gathered at point-of-care interventions or collected preemptively for patients as needed for decisions on drug choice and dose. In the former, genetic testing is reactive and often focused on a single drug or drug class. This is dependent on clinicians ordering the genetic testing, which may be more expensive and time-consuming than prospectively collecting and embedding data on multiple pharmacogenetic variants in EHRs [34]. Economic modeling must consider these alternative cost structures in evaluating the costs/benefits of PM interventions. It will also be important to identify ways of standardizing identified biomarkers and to establish methods for reliably sharing such information, so that when patients change providers, biomarkers do not need to be re-evaluated. There is considerable uncertainty on what tests or biomarkers will be reimbursed by insurance companies or government agencies, with substantial variability by location and payer.

The delivery of information to healthcare providers is a known challenge. Few medical professionals have extensive education in genomics or in the new biomarkers being developed. Specialized decision support tools that work hand-in-hand with EHRs used in doctors' offices and hospitals will be required for full implementation of PM. Similarly, we need to develop substantial additional and continuing education for healthcare providers, which may be particularly important for providers practicing in nonacademic centers or clinics. Some providers may also be sensitive to the use of PM biomarkers for the fear of being accused of racism or sexism, since biomarkers tend to vary by race and sex.

23.2 Methods in Precision Medicine and Applications in Health Disparities

23.2.1 Methods of Characterizing Individuals

23.2.1.1 Genetics Approaches: Genome-wide Association Study and Phenome-wide Association Study

Differences in human phenotypes (i.e., traits such as height, eye color, race, or some disease states) have varying levels of genetic and environmental components. Family studies conducted before the advent of modern molecular biology clearly showed traits are inherited, and diseases can sometimes be traced many generations back to distant ancestors. As deoxyribonucleic acid (DNA) markers replaced the modest number of blood group markers to characterize human variation, traits such as hypertension and some Mendelian diseases were mapped to certain regions of the genome using linkage analysis approaches. This advance, though informative, remained crude since only traits in which individual genetic variants accounted for most of the phenotypic variation in the population could be localized with linkage mapping. Nevertheless, some disease genes were discovered through this approach.

Ultimately, linkage mapping and small-scale candidate gene studies gave way to advances in technology that made it possible to genotype or sequence large parts of the genome in a relatively inexpensive way. Even the nongenotyped areas could be inferred through multiple imputation techniques [35]. The resulting genome variation could then be tested for association with a phenotype in a specific race/ethnic group so as to avoid spurious results due to population stratification and/or admixture. Such approaches, termed genome-wide association studies (GWAS), have greatly improved the identification of the genetic component to map common human diseases, quantitative biomarkers of interest, and pharmacogenomic phenotypes. Subsequent extensions of GWAS have moved beyond testing individual DNA variants and allow gene-based tests using single nucleotide polymorphisms (SNPs) local to a gene [36], or SNPs that are predictive of mRNA transcript levels measured for a gene [37, 38], all of which may contribute to explaining disparities in disease incidence and response to medical interventions.

As we have developed a better understanding of how genome variation affects human phenotypes, it has become clear that genome variation can be pleiotropic and affect multiple traits concurrently. This knowledge of genetic pleiotropy, together with existing databases of genotyped individuals with multiple traits (e.g., those in electronic medical records), led to development of the concept of phenome-wide association study (PheWAS) in which a given genetic variant is tested for association with multiple traits/phenotypes, making it possible to cluster traits. If genotypes are not available, one can indirectly infer genes or polymorphisms that might drive such traits. This approach is theoretically likely to identify diseases for which known treatments could work (e.g., in the context of off-label use).

For example, if a gene is associated with two or more diseases and a treatment is developed to inhibit that gene product in the treatment of one of the diseases, then the new treatment might also have value for treatment of the second disease. Thus, characterization of individuals at the genome and phenotype level is likely to continue to yield benefits, especially those that could help us understand disparities in disease incidence, prevalence, and response to treatment. Both genome variation and the medical phenome are big data, which creates a massive discovery space: many combinations of variables will be tested for association. Thus, discoveries must meet stringent criteria for significance and undergo replication and functional studies.

Advancements in molecular biology shifted the focus of research to characterizing patients based on the differences in their genomes. While the approach is appealing and was expected to explain a large proportion of the variation in disease susceptibility and treatment outcomes, genomic approaches continue to reveal the complexity of common human diseases and the response to drug therapies. The complex nature of the correlations among genome variants reveals how difficult it can be to know which of the many variants with high correlation may actually be affecting the phenotypes to which all of those variants show association. Discerning the function of noncoding variation is more challenging than for coding variation. Individual variants tend to have small effect sizes for common diseases. Even when multiples of the significantly and reproducibly associated variants are combined in a genetic score, the variance explained remains small, often less than 10%. This reflects both the limitation of including only the most well-established genetic risk factors when there are likely to be tens to thousands of variants with similar effect sizes that have not yet met those same criteria, and also the limitation of having conducted the discovery research using only the common variants that can be assayed on inexpensive genotyping chips. When larger numbers of associated SNPs are included in developing risk scores, a much larger proportion of the inter-individual variation may be explained, and, similarly, when the contribution of both rare and common variants are considered. Recent publications on height, for example, suggest that much of the heritability of height can be explained using both of these sources of variation [35].

The evidence to date on the effect sizes for rare coding variants contributing to common human diseases and related quantitative traits is mixed. Early onset diseases with a clear consequence on fitness may derive notable susceptibility from coding variants that are quite rare because of the selection against them. But there is little evidence for large-effect-size coding variants in late-onset common human diseases [36], even in tens of thousands of cases and controls. While it is widely recognized that nongenetic factors influence the risk of common diseases and both the adverse events and efficacy of drug therapies, we do not always know which aspects of diet and lifestyle are contributing to which phenotypes. We generally have little information on environmental exposures in the discovery research currently conducted. While summary measurements of exposure, such as DNA methylation studies, metabolomics studies, and studies of the microbiome, may ultimately provide more information on environmental exposures and their physiological consequences (which will improve the ability to detect the effects of genetic variation and potential interactions of genetic variation with exposures), large-scale studies of exposure remain rare and few of these technologies have been developed for routine clinical use.

Nonetheless, rigorous studies are ongoing to identify noncorrelated DNA-, ribonucleic acid- (RNA), and/or protein-based biomarkers that can be measured relatively noninvasively and with sufficient precision to be used in an integrated fashion to characterize individuals for disease susceptibility and response to interventions, such as treatment with pharmaceutical products. While identification of genetic biomarkers for PM has mainly been done within specific race/ethnic groups, newer approaches combine race/ethnic groups so as to leverage differences in haplotype structures across ancestries (e.g., Bayesian trans-ethnic meta-analyses) and maximize discovery of genetic biomarkers [39, 40].

23.2.1.2 Proteomics and Metabolomics

Recent work, driven by advances in high-performance liquid chromatography and mass spectrophotometer (ms/ms spec) technologies, has led to relatively inexpensive approaches that measure protein and metabolite profiles in human samples. Many diagnostic tests that characterize individuals have become available and contributed to improvements in PM. Recent examples include findings that individuals with higher concentrations of branched chain amino acids detected from metabolomics profiles are at increased risk for diabetes, a finding that has been replicated in various studies. High-sensitivity assays, such as C-reactive protein or cardiac troponins, have also been identified and used in disease diagnosis and disease susceptibility prediction models.

Despite the promise of metabolomics and proteomics approaches, several barriers still limit their use in routinely characterizing individuals for disparities. First, the proteins or metabolites identified tend to be unstable and are very sensitive to simple changes in patient characteristics, such as fasting state, diet, or medications. Second, large amounts of multidimensional, highly correlated data are produced, making them difficult to use without sophisticated training in statistical procedures. Third, protein biomarkers are difficult to replicate in their measurement, making them less desirable as biomarkers for studying disease disparities or as biomarkers for disease susceptibility. Use of the rapidly emerging radiomics technology will allow precise phenotyping based on quantitative data from magnetic imaging studies, computerized tomography, and nuclear medicine imaging [41].

23.2.1.3 Person-specific Modeling

Most work that has identified disease risk factors is based on prospective epidemiologic studies that model baseline characteristics of study participants for association with clinical outcomes, such as incidence of disease. While this approach has led to major medical breakthroughs, including approval of drugs in clinical trials, there are a number of limitations. First, changes in risk factor profiles between baseline and the time of disease onset are rarely known, making it difficult to accurately model causes of clinical outcomes. Second, risk factors identified at the population level and often based on a single baseline measurement, do not explain increased risk of disease in some population subgroups (e.g., typical CVD risk factors or poorly identify CVD risk in middle-aged women). One approach to circumvent this limitation and improve personalization of risk prediction is to collect dense longitudinal data from individuals before they develop the attribute of interest and use sophisticated computer algorithms to create risk factor trajectories. The latter are then tested for association with incident clinical events.

Various techniques, such as machine learning algorithms, growth curve modeling, and clustering approaches, can be used to perform person-specific modeling to determine risk factors that might be unique to an individual's disease risk. Person-specific changes in body mass index and blood pressure over long periods of time are being actively investigated to determine whether unique within-person variability in risk factors can effectively predict major disease events. Though person-specific modeling approaches have not been used extensively in medicine, they are being used in psychology to identify individual risk factors for events. Some commonly used person-specific modeling techniques are described elsewhere [42, 43] and have been used to identify individuals at risk for events such as suicide, loss of physical function, and other conditions. Person-specific modeling approaches, though very appealing in terms of personalizing risk profiles and accounting for interaction between social and nonsocial variables, are limited by their complex nature, since they rely mainly on Markov chain and Bayesian modeling techniques. They are also limited by lack of generalizability, since they are not easily transferable to population subgroups that were not included in the study. Nonetheless, person-specific modeling approaches could be extended to studies on disparities and PM for many conditions.

23.2.1.4 Data Integration

With advances in electronic data capture and genomic technologies, such as GWAS, deep DNA and RNA sequencing, metabolomics, and proteomics, researchers and healthcare providers gain large amounts of data that could potentially improve clinical outcomes. However, a major limitation exists in how such data could be integrated, analyzed, interpreted, and incorporated in patient care. Despite the great progress that has been made in applying machine learning techniques to cluster various facets of data and predict outcomes, current data integration models lack uniform standards in their application, and when data are integrated, several components are lacking. Other issues remain a problem for effective data integration. If data are generated from different sources and on different instruments, data validation, harmonization, storage, and ownership remain major concerns.

23.2.2 Methods of Predicting Disease Risk

The success of PM depends on effective identification of population subgroups that are likely to benefit the most from a given preventive or therapeutic approach. Traditionally, identification of individuals likely to experience harm (high-risk groups) or those less likely to respond (poor responders) has been based on association analyses. In this approach, data from cross-sectional studies or, preferably, from prospective cohort studies or clinical trials, are analyzed to determine whether certain exposures (e.g., lifestyle attributes like smoking, medications, or genetic polymorphisms) are associated with a given clinical outcome (e.g., incidence of myocardial infarction [MI], a binary outcome, or elevated LDL-C [low-density lipoprotein-cholesterol], a continuous outcome). For binary outcomes, logistic regression, or Cox-regression models, are typically used to estimate measures of disease risk, such as the relative risk or the hazard ratio in the case of prospective studies or the odds ratios in the case of cross-sectional studies. For continuous outcomes, disease risk is typically assessed using linear regression or analysis of variance models in which differences in the mean between those exposed and those who are not are used.

For instance, individuals with a certain SNP in the gene for the enzyme proprotein convertase subtilisin/kexin type 9 (PCSK9) will tend to have high levels of LDL-C and are at greater risk for developing myocardial infarction. To identify the increase in risk of CVD associated with high levels of PCSK9, linear regression models are used to estimate the mean difference in LDL-C concentrations between those who have the SNP compared to those who do not. Since the risk of MI associated with each $mg\,dl^{-1}$ increase in LDL-C concentration is known, the risk of MI can be computed indirectly. On the other hand, if data are from a prospective study with well-curated clinical events such as MI, then the risk of disease can be computed directly by estimating the relative risk or hazard ratio associated with carrying one or two alleles of a given genetic variant when compared to those who do not carry the variant allele.

More recent work on disease risk has utilized existing algorithms (e.g., the Framingham Risk Scores) or the American Heart Association/American College of Cardiology calculators to estimate 5-year, 10-year, or lifetime risk for CVD. Newer tools based on machine learning (e.g., artificial neural networks) are being developed for use with EHRs to automatically detect individuals at a greater risk for medication side effects or other clinical outcomes [44, 45].

While much progress has been made in detection of disease risk, the methods remain inadequate, and several individuals without typical risk factors used in disease risk models develop major clinical events such as MI or stroke. Thus, more studies that account for disparities in populations are needed to refine the existing disease risk algorithms.

23.3 Future Directions and Research Needs

In the following sections, we describe future directions and research needs that can improve current efforts in PM.

23.3.1 Engaging Minority Communities

The success of PM also depends on treatment response prediction models that are inexpensive, are easy to use, and result in health outcomes that are significantly better than when decisions are simply based on population averages or the clinical intuition of an experienced physician. Many existing prediction algorithms do not perform well in subpopulations that were underrepresented in studies that gave rise to the data used to develop the algorithms. For example, most US prospective studies that have adequate follow-up either have few or no minorities. When such studies are used for algorithm development, the resulting models are inadequate for use in underrepresented groups. Such observations have been made for atherosclerotic CVD models, which show poor performance among individuals with diabetes, since the latter were underrepresented in cohorts used to develop the models. Future efforts for PM will need to increase the numbers of underrepresented populations in studies. Furthermore, there is a need to demonstrate clear economic and clinical benefit of adding genomic information to standard clinical approaches to medical dosing when trying to improve clinical outcomes or patient-reported outcomes. Efforts to develop prediction models are needed for specific population subgroups, as well as efforts to prove that models perform equally well across subpopulations. This will be particularly important for models that rely heavily on genomic data, since allele frequency varies by genomic ancestry.

23.3.2 Engaging Health Providers

A retrospective cohort study among US veterans with diabetes shows that compared to non-Hispanic Whites, non-Hispanic Blacks are less likely to start statin therapy, a potent intervention for reducing CVD [46]. Precision medicine algorithms could inform prompts for healthcare providers to initiate statin therapy in underrepresented minorities that tend to have the largest burden of cardiometabolic conditions. This should go hand-in-hand with efforts to engage healthcare providers, especially minority practitioners and those working in rural areas, to engage their clients to participate in research studies.

23.3.3 Challenges Unique to Precision Medicine

A lack of disease prediction algorithms that can be used easily, economically, and universally in typical clinical practice settings (e.g., nonacademic medical centers or clinics) is a major challenge, yet there are other challenges that will need to be overcome. To begin with, the concept of PM is relatively new and its boundaries and expectations keep changing. There is a need to standardize its terminology and implementation. Precision medicine biomarkers that are very promising in terms of guiding selection of treatment regimens have mainly been evaluated in academic settings and lack thorough evaluation for economic viability and feasibility of implementation in real-world settings. For many conditions that cause untold suffering to populations there are not enough data to develop robust prediction models, especially in some population subgroups. Precision medicine biomarkers for many disease conditions are not yet eligible for reimbursement from insurance companies and are too expensive for patients to pay for out of pocket. Resources for characterizing individuals, especially in rural and underdeveloped countries or even non-tertiary hospitals in urban areas of developed countries, are not available. Thus, it is possible that healthcare providers may be reluctant to offer uninsured/underinsured patients an opportunity to be fully characterized (e.g., through genotyping) for tailored interventions. Furthermore, minority populations that tend to carry the largest burden of diseases and, therefore, are a natural target for PM efforts to reduce health disparities may be less likely to trust PM efforts that include considerations of race in their implementation. These attributes could greatly undermine PM efforts.

More research is needed to evaluate proposed and emerging PM approaches for their economic feasibility in the current or anticipated healthcare settings. As pointed out by Verbelen and others [47], several candidates for PM (e.g., pharmacogenetic markers) have not yet received a comprehensive, or indeed any, evaluation for their cost-effectiveness or cost-saving potential. Such evaluations on cost are urgently needed so that they can be used together with results from efficacy studies to enable implementation of PM.

23.4 Genomic Health Literacy

The field of genomics is broad and encompasses processes from traditional genomic variation to epigenetic processes in the genome. Enhanced training of healthcare providers in areas that are key to PM is required. About 50% of practicing physicians have never had a formal course in genetics, and 30% of medical schools do not include a course in genetics in their curriculum. Training providers on the usefulness of such information could be valuable. A detailed knowledge of genomics and its role in health is difficult to fully grasp in practical terms, even for highly educated individuals. Conveying genomic and PM information to patients is a challenge, especially in a typical 15–20-minute provider encounter. Nonetheless, educating populations on basic concepts in genomics and exploring basic links with disease incidence, response to interventions, and some disparities in health is important. While all population strata would likely benefit from health literacy focused on genomics, young people are likely to be more receptive and grow into these concepts. Pilot projects on developing curricula for genomic health literacy in middle schools, high schools, or colleges would likely be very useful.

23.4.1 Cross-training in Precision Medicine and Health Disparities

While the reasons for some health disparities are unknown, many disparities occur due to known unaddressed risk factors usually related to SDH. Practitioners of effective PM will require training in how to identify drivers of disparities and how such drivers affect medical interventions. Since health disparities affect PM, ability to identify and address disparities should be part of future medical education. Providers already in practice could undergo health disparities training. In the long term, creating a specialty in PM where practitioners can obtain certification could also help to enhance the number of providers with deep knowledge in disparities and PM. Producing a cadre of healthcare providers who have ample training and are certified in health disparities could have a great impact in enhancing development, evaluation, and implementation of PM in typical clinical practices.

23.4.2 Integration of Cultural, Social, and Environmental Data

It is well established that cultural, social, and environmental factors individually or in combination contribute to health disparities. If measured effectively, they could be great avenues for addressing disease disparities and could enhance PM efforts. However, valid instruments for assessing these attributes are not common and, when available, they may not be generalizable enough. Furthermore, their integration into PM algorithms is limited by a lack of studies showing their joint impacts on clinical outcomes. Healthcare providers will need to work closely with social scientists, psychologists, environmentalists, and other stakeholders to understand how variables could be integrated into disease prediction models so as to identify individuals who could better benefit from health interventions. Use of novel approaches such as machine learning tools could help the integration of traditional and nontraditional variables in identifying health disparities and avenues for PM.

23.5 Success in Precision Medicine

While global metrics for success in PM will be difficult to define since they will be context-dependent (e.g., in terms of type of disease, efficacy in classifying patients, whether cost-effective or cost-saving), measures of success can be envisaged for individual disease processes. We think PM will be considered successful if most of the criteria below are met for a given disease process and for an individual patient:

a) Important biological, behavioral, cultural, or socioeconomical markers are identified and when used individually or in combination to choose a treatment plan, lead to meaningful changes in clinical outcomes, such as improvement in quality of life (or other patient-reported outcomes), better disease control (e.g., controlled blood pressure, hemoglobin A1c [HbA1c], LDL-C), or reduction in rates for adverse events, hospital readmission, or recurrent events.
b) Use of PM markers is affordable and reimbursable by public or private payers and is acceptable and available to healthcare providers regardless of their institutional affiliation or the community they serve.
c) Identified PM markers can be measured at the point of care and made available in electronic medical records regardless of race, sex, geography, socioeconomic status of patients, and other indicators of discrimination/disparities.
d) A system is available to make it possible for standardized invariant PM markers (e.g., genotypes) to be shared across providers, even in different healthcare networks.

e) Both patients and providers, regardless of age or place of medical practice, attain a clear understanding and comfort in using PM without fear of litigation or of being targeted for discrimination (especially in minority patients).
f) The practice of PM leads to substantial reductions in health disparities.
g) Use of PM markers results in cost savings that improve overall access to and delivery of healthcare.

For an individual or combination of PM markers, success will be attained when failure to use them will be considered a departure from the norm and healthcare providers are required to justify their failure to use such markers. This will be similar to the current standard where blood typing and matching of blood type before blood transfusion is a requirement or norm and departure from this standard protocol (e.g., in an emergency situation) requires a scientific or clinical justification.

23.6 Conclusion

While health disparities have been described mainly as negative factors in terms of the efficiency of public health systems, knowledge of health disparities is important for discovery of what causes differences in disease incidence and response to interventions. If well studied, health disparities will inform PM approaches, which aim to improve benefits from a given treatment or preventive approach while minimizing the risk of harm. Effective PM will require stakeholders, including members of the general population and healthcare providers, to be educated on the key concepts and promise of PM. Successful development and implementation of PM will require the active participation of underrepresented groups (e.g., racial minorities or carriers of a genetic variant) in research studies and for researchers to develop disease or treatment response models that are accurate, inexpensive, and easy to use in typical healthcare settings and that clearly add value when used in addition to standard clinical algorithms. While most efforts on PM have focused on treating diseases, efforts are also needed for precision preventive medicine and precision public health, since the three are interrelated.

23.7 Key Points

- Precision medicine, which focuses on individual variation in disease incidence, effectiveness of preventative strategies, and treatment response, has the potential to reduce disparities in health outcomes.
- Novel approaches are needed to optimize genomic analyses in diverse populations.
- Effectively measuring and integrating clinical, biological, social, cultural, and environmental data is an important strategy to address health disparities and could enhance PM efforts.
- Strategies are needed to overcome barriers to implementation of PM, such as limited genomic health literacy and concerns about data privacy, which may be more common among racial and ethnic minorities.

Acknowledgments

The authors wish to thank Edmond K. Kabagambe, DVM, PhD, Vanderbilt University Medical Center, and Ana Palacio, MD, MPH, University of Miami Miller School of Medicine, for their significant contributions to this book chapter. Some of the work reported in this publication was supported by the National Institutes of Health (NIH) under Award Number U54MD010722. The content is solely the responsibility of the authors and does not necessarily represent the official views of the NIH.

Disclaimer

The views and opinions expressed in this chapter are those of the authors only and do not necessarily represent the views, official policy, or position of the U.S. Department of Health and Human Services or any of its affiliated institutions or agencies.

References

1 National Research Council (2011). Committee on a framework for developing a new taxonomy of disease. In: *Toward Precision Medicine: Building a Knowledge Network for Biomedical Research and a New Taxonomy of Disease*. Washington, DC: National Academies Press.

2 Li, L., Cheng, W.Y., Glicksberg, B.S. et al. (2015). Identification of type 2 diabetes subgroups through topological analysis of patient similarity. *Science Translational Medicine* 7: 311ra174.

3 Lyman, G.H. and Moses, H.L. (2016). Biomarker tests for molecularly targeted therapies – the key to unlocking precision medicine. *New England Journal of Medicine* 375: 4–6.

4 Dugger, S.A., Platt, A., and Goldstein, D.B. (2017). Drug development in the era of precision medicine. *Nature Reviews Drug Discovery* 17 (3): 183–196. https://doi.org/10.1038/nrd.2017.226.

5 Cavallari, L.H., Lee, C.R., Beitelshees, A.L. et al. (2017). Multisite investigation of outcomes with implementation of CYP2C19 genotype-guided antiplatelet therapy after percutaneous coronary intervention. *JACC: Cardiovascular Interventions* 11 (2): 181–191.

6 de Bono, J.S. and Ashworth, A. (2010). Translating cancer research into targeted therapeutics. *Nature* 467: 543–549.

7 Xue, Y. and Wilcox, W.R. (2016). Changing paradigm of cancer therapy: precision medicine by next-generation sequencing. *Cancer Biology & Medicine* 13: 12–18.

8 Buettner, R., Wolf, J., and Thomas, R.K. (2013). Lessons learned from lung cancer genomics: the emerging concept of individualized diagnostics and treatment. *Journal of Clinical Oncology* 31: 1858–1865.

9 Snyder, A., Makarov, V., Merghoub, T. et al. (2014). Genetic basis for clinical response to CTLA-4 blockade in melanoma. *New England Journal of Medicine* 371: 2189–2199.

10 Chalandon, Y., Tiercy, J.M., Schanz, U. et al. (2006). Impact of high-resolution matching in allogeneic unrelated donor stem cell transplantation in Switzerland. *Bone Marrow Transplantation* 37: 909.

11 Olson, M.V. (2017). A behind-the-scenes story of precision medicine. *Genomics, Proteomics & Bioinformatics* 15: 3–10.

12 Terry, S.F. (2015). Obama's precision medicine initiative. *Genetic Testing Molecular Biomarkers* 19: 113–114.

13 Collins, F.S. and Varmus, H. (2015). A new initiative on precision medicine. *New England Journal of Medicine* 372: 793–795.

14 Khoury, M.J., Iademarco, M.F., and Riley, W.T. (2016). Precision public health for the era of precision medicine. *American Journal of Preventive Medicine* 50: 398–401.

15 Meagher, K.M., McGowan, M.L., Settersten, R.A. et al. (2017). Precisely where are we going? Charting the new terrain of precision prevention. *Annual Review of Genomics and Human Genetics* 18: 369–387.

16 Dankwa-Mullan, I., Bull, J., and Sy, F. (2015). Precision medicine and health disparities: advancing the science of individualizing patient care. *American Journal of Public Health* 105 (Suppl 3): S368.

17 Adler, N.E. and Stead, W.W. (2015). Patients in context--EHR capture of social and behavioral determinants of health. *New England Journal of Medicine* 372: 698–701.

18 Palacio, A., Suarez, M., Tamariz, L., and Seo, D.A. (2017). Road map to integrate social determinants of health into electronic health records. *Population Health Management* 20: 424–426.

19 Beach, M.C., Gary, T.L., Price, E.G. et al. (2006). Improving health care quality for racial/ethnic minorities: a systematic review of the best evidence regarding provider and organization interventions. *BMC Public Health* 6: 104.

20 Ribaudo, H.J., Liu, H., Schwab, M. et al. (2010). Impact of CYP2B6, ABCB1 and CYP3A5 polymorphisms on efavirenz pharmacokinetics and treatment response: an AIDS Clinical Trials Group Study. *Journal of Infectious Disease* 202: 717–722.

21 Wechsler, M.E., Castro, M., Lehman, E. et al. (2011). Impact of race on asthma treatment failures in the asthma clinical research network. *American Journal of Respiratory and Critical Care Medicine* 184: 1247–1253.

22 Vardeny, O., Cavallari, L.H., Claggett, B. et al. (2013). Race influences the safety and efficacy of spironolactone in severe heart failure. *Circulation: Heart Failure* 6: 970–976.

23 IMS Institute for Healthcare Informatics (2016). *Medicines Use and Spending in the US. A Review of 2015 and Outlook to 2020*. Parsippany, NJ: IMS Health.

24 Maher, K.N., Tan, M., Tossell, J.W. et al. (2013). Risk factors for neutropenia in clozapine-treated children and adolescents with childhood-onset schizophrenia. *Journal of Child and Adolescent Psychopharmacoly* 23: 110–116.

25 Krakowski, M., Czobor, P., and Citrome, L. (2009). Weight gain, metabolic parameters, and the impact of race in aggressive inpatients randomized to double-blind clozapine, olanzapine or haloperidol. *Schizophrenia Research* 110: 95–102.

26 Patel, N.C., Crimson, L., Shafer, A. et al. (2006). Ethnic variation in symptoms and response to risperidone in youths with schizophrenia-spectrum disorders. *Soc Psychiatry and Psychiatric Epidemiology* 41: 341–346.

27 Tiwari, A., Rodgers, K., Sicard, R. et al. (2010). Association study of polymorphisms in cholecystokinin gene and its receptors with antipsychotic induced weight gain in schizophrenia patients. *Progress in Neuro-psychopharmacology and Biological Psychiatry* 34: 1484–1490.

28 Fijal, B.A., Kinon, B.J., Kapur, S. et al. (2009). Candidate-gene association analysis of response to risperidone in African-American and white patients with schizophrenia. *The Pharmacogenomics Journal* 9: 311–318.

29 Müller, D.J., Klempan, T.A., De Luca, V. et al. (2005). The SNAP-25 gene may be associated with clinical response and weight gain in antipsychotic treatment of schizophrenia. *Neuroscience Letters* 379: 81–89.

30 McGuire, T.G. and Miranda, J. (2008). New evidence regarding racial and ethnic disparities in mental health: policy implications. *Health Affairs (Millwood)* 27: 393–403.

31 Schwartz, R.C. and Blankenship, D.M. (2014). Racial disparities in psychotic disorder diagnosis: a review of empirical literature. *World Journal of Psychiatry* 4: 133–140.

32 Diler, R.S., Ladouceur, C.D., Segreti, A. et al. (2013). Neural correlates of treatment response in depressed bipolar adolescents during emotion processing. *Brain Imaging Behavior* 7: 227–235.

33 Kabagambe, E.K., Beasley, T.M., and Limdi, N.A. (2013). Vitamin K intake, body mass index and warfarin maintenance dose. *Cardiology* 126: 214–218.

34 Schildcrout, J.S., Denny, J.C., Bowton, E. et al. (2012). Optimizing drug outcomes through pharmacogenetics: a case for preemptive genotyping. *Clinical Pharmacology & Therapeutics* 92: 235–242.

35 Howie, B., Fuchsberger, C., Stephens, M. et al. (2012). Fast and accurate genotype imputation in genome-wide association studies through pre-phasing. *Nature Genetics* 44: 955–959.

36 Edwards, S.L., Beesley, J., French, J.D., and Dunning, A.M. (2013). Beyond GWASs: illuminating the dark road from association to function. *American Journal of Human Genetics* 93: 779–797.

37 GTEx Consortium (2013). The genotype-tissue expression (GTEx) project. *Nature Genetics* 45: 580–585.

38 Wheeler, H.E., Shah, K.P., Brenner, J. et al. (2016). Survey of the heritability and sparse architecture of gene expression traits across human tissues. *PLoS Genetics* 12: e1006423.

39 Hong, J., Lunetta, K.L., Cupples, L.A. et al. (2016). Evaluation of a two-stage approach in trans-ethnic meta-analysis in genome-wide association studies. *Genetic Epidemiology* 40: 284–292.

40 Liu, C.T., Raghavan, S., Maruthur, N. et al. (2016). Trans-ethnic meta-analysis and functional annotation illuminates the genetic architecture of fasting glucose and insulin. *American Journal of Human Genetics* 99: 56–75.

41 Limkin, E.J., Sun, R., Dercle, L. et al. (2017). Promises and challenges for the implementation of computational medical imaging (radiomics) in oncology. *Annals of Oncology* 28: 1191–1206.

42 Deleris, L.A., Aonghusa, P.M., and Shorten, R. (2015). Person-specific standardized vulnerability assessment in health and social care. *Studies in Health Technology and Informatics* 216: 462–466.

43 Molenaar, P.C.M., Sinclair, K.O., Rovine, M.J. et al. (2009). Analyzing developmental processes on an individual level using nonstationary time series modeling. *Developmental Psychology* 45: 260–271.
44 Podda, G.M., Grossi, E., Palmerini, T. et al. (2017). Prediction of high on-treatment platelet reactivity in clopidogrel-treated patients with acute coronary syndromes. *International Journal of Cardiology* 240: 60–65.
45 Palmerini, T., Calabro, P., Piscione, F. et al. (2014). Impact of gene polymorphisms, platelet reactivity, and the SYNTAX score on 1-year clinical outcomes in patients with non-ST-segment elevation acute coronary syndrome undergoing percutaneous coronary intervention: the GEPRESS study. *JACC: Cardiovascular Interventions* 7: 1117–1127.
46 Brennan, M.B., Huang, E.S., Lobo, J.M. et al. (2017). Longitudinal trends and predictors of statin use among patients with diabetes. *Journal of Diabetes and its Complications* 32 (1): 27–33. https://doi.org/10.1016/j.jdiacomp.2017.09.014.
47 Verbelen, M., Weale, M.E., and Lewis, C.M. (2017). Cost-effectiveness of pharmacogenetic-guided treatment: are we there yet? *The Pharmacogenomics Journal* 17: 395–402.

24

Recruitment, Inclusion, and Diversity in Clinical Trials

Bernadette Boden-Albala[1,2], Salina P. Waddy[3], Noa Appleton[4], Heather Kuczynski[5], Emily Nangle[6], and Nina S. Parikh[6]

[1] *Department of Population Health and Disease Prevention, University of California, Irvine, Irvine, CA, USA*
[2] *Department of Epidemiology, University of California, Irvine, Irvine, CA, USA*
[3] *Department of Neurology, Atlanta Veterans Administration, Decatur, GA, USA*
[4] *Department of Population Health, New York University School of Medicine, New York, NY, USA*
[5] *New York University, New York, NY, USA*
[6] *Department of Epidemiology, School of Global Public Health, New York University, New York, NY, USA*

24.1 Background/Context

Inadequate participation of underrepresented groups, such as women and racial/ethnic minorities, can negatively affect the scientific and ethical value of a clinical trial [1]. Specifically, nonrepresentative samples may limit a trial's safety, efficacy, and relevance to the general population [2]. Accordingly, the National Institutes of Health (NIH) Revitalization Act of 1993 established guidelines for appropriate representation of women and minorities in NIH-funded clinical research. Despite this legislative action 27 years ago, participation rates of minority groups remain well below representation in the US population.

A recent paper analyzing 2003–2016 cancer trial data reported in ClinicalTrials.gov, a database of publicly and privately funded clinical studies, found that non-Hispanic Whites were significantly more likely to be enrolled in clinical cancer trials compared to African Americans and Hispanics. Enrollment fractions (EFs), defined as the number of trial enrollees divided by cancer prevalence, were 1.2% for non-Hispanic Whites compared to 0.7% for African Americans and 0.4% for Hispanics [3]. Additionally, this study found that in trials of certain cancers (melanoma, lung cancer, pancreatic cancer), enrollment of women was at or below 40% [3]. An analysis of clinical trials funded by the National Institute of Neurological Disorders and Stroke (NINDS) reported that while participation rates of women and African Americans have increased since implementation of the NIH Revitalization Act, Hispanic participation has declined to just 5.8% [4]. Underrepresented of minorities in clinical research has also been documented across numerous other conditions, including pulmonary disease [5], bipolar disorder [6], and human immunodeficiency virus (HIV) [7]. Given

The Science of Health Disparities Research, First Edition. Edited by Irene Dankwa-Mullan, Eliseo J. Pérez-Stable, Kevin L. Gardner, Xinzhi Zhang, and Adelaida M. Rosario.

the documented racial/ethnic disparities in the incidence and prevalence of many diseases, representative clinical trial study samples are critical.

From a scientific standpoint, adequate inclusion in clinical trial samples is essential to determine the safety and efficacy of a treatment, as treatment effects can vary significantly by sex and race/ethnicity. Across the board, women have been found to be approximately 1.5 times more likely to experience adverse drug effects compared to men [8]. One recent example is the sleeping aid Zolpidem (Ambien). Zolpidem has lower clearance (rate of elimination from the body) in women than in men; however, this treatment difference was not initially detected in early trials. In 2013, in response to studies confirming reports of impaired next-day driving, particularly among female users, the FDA reduced the recommended dose for women [9]. Variations in medication metabolism and toxicity leading to differences in safety and efficacy have also been identified across race/ethnic groups in chemotherapy and immunosuppressant and cardiovascular drugs [10–12].

For example, the blood thinner Plavix (Clopidogrel), prescribed as an alternative to aspirin in preventing ischemic events, has limited efficacy among a large percentage of Asians and Pacific Islanders who possess a genetic variant that affects drug metabolism [13]. This has led to an increased risk of adverse outcomes and also prompted numerous high-profile individual and class action lawsuits, including from the state of Hawaii, which has a large Asian/Pacific Islander population.

An area of continuing discussion relates to the question of just how many participants constitutes a representative sample. As part of the NIH Revitalization Act policy, there is a requirement that phase III trials include appropriate designs to determine whether treatment may differentially impact population subgroups. While this policy does not stipulate that analyses have adequate power, it does suggest that there must be "sufficient" numbers from subgroups represented in order to perform a valid analysis. This very important area remains controversial in that in design of trials, investigators do not report having data to anticipate differential effect. There is also ambiguity as to what statistical considerations are associated with valid analysis.

Aside from the scientific importance of determining variations in drug safety and efficacy across demographic subgroups, inclusion in clinical trials is critical from an ethical and social justice perspective. Everyone has a right to be represented and participate in trials that shape medical knowledge and clinical guidelines. Additionally, involvement in clinical research often provides participants access, at low or no cost, to the newest medical care that may otherwise be inaccessible to them. Clinical trial participation also offers insight into areas beyond treatment safety and efficacy, such as issues with the logistics and feasibility of a given treatment, which may also vary across sex, age, and race/ethnicity.

For all of these reasons, appropriate and equitable inclusion in clinical trials has been identified as a fundamental issue in health disparities research and intervention. While this chapter focuses on inclusion of women and racial/ethnic minorities in clinical research, some of the issues discussed may apply equally to other underrepresented or vulnerable groups such as children, older adults, rural residents, and the uninsured.

24.2 Understanding Barriers to Inclusion in Clinical Research

The overall challenges to inclusion in clinical research can be characterized as relating to: (i) optimizing the pool of eligible patients screened for a study, (ii) converting those screened patients to enrolled subjects, and (iii) retaining enrolled subjects. This chapter will provide an overview of the specific barriers associated with each of these challenges as well as best practices and strategies to address them. Barriers exist at multiple levels, including the patient/community level, investigator level, and structural/institutional level. While certain challenges apply specifically to the recruitment and retention of women and racial/ethnic minority populations, many barriers affect overall recruitment and retention among all potential study subjects.

24.2.1 Patient-level Barriers

Previous research has explored lay-/patient-level recruitment and retention barriers in clinical trials. Many identified lay-level barriers relate to trial logistics, such as burdensome time commitments, challenges arranging transportation for follow-up visits, and anticipation of additional costs related to study procedures [14, 15]. A lack of childcare as well as difficulty taking time off from work pose additional obstacles, especially for women [14]. While these logistical challenges apply to all potential study participants, they have been found to be a particularly relevant burden in the recruitment of women and minorities.

Barriers related to language and education are also frequently cited as affecting minority participation. Those who do not speak English as their preferred language are often less likely to be enrolled in trials. Disparities in clinical trial awareness by race/ethnicity and education level persist, which likely also contribute to recruitment differences [16].

Finally, many studies have reported a fear of or mistrust toward the healthcare system among minority groups [14, 17]. In particular, minorities have noted fears of mistreatment, exploitation, and being treated as a guinea pig (e.g., Tuskegee Syphilis Study) [18] as prominent factors when considering clinical trial participation. Many potential minority participants also cite fear of a loss of privacy/confidentiality and apprehension surrounding the concept of randomization as reasons to avoid participation [15].

24.2.2 Investigator-level Barriers

Despite the patient-level barriers, research suggests that Hispanic and Black populations report being positively interested in trial participation; however, they are often neither asked nor eligible [19, 20]. Examples of eligibility criteria that may impact minority enrollment include trial inclusion requiring early-stage cancer diagnosis, as certain underrepresented populations, such as African Americans, Hispanics, and American Indians, are often diagnosed at more advanced stages compared to Whites [21]. Additionally, a number of studies set serum creatinine cut-offs for inclusion that systematically exclude Black participants [22].

Researchers' attitudes and behaviors, as well as trial procedures, may be directly related to the success or failure to recruit representative populations [17, 23]. Indeed, researchers stand at the forefront of enrollment procedures and act as intermediaries between patients and therapeutic options. Yet, the literature is sparse on investigator-level barriers to minority involvement in clinical trials.

In 2010, the National Initiative for Minority Involvement in Neurological Clinical Trials (NIMICT) was created by NIH, NINDS, and the National Institute on Minority Health and Health Disparities (NIMHD) to identify the constellation of investigator-level barriers and create and test a series of evidence-based toolkits that address minority recruitment and retention challenges in neurological clinical trials. The authors of this chapter were part of the NIMICT study team, including principal investigator (PI) Dr. Boden-Albala. As part of NIMICT's formative research, the team employed a mixed-method approach to identify investigator-level barriers to minority involvement in stroke clinical trials. These barriers included insufficient planning and budgeting, translation and communication challenges, issues with staff hiring and training, and researcher/health provider attitudes and biases [24, 25].

NIMICT's research found that nearly half of leading stroke PIs do not incorporate active planning for recruitment of minority populations [24]. In order to be able to perform valid subgroup analysis by gender and race/ethnicity, a requirement of the NIH Revitalization Act, a study must have a larger sample size than otherwise planned [26]. In addition to having to recruit and retain more participants, there are additional costs associated with ethics committees' requirements, such as translation of consent forms/informative material, hiring of translators, and hiring of staff with multilinguistic or cultural skills [26]. Many investigators are resistant to incurring these additional costs or do not budget adequately for them in planning their trials [24].

Translation can act as an even greater burden within acute clinical trials that have narrow eligibility timeframes. For example, some trials require that patients be enrolled within hours of experiencing an event. In such cases, study coordinators will often not approach someone that is non-English-speaking or who speaks English as a nonpreferred language, as they anticipate that the extra time required for translation would impede successful enrollment [25]. The need for translation also raises concerns regarding the ability for measurement scales to maintain their validity.

Beyond translation difficulties, PIs and study coordinators also cite general communication challenges as a barrier to minority enrollment. In NIMICT's survey of leading stroke researchers, respondents expressed difficulty in relaying the concept of prognostic uncertainty, explaining research concepts in lay terminology (e.g., therapeutic misconception), and communicating due to language, education, and cultural barriers [24]. Investigators also emphasized the difficultly of recruitment communication and, in particular, establishing trust.

NIMICT's focus groups with PIs also highlighted issues related to hiring and training of study staff. They referenced the young age at which research coordinators are hired, high turnover rates, lack of support, insufficient training, and general inexperience as significant issues. PIs noted that the research coordinators' lack of confidence in performing the job may have led to denial of potential participants [25]. Additionally, NIMICT's surveys revealed that roughly two-thirds of leading stroke PIs do not require any form of cultural competency or cultural humility training for research staff [24, 27]. This leaves study recruiters further unprepared for potentially difficult cross-cultural recruitment interactions.

Finally, biased attitudes (both conscious and unconscious) held by researchers and/or healthcare providers may impede recruitment of underrepresented groups. Research has shown that healthcare providers are, on average, less comfortable explaining clinical trials to non-English-speaking patients and patients from low-socioeconomic groups, and assume these patients would be less interested in participating in clinical trials [28, 29]. NIMICT's focus groups with research coordinators found that their anticipation of a lack of trust in clinical trials among racial/ethnic minorities caused them to be less likely to engage those patients when recruiting for studies [25].

24.2.3 Structural and Institutional Barriers

Structural and institutional policies and resources shape investigators' recruitment strategies as well as patient enrollment patterns. While the NIH Inclusion Policy has changed standards for representation of racial/ethnic minorities and women in clinical trials, these standards have not translated directly to researchers' practices. In NIMICT's surveys, investigators expressed confusion about the practical application of the NIH Inclusion Policy for valid analysis through adequate representation. In a discussion of adequate power for multiple racial-ethnic group comparisons, several questioned operational parameters and denominator definitions for adequate representation. One respondent asked, "If the African American community makes up 12% of the population and you have 12% of your study patients who are African American, is that adequate? Will you be able to do anything with that analysis?" [24] Further clarification and training is needed to allow investigators to plan adequately to meet inclusion standards.

Inter- and intra-institutional challenges have also been cited as barriers to minority recruitment [24]. For example, NIMICT's survey respondents noted examples of geographical and philosophical disconnect between research institutions and organizations serving minority populations. One investigator explained, "[An academic institution] is two miles from the clinic, and the patients are there, the diversity is there but the patients are not traveling to [the institution] and vice versa. I think that is one of the biggest challenges... bridg[ing] trials to communities" [24]. Additionally, investigators reported a lack of meaningful community partnerships and a lack of institutional support to accommodate challenges accompanying minority recruitment.

Finally, documented racial/ethnic disparities in access to healthcare services further prevent patients from accessing clinical research. These disparities result from factors such as lack of insurance, noncitizen status, geographic location, and structural racism. Those who do not have health insurance or who have difficulty affording copayments or deductibles are likely to have fewer visits with their physicians, and therefore may miss opportunities for referral to clinical trials [26]. Limited access to healthcare also often results in nontimely diagnosis, which poses a barrier to trial participation, particularly in cancer and stroke research, when timely diagnosis is an inclusion requirement (Table 24.1).

Table 24.1 Investigator-identified best practices and recommendations.

Barrier	Theme	Illustrative quote(s)	Best practice, recommendation(s)
Structural and institutional	NIH inclusion policy clarification	If the African American community makes up 12% of the population and you have 12% of your study patients who are African American, is that adequate? Will you be able to anything with that analysis?	• Guidance on criteria for optimal inclusion rates • A priori minority recruitment and retention budget parameter
	Trial design and site selection	What is the prevalence of the disease? Are you mimicking the prevalence or the racial breakdown of wherever you are conducting the trial?	• Best practices on clinical trial design/analysis
	Value of community engagement and partnerships	You have to go to the community to develop those relationships [An academic institution] is 2 miles from the clinic, and the patients are there, the diversity is there but the patients are not traveling to [the institution] and vice versa. I think that is one of the biggest challenges. . .bridg[ing] trials to communities.	• Tips/tools on fostering and maintaining equitable community partnerships
	Physician relationships	If the physician believes it is a good choice for [the patient then] he/she is usually more onboard.	• Primary physician toolkit with active trial information
Recruitment communication	Effective training	I would support training to recruit minorities or recruit anyone, recruitment is training people about what are these people's concerns, how do I establish trust?, what are the patients' needs? Recruitment is not based on the recruiter but it is really the psychologies, motivational interviewing, the "human factor."	• Communication education • Cultural competency training • Motivational interviewing
Unique challenges	Clinical environment and patient population	Because neurologic injury occurs, a proxy now has to make a decision about a loved one to enroll in a clinical trial. . .It's not your own risk; it's the risk for a loved one.	• Navigating acute clinical trials • Guidance on patient denial/stigma

Source: Reprinted from Boden-Albala et al. [24].

24.3 Best Practices for Optimizing Inclusion in Clinical Research

Strategies to enhance inclusion in clinical trials must address the patient, investigator, and institutional-level barriers described above in order to expand and diversify the pool of patients invited to enroll in studies, gain trust to provide consent, and retain participants for the duration of a study. Over the past two decades, researchers across disciplines and diseases have identified many strategies and best practices to overcome these barriers. Best practices related to site selection, screening procedures, community engagement, communication strategies, and participant retention are outlined below.

24.3.1 Site Selection

The sites selected to participate in a trial will often determine, to a large extent, the demographics of the study sample. Careful and intentional site selection, therefore, is a critical tool for optimizing a study's screening pool. Before selecting sites, investigators should evaluate the demographic profile of potential sites. National or regional data can be used to assess each institution's surrounding communities. Investigators can also access the racial/ethnic and sex breakdown of admitted patients in order to make reasonable projections and evaluate each site's enrollment feasibility. Sources of data include International Classification of Disease (ICD-9) codes, quality outcome measures (e.g., American Heart Association Get with the Guidelines metrics), and real-time experience. Other site selection considerations that can maximize inclusion include site location and accessibility, to encourage compliance with follow-up appointments, and satellite sites in close proximity to communities of interest. Once sites are selected, PIs and research staff should be encouraged to carefully assess referral patterns, taking into consideration each site's racial/ethnic makeup, socioeconomic status, sex, language, and education/literacy levels. This may help identify challenges and suggest strategies to work through them early in the outreach efforts.

24.3.2 Screening

Determining the patient population available for trials is critical to ensuring representative enrollment. Trial PIs and site PIs should work in tandem to establish screening procedures before trial enrollment begins. Consistent patient screening allows for assessment of challenges and barriers to enrollment. In general, screening approaches should aim to include a large pool of patients and narrow down to identify those eligible, rather than screening within a narrow set of criteria, as more comprehensive screening procedures tend to yield a wider pool of participants. For inpatient screening, use of grand and attending rounds often enhances information from electronic medical records. Other proposed strategies to optimize screening include a generic consent for research participation, which allows for broader screening.

24.3.3 Community Engagement

Researcher engagement with target communities is critical to the success of recruiting and retaining underrepresented groups in clinical trials [30]. In community-engaged research, community members should help set the research agenda, share resources equitably through collaborative leadership models, and be involved at all stages of the research process. This includes dissemination of research proposals, progress, and results. Investigators can work with community leaders and/or use focus groups to formulate and test hypotheses, develop mutual goals, and define the roles of stakeholders. A community advisory board and/or community information sessions can be effective ways to formalize involvement with the community and potentially build even stronger relationships [24, 25]. Site investigators can also work with their hospitals' public/community relations departments to identify patient engagement groups or community activities and advertising that might be utilized for recruitment.

Another key community engagement strategy that can increase minority patient referrals is to engage the larger nonresearch community (i.e., primary care physicians) in study recruitment [14, 24]. There is often a disconnect between the investigators conducting trials and the clinics and community practices that serve racial/ethnic minority patients. To address this barrier, clinical investigators can strengthen and maintain relationships with surrounding clinics and primary care settings and provide them with active and updated trial information. Nonresearch physicians should be treated as partners in the research process, serving as coinvestigators or subinvestigators. Any local concerns, such as assuring that patients will not get lost or billed for procedures associated with the trial, must also be addressed.

A number of recent interventions and initiatives have specifically focused on engaging minority-serving physicians in research. The Randomized Recruitment Intervention Trial (RECRUIT) at The University of Texas is testing a program in 50 specialty clinics to help investigators and coordinators build relationships with community clinicians in order to increase minority patient referrals [31]. Another program, through the National Medical Association's (NMA) Project IMPACT (Increase Minority Participation and Awareness of Clinical Trials), aims to encourage greater participation of African American physicians and patients in clinical research [32]. Their activities include education and training workshops aimed at minority physicians and the establishment of a physician-investigator database.

24.3.4 Communication Strategies

Researchers in recent years have sought to understand how recruitment communication behaviors contribute to building positive relationships with potential study participants [33]. One of NIMICT's main findings was that researchers request additional training for study staff to improve patient engagement and communication skills [24]. In response, NIMICT developed an online training curriculum for researchers in motivational interviewing, a set of proven collaborative communication strategies to facilitate behavior change. Motivational interviewing incorporates strategies such as asking open-ended questions, affirming where the individual is in their thinking, using reflective listening, and exploring the positives and negatives of making a given decision [34]. The goal of NIMICT's motivational interviewing training is to make study coordinators more comfortable during often stressful situations when called to speak with potential participants who have different cultural backgrounds or limited exposure to clinical research. Recently, the NIMICT team successfully piloted this curriculum in a group of research coordinators. Given the feasibility of online training modules and the ability to implement and monitor the training curriculum remotely, applicability across nationwide clinical trials appears promising.

Trial recruitment materials must also utilize communication best practices. Messaging materials, such as fliers, posters, brochures, and online advertisements, should be clear, accessible, health literate, and culturally appropriate. Recruitment materials should be tailored to specific target populations. Technology, and specifically video series and narratives, can be an effective medium to communicate study information to patients, families, and caregivers. Technology can also be used by investigators to help overcome language barriers; for example, by developing recruitment videos in multiple languages [35]. Finally, some studies have reported recruitment success through the use of patient navigators who can help bridge communication barriers with target populations [36].

24.3.5 Retention

Participant attrition is a major issue affecting the viability of clinical trials. Additional accommodations are often necessary to ensure participant retention and should be considered when creating study budgets. Retention strategies should seek to promote patient engagement and minimize the logistical burdens associated with trial participation. Flexible hours can be offered to accommodate the busy schedules of working individuals and those with caregiver issues [33]. Incentives and reimbursements can be offered to minimize financial burdens, such as transportation or parking vouchers, meal reimbursement, and gift cards. Many studies support the importance of maintaining regular and repeated contact with study participants to improve retention [30, 33]. Contact can take the form of birthday and/or holiday cards, appointment reminders, and study newsletters. Providing a designated study number or email account for participants to reach research staff can also promote active communication. Finally, subjects will be more likely to remain in a study if they feel that their time and effort are appreciated [33]. This can be accomplished through providing certificates, thank-you notes, and study-branded gifts, such as pens, calendars, bags, and magnets. Beyond these accommodations, continued engagement with community partners throughout all stages of the research process will help to maximize retention.

24.4 Future Directions and Research Needs

24.4.1 Federal Policies

Solutions are not one-dimensional and must be implemented on all levels. Policy changes and enforcement must come from the government at early stages in the grant review process and continue throughout study setup, implementation, and community dissemination.

One recent approach is through the use of disease-focused clinical trial networks within NIH, such as NIH StrokeNet [37]. It is possible that these clinical research networks will allow for greater communication between researchers regarding effective recruitment strategies. Also, as part of the paradigm shift at NIH toward inclusion in clinical research, NIH StrokeNet has required that minority recruitment procedures be outlined and reviewed before all grants are submitted. This kind of structural-level mandate may be as useful in improving levels of inclusion as additional research and innovation. Another recent suggestion is that gold-standard journals should require reporting of information on subgroup baseline demographics and treatment effects, as this would create an impetus for investigators to examine and include this information.

Despite NIH's efforts, the pharmaceutical industry funds the vast majority of clinical trials in the United States, and, until recently, the inclusion of population subgroups was not required. In 2013, the FDA implemented a policy requiring the reporting of age, sex, and race/ethnicity for all drugs submitted for regulatory approval. They also created a website for the public to access this information. In response, the pharmaceutical industry is currently thinking through strategies to expand their participant base. Working in partnership with the research community, industry and government must find solutions to ensure equitable representation in clinical research.

24.4.2 Interactive Online Toolkits

A number of online toolkits have been developed in recent years that provide investigators and research staff with interactive, responsive resources for diagnosing recruitment barriers and supporting the adoption of best practices. In the area of neurological research, the NIMICT team has launched NIMICT.com, an online platform for principal investigators, research staff, and clinical trial participants and their friends/family members to access critical information and training resources. These resources include budgeting templates and diagnostic assessments for PIs, recruitment and retention best practices for coordinators, guidelines for improved screening, research staff training materials, as well as videos and infographics presenting clinical research concepts for lay audiences (Figure 24.1).

Another NIMHD-supported initiative, the Enhancing Minority Participation in Clinical Trials (EMPaCT) consortium, has created an online toolkit targeting cancer researchers at empactconsortium.com. The EMPaCT toolkit came out of a collaboration across five regional cancer centers throughout the United States and was developed in partnership with the American Cancer Society [38]. Their website offers online training courses, training videos, and resource collections and reviews. Courses cover topics such as cultural competency training, building trust and establishing rapport, and successful models for minority recruitment (Figure 24.2).

A third toolkit was created by the Building Trust Between Minorities and Researchers project, a bioethics research initiative run through the University of Maryland Center of Health Equity that is also supported by NIMHD. The website[1] contains tailored curricula and multimedia resource collections for researchers and minority communities aimed at enhancing minority engagement in research (Figure 24.3).

The proliferation of high-quality, interactive online toolkits has greatly expanded the resources and training opportunities available to investigators to support women and minority inclusion in clinical research. These websites also provide valuable and accessible information to patients and community members, which can help elucidate the importance and relevance of clinical research and demystify the processes involved with participation (Table 24.2).

[1] www.buildingtrustumd.org

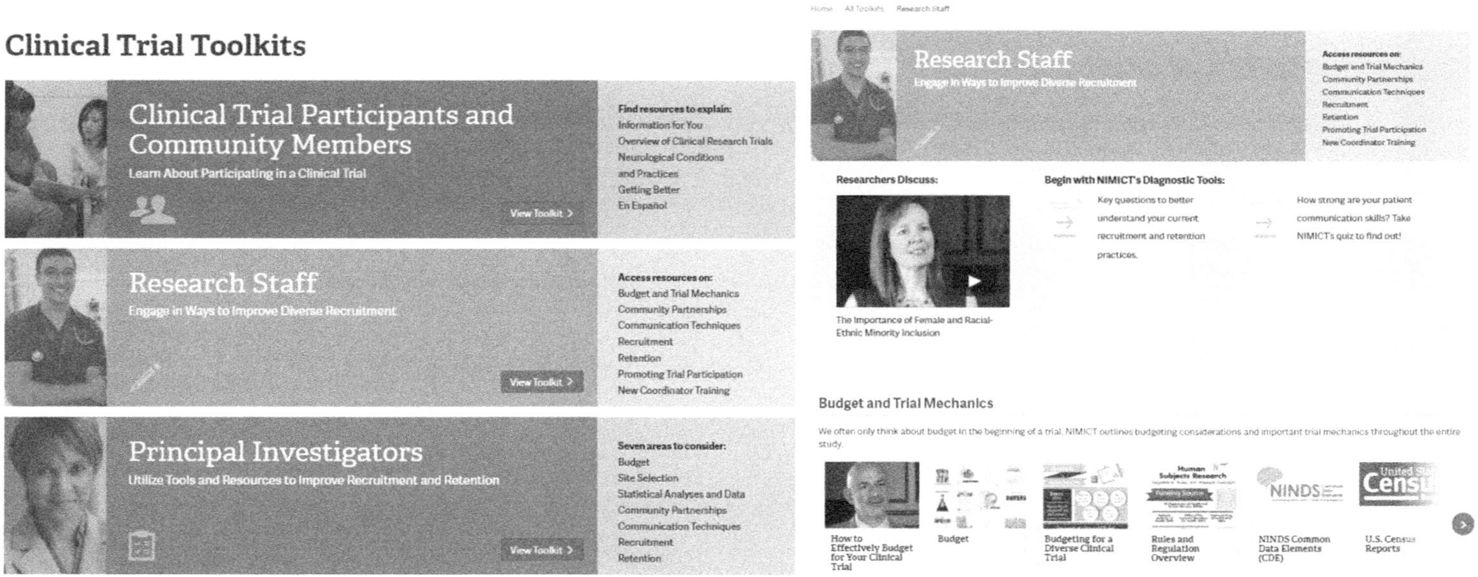

Figure 24.1 Screenshots of NIMICT.com online toolkits. (See color representation of this figure in color inserts section.)

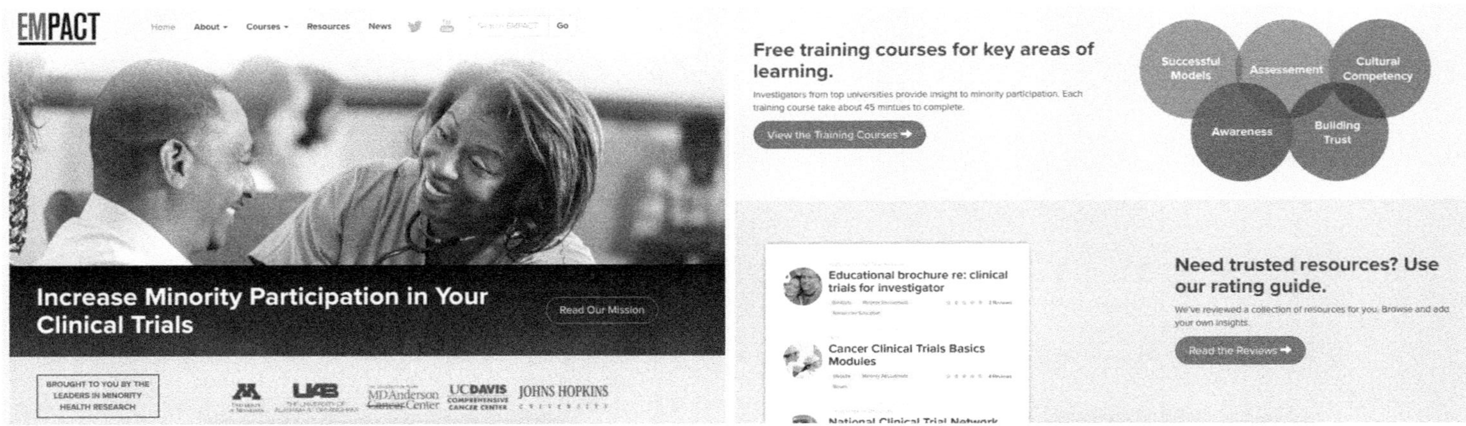

Figure 24.2 Screenshots of EMPaCT consortium online toolkit. *Source:* empactconsortium.com. (See color representation of this figure in color inserts section.)

Figure 24.3 Screenshots of building trust between minorities and researchers online toolkit. *Source:* buildingtrustumd.org. (See color representation of this figure in color inserts section.)

Table 24.2 Minority recruitment online toolkit features and content.

Title/website	Funder	Target audience	Types of tools	Selected topics
National initiative for minority involvement in neurological clinical trials (NIMICT)	NIH, NIMHD, NINDS	Neurological principal investigators	Video series	Budgeting
nimict.com		Research staff	Infographics	Community partnerships
		Clinical trial participants and community members	Resource collections	Communication techniques
			Diagnostics	Site selection
			Case studies	Statistical analysis
			Templates	Recruitment and retention
Enhancing minority participation in clinical trials (EMPaCT)	NIMHD	Cancer principal investigators	Free training courses	Building trust
Vickers and Fouad [38]		Research staff	Training videos	Cultural competency training
		Patient navigators	Resource collection with ratings	Successful models for minority recruitment
		Referring physicians		Increasing clinical trial awareness among minorities
Building trust between minorities and researchers	NIH, NIMHD	Investigators	Multimedia resource center	Importance of research
buildingtrustumd.org		Research staff	Curricula targeting researchers and community members	Informed decision making
		IRB members		Research, community and you
		Community members		
Project increase minority participation and awareness of clinical trials (IMPACT)	National Medical Association	African American physicians and community members	Physician education and training programs	Why physicians should care
impact.nmanet.org			Consumer information and materials	Clinical research overview and training
			Blog posts	Is a trial right for you?
			Resource links	Benefits and risks
			Physician-investigator database	Safety and patient rights

24.5 Conclusion

The US demographic landscape is becoming more multiethnic, and the lack of balanced clinical trial participation by diverse racial/ethnic groups hinders our ability to generalize scientific findings at the population level. Research suggests that a multidimensional approach is required to achieve equitable rates of participation. This chapter highlights patient-, investigator-, and structural-level barriers and best practices and identifies future directions for creating evidence-based tools and solutions.

24.6 Key Points

- Inadequate participation of under-represented groups, such as women and racial/ethnic minorities, can negatively affect the scientific and ethical value of a clinical trial.
- Recruitment barriers exist at multiple levels, including the patient/community level, investigator level, and structural/institutional level. Recruitment barriers exist at multiple levels, including the patient/community level, investigator level, and structural/institutional level.
- In recent years, many best practices have been identified relating to site selection, screening procedures, community engagement, communication strategies, and participant retention.
- A multi-dimensional approach involving collaboration among government, industry and the research community will be required to ensure equitable representation in clinical research.

Disclaimer

The views and opinions expressed in this chapter are those of the authors only and do not necessarily represent the views, official policy, or position of the U.S. Department of Health and Human Services or any of its affiliated institutions or agencies.

References

1 Roberts, J., Waddy, S., and Kaufmann, P. (2012). Recruitment and retention monitoring: facilitating the mission of the National Institute of Neurological Disorders and Stroke (NINDS). *Journal of Vascular and Interventional Neurology* 5 (supp): 14–19.

2 Geller, S.E., Koch, A., Pellettieri, B., and Carnes, M. (2011). Inclusion, analysis, and reporting of sex and race/ethnicity in clinical trials: have we made progress? *Journal of Women's Health (Larchmt)* 20 (3): 315–320.

3 Duma, N., Aguilera, J.V., Paludo, J. et al. (2018). Representation of minorities and women in oncology clinical trials: review of the past 14 years. *Journal of Oncology Practice* 14 (1): e1–e10.

4 Burke, J.F., Brown, D.L., Lisabeth, L.D. et al. (2011). Enrollment of women and minorities in NINDS trials. *Neurology* 76 (4): 354–360.

5 Burchard, E.G., Oh, S.S., Foreman, M.G., and Celedon, J.C. (2015). Moving toward true inclusion of racial/ethnic minorities in federally funded studies. A key step for achieving respiratory health equality in the United States. *American Journal of Respiratory and Critical Care Medicine* 191 (5): 514–521.

6 Akinhanmi, M.O., Biernacka, J.M., Strakowski, S.M. et al. (2018). Racial disparities in bipolar disorder treatment and research: a call to action. *Bipolar Disorders* 20 (6): 506–514.

7 Gifford, A.L., Cunningham, W.E., Heslin, K.C. et al. (2002). Participation in research and access to experimental treatments by HIV-infected patients. *New England Journal of Medicine* 346 (18): 1373–1382.

8 Zopf, Y., Rabe, C., Neubert, A. et al. (2008). Women encounter ADRs more often than do men. *European Journal of Clinical Pharmacology* 64 (10): 999–1004.

9 Food and Drug Administration (2013). *FDA Drug Safety Communication: Risk of Next Morning Impairment after Use of Insomnia Drugs; FDA Requires Lower Recommended Doses for Certain Drugs Containing Zolpidem (Ambien, Ambien CR, Edluar, and Zolpimist).* Rockville, MD: US FDA.

10 Schaefer, B.M., Caracciolo, V., Frishman, W.H., and Charney, P. (2003). Gender, ethnicity and genetics in cardiovascular disease: part 1: basic principles. *Heart Disease* 5 (2): 129–143.

11 Dirks, N.L., Huth, B., Yates, C.R., and Meibohm, B. (2004). Pharmacokinetics of immunosuppressants: a perspective on ethnic differences. *International Journal of Clinical Pharmacology and Therapeutics* 42 (12): 701–718.

12 Phan, V.H., Moore, M.M., McLachlan, A.J. et al. (2009). Ethnic differences in drug metabolism and toxicity from chemotherapy. *Expert Opinion on Drug Metabolism & Toxicology* 5 (3): 243–257.

13 Hasan, M.S., Basri, H.B., Hin, L.P., and Stanslas, J. (2013). Genetic polymorphisms and drug interactions leading to clopidogrel resistance: why the Asian population requires special attention. *The International Journal of Neuroscience* 123 (3): 143–154.

14 Barrett, N.J., Ingraham, K.L., Vann Hawkins, T., and Moorman, P.G. (2017). Engaging African Americans in research: the recruiter's perspective. *Ethnicity & Disease* 27 (4): 453–462.

15 Schmotzer, G.L. (2012). Barriers and facilitators to participation of minorities in clinical trials. *Ethnicity & Disease* 22 (2): 226–230.

16 Leiter, A., Diefenbach, M.A., Doucette, J. et al. (2015). Clinical trial awareness: changes over time and sociodemographic disparities. *Clinical Trials* 12 (3): 215–223.

17 Durant, R.W., Wenzel, J.A., Scarinci, I.C. et al. (2014). Perspectives on barriers and facilitators to minority recruitment for clinical trials among cancer center leaders, investigators, research staff, and referring clinicians: enhancing minority participation in clinical trials (EMPaCT). *Cancer* 120 (Suppl 7): 1097–1105.

18 Scharff, D.P., Mathews, K.J., Jackson, P. et al. (2010). More than Tuskegee: understanding mistrust about research participation. *Journal of Health Care for the Poor and Underserved* 21 (3): 879–897.

19 US Institute of Medicine (2012). *Envisioning a Transformed Clinical Trials Enterprise in the United States: Establishing an Agenda for 2020: Workshop Summary*. Washington, DC: National Academy of Sciences.

20 Wendler, D., Kington, R., Madans, J. et al. (2006). Are racial and ethnic minorities less willing to participate in health research? *PLoS Medicine* 3 (2): e19.

21 Miller, B.A., Hankey, B.F., and Thomas, T.L. (2002). Impact of sociodemographic factors, hormone receptor status, and tumor grade on ethnic differences in tumor stage and size for breast cancer in US women. *Ameican Journal of Epidemiology* 155 (6): 534–545.

22 Hsu, J., Johansen, K.L., Hsu, C.Y. et al. (2008). Higher serum creatinine concentrations in black patients with chronic kidney disease: beyond nutritional status and body composition. *Clinical Journal of the American Society of Nephrology* 3 (4): 992–997.

23 Corbie-Smith, G.M., Durant, R.W., and St George, D.M. (2006). Investigators' assessment of NIH mandated inclusion of women and minorities in research. *Contemporary Clinical Trials* 27 (6): 571–579.

24 Boden-Albala, B., Carman, H., Southwick, L. et al. (2015). Examining barriers and practices to recruitment and retention in stroke clinical trials. *Stroke* 46 (8): 2232–2237.

25 Haley, S.J., Southwick, L.E., Parikh, N.S. et al. (2017). Barriers and strategies for recruitment of racial and ethnic minorities: perspectives from neurological clinical research coordinators. *Journal of Racial and Ethnic Health Disparities* 12 (6): 1225–1236.

26 Hussain-Gambles, M., Atkin, K., and Leese, B. (2004). Why ethnic minority groups are under-represented in clinical trials: a review of the literature. *Health & Social Care in the Community* 12 (5): 382–388.

27 Yeager, K.A. and Bauer-Wu, S. (2013). Cultural humility: essential foundation for clinical researchers. *Applied Nursing Research* 26 (4): 251–258.

28 Stone, V.E., Mauch, M.Y., and Steger, K.A. (1998). Provider attitudes regarding participation of women and persons of color in AIDS clinical trials. *Journal of Acquired Immune Deficiency Syndromes and Human Retrovirology* 19 (3): 245–253.

29 Thompson, E.E., Neighbors, H.W., Munday, C., and Jackson, J.S. (1996). Recruitment and retention of African American patients for clinical research: an exploration of response rates in an urban psychiatric hospital. *Journal of Consulting and Clinical Psychology* 64 (5): 861–867.

30 Nicholson, L.M., Schwirian, P.M., and Groner, J.A. (2015). Recruitment and retention strategies in clinical studies with low-income and minority populations: progress from 2004-2014. *Contemporary Clinical Trials* 45 (Pt A): 34–40.

31. Amorrortu, R.P., Arevalo, M., Vernon, S.W. et al. (2018). Recruitment of racial and ethnic minorities to clinical trials conducted within specialty clinics: an intervention mapping approach. *Trials* 19 (1): 115.

32 Powell, J.H., Fleming, Y., Walker-McGill, C.L., and Lenoir, M. (2008). The project IMPACT experience to date: increasing minority participation and awareness of clinical trials. *Journal of the National Medical Association* 100 (2): 178–187.

33 Morgan, S.E., Occa, A., Potter, J. et al. (2017). "You need to be a good listener": recruiters' use of relational communication behaviors to enhance clinical trial and research study accrual. *Journal of Health Communication* 22 (2): 95–101.

34 Resnicow, K. and McMaster, F. (2012). Motivational interviewing: moving from why to how with autonomy support. *International Journal of Behavioral Nutrition and Physical Activity* 9: 19.

35 Huang, H.Y., Ezenwa, M.O., Wilkie, D.J., and Judge, M.K. (2013). ResearchTracking: monitoring gender and ethnic minority recruitment and retention in cancer symptom studies. *Cancer Nursing* 36 (3): E1–E6.

36 Cook, E.D., Yeager, K.A., Cecchini, R.S. et al. (2018). Recruitment practices for US minority and underserved populations in NRG oncology: results of an online survey. *Contemporary Clinical Trials Communications* 10 (6): 100–104.

37 Broderick, J.P., Palesch, Y., and Janis, S. (2016). The NIH StrokeNet: a User's guide. *Stroke, a Journal of Cerebral Circulation* 47 (2): 301–303.

38 Vickers, S.M. and Fouad, M.N. (2014). An overview of EMPaCT and fundamental issues affecting minority participation in cancer clinical trials. *Cancer* 120 (0 7): 1087–1090.

25

Sexual and Gender Minority Health Disparities

Concepts, Methods, and Future Directions

Mark L. Hatzenbuehler[1] *and John E. Pachankis*[2]

[1] *Department of Psychology, Harvard University, Cambridge, MA, USA*
[2] *School of Public Health, Yale University, New Haven, CT, USA*

25.1 Introduction to the Topic, Including Key Definitions

Defining sexual and gender minority (SGM)[1] populations is a complex task given a lack of objective markers of SGM status, diverse manifestations of SGM expression, and the challenges of parameterizing any socially constructed phenomenon. Yet all definitions of SGM status ultimately rely on definitions of gender and sexual orientation.[2] *Gender*, a socially constructed notion of masculinity or femininity, describes the roles, behaviors, activities, and attributes that society considers appropriate for men and women. Gender can be fluid and is not always aligned with biological sex. For transgender individuals, *gender identity* (which refers to a person's perception of themselves as male, female, both or neither) does not conform to conventional, dichotomous notions of male or female or to the biological sex to which one was assigned at birth (*cisgender* refers to individuals for whom biological sex and gender identity are in alignment). *Sexual orientation* refers to an individual's stable pattern of identity, sexual behavior, or attraction in relation to the gender or sex to which he or she is attracted. Sexual minority status is typically defined as separate from gender minority status, with any individual occupying one status, both, or neither. In this chapter, we define SGM as lesbian, gay, bisexual, and transgender populations as well as individuals whose sexual orientation, gender identity and expressions, or reproductive development varies from traditional, societal, cultural, or physiological norms.

---

1 The National Institutes of Health (NIH) utilizes the umbrella term sexual and gender minority to encompass lesbian, gay, bisexual, and transgender populations as well as those whose sexual orientation, gender identity and expression, or reproductive development varies from traditional, societal, cultural, or physiological norms. The NIH moved to this language in 2015 in an effort to be more inclusive of emerging identities within the community and to remain consistent with the scientific literature.

2 NIH also includes persons with disorders or differences in sex development (DSD), sometimes referred to as "intersex," in their definition of SGM. Given our focus on stigma as it relates to social identities, and given that identities vary widely among those with DSD conditions, this group was excluded from the scope of this chapter.

The Science of Health Disparities Research, First Edition. Edited by Irene Dankwa-Mullan, Eliseo J. Pérez-Stable, Kevin L. Gardner, Xinzhi Zhang, and Adelaida M. Rosario.

Health disparities researchers operationalize SGM status according to their specific research purposes. For instance, studies on human immunodeficiency virus (HIV) disparities often measure sexual orientation in terms of sexual behavior (e.g., men who have sex with men) given the clear association between sexual behavior and HIV transmission. On the other hand, studies into social determinants of disease, such as stigma, often assess sexual orientation in terms of personal identity (e.g., identifying as lesbian, gay, or bisexual [LGB]) given that identity provides the psychological lens through which stigma is ultimately filtered to impact health. Assessing sexual orientation in terms of attraction is a common approach, particularly when studying sexual minority youth. Given that attraction can precede identification and behavior by several years [1], using this approach can capture a larger proportion of the young SGM population.

Each approach to defining sexual orientation is associated with important caveats. For instance, given that about half of men who ever have had sex with men report that this behavior was limited to their adolescence, researchers measuring sexual orientation in terms of behavior are careful to specify the time frame of this behavior (e.g., past 12 months versus lifetime). Relying on self-reported identity as a measure of sexual orientation is limited by the fact that some sexual minorities, perhaps due to strong social pressures, do not adopt sexual minority identities. Defining sexual orientation in terms of attraction is limited by its requirement that attractions in counterfactual situations be verifiable [2]. The importance of appropriate selection of definitions is underscored by the fact that these different definitions give rise to different prevalence rates of the SGM population and of the size of SGM health disparities [3]. For current population estimates, see Figure 25.1.

25.2 How and Why This Topic is Important to Minority Health and Health Disparities Research

In the past two decades, probability-based health studies have begun to assess sexual orientation, allowing researchers to determine the population prevalence of health conditions by sexual orientation status. Across these studies, consistent evidence indicates that LGB

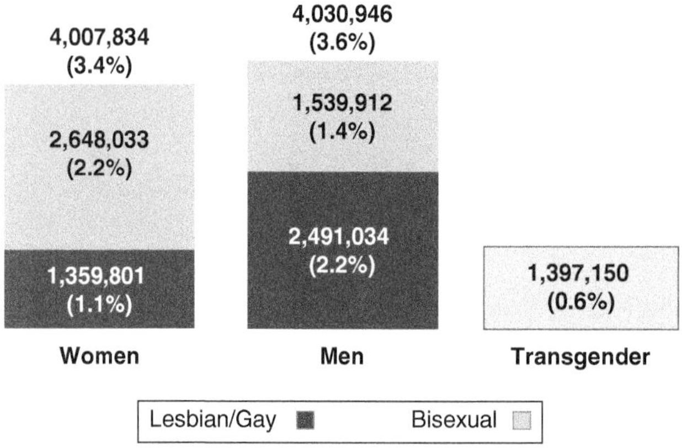

Figure 25.1 Sexual and gender minority population estimates. *Source:* [4] (Data on gay, lesbian, and bisexual estimates); [5] (Data on transgender estimates).

individuals report higher levels of mental and physical health problems compared to heterosexuals [6]. Sexual minorities face among the highest odds of depression, anxiety, and suicide of any risk group. These disparities emerge early in life and persist across some age groups. Comparatively less research has been conducted in the area of physical health disparities, with the exception of HIV/AIDS (acquired immune deficiency syndrome). Emerging research, however, suggests that there may be other sexual orientation-based health disparities for some physical health conditions and for some subgroups of LGB populations [7]. Future research needs to include a broader range of physical health conditions as well as additional markers of health that do not rely on self-reports, such as indicators of allostatic load, to more comprehensively characterize the nature of physical health disparities related to sexual orientation [7].

Strikingly, until recently, there were no population-based health datasets that included measures of gender identity. Thus, research on health disparities related to gender identity has relied on data from nonprobability samples, limiting generalizability. A recent review of 116 studies of the health burden experienced by transgender populations found that very few studies compared transgender and cisgender samples, largely precluding estimates of health disparities by gender identity [8]. Nevertheless, existing studies show a high prevalence of mental, sexual, reproductive, as well as substance-related health burdens among transgender populations, with stigma and violence being strongly related to these outcomes. The few existing comparative studies suggest substantial disparities in these outcomes according to gender identity [6]. For instance, in a cohort of youth attending a community health center, transgender patients had a two-to-three times greater risk of depression, anxiety, suicidality, and inpatient and outpatient mental health treatment than matched cisgender controls [9]. The incorporation of gender identity into additional population-based datasets is urgently needed to produce more reliable estimates of the prevalence and scope of health disparities between transgender and cisgender individuals.

In recognition of the magnitude of these disparities, in October of 2016 the National Institutes of Health designated SGM as a Health Disparity population for research purposes. The study of SGM health disparities is important not only because of the size of the disparity and the necessity of a commensurate public health response, but also because the study of SGM disparities at this point in history provides an opportunity to better understand health determinants affecting all disparity populations.

For example, the rapidly changing legal and social climate surrounding SGM populations has allowed researchers to understand how these climates affect health disparities. Whereas stigma toward other disparity groups, such as racial minorities, might be less obvious because legal discrimination toward those groups has been outlawed for decades, studying structural stigma toward SGM communities can provide insight into how stigma affects populations that have been similarly discriminated against in the past and those that might be into the future. Further, whereas SGM status differs from certain other disparity populations, it also shares important features with some stigmatized populations. For instance, SGM status is often concealable, emerges relatively early in development, and is spread across the population. Knowing how these dimensions relate to SGM health can inform research and intervention for populations similarly located along these dimensions, e.g., those with mental illness, physical disability, religious minority status, or obesity [10].

25.3 Most Relevant Measures and Research Methods Used and Recommended

Having established the existence of SGM health disparities, researchers turned to identifying determinants that can explain them. Stigma has been the predominant theoretical framework for understanding the cause of these disparities. Link and Phelan put forward a widely used conceptualization that defines stigma as the co-occurrence of several interrelated components:

> In the first component, people distinguish and label human differences. In the second, dominant cultural beliefs link labeled persons to undesirable characteristics – to negative stereotypes. In the third, labeled persons are placed in distinct categories so as to accomplish some degree of separation of "us" from "them". In the fourth, labeled persons experience status loss and discrimination that lead to unequal outcomes. Stigmatization is entirely contingent on access to social, economic and political power that allows the identification of differentness, the construction of stereotypes, the separation of labeled persons into distinct categories and the full execution of disapproval, rejection, exclusion and discrimination. Thus, we apply the term stigma when elements of labeling, stereotyping, separation, status loss and discrimination co-occur in a power situation that allows them to unfold [11]. (p.367)

According to this definition, discrimination is a necessary component of stigma; indeed, Link and Phelan [11] argue that stigma "cannot hold the meaning we commonly assign to it" (p.370) when discrimination is not included. Accordingly, the stigma literature shares important commonalities with the larger empirical literature on discrimination (e.g., within the context of race), but the stigma concept is considered broader than discrimination because stigma processes involve additional elements (e.g., labeling, stereotyping, status loss). Stigma theory also points to several processes and mechanisms through which stigma contributes to adverse health outcomes, including the availability of health-promoting resources (e.g., money, knowledge, power, prestige), beneficial social relationships, psychological and behavioral responses (e.g., self-stigmatization, emotion regulation difficulties), and stress [12].

Researchers have conceptualized stigma as a multilevel construct that exists at individual (e.g., self-stigma), interpersonal (e.g., discriminatory treatment), and structural (e.g., institutional policies) levels [13]. Figure 25.2 depicts these different levels, their interrelationships, and the pathways through which they may operate to affect SGM health. Consequently, measures have been developed to examine stigma at each of these levels of analysis. Below, we discuss examples of these measures and evaluate their strengths and limitations. Ultimately, the decision of which measure to use should be guided by a thorough understanding of these strengths and limitations, as well as the specific research question [14].

25.3.1 Individual/Interpersonal Measures

While researchers typically conceptualize stigma as a social process, it has most frequently been measured at the individual/interpersonal level. These measures either assess

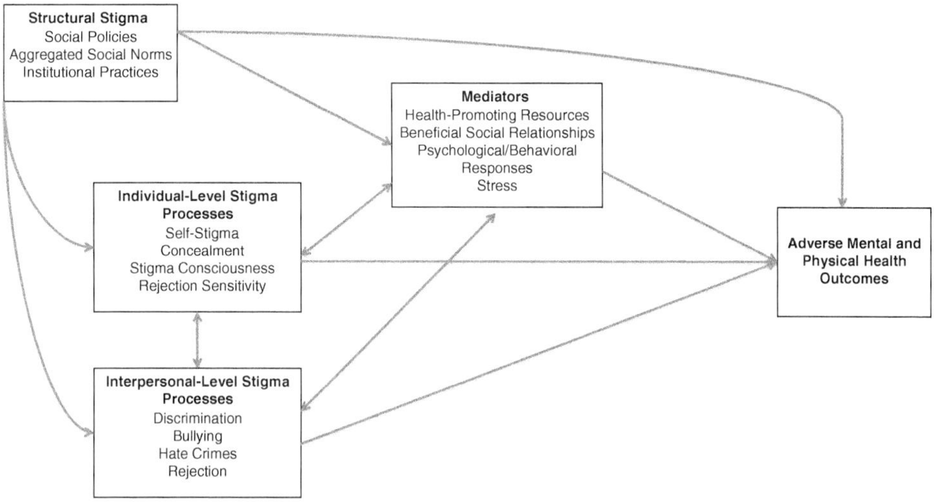

Figure 25.2 Levels of stigma and their relationships.

intrapersonal responses to stigma (e.g., internalized stigma) or perceptions of interpersonal discrimination, which query respondents about whether they have been discriminated against in a variety of major life domains (e.g., healthcare, employment) or in everyday circumstances (e.g., called names or insulted) as a result of their SGM status.

These measures have a number of strengths, including ease of administration, the ability to capture intrapersonal reactions to stigma, and empowering SGM's perspectives. At the same time, they are also subject to numerous limitations [14]. One of the most important is the potential for confounding between exposure and outcome. That is, mental health can affect the reporting of stigma experiences (e.g., those who are depressed are more likely to report stigma), which poses significant obstacles for establishing causal relations. Additionally, there are several psychological, identity, and motivational characteristics that affect the reporting of stigma. In many instances, these same characteristics have also been investigated as risk factors for poor health. If these same factors simultaneously lead to both under-reporting of discrimination and to adverse outcomes, this would create a negative bias in studies.

25.3.2 Structural Measures

Structural forms of stigma refer to "societal-level conditions, cultural norms, and institutional policies and practices that constrain the opportunities, resources, and well-being of the stigmatized" [15] (p. 2). Measures of structural stigma have tended to come in two forms [13]. In the first measurement approach, the content of public policies (at the country, state, and/or municipal level) is coded to determine the presence of structural stigma in institutions. The main advantage of this approach is that it relies on objective data sources to code the policies. The primary limitation is that policies often do not capture the unwritten customs or procedures that undergird more informal institutional practices.[3]

3 www.mentalhealthcommission.ca

An additional limitation is that some policies can represent a ubiquitous exposure [14]. For instance, if enacted at the federal level (as was the case with the Defense of Marriage Act), it applies to all 50 states and thus there is no variation in which to detect an effect on SGM health – although variability could be detected with a cross-national comparison.

In the second measurement approach, researchers obtain data on individual attitudes toward members of stigmatized groups [16] and then aggregate these individual responses up to the community level (defined at various geographic scales, such as counties), so that the level of structural stigma can be compared across communities [17]. A methodological strength of this approach is that members of stigmatized groups are not asked about their perceptions of community attitudes. Instead, the community residents themselves report on their own attitudes, thereby overcoming the limitations of self-reported perceptions of stigma among the stigmatized. One limitation, however, is that this approach can potentially underestimate levels of structural stigma, given that self-reported attitudes are subject to social desirability biases.

25.4 Selected Examples of the State of the Science to Illustrate Best Practices

Next, we review illustrative examples of methodological approaches that researchers have used to examine stigma as a determinant of SGM health disparities [7, 13, 18]. Each method has its strengths and limitations. Consequently, researchers seek to obtain triangulation of evidence across different methods to improve causal inferences about the relationships between stigma and SGM health disparities.

25.4.1 Cross-sectional Designs

Most studies examining the stigma–health relationship have used cross-sectional designs that measure stigma (via self-reported surveys) at one point in time. In one example of this method, Mays and Cochran [19] used data from the Midlife in the United States (MIDUS) study, a nationally representative sample of adults aged 25 to 74. Respondents completed several self-reported survey items regarding their perceptions of experiences with discrimination across four domains (e.g., lifetime discrimination, day-to-day experiences of discrimination). For health measures, a semi-structured diagnostic interview was used to generate psychiatric diagnoses defined according to the *Diagnostic and Statistical Manual of Mental Disorders* (DSM). Self-reported measures of self-rated health and psychological distress were also assessed.

Results indicated that after controlling for potential confounders, LGB respondents were almost twice as likely as heterosexuals to report at least one lifetime discriminatory event, and were nearly two-and-a-half times more likely than heterosexuals to report day-to-day experiences with discrimination. In turn, controlling for discrimination significantly attenuated the relationship between sexual orientation and each of the health indicators (psychiatric diagnosis, self-rated health, and psychological distress), indicating that perceptions of discrimination at least partially account for sexual orientation health disparities [19].

25.4.2 Daily Diary Studies

Cross-sectional studies offer important initial insights about associations between stigma and SGM health. At the same time, there are a number of significant limitations to this method. In part to overcome these limitations, researchers have used alternative methods, such as experience sampling methods (also known as ecological momentary assessments and daily diary studies), which offer a number of methodological strengths, including: capturing reported events and psychological reactions longitudinally in their natural, spontaneous context, thereby permitting the examination of person-by-situation interactions; reducing recall bias (because the approach minimizes the amount of time that elapses between an experience and the reporting of the experience); improving the validity of modeling within individual changes (because of the much larger number of assessments that are possible with this design); and affording the opportunity to examine the temporal sequence of events and to control for third variables by using individuals as their own controls, thereby improving causal inferences [20]. At the same time, experience sampling studies are not without their limitations, [20] which include a high level of participant involvement and commitment (which can lead to participant burden and attrition); the limited nature of what can be assessed (to counteract participant burdens); and certain threats to validity (e.g., habituation, reactivity).

Daily diary studies are increasingly being used to study stigma and SGM health. Frost and Fingerhut [21] used this design to obtain daily reports on health and stress exposure from 62 same-sex couples from four states in the month before citizens of these states voted on same-sex marriage in voter referenda. Daily exposure to negative campaign messages was associated with increased negative affect as well as decreased positive affect and relationship satisfaction, controlling for baseline measures of depression and daily fluctuations in general stress among both members of the couple [21]. Thus, even though all four states went on to vote in favor of same-sex marriage, the toxic social environment created by public debates about the rights of SGM individuals affected the mental health of same-sex couples.

25.4.3 Quasi-experiments

Researchers have complemented these observational designs through the use of several types of experiments (quasi, field, and laboratory), which permit stronger inferences regarding the relationship between stigma and health. Quasi-experiments are used in situations where it is not possible or ethical to randomly assign individuals to a particular condition, as is the case in studying structural forms of stigma. In such cases, researchers examine whether SGM health changes following an increase or decrease in structural stigma.

One example of this work [22] took advantage of the fact that leading up to and during the 2004 election, several states passed constitutional amendments banning same-sex marriage. These events occurred between two waves of data collection of the National Epidemiologic Survey on Alcohol and Related Conditions (NESARC), a nationally representative survey of US adults. Respondents were first interviewed in 2001 and then were reinterviewed in 2005, following the passage of the same-sex marriage bans. This provided a natural experiment that enabled researchers to examine changes in the prevalence of psychiatric disorders among LGB respondents who had been assessed before and after the bans were passed. LGB adults who lived in states that passed same-sex marriage bans experienced a 37% increase in mood disorders, a 42% increase in alcohol use disorders, and a 248% increase in generalized anxiety disorder between the two waves [22]. In contrast, LGB respondents in states without these bans did not experience a significant increase in psychiatric disorders during the study period. Moreover, the mental health of heterosexuals in states that passed the bans was largely unchanged between the two waves.

This quasi-experimental design was particularly strong because it included two comparison groups (i.e., LGB respondents in states that did not pass amendments and heterosexuals in states that did pass amendments) and a pre-post assessment of mental health outcomes. Nevertheless, quasi-experiments do have several limitations, in particular the possibility that other factors that occurred contemporaneous to the change in structural stigma are responsible for the results (i.e., serve as confounders). Researchers can address this in part through providing evidence for why these alternative explanations are unlikely to bias the results.

25.4.4 Field Experiments

One of the strengths of field experiments is that they retain the internal validity of a traditional randomized experiment but improve external validity by examining stigma processes in real-world settings. One particular type of field experiment is the audit experiment, which has been used in several studies on discrimination. An innovative example of this approach was conducted by Tilcsik [23], who explored sexual orientation-based discrimination in employment outcomes among men. The researcher submitted a pair of fictitious, but ostensibly real, résumés to actual job postings of white-collar, entry-level jobs in seven states, chosen based on whether they had employment nondiscrimination laws that included sexual orientation as a protected class. The sexual orientation of the applicant was randomly assigned to each résumé pair before they were sent out for the position. The sexual orientation of the applicant was signaled through the applicant's membership in a campus organization during college. Although the résumés differed slightly to avoid raising suspicion, there was no systematic relationship between résumé quality and sexual orientation; as such, any difference that was observed in call-back rates (the dependent variable of interest) could be attributed to the sexual orientation of the applicant.

Gay men were approximately 40% less likely to be offered a job interview than similarly qualified heterosexual men, an effect that is similar to previous audit studies on Black/White disparities in employment outcomes [23]. There was also substantial variation in the level of hiring discrimination across the seven states. The states with higher rates of employment discrimination were those that did not have employment nondiscrimination policies that protected gay men. This study not only provided experimental evidence that both interpersonal and structural discrimination influence employment outcomes, but also that these forms of discrimination interact to produce adverse outcomes for gay men.

While this research focused on employment outcomes, the results have important implications for health, given robust relationships between employment and poor health [24]. As with all methods, field experiments (including audit experiments) have certain limitations. For instance, inferences are only valid if researchers can ensure that there are no systematic differences in the applicants (in this case, in their résumés), which can prove challenging. Researchers can, however, take numerous steps through the study design to minimize the likelihood that these systematic differences occur.

25.4.5 Interventions

Randomized controlled trials represent an experimental approach to determining if psychosocial interventions can reduce stigma's negative health effects. One of the few such trials applied to stigma and SGM health tested the efficacy a 10-session skills training intervention focused on teaching gay and bisexual young men cognitive, affective, and behavioral skills for overriding the psychosocial damage of early and ongoing exposure to stigma. Compared to a waitlist control, the intervention significantly reduced depression, substance use, and HIV-transmission risk, as well as the psychosocial mechanisms, such as rejection sensitivity and social isolation, through which stigma is purported to produce adverse health outcomes [25]. Adaptations of the treatment for other SGM populations (e.g., lesbian and bisexual women) are underway.

25.5 Challenges and Future Opportunities

In this section, we discuss methodological and ethical challenges to conducting research on SGM health disparities, as well as opportunities for advancing research on this topic.

25.5.1 Methodological Challenges Related to Sampling SGM Populations

Perhaps the greatest challenge to conducting research on SGM health disparities involves the lack of a sampling frame for SGM populations. Because SGM status is not routinely assessed as part of national censuses or national health registries, researchers lack accurate estimates of even the size of the SGM population.[4] Nonetheless, the assessment of sexual orientation in population-based health surveys in the past two decades, and the recent inclusion of gender identity in a handful of health surveys, have been and remain an essential source of information regarding SGM health disparities. To the extent that future population-based surveys continue assessing SGM status and add measures of potential mechanisms underlying SGM health disparities, such as stigma exposure, this will greatly advance research on SGM health, which has historically relied heavily on nonrepresentative sampling. However, population-based sampling is often limited by small sample sizes of SGM and a lack of measures specific to this group. Therefore, the development of additional methods and improvements upon existing methods (e.g., respondent driven sampling) for sampling SGM populations are needed to accurately assess the health needs and determinants affecting this population.

4 http://news.gallup.com/poll/201731/lgbt-identification-rises.aspx

25.5.2 Ethical Challenges and Considerations

Several scholars have discussed the distinct ethical issues that are raised in conducting research with SGM youth – in particular, requiring parental notification for participating in research, which could potentially lead to victimization and rejection if the parents are unaware of or hostile toward their child's sexual orientation or gender identity. Research indicates [26] that requiring parental consent would also under-represent particular subgroups of SGM youth (e.g., those who have less family support, are racial/ethnic minorities, and identify as bisexual). A number of specific strategies for minimizing risks to SGM youth who participate in research studies have been suggested, such as enrolling a youth advocate at the research site who is unaffiliated with the study and who can provide guidance to SGM youth about their participation, as well as obtaining a Federal Certificate of Confidentiality to further protect participant privacy [26]. For a discussion of specific recommendations for addressing Institutional Review Board concerns in studies with SGM youth, we refer readers to a thoughtful analysis provided by Mustanski [26].

Of course, ethical issues are not only relevant for SGM youth. For instance, anti-SGM ideological arguments have long used evidence of SGM health disparities to advance notions that being SGM itself causes these disparities or that unhealthy environmental exposures cause both SGM status and poor health, with the corollary conclusion being that SGM status can and should be changed to reduce these disparities. Such suggestions were historically reflected in official diagnostic nomenclatures (e.g., *DSM*). Yet compelling contemporary evidence suggests that the root of these disparities lies largely in stigma, leading professional bodies to agree that the clearest ethical public health response to SGM health disparities involves reducing stigma toward SGM populations [27].

25.5.3 Directions for Future Research

In the past two decades, there have been important advancements in the study of stigma and SGM health disparities. At the same time, there are a number of areas that require additional research in order to advance this literature. As a comprehensive discussion of future directions is outside the scope of this chapter, we highlight three illustrative areas and refer readers to more in-depth discussions of this topic [7, 13].

25.5.4 Methodological and Measurement Innovations

As is evident in the work reviewed in this chapter, researchers have employed increasingly sophisticated methods and measures to study the role of stigma in shaping SGM health disparities. Nevertheless, the field would benefit from additional advancements in both areas.

One particularly important area for innovation is the development of novel measures for assessing exposure to stigma at different levels of analysis. A recent example is the use of Google searches to create area-level, or structural, measures of prejudice. In one study, Chae and colleagues [28] developed a measure of community-level racial prejudice by aggregating Google searches of racial epithets. They then linked this measure to a health dataset and found that Blacks who lived in communities characterized by greater racial animus (via the Google measure) had increased mortality risk compared to Blacks living in low-prejudice communities. One of the advantages of this measurement approach is that it minimizes social desirability biases, because respondents are less likely to censor their socially deviant attitudes on the Internet. While this study focused on area-level measures of racism, similar approaches could be used to study structural forms of stigma related to sexual orientation, for example through the use of Google searches for homophobic terms (e.g., “fag”).

Researchers are also beginning to use previously developed measures and apply them in innovative ways to the study of stigma processes at the individual and interpersonal levels. For instance, while originally developed to capture implicit attitudes of potential stigmatizers, the Implicit Association Test (IAT) has recently been used to measure stigma processes among the stigmatized, such as out-group bias [29] and exposure to discrimination [30]. In these studies, the IAT was associated with psychological distress [29] and hypertension [30] among LGB and Black respondents, respectively. One of the advantages of the IAT is that it does not rely on self-reported perceptions of stigma processes, which is particularly important in studies that also measure health outcomes that are self-reported. In these instances, the use of implicit tasks overcomes the limitations associated with same-source bias, which can occur when the exposure and outcome are measured via the same method.

In addition to novel measurement approaches to capture stigma processes, the field also requires the development and application of innovative methods to study stigma and SGM health. Agent-based models (ABMs), a tool for exploring complex systems across multiple levels of influence [31], represents one potential example. Despite calls for treating stigma within a complex systems framework, ABMs have thus far not been used to address the dynamic relationships among multiple forms of stigma (individual, interpersonal, structural) simultaneously in shaping the health of stigmatized individuals, including SGM populations.

25.5.5 Intersectional Perspectives

Important questions remain about the additive or multiplicative effects of intersections of multiple stigmatized identities within SGM populations. Given its ability to highlight the health issues posed by multiple overlapping systems of oppression, intersectionality theory offers a potentially fruitful template to begin studying these issues [32]. Indeed, the 2011 Institute of Medicine report on SGM health disparities [6] highlighted four theoretical approaches for advancing the science and treatment of SGM health, one of which was intersectionality. The theory of intersectionality proposes that social experiences differ based upon the unique intersections of various identities. For example, a Black lesbian may not simply experience the stigma related to being Black plus the stigma related to being a sexual minority. Instead, she may face unique stigma related to the *intersection* of these two stigmatized identities. Thus far, evidence that multiple stigmatized identities confer additional risk has been somewhat mixed. Datasets that include more variability in terms of sex, age, race, socioeconomic status, and other axes of social stratification are needed to more comprehensively assess how intersectional identities predict health outcomes within SGM populations.

25.5.6 Interventions to Reduce Stigma and Its Negative Consequences

Additional research is needed to further design, develop, and evaluate evidence-based interventions that both reduce stigma related to sexual orientation and gender identity, as well as address the negative consequences associated with stigma among SGM populations. For instance, despite some progress, most stigma coping interventions are in the early stages of efficacy testing, relying on case studies or pre-post designs without control groups [33]. Stronger tests of these interventions are needed to establish stronger causal inferences [25]. Moreover, although prejudice among healthcare providers, as well as structural aspects of the medical institution itself, may compromise the quality of care that SGM populations receive, few studies have examined the impact of stigma-reduction efforts specifically in healthcare settings, an important avenue for future intervention work [34].

25.5.7 Research on Other Determinants of SGM Health Disparities

We have focused in this chapter on stigma, because it has been the predominant framework for understanding SGM health disparities. At the same time, there is a small but growing area of research on determinants of SGM health disparities that do not draw on stigma theory. We briefly discuss examples of this work, organized by explanations at the biological, psychological, and social levels.

One set of explanations focuses on biological mechanisms underlying specific diseases or on disease comorbidity. For example, anal sex is common among gay men and transgender women and is also a risk factor for human papilloma virus, anal cancer, and HIV infection. When controlling for HIV infection, sexual orientation disparities in some adverse health outcomes for men are attenuated [35], suggesting that disease comorbidity might partially explain certain sexual orientation health disparities. Nonmedically supervised hormone or silicone injection represents another biological route of HIV transmission for some transgender individuals.

Another set of explanations focuses on universal psychological processes, which refer to established cognitive, affective, and social risk factors for mental health outcomes [36]. Rather than assessing characteristics that distinguish sexual minorities from heterosexuals (e.g., stigma experiences), this research examines the psychological processes common across sexual/gender identities and seeks to determine whether these factors explain SGM health disparities. For example, Hatzenbuehler et al. [37] showed that sexual minority youth were significantly more likely to ruminate – a general emotion regulation risk factor for depression – than their heterosexual peers. In turn, rumination mediated the relationship between sexual orientation and depressive and anxious symptoms, suggesting that rumination explains, at least in part, sexual orientation disparities in these mental health outcomes. Similarly, compared to their nontransgender siblings, transgender adults report lower social support, a universal risk factor for mental and physical illness [38].

Finally, researchers have also evaluated social determinants of SGM health disparities other than stigma. Differences in the family structures of sexual minority individuals (such as being less likely than heterosexuals to be parents), for instance, has been raised as a possible explanation for heightened risk for adverse health outcomes ranging from breast cancer (due to nulliparity) to substance use (due to more permissive social norms) among sexual minority women [39].

None of these explanations has accumulated the level of empirical support that stigma has. Nevertheless, their existence points to the importance of additional scholarly attention to factors other than stigma that may contribute to SGM health disparities.

25.6 Summary

In this chapter, we have provided a general introduction to the emerging field of research on SGM health disparities. We briefly reviewed evidence that SGM populations are at heightened risk for numerous adverse mental and physical health outcomes relative to cisgender heterosexual populations. We also noted that future research should (i) include measures of gender identity in a wider range of population-based datasets in order to generate nationally representative estimates of health outcomes among transgender populations and (ii) incorporate a broader range of physical health conditions and biological markers of health into studies of SGM health. In addition, we discussed several methodological approaches – including cross-sectional designs, daily diary studies, quasi-experiments, field experiments, and randomized controlled trials – that have been used to evaluate the role that stigma plays in contributing to SGM health disparities. While each of these methods has its own strengths and limitations, researchers have been able to amass a robust body of evidence across these different approaches indicating that stigma is a social determinant of SGM health disparities. Finally, we highlighted several areas for future research, including: (i) identifying additional methods, and improving upon existing methods, for sampling SGM populations; (ii) developing novel measures for assessing stigma at individual, interpersonal, and structural levels, and for examining interrelationships among these levels; (iii) adopting an intersectional approach to the study of SGM health; (iv) developing and testing evidence-based interventions to reduce stigma and its negative consequences; and (v) uncovering new mechanisms underlying SGM health disparities.

25.7 Key Points

- SGM populations are at heightened risk for numerous adverse mental and physical health outcomes relative to cisgender heterosexual populations.
- The field has used multiple methods and measures to demonstrate that stigma is a robust social determinant of SGM health disparities.
- This chapter identified a number of important future research directions that are needed to advance knowledge about the causes of SGM health disparities and about effective prevention interventions for reducing stigma and its negative consequences among SGM populations.

Disclaimer

The views and opinions expressed in this chapter are those of the authors only and do not necessarily represent the views, official policy, or position of the U.S. Department of Health and Human Services or any of its affiliated institutions or agencies.

References

1 Calzo, J.P., Antonucci, T.C., Mays, V.M., and Cochran, S.D. (2011). Retrospective recall of sexual orientation identity development among gay, lesbian, and bisexual adults. *Developmental Psychology* 47 (6): 1658–1673.

2 Stein, E. (1999). *The Mismeasure of Desire: The Science, Theory and Ethics of Sexual Orientation*. New York: Oxford University Press.

3 Bostwick, W.B., Boyd, C.J., Hughes, T.L., and McCabe, S.E. (2010). Dimensions of sexual orientation and the prevalence of mood and anxiety disorders in the United States. *American Journal of Public Health* 100 (3): 468–475.

4 Flores, A.R., Herman, J.L., Gates, G.J., and Brown, T.N.T (2016). How many adults identify as transgender in the United States? The Williams Institute. https://williamsinstitute.law.ucla.edu/wp-content/uploads/How-Many-Adults-Identify-as-Transgender-in-the-United-States.pdf

5 Gates, G.J. (2011). How Many People are Lesbian, Gay, Bisexual, and Transgender? The Williams Institute. https://williamsinstitute.law.ucla.edu/wp-content/uploads/Gates-How-Many-People-LGBT-Apr-2011.pdf

6 Institute of Medicine (2011). *The Health of Lesbian, Gay, Bisexual, and Transgender People: Building a Foundation for Better Understanding*. Washington, DC: National Academies Press.

7 Pachankis, J.E. and Lick, D.J. (2017). Sexual minority stigma and health. In: *The Handbook of Stigma, Discrimination, and Health* (eds. B. Major, J.F. Dovidio and B.G. Link). New York: Oxford University Press chapter 26.

8 Reisner, S.L., Poteat, T., Keatley, J. et al. (2016). Global health burden and needs of transgender populations: a review. *The Lancet* 388 (10042): 412–436.

9 Reisner, S.L., Vetters, R., Leclerc, M. et al. (2015). Mental health of transgender youth in care at an adolescent urban community health center: a matched retrospective cohort study. *Journal of Adolescent Health* 56 (3): 274–279.

10 Pachankis, J.E., Hatzenbuehler, M.L., Wang, K. et al. (2018). The burden of stigma on health and wellbeing: a taxonomy of concealment, course, disruptiveness, aesthetics, origin and peril across 93 stigmas. *Personality and Social Psychology Bulletin* 44 (4): 451–474.

11 Link, B.G. and Phelan, J. (2001). Conceptualizing stigma. *Annual Review of Sociology* 27: 363–385.

12 Hatzenbuehler, M.L., Phelan, J.C., and Link, B.G. (2013). Stigma as a fundamental cause of population health inequalities. *American Journal of Public Health* 103: 813–821.

13 Hatzenbuehler, M.L. (2016). Structural stigma: research evidence and implications for psychological science. *American Psychologist* 71 (8): 742–751.

14 Meyer, I.H. (2003). Prejudice as stress: conceptual and measurement problems. *American Journal of Public Health* 93 (2): 262–265.

15 Hatzenbuehler, M.L. and Link, B.G. (2014). Introduction to the special issue on structural stigma and health. *Social Science & Medicine* 103: 1–6.

16 Dodge, B., Herbenick, D., Friedman, M.R. et al. (2016). Attitudes toward bisexual men and women among a nationally representative probability sample of adults in the United States. *PLoS One* 11 (10): e0164430.

17 Hatzenbuehler, M.L., Flores, A.R., and Gates, G.J. (2017). Social attitudes regarding same-sex marriage and LGBT health disparities: results from a national probability sample. *Journal of Social Issues* 73: 508–528.

18 Meyer, I.H. (2003). Prejudice, social stress, and mental health in lesbian, gay, and bisexual populations: conceptual issues and research evidence. *Psychological Bulletin* 129 (5): 674–697.

19 Mays, V.M. and Cochran, S.D. (2001). Mental health correlates of perceived discrimination among lesbian, gay, and bisexual adults in the United States. *American Journal of Public Health* 91 (11): 1869–1876.

20 Bolger, N., Davis, A., and Rafaeli, E. (2003). Diary methods: capturing life as it is lived. *Annual Review of Psychology* 54 (1): 579–616.

21 Frost, D.M. and Fingerhut, A.W. (2016). Daily exposure to negative campaign messages decreases same-sex couples' psychological and relational well-being. *Group Processes & Intergroup Relations* 19 (4): 477–492.

22 Hatzenbuehler, M.L., McLaughlin, K.A., Keyes, K.M., and Hasin, D.S. (2010). The impact of institutional discrimination on psychiatric disorders in lesbian, gay, and bisexual populations: a prospective study. *American Journal of Public Health* 100 (3): 452–459.

23 Tilcsik, A. (2011). Pride and prejudice: employment discrimination against openly gay men in the United States. *American Journal of Sociology* 117 (2): 586–626.

24 Link, B.G. and Phelan, J. (1995). Social conditions as fundamental causes of disease. *Journal of Health and Social Behavior* 35: 80–94.

25 Pachankis, J.E., Hatzenbuehler, M.L., Rendina, H.J. et al. (2015). LGB-affirmative cognitive-behavioral therapy for young adult gay and bisexual men: a randomized controlled trial of a transdiagnostic minority stress approach. *Journal of Consulting and Clinical Psychology* 83 (5): 875–889.

26 Mustanski, B. (2011). Ethical and regulatory issues with conducting sexuality research with LGBT adolescents: a call to action for a scientifically informed approach. *Archives of Sexual Behavior* 40 (4): 673–686.

27 Anton, B.S. (2010). Proceedings of the American Psychological Association for legislative year 2009: minutes of the annual meeting of the council of representatives and minutes of the meetings of the board of directors. *American Psychologist* 65: 385–475.

28 Chae, D.H., Clouston, S., Hatzenbuehler, M.L. et al. (2015). Association between an internet-based measure of area racism and black mortality. *PloS One* 10 (4): e0122963.

29 Hatzenbuehler, M.L., Dovidio, J.F., Nolen-Hoeksema, S. et al. (2009). An implicit measure of anti-gay attitudes: prospective associations with emotion regulation strategies and psychological distress. *Journal of Experimental Social Psychology* 45 (6): 1316–1320.

30 Krieger, N., Carney, D., Lancaster, K. et al. (2010). Combining explicit and implicit measures of racial discrimination in health research. *American Journal of Public Health* 100 (8): 1485–1492.

31 El-Sayed, A.M., Scarborough, P., Seemann, L. et al. (2012). Social network analysis and agent-based modeling in social epidemiology. *Epidemiologic Perspectives and Innovations* 9 (1): 1–9.

32 Cole, E.R. (2009). Intersectionality and research in psychology. *American Psychologist* 64 (3): 170.

33 Chaudoir, S.R., Wang, K., and Pachankis, J.E. (2017). What reduces sexual minority stress? A review of the intervention “toolkit”. *Journal of Social Issues* 73 (3): 586–617.

34 Hatzenbuehler, M.L. and Pachankis, J.E. (2016). Stigma and minority stress as social determinants of health among lesbian, gay, bisexual, and transgender youth: research evidence and clinical implications. *Pediatric Clinics of North America* 63: 985–997.

35 Cochran, S.D. and Mays, V.M. (2007). Physical health complaints among lesbians, gay men, and bisexual and homosexually experienced heterosexual individuals: Results from the California Quality of Life Survey. *American Journal of Public Health* 97 (11): 2048–2055.

36 Hatzenbuehler, M.L. (2009). How does sexual minority stigma “get under the skin”? A psychological mediation framework. *Psychological Bulletin* 135 (5): 707–730.

37 Hatzenbuehler, M.L., McLaughlin, K.A., and Nolen-Hoeksema, S. (2008). Emotion regulation and internalizing symptoms in a longitudinal study of sexual minority and heterosexual adolescents. *Journal of Child Psychology and Psychiatry* 49 (12): 1270–1278.

38 Factor, R.J. and Rothblum, E.D. (2008). A study of transgender adults and their non-transgender siblings on demographic characteristics, social support, and experiences of violence. *Journal of LGBT Health Research* 3 (3): 11–30.

39 Cochran, S.D., Grella, C.E., and Mays, V.M. (2012). Do substance use norms and perceived drug availability mediate sexual orientation differences in patterns of substance use? Results from the California quality of life survey II. *Journal of Studies on Alcohol and Drugs* 73 (4): 675–685.

26

Workforce Diversity and Capacity Building to Address Health Disparities

Hannah A. Valantine and Alison F. Davis

Scientific Workforce Diversity, Office of the Director, National Institutes of Health, Bethesda, MD, USA

26.1 Background

This chapter introduce readers to emerging research at the interface of scientific workforce diversity and health disparities. It describes the various types of diversity affecting biomedicine and includes the most up-to-date data on the effects of scientific workforce diversity on health outcomes, focusing on health disparities. Among topics explored are concordance – the nature of the relationship between researchers and research participants – and the study of scientists' choice of specialization and research topics. We also provide potential methodological approaches to enrich scholarship in areas central to scientific workforce diversity, as applied through a health disparities lens. These include: understanding how researcher/provider diversity affects the knowledge and products of biomedical inquiry and care, promoting rigorous data collection and evaluation in investigations of diversity practices and programs and care environments, and methods to interrogate the effect of sociocultural factors on researchers/research participants and providers/patients.

Overall, this body of research is in its infancy, and much work remains to be done to determine the impact of various domains of workforce diversity on health outcomes, and in particular on health disparities.

26.2 Introduction

Scientific workforce diversity drives biomedical innovation, facilitates translation of advances to enhance health, and prepares scientists and healthcare professionals to better serve an increasingly diverse US population [1]. To date, most of the evidence to support the value of workforce diversity exists in scholarship outside biomedicine, pointing to the need for further study, analyses, and application within scientific and health-related settings. Developing and retaining a diverse scientific workforce has been encouraged as a key

The Science of Health Disparities Research, First Edition. Edited by Irene Dankwa-Mullan, Eliseo J. Pérez-Stable, Kevin L. Gardner, Xinzhi Zhang, and Adelaida M. Rosario.

strategy for resolving health disparities. This chapter reviews evidence to support this claim, cites examples, and points to gap areas where more scholarship is needed.

To demonstrate the effects of workforce diversity on health disparities, sound conceptual and methodological approaches must interrogate this question to identify the positive effects and challenges of workforce diversity on health outcomes in under-represented populations. Moreover, as health outcomes are highly contextual, integrated methods of study are needed.

Our search strategy included the following databases: PubMed/MEDLINE, Scopus, and PsycINFO/APA Psycnet. We selected and used a variety of synonyms to represent each concept of the research question at hand: What are the effects of scientific workforce diversity on health disparities? These terms included diversity, scientific research workforce, and health disparities.

26.3 Diversity in the US Scientific and Health Workforce

The US federal government stratifies the nation's population according to several demographic categories. These include African American or Black, Asian (spanning more than 20 countries), American Indian or Alaska Native, Native Hawaiian or other Pacific Islander, Latino or Hispanic (spanning 20 countries), and White.[1] Much more broadly, scientific workforce diversity is characterized by the inclusion of a variety of backgrounds, life experiences, ability and disability, and socioeconomic disadvantage, and encompasses groups that are under-represented in science. Diversity of thought, or cognitive diversity, is often approximated by differences in people's experiences and perspectives.

Extensive National Institutes of Health (NIH)-funded research has documented a persistent lack of diversity in the scientific workforce based on gender, race/ethnicity, physical ability, socioeconomic status, and sex/gender orientation. Research shows that individuals whose identity encompasses two or more of these groups (intersectionality) face unique challenges due to double-bind effects [2]. Across the United States, women comprise 42% of tenure-track faculty and 26% of tenured faculty, and even fewer hold leadership positions.[2] Under-representation is more pronounced for individuals from certain racial/ethnic groups, who comprise 13.5% of biomedical PhD recipients but only 7.8% of research faculty.[3]

NIH-funded efforts to diversify the scientific workforce over recent decades have been successful, but they have been largely focused on early-career stages. The result has been an increased number of biomedical doctorates but less impact on diversifying academic faculty and senior leadership [3, 4]. Although similar pipeline approaches have been pursued in the health provider arena, there has been less attention paid to understanding the effects of workforce diversity on patient care, and in turn, on health outcomes [5]. Recent studies, however, have delineated the characteristics of medical school graduates that

1 http://www.census.gov/topics/population/race/about.html

2 https://www.aamc.org/data-reports/workforce/interactive-data/figure-15-percentage-full-time-us-medical-school-faculty-race/ethnicity-2018

3 https://www.aamc.org/system/files/2020-01/2019Table15.pdf

predict an increased likelihood they will work in under-served communities that bear the greatest burdens of health disparities [6].

A diverse healthcare workforce should reflect the community it serves, yet in the United States, most of the nation's health workers are not culturally, racially, or ethnically similar to individuals for whom they provide care. Moreover, reliance upon aggregate descriptors as lenses for evaluating the effects of diversity on healthcare risks masks important nuances. For example, a 2010 study of nurses, the largest US subgroup of healthcare workers, showed that racial/ethnic similarity does not coincide with cultural similarity. Similar findings have been reported internationally [7]. This suggests that the broader concept of cultural competency is particularly relevant for: (i) considering the interplay between diversity and care, (ii) addressing awareness of existing racial and ethnic health disparities, (iii) understanding the incidence and prevalence of health conditions among diverse populations, and (iv) recognizing the effects of stereotype and bias.

26.4 Diversity and Health Disparities: What Are the Links?

Increasing the number of health providers from under-represented groups has the potential to broaden the nation's health disparities research agenda [8]. Evidence suggests that individuals from under-represented groups may prefer certain types of research that align with externally focused values, such as applying research interests to community health needs [9]. Empirically, these tendencies are manifest, yet further study is needed to understand what drives topic choice in biomedical research.

One strategy to enhance workforce diversity has been to create cohorts of investigators who are engaged in health disparities research, scholarship, and practice by encouraging scientists from under-represented groups to work in these areas of study [10]. Partnerships between research-intensive and minority-serving institutions have been established to diversify the biomedical research workforce[4] as well as to support individuals from under-represented groups to pursue health disparities research [11]. Other approaches have embedded researchers in public health departments in major urban settings to provide trainees with practical experience in applied health disparities research. Such programs have as a stated goal the desire to increase racial/ethnic and socioeconomic diversity in the public health workforce [12].

The above-noted reports offer indirect evidence of the effects of enhanced scientific workforce diversity on health disparities, including setting researcher priorities, improving the number of providers from under-represented groups, and providing cultural competency education to biomedical trainees.

The goal of many academic-government partnerships that combine research education and mentoring is to enhance workforce diversity and address regional health disparities. One such partnership is the NIH-funded Programs to Increase Diversity among Individuals Engaged in Health-Related Research (PRIDE), a longitudinal, summer-based initiative in cooperation with US-based academic institutions [13]. The research training elements of

4 http://www.diversityprogramconsortium.org

the Jackson Heart Study are another [14]. Diversifying emerging areas of biomedical research, such as data science, may reap benefits for bringing the power of technology to bear on understanding the complexity of health disparities, which result from a multitude of overlapping factors and forces [15]. Programs such as these are works in progress, and their impact on creating and sustaining workforce diversity and reducing health disparities awaits future testing.

26.5 Methodological Approaches to Understanding Workforce Diversity and Health Disparities

The challenges of enhancing scientific workforce diversity, resolving health disparities, and understanding the relationship between the two call for a rigorous scientific approach akin to the biomedical discovery process. NIH has articulated four cross-cutting challenges that guide efforts to diversify the workforce [1]. This provides an excellent framework for setting a research agenda to understand health disparities.

26.5.1 Expand Scholarship of the Science of Diversity as Related to Health Outcomes

Much has been written about the effects of workforce diversity on creativity, innovation, productivity, and other metrics. However, as noted, little is known about the effects of diversity on health outcomes, including disparities. In addition, most published work addressing this topic has focused on racial/ethnic and gender diversity.

Diversity arises from many domains, such as physical ability, socioeconomic status, sex/gender orientation, geography, age, and career focus. A person with a disability faces numerous challenges that may be difficult to approximate without a vivid understanding of that individual's daily life, preferences, and expectations. This situation may be improved by increasing the number of health providers with disabilities, but doing so would require ensuring reasonable accommodations for both providers and patients. To date, this hypothesis remains untested and may be difficult to test given the very wide range of disabilities [16].

The inclusion of sex gender minority (SGM) individuals constitutes another underdeveloped talent pool within the health provider workforce. Will enhancing representation of this population sector improve healthcare, and health outcomes, for this population of individuals? This is an urgent question, since SGM individuals suffer many documented health disparities [17] and are frequent targets of stigmatization [18], yet they have been vastly understudied in biomedicine. There is a unique opportunity to design studies and interventions to address the effect of workforce diversity on health outcomes in this critical context [19].

26.5.2 Promote Rigorous Data Collection and Evaluation

Scientific workforce diversity is a complex system influenced by a set of inputs and outputs. Both institutional programming and personal choice affect the demographic distribution and other types of diversity of researchers and healthcare workers.

Data-driven approaches offer the best opportunity to manipulate and observe system alterations. NIH has applied this strategy to try to understand the factors driving recruitment and retention of undergraduate students from under-represented groups, as well as to study various approaches to mentoring.[5]

The Diversity Program Consortium consists of three integrated elements: the Building Infrastructure Leading to Diversity program, the National Research Mentoring Network, and a Coordination and Evaluation Center. Evaluation of the results of these experimental programs will guide future federal investments in diversity programs as well as collect and disseminate lessons learned toward private sector investments at institutions and in industry. Early data toward understanding factors that affect career development of individuals from under-represented groups is exciting and promising. For example, a stereotype-threat intervention improved the academic outcomes of students from under-represented groups and decreased their worries about confirming ethnic/racial stereotypes [20]. Another study showed that the quality of faculty mentorship translated to more student success in undergraduate research experiences at a minority-serving institution [21]. Data-driven knowledge may also guide in-depth analyses of relationships between researcher diversity and health disparities.

Additional systems-based approaches to workforce modeling offer promise for understanding the effects on diversity generally, and are in particular a practical framework for dynamically probing the interplay between workforce diversity and health disparities [22]. System-based approaches view workforce stability as a system of hires and departures and reveal how the steady state level of diversity can be stubbornly resistant to change if inclusion efforts are not put into place to retain talent and build a critical mass of minority-group participants.

26.5.3 Expand Study of Sociocultural Factors as Related to Health Outcomes

Culture is a social domain that emphasizes practices, discourses, and material expressions, which, over time, express continuities and discontinuities of meaning in common experience. Culture assumes a critical role for individuals from under-represented groups in systems (such as healthcare) generally designed and populated by individuals from majority populations.

Thinking broadly across diversity domains invites consideration of cultural competency and acknowledges and incorporates cultural differences that may arise related to knowledge, communication, difference, and services as related to research and/or healthcare. As with other subareas discussed in this chapter, there is scant published literature on the effects of cultural competency on health outcomes [23]. Many important and interesting questions remain unasked in the realm of cultural competency – in particular, its effect on health outcomes in a vast range of socioeconomic and geographic arenas. Moreover, the vast majority of health providers work in teams within hospitals or other large healthcare organizations. Thus, any measures to affect cultural competency must be tested, validated,

5 www.diversityprogramconsortium.org

and implemented within institutional settings. Ultimately, those that prove effective must be endorsed by institutional leaders. This is no different from the case of measures for enhancing scientific workforce diversity and inclusion within university and medical center settings.[6]

Much work remains, and several hypotheses are ripe for testing, such as carefully designed interventions to investigate the value of a range of cultural competency techniques on health outcomes in disadvantaged groups. These encompass many realms, including language, recruitment and retention techniques, training, alliances with culturally relevant health customs and preferences, communication, and the inclusion of family/community members in decision making [24].

Concordance describes the nature of the relationship between a physician and his or her patient and can influence the nature and quality of care. Lack of concordance between patients and physicians may result in unequal treatment for individuals from under-represented groups. Research has shown that sociocultural factors affect concordance, which in turn may contribute to health disparities. For example, it is known that physicians' implicit biases affect the use of medical procedures, such as thrombolytic therapy [25] and cardiac catheterization [26]. Other potentially harmful consequences of discordance between providers and patients include missed opportunities for screening, lack of awareness about differing responses to medication and traditional remedies that could provoke drug interactions, and medical errors resulting from miscommunication.

To remediate such effects, it has been proposed that enhancing health-provider workforce diversity may reduce health disparities through matching patients with medical professionals who share their linguistic and cultural backgrounds [27]. More research is needed to understand this interaction and any potential impact. However, there is evidence to suggest that Black and Latino patients – when they have a choice – are more likely to choose physicians based on race/ethnicity, especially in the context of limited English fluency [28]. Conversely, a physician's race/ethnicity is a strong predictor of subspecialty choice and whether he or she provides care to patients from under-represented groups and those who are under- or uninsured [29]. The nation's uneven geographic distribution of primary care physicians, combined with physicians' willingness to care for patients with limited insurance, most certainly contributes to health disparities [30]. The overall number of trained and practicing physicians from under-represented groups remains disproportionately low, approximately 9–10% in aggregate, compared to the US population.[7] However, physicians represent only a relatively small component of the entire health-provider population. Diversity efforts must address the many other elements of the workforce, including nursing, which according to the US Census Bureau is predominantly female (91%) and White (75%) [31].

A broader discussion of the effects of discordant relationships between physician and patient is beyond the scope of this chapter, but it is worth noting that concordance and cultural competency also likely affect researcher–research participant interactions, and

---

6 http://www.nsf.gov/crssprgm/advance/advance_Two-Page_Summaries_2007.pdf
7 https://www.aamc.org/data-reports/workforce/interactive-data/figure-18-percentage-all-active-physicians-race/ethnicity-2018

potentially, research outcomes [32]. A detailed discussion of inclusion of research participants from under-represented groups appears in Chapter 24 of this volume.

26.5.4 Can Scientific Workforce Diversity Partnerships with the Private Sector and Community Organizations Reduce Health Disparities?

Some tools have been developed to promote and enhance scientific workforce diversity, and implementation may be facilitated through collaboration outside academia and government. The NIH Scientific Workforce Diversity office has developed evidence-based tools to address enhancing faculty diversity.[8] We know that an integrated approach is necessary, and these tools include search strategies for identifying a diverse candidate pool, education on how to mitigate implicit bias, strategies for conducting person-centered outreach, and evidence-based approaches on mentoring and networking.

Our current challenge is to focus on efforts that help transition researchers from under-represented groups from training into independent careers in academia and elsewhere. Those efforts should be a shared responsibility of academia, government, and industry, because the nation benefits from diverse perspectives driving biomedical inquiry in various areas of science, including health disparities. Innovative technology may also provide new opportunities for matching researchers and providers with under-served individuals. The University of Hawaii child and adolescent psychiatry telemental health program trains child and adolescent psychiatry fellows to provide culturally relevant behavioral health services for children living in rural Hawaii and the Pacific Islands. The program incorporates team building and workforce development through interactive videoconferencing that can bridge geographic and cultural distances [33]. This is one example of efforts to enhance workforce diversity while addressing health disparities. Long-term impact might provide indication of efficacy toward that goal.

26.6 Closing Thoughts

It is not a coincidence that individuals and groups affected by health disparities are themselves less represented among the ranks of scientists and health professionals who acquire biomedical knowledge and deliver care. A national effort to train individuals from under-represented groups to address health disparities will enable researchers to better assess the link between diversity and health disparities using the scientific method. We have provided examples of programs aiming to enhance workforce diversity as a key goal toward resolving health disparities. Such approaches are needed to identify and disseminate successful strategies for addressing health disparities that hinge on collaboration between biomedical researchers, healthcare providers, and research participants and patients.

In this chapter, we have provided a brief review and discussion of what is known about the relationship between scientific workforce diversity and health disparities. We conclude that much more research is needed to fully understand the potential impacts of scientific

8 https://diversity.nih.gov/sites/coswd/files/images/2018-01/SWD_Toolkit_Interactive-updated_508.pdf

workforce diversity on resolving health disparities. This topic is highly worthy of exploration, including deeper study of how concordance between patients and providers, and between researchers and research participants, affects scientific knowledge and health outcomes.

An especially intriguing but underdeveloped area of inquiry is topic choice: How do scientists decide what to study and why? Precious little is known about this important issue that rests at the heart of how our nation sets research priorities. Further study of relationships between social identity, life experiences, and other defining factors of individuals may shed light.

We assert that investigation of the role of sociocultural influences undergirds the study of workforce diversity, health outcomes, and the interactions between the two. Overall, this body of research is in its infancy, and much work remains to be done. Solid, data-driven approaches that carefully weigh context are needed to determine the effect of various domains of workforce diversity on health outcomes.

26.7 Key Points

- Most of the nation's health workers are not culturally, racially, or ethnically similar to individuals for whom they provide care.
- The challenges of enhancing scientific workforce diversity, resolving health disparities, and understanding relationships between the two call for a rigorous scientific approach akin to the biomedical discovery process.
- More research is needed to fully understand potential impacts of scientific workforce diversity on resolving health disparities.
- Investigation of the role of sociocultural influences undergirds the study of workforce diversity, health outcomes, and any interactions between the two.

Acknowledgments

The authors graciously acknowledge the contributions of Alicia Livinski and Loan Nguyen, of the NIH Library, for help with literature searching and curation of results.

Disclaimer

The views and opinions expressed in this chapter are those of the authors only and do not necessarily represent the views, official policy, or position of the U.S. Department of Health and Human Services or any of its affiliated institutions or agencies.

References

1 Valantine, H.A. and Collins, F.S. (2015). National Institutes of Health addresses the science of diversity. *Proceedings of the National Academy of Sciences* 112 (40): 12240–12242.

2 Ginther, D.K., Kahn, S., and Schaffer, W.T. (2016). Gender, race/ethnicity, and National Institutes of Health R01 research awards: is there evidence of a double bind for women of color? *Academic Medicine* 1 (8): 1098–1107.

3 Valantine, H.A., Lund, P.K., and Gammie, A.E. (2016). From the NIH: a systems approach to increasing the diversity of the biomedical research workforce. *CBE-Life Sciences Education* 15 (3): fe4.

4 Gibbs, K.D., Basson, J., Xierali, I.M., and Broniatowski, D.A. (2016). Decoupling of the minority PhD talent pool and assistant professor hiring in medical school basic science departments in the US. *eLife* 5: e21393.

5 Peek, M.E., Wilson, S.C., Bussey-Jones, J. et al. (2012). A study of national physician organizations' efforts to reduce racial and ethnic health disparities in the United States. *Academic Medicine* 87 (6): 694–700.

6 Garcia, A.N., Kuo, T., Arangua, L., and Pérez-Stable, E.J. (2017). Factors associated with medical school graduates' intention to work with underserved populations: policy implications for advancing workforce diversity. *Academic Medicine* 93 (1): 82–89.

7 McGinnis, S.L., Brush, B.L., and Moore, J. (2010). Cultural similarity, cultural competence, and nurse workforce diversity. *Western Journal of Nursing Research* 32 (7): 894–909.

8 Oh, S.S., Galanter, J., Thakur, N. et al. (2015). Diversity in clinical and biomedical research: a promise yet to be fulfilled. *PLoS Medicine* 12 (12): e1001918.

9 Gibbs, K.D. Jr. and Griffin, K.A. (2013). What do I want to be with my Ph.D.? The roles of personal values and structural dynamics in shaping the career interests of recent biomedical science PhD graduates. *CBE-Life Sciences Education* 12 (4): 711–723.

10 Daley, S.P., Broyles, S.L., Rivera, L.M., and Reznik, V.M. (2009). Increasing the capacity of health sciences to address health disparities. *Journal of the National Medical Association* 101 (9): 881–885.

11 Powell, D.L. and Gilliss, C.L. (2003). Partnering in the cultivation of the next generation of ethnic minority nurse scientists: responding to a compelling national agenda. *Journal of the National Black Nurses Association* 14 (2): 34–43.

12 Blake, J., Choden, T., Hemans-Henry, C. et al. (2011). NYC Epi scholars program: promoting applied health disparities research in an urban public health department-a program model. *Journal of Public Health Management & Practice* 17 (4): 313–315.

13 Boyington, J.E., Maihle, N.J. et al. (2016). A perspective on promoting diversity in the biomedical research workforce: the National Heart, Lung, and Blood Institute's PRIDE program. *Ethnicity & Disease* 26 (3): 379–386.

14 Taylor, H.A. Jr. (2005). The Jackson heart study: an overview. *Ethnicity & Disease* 15 (4 Suppl 6): S6-1–S6-3.

15 Canner, J.E., McEligot, A.J., Rice, T.K. et al. (2017). Enhancing diversity in biomedical data science. *Ethnicity & Disease* 27 (2): 107–116.

16 Iezzoni, L.I. (2016). Why increasing numbers of physicians with disability could improve care for patients with disability. *AMA Journal of Ethics* 18 (10): 1041–1049.

17 Operario, D., Gamarel, K.E., Grin, B.M. et al. (2015). Sexual minority health disparities in adult men and women in the United States: national health and nutrition examination survey, 2001-2010. *American Journal of Public Health* 105 (10): e27–e34.

18 Meyer, I.H. (2003). Prejudice, social stress, and mental health in lesbian, gay, and bisexual populations: conceptual issues and research evidence. *Psychological Bulletin* 129 (5): 674–697.

19 Mansh, M., Garcia, G., and Lunn, M.R. (2015). From patients to providers: changing the culture in medicine toward sexual and gender minorities. *Academic Medicine* 90 (5): 574–580.

20 Ben-Zeev, A., Paluy, Y., Milless, K.L. et al. (2017). “Speaking truth” protects underrepresented minorities’ intellectual performance and safety in STEM. *Education Sciences* 7 (2): 65.

21 Daniels, H., Grineski, S.E., Collins, T.W. et al. (2016). Factors influencing student gains from undergraduate research experiences at a Hispanic-serving institution. *CBE-Life Sciences Education* 15 (3): ar30.

22 O’Brien, K.R., Scheffer, M., van Nes, E.H., and Van Der Lee, R. (2015). How to break the cycle of low workforce diversity: a model for change. *PLoS One* 10 (7): e0133208.

23 Horvat, L., Horey, D., Romios, P., and Kis-Rigo, J. (2014). Cultural competence education for health professionals. *The Cochrane Database System Reviews* 5 (5): CD009405.

24 Brach, C. and Fraserirector, I. (2000). Can cultural competency reduce racial and ethnic health disparities? A review and conceptual model. *Medical Care Research and Review* 57 (Suppl 1): 181–217.

25 Green, A.R., Carney, D.R., Pallin, D.J. et al. (2007). Implicit bias among physicians and its prediction of thrombolysis decisions for black and white patients. *Journal of General Internal Medicine* 22 (9): 1231–1238.

26 Schulman, K.A., Berlin, J.A., Harless, W. et al. (1999). The effect of race and sex on physicians’ recommendations for cardiac catheterization. *New England Journal of Medicine* 340 (8): 618–626.

27 Saha, S., Taggart, S.H., Komaromy, M., and Bindman, A.B. (2000). Do patients choose physicians of their own race? *Health Affairs (Millwood)* 19 (4): 76–83.

28 Traylor, A.H., Schmittdiel, J.A., Uratsu, C.S. et al. (2010). The predictors of patient-physician race and ethnic concordance: a medical facility fixed-effects approach. *Health Services Research* 45 (3): 792–805.

29 Moy, E. and Bartman, B.A. (1995). Physician race and care of minority and medically indigent patients. *Journal of the American Medical Association* 273 (19): 1515–1520.

30 Mensah, M.O. and Sommers, B.D. (2016). The policy argument for healthcare workforce diversity. *Journal of General Internal Medicine* 31 (11): 1369–1372.

31 Gilliss, C.L., Powell, D.L., and Carter, B. (2010). Recruiting and retaining a diverse workforce in nursing: from evidence to best practices to policy. *Policy, Politics, & Nursing Practice* 11 (4): 294–301.

32 George, S., Duran, N., and Norris, K. (2014). A systematic review of barriers and facilitators to minority research participation among African Americans, Latinos, Asian Americans, and Pacific Islanders. *American Journal of Public Health* 104: e16–e31.

33 Alicata, D., Schroepfer, A., Unten, T. et al. (2016). Telemental health training, team building, and workforce development in cultural context: the Hawaii experience. *Journal of Child and Adolescent Psychopharmacology* 26 (3): 260–265. Epub 2015 Oct 14.

Index

The Science of Health Disparities Research, First Edition. Edited by Irene Dankwa-Mullan, Eliseo J. Pérez-Stable, Kevin L. Gardner, Xinzhi Zhang, and Adelaida M. Rosario.

d

e

m

National Institute on Minority Health and Health Disparities Research Framework

Domains of Influence (Over the Life course)	Levels of Influence*: Individual	Interpersonal	Community	Societal
Biological	Biological Vulnerability and Mechanisms	Caregiver–Child Interaction Family Microbiome	Community Illness Exposure Herd Immunity	Sanitation Immunization Pathogen Exposure
Behavioral	Health Behaviors Coping Strategies	Family Functioning School/Work Functioning	Community Functioning	Policies and Laws
Physical/Built Environment	Personal Environment	Household Environment School/Work Environment	Community Environment Community Resources	Societal Structure
Sociocultural Environment	Sociodemographics Limited English Cultural Identity Response to Discrimination	Social Networks Family/Peer Norms Interpersonal Discrimination	Community Norms Local Structural Discrimination	Social Norms Societal Structural Discrimination
Healthcare System	Insurance Coverage Health Literacy Treatment Preferences	Patient–Clinician Relationship Medical Decision Making	Availability of Services Safety Net Services	Quality of Care Healthcare Policies
Health Outcomes	**Individual Health**	**Family/ Organizational Health**	**Community Health**	**Population Health**

National Institute on Minority Health and Health Disparities, 2018
*Health Disparity Populations: Race/Ethnicity, Low SES, Rural, Sexual/Gender Minority
Other Fundamental Characteristics: Sex/Gender, Disability, Geographic Region

Figure 1.3 The NIMHD Research Framework. *Source:* National Institute on Minority Health and Health Disparities. NIMHD Research Framework. 2017. Public Domain Available at: https://www.nimhd.nih.gov/about/overview/research-framework.html. Accessed March 2, 2018.

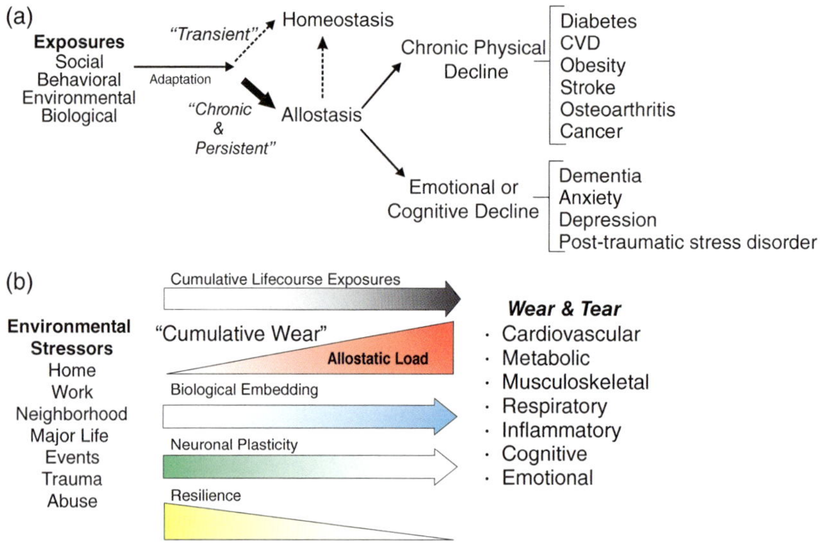

Figure 2.1 Schematic representation of the concept of Allostatic Load. (a) Graphic illustration of conceptual linkages between allostasis and environmental stress and chronic disease; (b) Graphic representation of the role and components of allostatic load over the life course illustrating influences of biological embedding, neuronal plasticity, and cumulative "wear and tear" in response to environmental stressors.

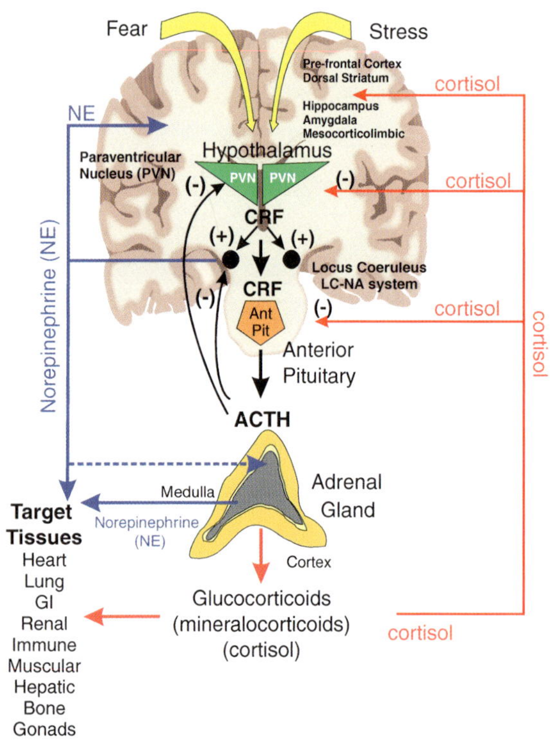

Figure 2.2 Schematic presentation of stress pathway outputs and inputs to the hypothalamic-pituitary-adrenal (HPA) axis in response to stress. NE, norepinephrine; CRF, corticotropin releasing factor; PVN, paraventricular nucleus; ACTH, adreno-corticotropin. *Source:* Derived in part from Spencer and Deak [9].

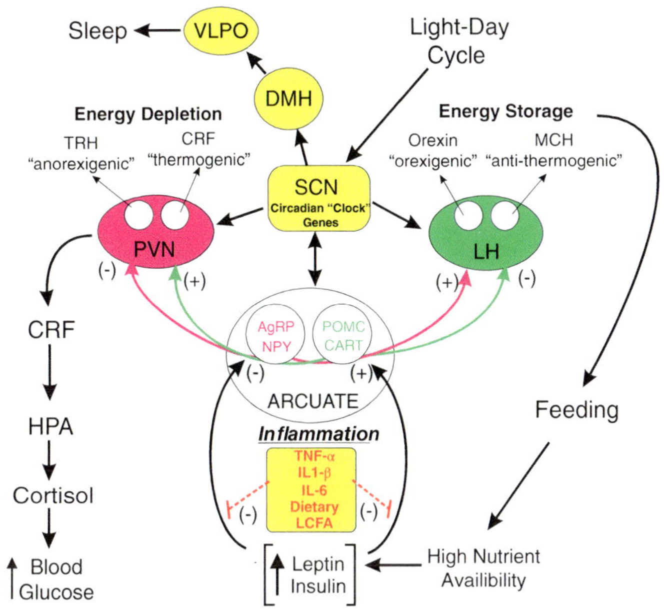

Figure 2.3 Schematic representation of pathways important in hypothalamic control of energy balance and sleep. VLPO, ventrolaterol preoptic nucleus; DMH, dorsomedial hypothalamus; TRH, thyrotropin releasing hormone; CRF, corticotrophin releasing factor; PVN, paraventricular nucleus; SCN, suprachiasmic nucleus; LH, lateral hypothalamus; HPA, hypothalamus-pituitary-adrenal axis; AgRP, agouti-related peptide; NPY, neuropeptide Y; POMC, pro-opiomelanocortin; CART, cocaine and amphetamine related transcript; MCH, melanin-concentrating hormone; TNF-α, tumor necrosis factor alpha; IL1-β, interleukin 1 beta; IL-6, interleukin-6; LCFA, long chain fatty acids.

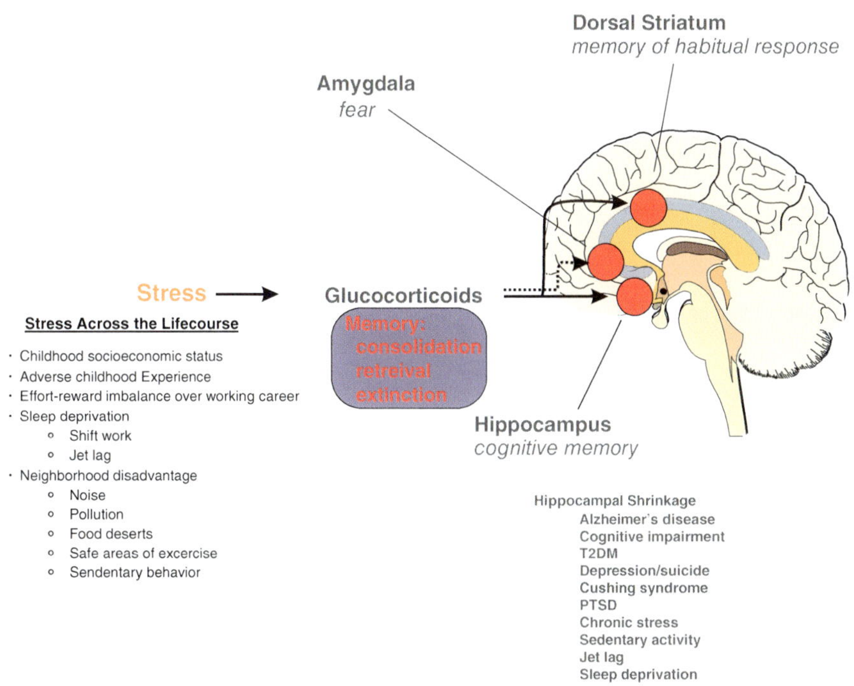

Figure 2.4 Schematic diagram of the influence of glucocorticoids and stress on memory, learning and mood. T2DM, Type 2 diabetes mellitus; PTSD, post-traumatic stress syndrome.

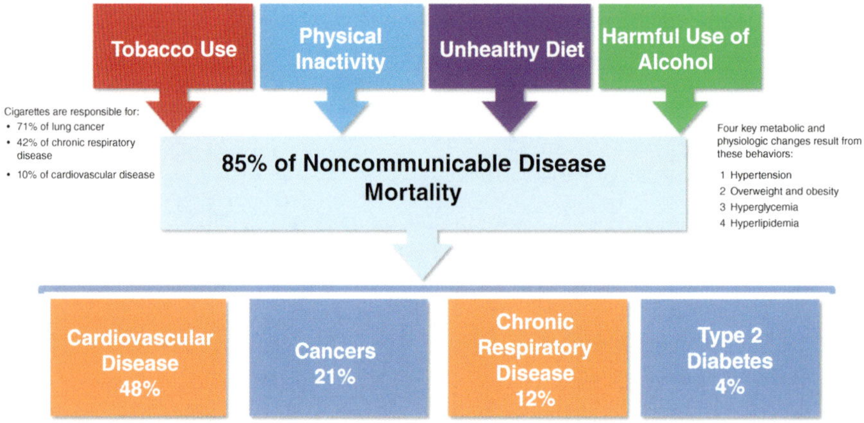

Figure 4.1 In this illustration of behavioral determinants for noncommunicable disease mortality in low- and middle-income countries, identified by the World Health Organization, four risk factors are linked to 85% of the noncommunicable disease mortality, which is caused by cardiovascular disease, cancer, chronic respiratory disease, and type 2 diabetes. *Source:* World Health Organization [17].

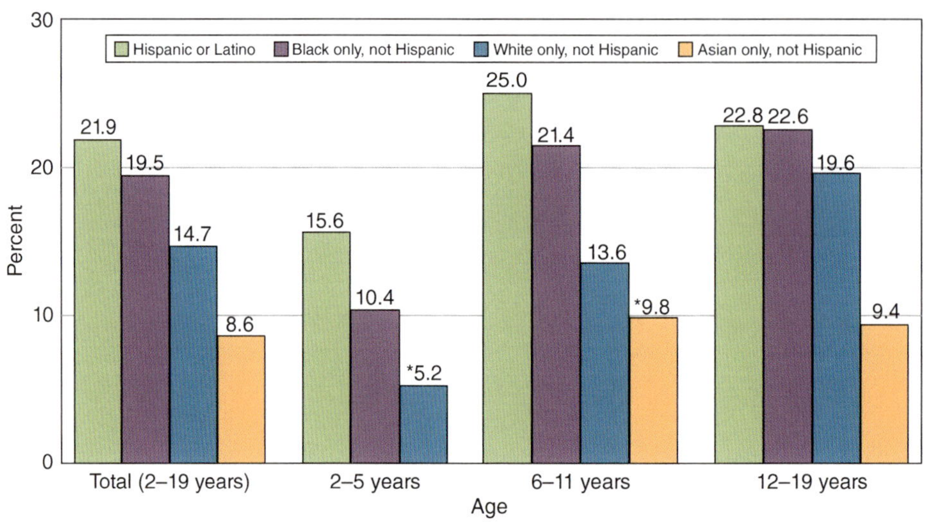

Figure 4.2 Children and adolescents with obesity, overall (2–19 years of age; top, left), and in (following, left to right), generally, preschool, elementary, and secondary school age groups. *Source:* US Centers for Disease Control and Prevention [11].

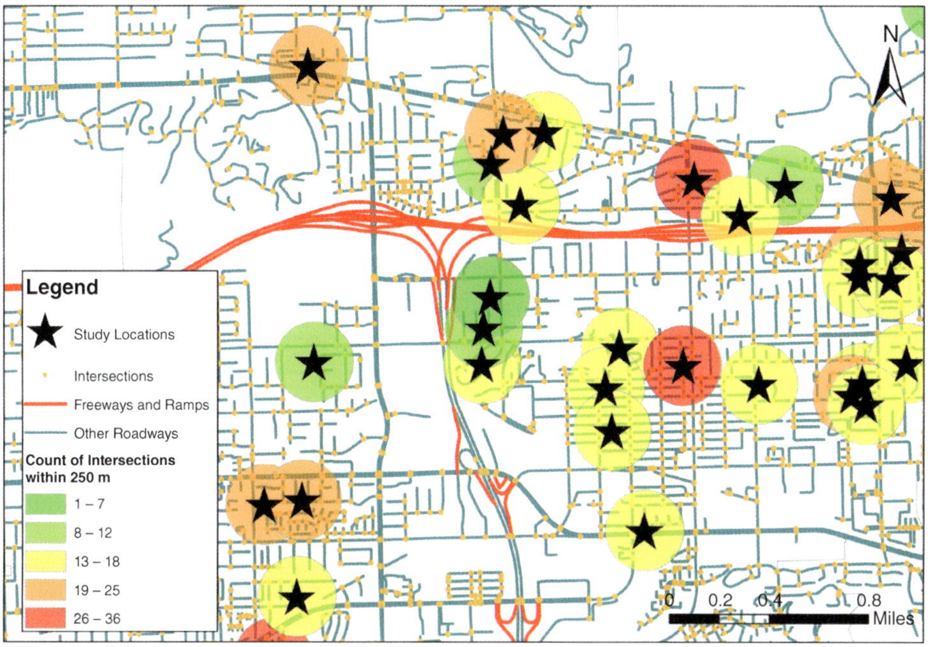

Figure 6.1 An illustration of the use of GIS to represent locations (points) in a study in relation to the road network (freeways, ramps, and other roadways as lines). The points of roadway intersections can be calculated and summarized within a 250 m buffer around a location. Color scales (green to red) can be used to represent the count of intersections on a continuous scale. *Source:* Figure created by Dr. Rima Habre, ScD, 2018, University of Southern California © OpenStreetMap (and) contributors, CC-BY_SA.

Figure 6.2 This infographic is an example of how environmental health research is presented to communities in a way that can be easily understood. It helps to create awareness and engage communities to learn about their physical environment. *Source:* Gutschow, W. (2016) Children's health and urban parks [infographic]. University of Southern California Environmental Health Centers.

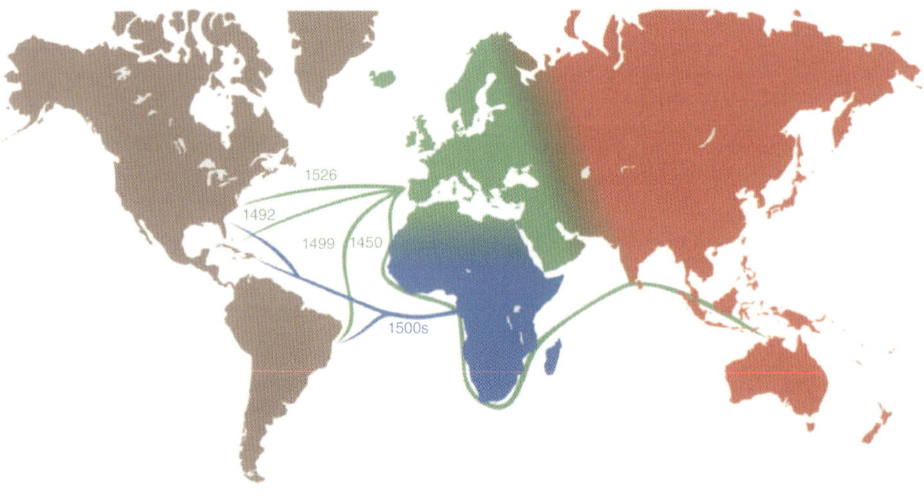

Figure 7.1 Major migrations and diasporas, 1400–1800, that are the sources of admixed populations in the Americas and elsewhere. The location of the ancient populations of Eurasia, Asia, sub-Saharan Africa, and the Americas are shown in distinct colors. *Source:* Adapted from Winkler et al. [19].

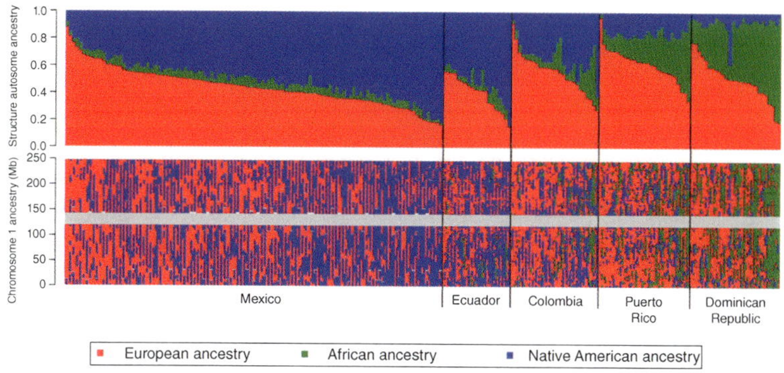

Figure 7.2 The top panel presents genome-wide ancestry of all autosomes for individuals from five Latino populations, showing the proportions of African, European, and Native American ancestry. The colors represent ancestry linkage disequilibrium blocks derived from the three ancestral populations contributing to the admixture. Each vertical slide represents a single individual, and the data for these individuals in each population are arranged from high to low European ancestry. The bottom panel presents the locus-specific (local) ancestry for chromosome 1, with each bar representing one individual. The two arms of the chromosomes are separated by a horizontal gray line representing the centromeric region. This figure makes the point that in an admixed population, each individual genome is a patchwork quilt of ancestral genomic segments. *Source:* Adapted from Bryc et al. [21] © 2010, Katarzyna Bryc.

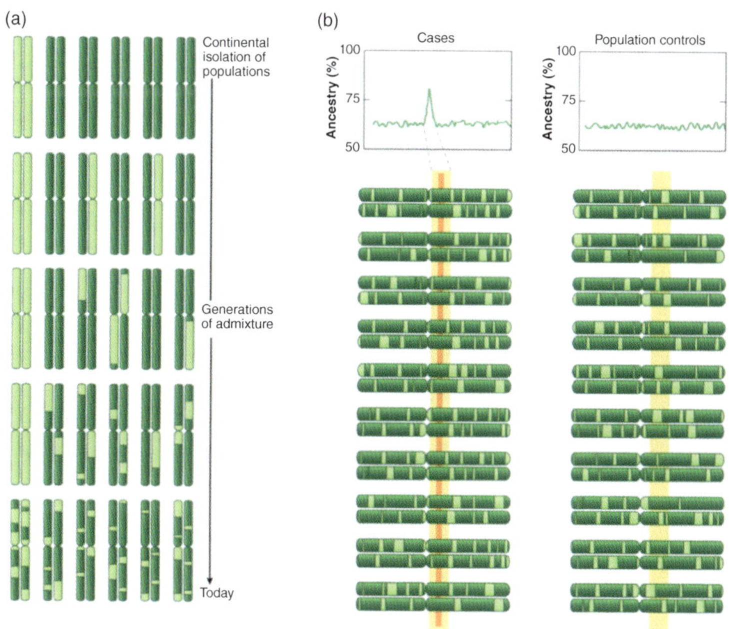

Figure 7.3 Shown is the process of admixture resulting from a moderate number of generations [4–18, 20, 21] since the initial admixture event. The boundary between light and dark green represents a recombination switch point. (a) The dark and light green color blocks represent chromosomal segments derived from matings between two parental (ancestral) populations. Starting with the second generation, meiotic cross-over (recombination) events produce chromosomal blocks of the two parental ancestries. Present-day admixed populations vary both in the extent of overall ancestry and block lengths due to differences in recombination events among individuals and because among individuals, original chromosomes have been subject to recombination for varying numbers of generations. (b) The graphs show the percentage of ancestry derived from the dark green segment of chromosome (~60%). In the 95% credible interval harboring the disease locus (yellow bar), there is an excess of majority ancestry blocks among cases, revealed as a spike in a graph of average ancestry for cases along the chromosome to ~80%. The orange bar indicates the location of the causal disease gene. Among population controls, the distribution of ancestry blocks is random across the ancestry. The spike of ancestry can be quantified either by comparing case ancestry versus control ancestry at the same locus (case-control statistic) or by comparing peak case local ancestry with case genome-wide average ancestry across all chromosomes (locus-genome statistic). *Source:* Adapted from Winkler et al. [19] © 2010, Elsevier.

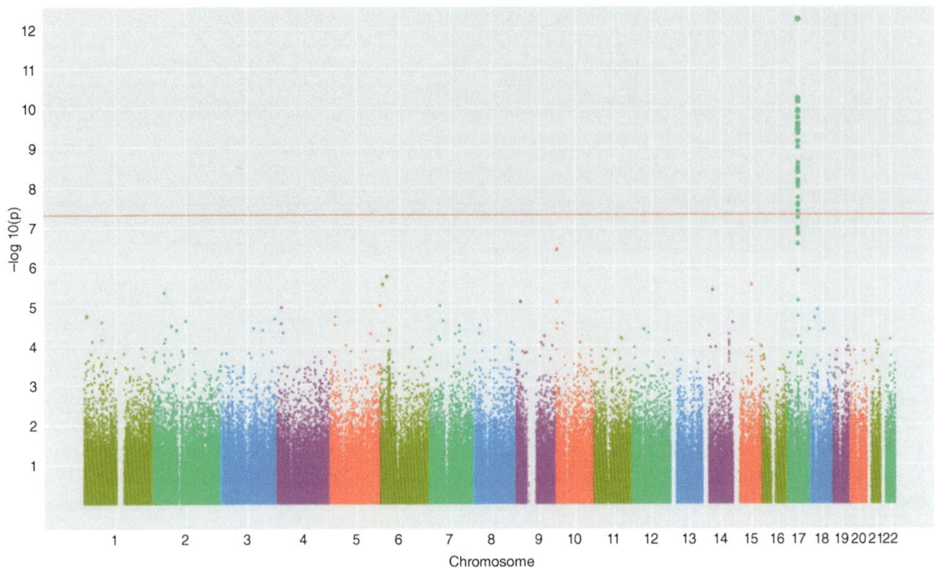

Figure 7.4 A Manhattan plot showing a peak of SNP association on chromosome 17q21 in a study of 1,893 Latinos with asthma and 1,881 Latino controls. The genomic coordinates, with chromosomes indicated, are displayed on the *X* axis, the negative log of the p values are displayed on the *Y* axis, and SNPs are represented by a point. The red line represents the genome-wide significance threshold, 5×10^{-8}. *Source:* Adapted from Galanter et al. [26].

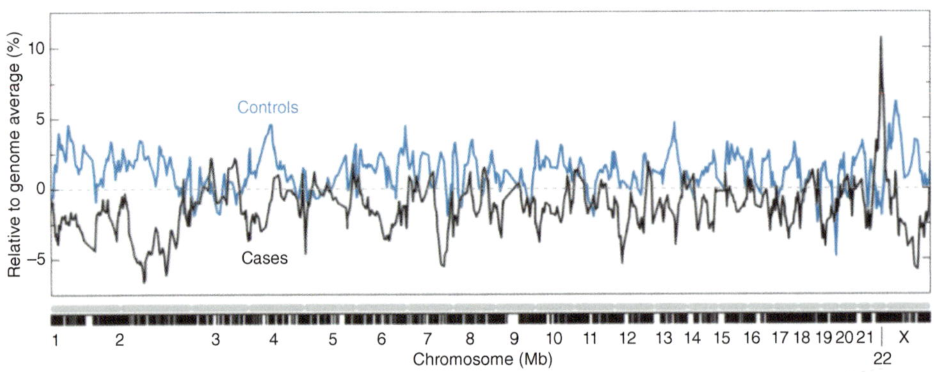

Locus-genome LOD score

	Genome-wide	Peak
Initial screen (1290 markers)	9.2	12.4
Even markers	8.7	10.1
Odd markers	7.7	8.8
Dense markers at peak	10.5	13.7

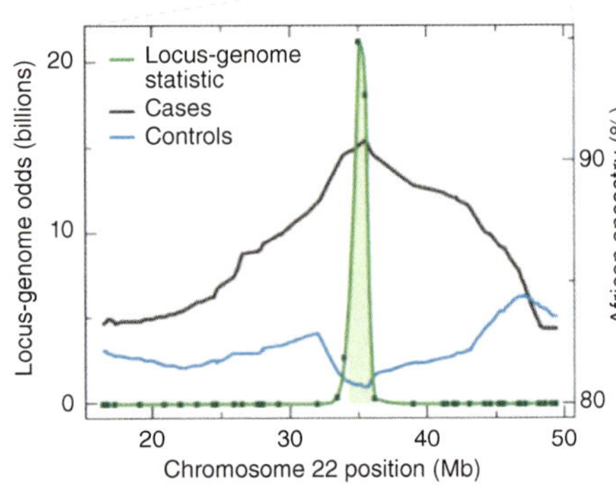

Figure 7.5 An admixture scan for focal segmental glomerulosclerosis and HIV-associated nephropathy, which are, respectively, fourfold and 60-fold more frequent in African Americans compared to European Americans [43]. The sharp peak of African ancestry occurs on chromosome 22 centered on the plausible candidate gene MYH9 and 32 other genes. Subsequently, the ancestry signal was shown to be due to the presence of two African-specific coding variants in the APOL1 gene encoding apolipoprotein L1 [9]. *Source:* Adapted from Winkler et al. [19].

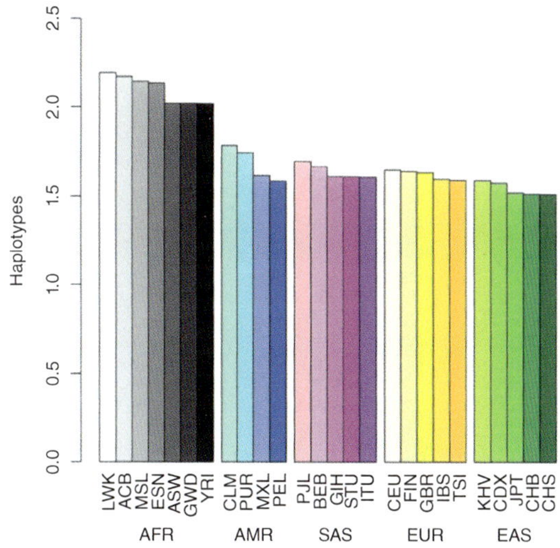

Figure 9.1 Haplotype diversity in the 1000 Genomes Project. Haplotypes were counted over windows of six consecutive diallelic SNPs, with a slide of one SNP. For each sample, all counted haplotypes have a frequency ≥1%. See Supplementary Material, table S1 [3] for sample descriptors. *Source:* Rotimi et al. [5] © 2017, Oxford University Press.

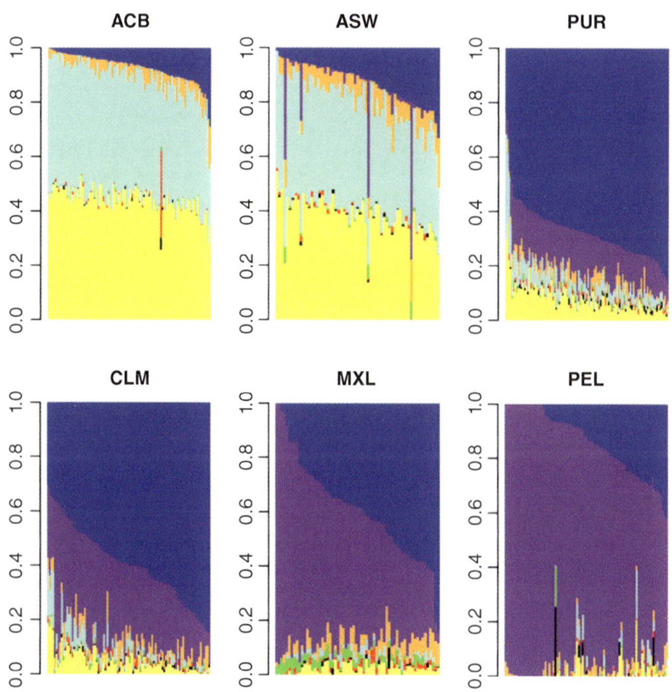

Figure 9.2 Individual admixture proportions from New World samples. The six samples are African Caribbean in Barbados (ACB), People with African Ancestry in Southwest USA (ASW), Puerto Ricans in Puerto Rico (PUR), Colombians in Medellín, Colombia (CLM), People with Mexican Ancestry in Los Angeles, California (MXL), and Peruvians in Lima, Peru (PEL) [3]. Conditional on eight ancestries, yellow corresponds to Western African ancestry, aquamarine corresponds to West-Central African ancestry, blue corresponds to Southern European ancestry, orange corresponds to Northern European ancestry, purple corresponds to Native American ancestry, red corresponds to South Asian ancestry, green corresponds to East Asian ancestry, and black corresponds to Southeastern Asian ancestry. Summing the Western and West-Central African ancestries, the range of individual admixture proportion across these six samples goes from 0 to 97.8%. Similarly, the range of individual admixture proportion for Native American ancestry goes from 0 to 100%. *Source:* Rotimi et al. [2] © 2016, Elsevier.

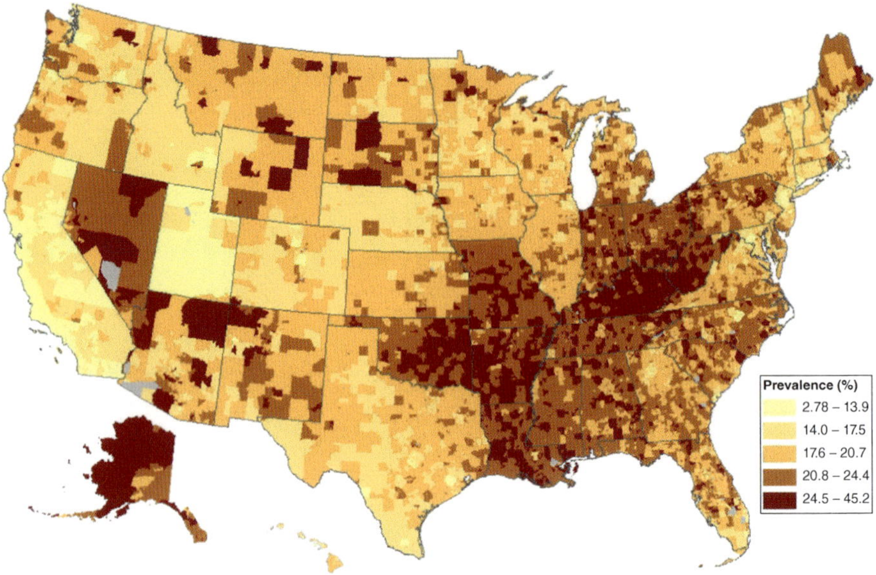

Figure 13.1 BRFSS model-prevalence estimates of current cigarette smoking among adults aged 18 and over in the United States, 2011 by 2010 census tract.

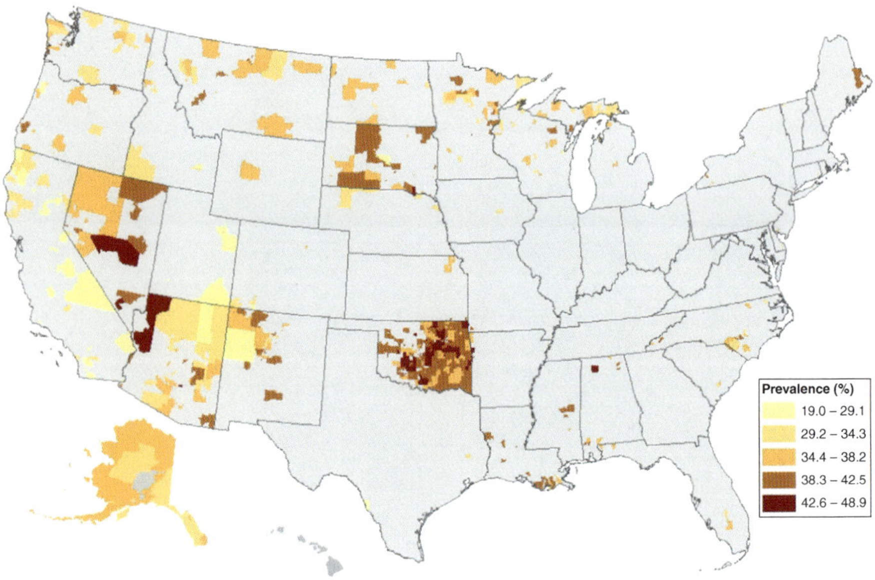

Figure 13.2 BRFSS model-prevalence estimates of current cigarette smoking among non-Hispanic American Indian and Alaska Native (NH-AIAN) adults aged 18 and over in the United States, 2011 by 2010 census tract with at least 100 NH-AIAN adults.

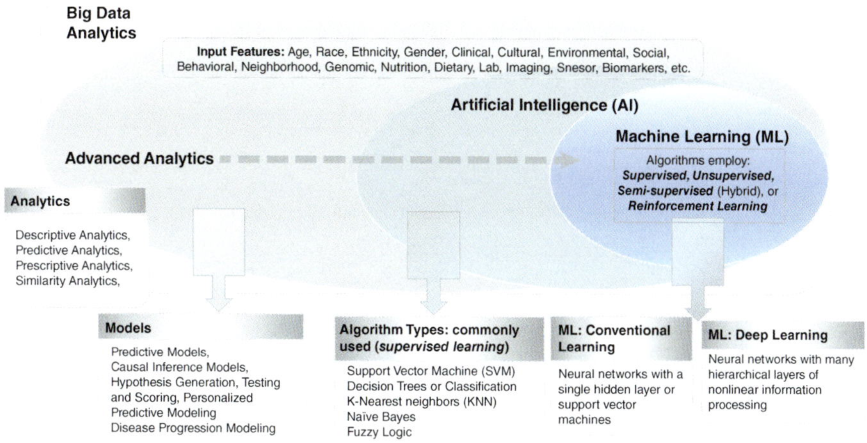

Figure 14.3 Spectrum of applications and techniques to analyzing big data.

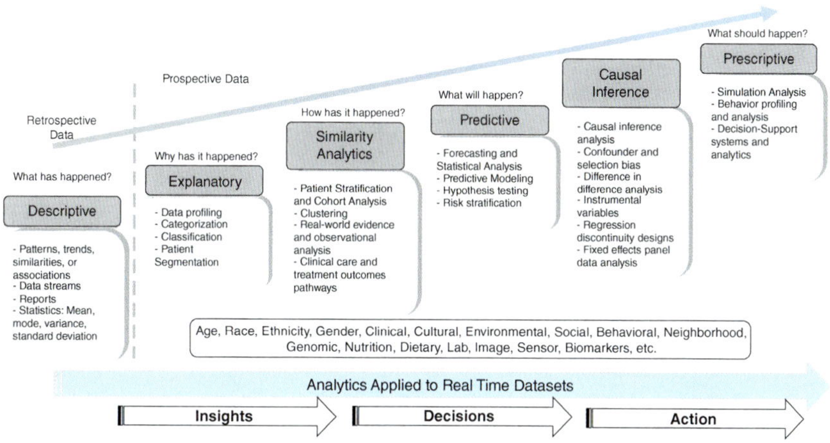

Figure 14.4 Schematic framework of advanced analytics for health disparity research.

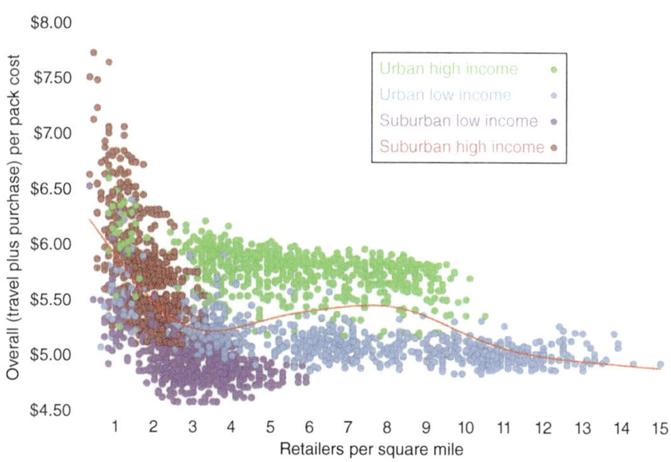

Figure 15.1 Relationship between tobacco retailer density and average, total per-pack cigarette cost (i.e., direct and indirect costs) across community contexts. *Source:* Luke et al. [25]. © 2017 American Public Health Association.

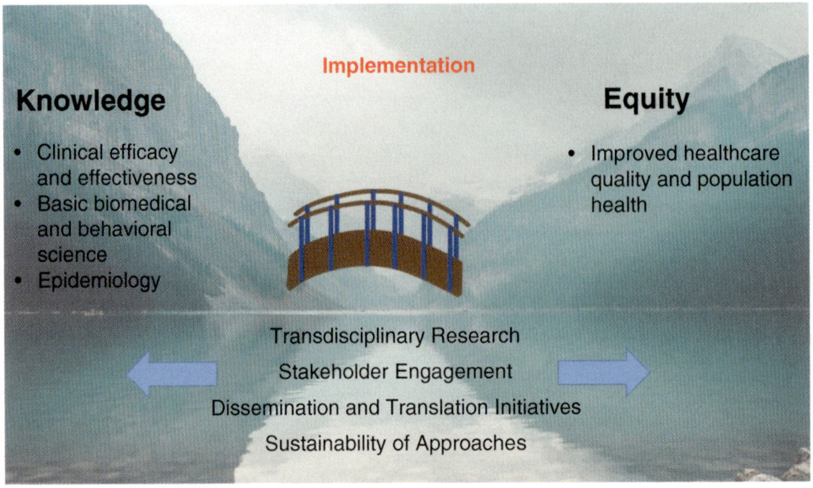

Figure 17.1 Population health and health disparities research: where we are and where we need to be. *Source:* Modified from The Colebrooke Centre for Evidence and Implementation: https://www.slideshare.net/TheSRAOrg/jane-lewis.

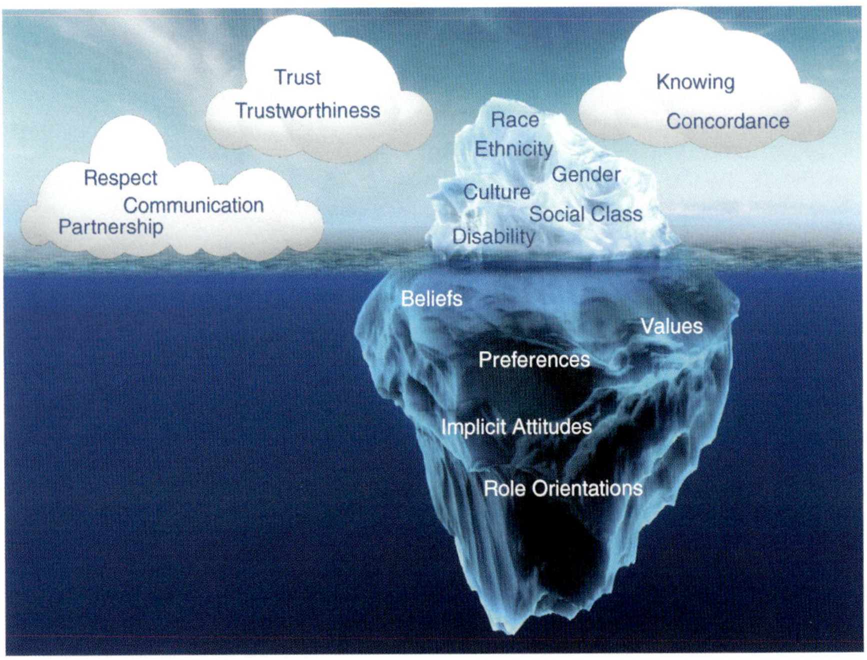

Figure 17.3 Iceberg concept of culture applied to relationship-centeredness in health disparities research. *Source:* Modified from Cooper et al. [29].

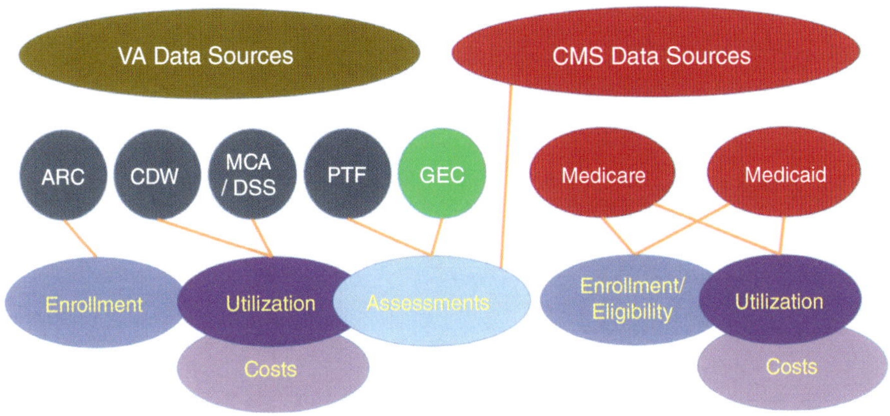

Figure 18.1 Distinct VA and CMS data sources needed to construct the VA HBPC model. ARC, Allocation Resource Center; CDW, Corporate Data Warehouse; CMS, Center for Medicare and Medicaid Services; GEC, VA Office of Geriatrics & Extended Care; MCA/DSS, Managerial Cost Accounting and Decision Support System; PTF, Patient Treatment File; VA, Veterans Affairs. *Source:* Courtesy of VA GEC IAGG Symposium Presentation, July 2017.

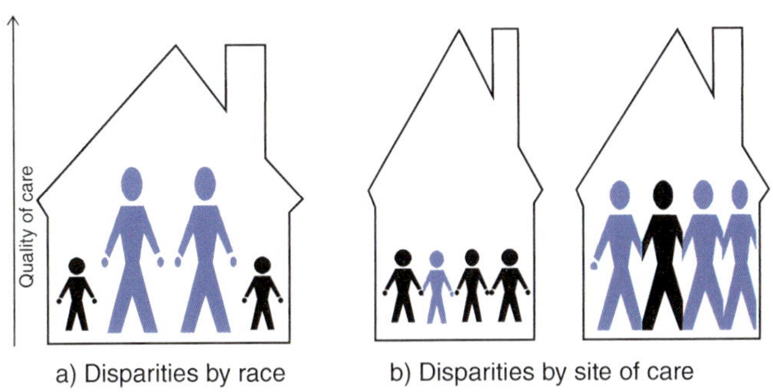

Figure 19.1 Healthcare disparities by race or site of care.

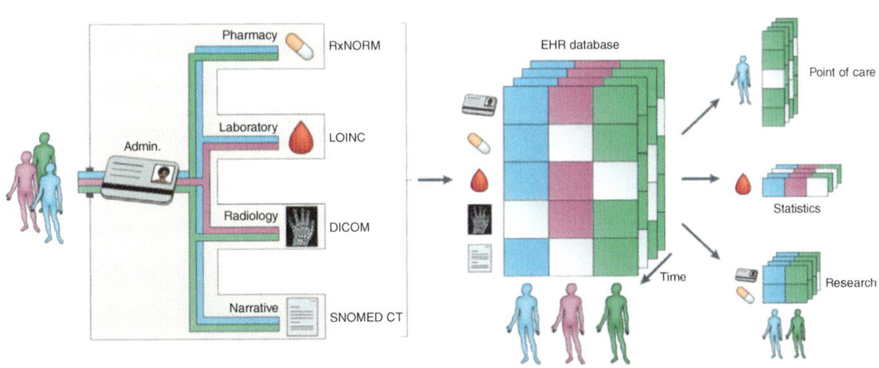

Figure 22.1 Electronic health records. *Source:* Adapted from Jensen et al. 2012.

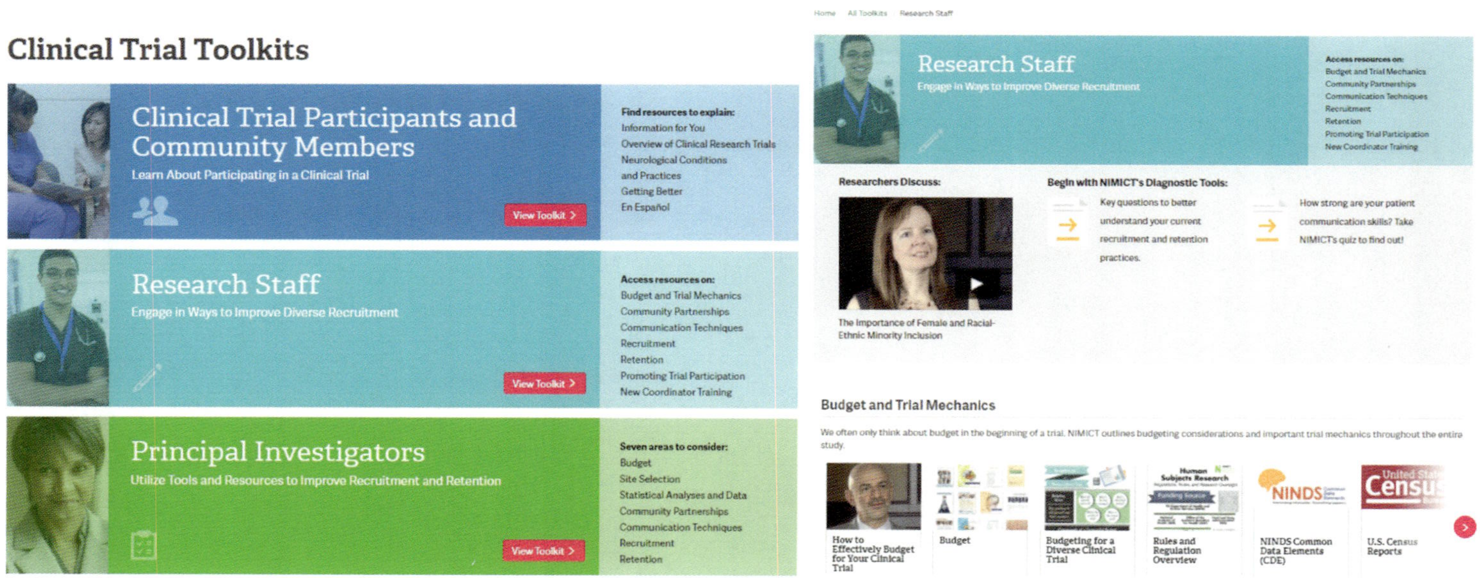

Figure 24.1 Screenshots of NIMICT.com online toolkits.

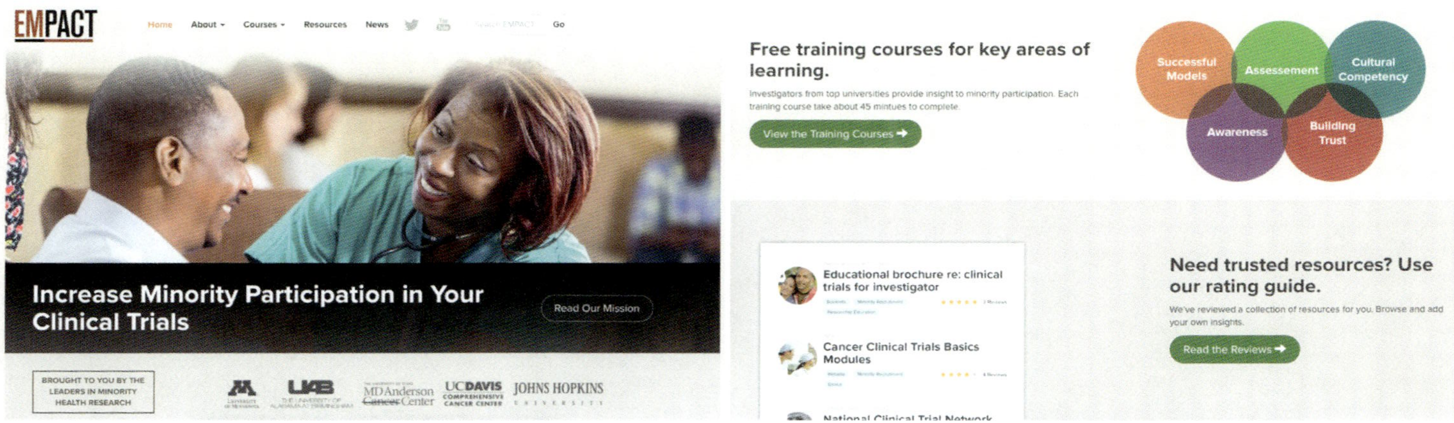

Figure 24.2 Screenshots of EMPaCT consortium online toolkit. *Source:* empactconsortium.com.

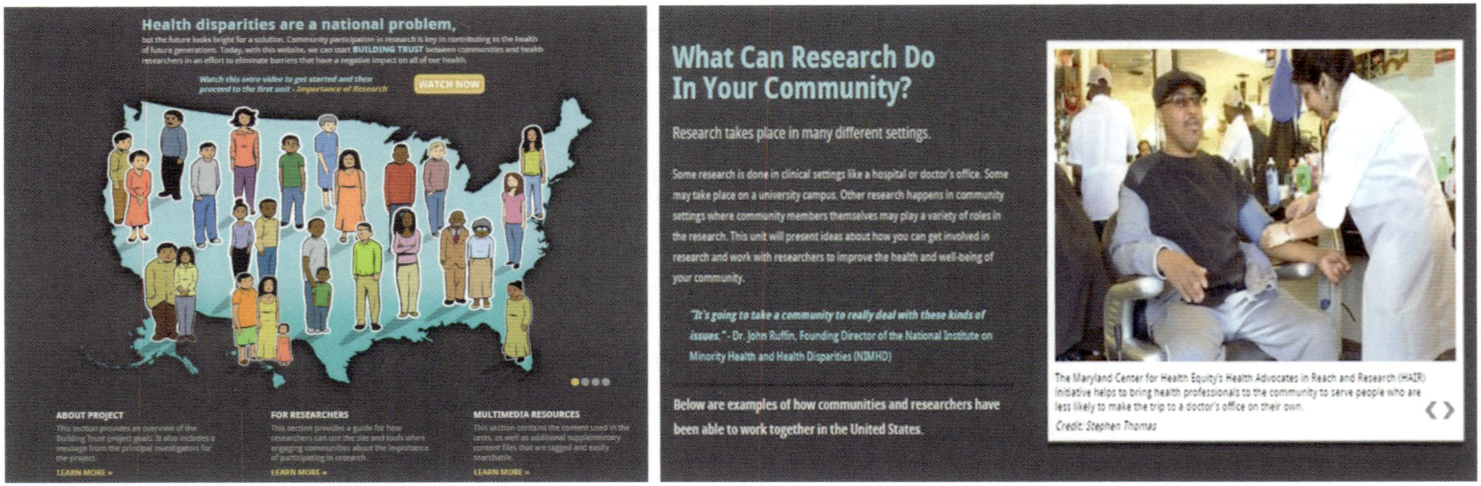

Figure 24.3 Screenshots of building trust between minorities and researchers, online toolkit. *Source:* buildingtrustumd.org.